Mind the Gap

AF709843

Peter M. Kappeler • Joan B. Silk
Editors

Mind the Gap

Tracing the Origins of Human Universals

Springer

Editors
Prof. Dr. Peter M. Kappeler
Deutsches Primatenzentrum
Abt. Verhaltensökologie & Soziobiologie
Kellnerweg 4
37077 Göttingen
Germany
pkappel@gwdg.de

Dr. Joan B. Silk
University of California,
Los Angeles
Dept. Anthropology
Los Angeles CA 90095
USA
jsilk@anthro.ucla.edu

ISBN: 978-3-642-02724-6 e-ISBN: 978-3-642-02725-3
DOI 10.1007/978-3-642-02725-3
Springer Heidelberg Dordrecht London New York

Library of Congress Control Number: 2009932123

© Springer-Verlag Berlin Heidelberg 2010
This work is subject to copyright. All rights are reserved, whether the whole or part of the material is concerned, specifically the rights of translation, reprinting, reuse of illustrations, recitation, broadcasting, reproduction on microfilm or in any other way, and storage in data banks. Duplication of this publication or parts thereof is permitted only under the provisions of the German Copyright Law of September 9, 1965, in its current version, and permission for use must always be obtained from Springer. Violations are liable to prosecution under the German Copyright Law.
The use of general descriptive names, registered names, trademarks, etc. in this publication does not imply, even in the absence of a specific statement, that such names are exempt from the relevant protective laws and regulations and therefore free for general use.

Cover: Photo by Etsuko Nogami, Kyoto University
A mother chimpanzee uses a pair of stones to crack open oil-palm nuts; watched by the son, 7 years old, and the daughter, 1.5 years old.

Cover design: WMXDesign GmbH, Heidelberg, Germany

Printed on acid-free paper

Springer is part of Springer Science+Business Media (www.springer.com)

Preface

This volume features a collection of essays by primatologists, anthropologists, biologists, and psychologists who offer some answers to the question of what makes us human, i.e., what is the nature and width of the gap that separates us from other primates? The chapters of this volume summarize the latest research on core aspects of behavioral and cognitive traits that make humans such unusual animals. All contributors adopt an explicitly comparative approach, which is based on the premise that comparative studies of our closest biological relatives, the nonhuman primates, provide the logical foundation for identifying human universals as well as evidence for evolutionary continuity in our social behavior. Each of the chapters in this volume provides comparative analyses of relevant data from primates and humans, or pairs of chapters examine the same topic from a human or primatological perspective, respectively. Together, they cover six broad topics that are relevant to identifying potential human behavioral universals.

Family and social organization. Predation pressure is thought to be the main force favoring group-living in primates, but there is great diversity in the size and structure of social groups across the primate order. Research on the behavioral ecology of primates and other animals has revealed that the distribution of males and females in space and time can be explained by sex-specific adaptations that are sensitive to factors that limit their fitness: access to resources for females and access to potential mates for males. The interaction of these selective pressures has favored the formation of stable social groupings, which range from pair-bonded family groups to large multi-male, multi-female groups. In Chap. 2, *Bernard Chapais* explores and reconstructs the deep social structure of human societies based on his extensive knowledge of primate social systems. He provides a convincing scenario for the transition from a chimpanzee-like system to one that characterizes all human societies. According to his analyses, the development of weapons has broken the polygyny potential of males, leading to the formation of stable breeding bonds. Reciprocal exchange of (female) mates among neighboring groups subsequently created a new dimension of kinship-mediated bonds across groups, which uniquely characterize human societies.

In Chap. 3, *Ryne Palombit* takes a broad look at bonding and conflict between the sexes, emphasizing commonalities between humans and other primates. He identifies sexual conflict as one constant in both human and primate societies and explores its consequences in depth. The susceptibility to infanticide is one important underlying feature of intersexual relations in primates with important consequences for their physiology and behavior. Sexual coercion and other forms of aggression between the sexes also have their origins in sexual conflict. This evolutionary approach can, therefore, explain some aspects of human behavior at least as well as culture-based alternatives.

Socio-ecological theory is based on the assumption that fundamental sex differences related to parental investment play an important role in shaping mammalian reproductive strategies. Differences in sexual and reproductive strategies between human males and females are, therefore, expected and have been documented in many cases. In Chap. 4, *Monique Borgerhoff Mulder* uses data from a rural forager-horticultural population in Tanzania to emphasize the fact that humans are not invariably restrained by deep physiological constraints into stereotypical gender roles. Her insightful analyses of the human pair bond reveals much more flexibility than previously assumed, raising interesting new questions about other prominent related aspects of human social behavior, including paternal care, female bonding, and multiple mating by females.

Politics and power. Apart from reproduction, much social behavior of primates revolves around dominance and power. Competition for access to resources or mates is ubiquitous in the animal world, and in long-lived species such as primates, where individuals interact on a daily basis, dominance offers a mechanism to minimize the immediate costs of competition. In Chap. 5, *David Watts* reviews our current understanding of dominance and power in primate societies. Following a most welcome critical discussion of the various terminologies used to describe agonistic asymmetries in primatology, he reviews the assumptions and predictions of the socioecological model that aims at explaining species differences in dominance styles. Watts argues that power and dominance are widespread among primates, but that politics, i.e., polyadic coalitions and social manipulations that require third-party awareness, are limited to great apes and differ importantly in kind from human politics.

In Chap. 6, *Aimée Plourde* turns to human power asymmetries. She argues that dominance is also pervasive in human societies, albeit based on a much richer repertoire of coercive behaviors. In addition, she identifies prestige as a unique source of social power in humans. She discusses the evolutionary origins of prestige, focusing on the hypothesis that prestige arose in parallel with the increasing importance of transmitting complex information culturally because individuals who successfully acquired and applied these vast sources of knowledge were treated with respect and admiration. Subsequently, signaling wealth and success took on an important role in reinforcing prestige asymmetries. Plourde goes on to depict how prestige has pervaded human social life in a unique manner, ranging from group competition to politics.

Laura Betzig focuses on a particular example to illustrate the consequences of power asymmetries in human societies in Chap. 7. Her detailed analysis of one period in Roman history documents pronounced skew in reproductive opportunities as a result of socially imposed and politically controlled power differences between emperors and eunuchs at the extreme ends of the social hierarchy. As such, this example attests to the social flexibility of our species already emphasized in the first section.

Intergroup relationships. Interactions with neighboring social units play an important part in the daily lives of most animals. In territorial species, groups defend resources located within their territories against covetous neighbors. Intergroup relations are, therefore, primarily characterized by mutual aggression. In Chap. 8, *Margaret Crofoot* and *Richard Wrangham* take a closer look at the nature and function of primate intergroup aggression. They show that the essential functional reasons for competition between groups are very similar across species, despite variable feeding ecologies and modes of interactions, and that numerical superiority is the best predictor of long-term success. In contrast to chimpanzees and humans, however, deliberate planning and regular lethal violence do not characterize intergroup violence in other primate species.

In Chap. 9, *Azar Gat* zooms in on human warfare as a form of collective, organized intergroup aggression that is not found in this form in other primates. He uses examples from various cultures and periods to illustrate the factors that favor this form of collective aggression. Again, competition for resources and reproductive opportunities loom large as important incentives, but religion and other supernatural beliefs also generate conflict among groups of humans. Gat also examines the role of proximate mechanisms, such as prestige, retaliation, and ecstasy. His informed evolutionary interpretation of the different aspects of warfare provides a compelling example of how a (near) human universal is shaped by general evolutionary processes.

Foundations of cooperation. One striking feature of human intragroup social behavior is the frequency and scope of cooperation. This problem is of special interest to students of behavior because cooperation is, by definition, associated with a cost for the actor and a benefit for the recipient, and therefore counterintuitive for both evolutionary and economic analyses of behavior. In Chap. 10, *Joan Silk* and *Robert Boyd* use a comparative approach to identify ways and mechanisms in which human cooperation differs from that of other primates. They review how kinship, reciprocity, and mutualism structure cooperative behavior of nonhuman primates in different functional domains. The same processes can be identified in humans, but Silk and Boyd argue that cultural evolution has created new opportunities for group-level cooperation.

In Chap. 11, *Venkat Lakshminarayanan* and *Laurie Santos* focus on the seemingly irrational aspects of cooperative behavior: foregoing individual pay-off maximization or, in other words, violating the norms of economic decision-making. They identify several key features of apparently irrational economic behavior in humans and explore whether similar features exist in other primates. They find similarities in aspects of this "economic cognition," such as loss aversion and inequity aversion. Their conclusions put claims about human economic irrationality into a broader perspective and identify new and exciting avenues for experimental work on primates.

Some of the seemingly irrational decisions of individuals, and prosocial acts in particular, may be proximately governed by emotions. In Chap. 12, *Daniel Fessler* and *Matthew Gervais* take a closer look at these emotions and their evolutionary origin. They broaden this approach by expanding their inquiry to all major emotions. Their comprehensive review reveals that many emotions have an evolutionary origin well outside the primate order. On the other hand, this broad comparative perspective helps identify likely universally human emotions, such as shame and norm-based guilt.

Language, thought, and communication. The unusually large human brain harbors the hardware for our cognitive abilities. After all, these abilities underlie our intelligence and behavioral flexibility. Language is the most salient aspect of human social cognition and communication. In Chap. 13, *Dorothy Cheney* and *Robert Seyfarth* examine the continuities and discontinuities between human language and vocal communication in nonhuman primates by focusing on vocal production and perception. Their review shows that humans are unique in the flexibility with which they produce learned, modifiable sounds, whereas fundamental differences with respect to call usage and perception are less clearly pronounced. Their experiments with wild baboons also reveal that a relatively small repertoire of fixed calls with specific meaning can, nonetheless, generate a formidable communication system.

In Chap. 14, *Chris Knight* takes a semantic look at the evolution of language. He considers fundamental aspects of digital and analog communication systems to illuminate the possible transition from primate vocal communication to language. Animal play provides an interesting situation in which animals modify their signals in a way that shares fundamental features with language. With this theoretical background, Knight develops a hypothesis about the origin of language in which menstruation and sexual conflict play a pivotal role in the socio-cultural evolution of human language.

In Chap. 15, *Robin Dunbar* examines the origin and functions of human brains within the broader context of primate brain evolution. After all, primates, as a group, are distinct from other mammals because of their large brains for their body size. He discusses three hypotheses about primate brain evolution that focus on ecological, life history, and social explanations. He stresses the often overlooked fact that these hypotheses imply different benefits and constraints so that a comprehensive approach that specifies causes, constraints, and emergent properties is required. The summary of his earlier empirical analyses strongly implicates group size as the main driving force in primate brain evolution. The special position of humans (and to a lesser extent of chimpanzees and bonobos) can be explained by the special cognitive demands of the dispersed, nested structure of their social groups.

In Chap. 16, *Michael Tomasello* and *Henrike Moll* outline their view of the uniqueness of human cognition. Accordingly, individual cognitive abilities, while impressive compared with other primates, are not what distinguish us ultimately from our close relatives. Instead, the synergies resulting from the combination of many individual brains in constructing and culturally transmitting social institutions and rules constitute a truly unique human achievement. The psychological mechanism facilitating these effects is shared intentionality. Tomasello and Moll compile an impressive array of experiments with great apes and human children to muster support for their claim. Their work also provides a compelling case for the coevolution of cognition and culture that can help to understand problems discussed in the last section (Innovation and Culture).

In Chap. 17, *David Bjorklund, Kayla Causey*, and *Virginia Periss* take up the theme of shared intentionality and focus on its developmental aspects. They emphasize the importance of mothers, in particular, in the development of the necessary cognitive abilities and mechanisms, such as gaze following and empathy. These authors chose chimpanzees raised by human caretakers as another interesting model for exploring the gap between humans and chimpanzees. They broaden their comparative analyses to social learning in general as well as the necessary theory of mind and conclude that, despite several striking similarities between human children and enculturated chimpanzees, the active role of human mothers in these interactions is unique.

In Chap. 18, *Robert Trivers* devotes his attention to two phenomena of the mind that are closely related but differ enormously in what we know about their effects on our daily behavior and decision-making. While deception is known to exist at all levels of life and expressed in the behavior and other traits of organisms, self-deception is only known from humans. Trivers presents evidence from several experiments that can only be explained by assuming that our unconscious mind hides true information from the conscious mind, thereby affecting the latter's performance. This constellation provides a fascinating playground for studying the interactions among deception, its detection, and self-detection, as well as a stimulating source of questions about the nature and organization of our minds.

This section concludes with yet another comparative perspective on primate cognition. In Chap. 19, *Claudia Fichtel* and *Peter Kappeler* look at the other side of the coin of human universals by asking which cognitive abilities humans share with other primates because they represent shared ancestral traits. They review the literature on cognitive abilities and the social behavior of strepsirrhine primates, which represent the best living models of ancestral primates. The available evidence suggests that strepsirrhines are by no means inferior to New World primates on a number of tests of technical intelligence, so that there appears to be a solid cognitive baseline common to all primates. In the realm of social behavior, however, group-living lemurs differ in a number of details, including coalition formation. Whether these discrepancies reflect a lack of social intelligence in lemurs or adaptations to particularly competitive regimes remains an open question for the time being.

Innovation and culture. The existence of multiple traditions that are transmitted via social learning is a hallmark of human societies. Once thought to be one of the main human universals, culture is now also known to have deep roots among the common ancestors of humans and other great apes. In Chap. 20, *Andrew Whiten* explores the depth and nature of these roots. He provides a useful summary of the culture debate and discusses the results of natural observations and clever experiments with chimpanzees to sharpen the distinction between different components and mechanisms of culture in these two species. He shows that there are many similarities in the patterns and mechanisms of cultural behaviors, but finds major differences in the complexity of human culture.

In Chap. 21, *Richard McElreath* describes human culture as an effective inheritance system for ecological and social information. In addition to genetic information and individual learning during development, socially mediated transfer of information provides a very flexible mechanism to accumulate locally relevant knowledge. He is interested in understanding how and why the human genome has learned to extract and transmit environmental information in such a unique and complex manner. His focus is on the origins of culture. How could it get started initially in the absence of large amounts of information, and, hence, immediate benefits? In this chapter, he introduces a model that suggests a possible scenario and highlights the important role of innovations in the initial process.

In Chap. 22, *Carel van Schaik* and *Judith Burkhart* develop the hypothesis that the need for assistance in rearing offspring and the development of cooperative or communal breeding systems has favored the evolution of the suite of derived traits that distinguish humans from other primates. Their chapter integrates the findings of many of the previous chapters in the volume, and provides a powerful example of the value of the comparative approach for understanding what makes us human.

Taken together, the chapters in this book provide the context for understanding the similarities and differences between humans and other primates. The data reviewed here provide fertile ground for developing and testing additional hypotheses about the origins and adaptive value of universal human traits, and for evaluating competing claims about the significance of the traits that distinguish us from other primates. The chapters in this book illuminate the magnitude and historical depth of the gap between humans and other primates, and help us to understand why and how our ancestors traversed the particular historical path that brings us to the present.

Acknowlegements

This volume is largely based on contributions to a conference held in Göttingen (Germany), in December 2007, the *VI. Göttinger Freilandtage*. We thank the Deutsche Forschungsgemeinschaft (DFG), the Deutsches Primatenzentrum (DPZ), the City of Göttingen, and the University of Göttingen for their support of this conference.

We subsequently solicited several additional contributions for this volume to cover topics not addressed during the conference. We are particularly grateful to those colleagues for contributing chapters to this volume at much shorter notice, and we appreciate their professional collegiality.

Every chapter of this volume was peer-reviewed, and we thank the following colleagues for providing helpful and constructive comments on at least one chapter: Filippo Aureli, Josep Call, Bernard Chapais, Dorothy Cheney, Lee Cronk, Jean-Louis Desalles, Frans de Waal, Charles Efferson, Nathan Emery, Azar Gat, Sue Healy, Karin Isler, Richard McElreath, Joseph Manson, John Mikhail, Susan Perry, Markus Port, Hannes Rakoczy, Laurie Santos, Brooke Scelza, Robert Seyfarth, Bernard Thierry, Michael Tomasello, Peter Turchin, Carel van Schaik, Andrew Whiten, Michael Wilson, and Richard Wrangham.

Special thanks to Ulrike Walbaum for formatting chapters, figures, and tables, and for carefully double-checking all references. P.M. Kappeler thanks Claudia, Theresa, and Jakob for moral support. J.B. Silk thanks P.M. Kappeler for the opportunity to help shape this volume.

Göttingen and Los Angeles
May 2009

Peter Kappeler & Joan Silk

Contents

Contributors

Laura Betzig The Adaptationist Program, Ann Arbor, MI, USA, lbetzig@gmail.com

David F. Bjorklund Department of Psychology, Florida Atlantic University, Boca Raton, FL, USA, dbjorklu@fau.edu

Monique Borgerhoff Mulder Department of Anthropology, University of California at Davis, Davis, CA, USA, mborgerhoffmulder@ucdavis.edu

Robert Boyd Department of Anthropology, University of California, Los Angeles, Los Angeles, CA, USA, rboyd@anthro.ucla.edu

Judith M. Burkart Anthropological Institute and Museum, University of Zürich, Zürich, Switzerland, judith.burkart@access.uzh.ch

Kayla Causey Department of Psychology, Florida Atlantic University, Boca Raton, FL, USA, klabeth@mac.com

Bernard Chapais Department of Anthropology, University of Montreal, Montreal, Canada, bernard.chapais@Umontreal.Ca

Dorothy L. Cheney Department of Biology, University of Pennsylvania, Philadelphia, PA, USA, cheney@sas.upenn.edu

Margaret C. Crofoot Department of Anthropology, Harvard University, Cambridge, MA, USA, crofoot@fas.harvard.edu

Robin I. M. Dunbar Institute of Cognitive & Evolutionary Anthropology, University of Oxford, Oxford, England, robin.dunbar@anthro.ox.ac.uk

Daniel M. T. Fessler Center for Behavior, Evolution, & Culture, Department of Anthropology, University of Anthropology, Los Angeles, CA, USA dfessler@anthro.ucla.edu

Claudia Fichtel Department of Behavioral Ecology & Sociobiology, German Primate Center, Göttingen, Germany, claudia.fichtel@gwdg.de

Azar Gat Department of Political Sciences, Tel-Aviv University, Tel-Aviv, Israel, Azargat@post.tau.ac.il

Matthew Gervais Department of Anthropology, Center for Behavior, Evolution, & Culture, University of California Los Angeles, Los Angeles, CA, USA, mgervais@ucla.edu

Peter M. Kappeler Department of Sociobiology/Anthropology, University of Göttingen, Göttingen, Germany, pkappel@gwdg.de

Chris Knight Department of Anthropology, School of Social Sciences, Media, and Cultural Studies, University of East London, London, England Chris.Knight@uel.ac.uk

Venkat Lakshminarayanan Department of Psychology, Yale University, New Haven, CT, USA, venkat.lakshminarayanan@yale.edu

Richard McElreath Department of Anthropology, University of Utah, Salt Lake City, UT, USA, rmcelreath@gmail.com

Henrike Moll Max Planck Institute for Evolutionary Anthropology, Leipzig, Germany, moll@eva.mpg.de

Ryne A. Palombit Department of Anthropology, Center for Human Evolutionary Studies, Rutgers University, New Brunswick, NJ, USA, rpalombit@anthropology.rutgers.edu

Virginia Periss Department of Psychology, Florida Atlantic University, Boca Raton, FL, USA, vperiss@gmail.com

Aimée M. Plourde AHRC Centre for the Evolution of Cultural Diversity, Institute of Archaeology, University College London, London, England, aimee.plourde@ucl.ac.uk

Laurie R. Santos Department of Psychology, Yale University, New Haven, CT, USA, laurie.santos@yale.edu

Robert M. Seyfarth Department of Psychology, University of Pennsylvania, Philadelphia, PA, USA, seyfarth@psych.upenn.edu

Joan B. Silk Department of Anthropology, University of California, Los Angeles, Los Angeles, CA, USA, jsilk@anthro.ucla.edu

Michael Tomasello Max Planck Institute for Evolutionary Anthropology, Leipzig, Germany, tomasello@eva.mpg.de

Robert L. Trivers Institute for Advanced Studies, Berlin, Germany, Department of Anthropology, Rutgers University, Rutgers, NJ, USA, trivers@rci.rutgers.edu

Carel P. van Schaik Anthropological Institute and Museum, University of Zürich, Zürich, Switzerland, vschaik@aim.unizh.ch

David Watts Department of Anthropology, Yale University, New Haven, CT, USA, david.watts@yale.edu

Andrew Whiten Centre for Social Learning & Cognitive Evolution and Scottish Primate Research Group, School of Psychology, University of St. Andrews, St. Andrews, Scotland, aw2@st-andrews.ac.uk

Richard W. Wrangham Department of Anthropology, Harvard University, Cambridge, MA, USA, wranghamrichard@gmail.com

Part I
Introduction

Chapter 1
Primate Behavior and Human Universals: Exploring the Gap

Peter M. Kappeler, Joan S. Silk, Judith M. Burkart, and Carel P. van Schaik

1.1 Introduction

What makes us human? This question has occupied people for millennia. A conclusive answer continues to elude us as we learn more about ourselves and other animals. A series of important discoveries over the last 50 years have led us to largely abandon the search for single traits that are unique to humans. We now know that tool use, language-like communication, lethal intergroup aggression, and an ability to anticipate future events can also be found in other species. However, humans are still quite different from other animals. So, the principal question has become: "What is the nature and the width of the gap that separates humans from primates and other animals?" This edited volume features a collection of essays by primatologists, anthropologists, biologists, and psychologists, who offer some partial answers to this question. In this introductory chapter, we briefly outline the background of this fundamental question about human universals and explain our emphasis on behavioral traits.

P.M. Kappeler (✉)
Department of Sociobiology/Anthropology, University of Göttingen, Göttingen, Germany
e-mail: pkappel@gwdg.de

J.S. Silk
Department of Anthropology, University of California, Los Angeles, CA, USA
e-mail: jsilk@anthro.ucla.edu

J.M. Burkart and C.P. van Schaik
Anthropological Institute and Museum, University of Zürich, Zürich, Switzerland
e-mail: judith.burkart@access.uzh.ch, vschaik@aim.unizh.ch

P.M. Kappeler and J.B. Silk (eds.), *Mind the Gap*,
DOI 10.1007/978-3-642-02725-3_1, © Springer-Verlag Berlin Heidelberg 2010

1.2 The Gap is Behavioral

An obvious place to start an inquiry into the distinctiveness of *Homo sapiens*, at least for anthropologists, is the formal description of our species. When Carolus Linnaeus provided the first scientific description of humans in 1758, he deviated from the rules he had developed for the definition of other species of plants and animals in two ways. First, he did not designate a holotype. Instead, his own remains were declared as a lectotype 200 years later (Stearn 1959). Second, and more importantly, instead of presenting the usual differential diagnosis of anatomical traits, he provided only a prompt: "*Nosce te ipsum!*" ("know thyself"). It was, therefore, left to later anatomists to identify and describe the few anatomical synapomorphies that distinguish our species from the other hominids, such as bipedalism, lack of an opposable big toe, an enlarged neocortex, and permanent breasts (Lovejoy 1981). Most other anatomical and physiological traits that make up the human body can be explained as homologies, reflecting our biological past as chordates, vertebrates, tetrapods, amniota, mammals, and primates, respectively.

Similarly, the analysis of the hominid fossil record provides a rough outline of the timing and sequence of anatomical changes leading up to the emergence of *H. sapiens* about 160,000 years ago, but most of the details have to do with bipedalism, changes in dentition and cranial volume, or they reflect changes in degree (e.g., in skull shape or brain size) rather than fundamental innovations (Henke and Tattersall 2007). Thus, comparative anatomists and paleoanthropologists can clearly identify a human being and distinguish it unequivocally from our closest biological relatives in the present or past, but their list of criteria does not answer the big question in a manner that would satisfy scholars of other disciplines.

The first complete sequencing of the chimpanzee genome (Mikkelsen et al. 2005) revealed that the genetic blueprints of humans and our closest living relatives are 98.77% identical. This result raises two important questions. First, given larger genetic differences between some other primate species, one can ask whether the separation of humans and chimpanzees at the species and, especially, at the genus level, is justified (cf. Diamond 1992). However, species concepts and definitions continue to be in flux (de Queiroz 2005), so the question about the formal taxonomic status of *H. sapiens* is, perhaps, only an academic one.

Second, the comparison of the human and chimpanzee genome suggests that everything that distinguishes us from chimpanzees must be encoded in the very small amount of uniquely human DNA. This hypothesis is based on the assumption that all morphological, physiological, and behavioral traits are controlled by the genes that we can sequence. If we assume that the majority of the interesting and fundamental human universals are directly or indirectly linked to our behavior (see below), this explanation could only be correct if relatively small genetic differences correspond to big behavioral differences. In voles (*Microtus* spp.), for example, relatively minor genetic differences in a vasopressin receptor gene correspond to major species differences in social organization and the mating system (Hammock and Young 2005). However, a recent study of the same vasopressin receptor gene in 12 Old World primates with variable mating systems revealed no covariation (Rosso et al. 2008).

Another potential example for minor genetic differences with major behavioral consequences is provided by the FOXP2 gene, which is critically involved in the control of the neural circuitry controlling speech and language (Vargha-Khadem et al. 2005). Its present form in humans, which differs from that of other great apes by only a few mutations, has been present for about 200,000 years, roughly coinciding with the emergence of modern humans (Enard et al. 2002). Similarly, *Microcephalin*, a gene involved in the regulation of brain growth, is more variable in humans than in other primates, and it has been under positive selection since the origin of the last common ancestor of humans and great apes (Wang and Su 2004). Moreover, one genetic variant of this gene in humans has been under positive selection in the past 40,000 years (Evans et al. 2005). Thus, some important behavioral innovations of humans appear to have a genetic underpinning, but, crucially, it remains largely unclear to what extent which aspects of human behavior are under direct genetic control.

Since humans do not differ qualitatively in their anatomy from great apes, except for the adaptations related to bipedalism, and because the genetic differences between these taxa – to the extent that we understand them beyond the primary sequences – do not appear to be tremendous, the main difference must and does exist in the realm of behavior and cognition. There is indeed little doubt that *H. sapiens* is the most intelligent and socially complex animal. Human cultural and technological achievements, powered by our large brains and capacities for language, are astounding. Within a few thousand years, we have come to build spacecraft that explore the solar system, work with nuclear power, manipulate the genome of plants and animals, eradicate and heal diseases, and transmit information instantaneously around the globe via computer technologies. It is widely accepted that intelligence and rationality are the salient driving forces of human behavior, which facilitate all those achievements. However, the very same rational and intelligent individuals engage in futile contests over social status, discriminate against members of other social groups, and rape, torture, or loot whenever social control mechanisms fail. But humans also donate money to support common welfare, help strangers, respond in predictable ways to particular stimuli of beauty or emotion, and consistently exhibit sex differences in many aspects of social behavior across cultures. Evolutionary processes, therefore, have also profoundly shaped the patterns of human social organization and behavior. How exactly evolutionary and cultural mechanisms interact in shaping human social behavior is still to be discovered. What is clear, however, is that the gap is behavioral and cognitive; what is less clear is how wide it is.

1.3 A Brief History of the Gap

The questions as to how and why humans differ from (other) animals have occupied philosophers, theologians, psychologists, and anthropologists long before the genetic basis of adaptations was discovered. Their reflections have been summarized and discussed at length, so that only a few examples may suffice to illustrate pre-Darwinian attempts to address this question. Aristotle, for example, pioneered a broad comparative approach by carefully comparing details of anatomy, reproduction, and behavior of the animals familiar to him. Humans were explicitly included, as they represented after all the best-known species, and he observed that, in contrast to animals, "*humans are not exclusively occupied with the two basic purposes of life: to maintain themselves and to perpetuate their kind.*" This first indirect allusion to human culture was to dominate this debate 2,400 years later.

Other eminent philosophers focused more specifically on the human psyche. Immanuel Kant (1797/1798), for example, concluded that "*the ability to set themselves any kind of purpose is what sets humans apart from animals.*" This ability exemplifies one aspect of human rationality, long thought to be our main distinguishing feature as a species. Ever since Plato, it had been held that the human mind and matter are two ontologically separate categories, giving rise to a philosophy of dualism (Descartes 1641) that is also in accordance with many religious beliefs, whose influences have dominated this discussion for centuries. The scientific study of the mind of animals began in earnest only in the late twentieth century (Griffin 1976), however, so that these assertions about (unique) qualities of the human mind remained unchallenged for a long time.

Charles Darwin revolutionized human self-conception. He not only developed a theory that firmly established man's place in nature (1859), but he also made influential speculations about unique human traits and their origin (1871). He realized the importance of our intellectual abilities in explaining our success as a species "*...the intellect must have been all-important to [man], even at a very remote period, enabling him to use language, to invent and make weapons, tools, traps etc; by which means, in combination with his social habits, he long ago became the most dominant of all living creatures.*" He also noted that we have "special social instincts," which underlie our unusual cooperative tendencies: "*These instincts (of moral qualities), are of a highly complex nature, and in the case of the lower animals give special tendencies towards certain definite actions; but the more important elements for us are love, and the distinct emotion of sympathy.*" While arguing that these instincts have "*in all probability been acquired through natural selection,*" he also maintained that some aspects of our social behavior must have a different origin: "*man resembles those forms, called by naturalists protean or polymorphic, which have remained extremely variable, owing, as it sees, to their variations being of an indifferent nature, and consequently to their having escaped the action of natural selection.*"

The notion that the capacity for culture has profoundly transformed evolutionary dynamics within the human species gained momentum among social and natural scientists alike. Sigmund Freud ((1927) 2005), for example, was even more specific and considered human culture as the hallmark of humans: "*Human culture – and I mean all that in which human life extolled over its animalistic conditions and distinguishes itself from the life of animals.*" The influential social anthropologist Leslie White (1949) made Darwin's second point explicit: "*Culture may thus be considered as a self-contained, self-determined process; one that can be explained only in terms of itself.*" But also eminent evolutionary biologists, such as Theodosius Dobzhansky, saw human culture as a unique phenomenon or process that is beyond the influence of natural selection: "*Man is a unique product of evolution in that he, far more than any other creature, has escaped from the bondage of the physical and biological into the multiform social environment*" (Dobzhansky and Montagu 1947) and "*culture is an adaptive mechanism supplemental to, but not incompatible with biological adaptation*" (Dobzhansky 1961).

This question about the control of human behavior has played both a central and crucial role in the debate about human uniqueness. The humanities and social sciences in western society developed a *Weltanschauung*, in which humans were completely isolated from the rest of nature and all psychological traits beyond our sensory abilities and a small number of basic, general-purpose rules guiding homeostatic behaviors were considered the product of socially constructed learning, socialization processes, and conscious reasoning. In following the behaviorists' paradigm, many social scientists maintained that humans are emancipated from the genetic control of behavior, and the resulting ability to create and perpetuate complex culture sets us apart from primates and all other animals (Lewontin et al. 1984). This view assumed that "*Man is not committed in detail by his biological constitution to any particular variety of behavior,*" and as a result, "*culturally conditioned responses make up the greater part of our huge equipment of automatic behavior*" (Benedict (1934) 2005). Thus, "*evolved psychological components place only the broadest constraints on what a human mind can become*" (Mameli 2007).

The aversion of the social sciences and humanities to evolutionary analyses of human behavior had much to do with early attempts to biologize human behavior, which emphasized the genetic determinism of eugenics and reified racial classifications and their presumed mental correlates. This work was eventually used to justify restrictive immigration policies, unpalatable social policies, and even genocide. This sordid history induced a collective denial of any biological influence on the mind and behavior of humans, and was facilitated by the notion that to explain human behavior we only needed to turn to culture, which hermetically sealed human behavior against any biological influences. In Ridley's (1996) words: "*today [cultural] anthropologists demand that the existence of culture, reason or language exempt us from biology.*" Thus, the isolation of the social sciences and humanities from biology was understandable, if unfortunate.

1.4 Explaining the Gap

One might argue that the social sciences have been considerably less successful in building a strongly predictive, integrative theoretical structure than their counterparts in the life sciences. The proper infusion of biological thinking might improve their explanatory power. To a modern biologist, the so-called "standard social sciences model" (Tooby and Cosmides 1992) almost reads like a creationist manifesto. It produced positions such as the Seville Statement on Violence (1986, see de Waal 1993), which in retrospect are remarkable in their denial of any biological influences on human behavior.

All this began to change after 1975, when E.O. Wilson published *Sociobiology: The New Synthesis*. Wilson generated vehement controversy over his last chapter, in which he boldly extended the evolutionary approach to the study of humans: "*In this macroscopic view, the humanities and social sciences shrink to specialized branches of biology; history, biography and fiction are the research protocols of human ethology; and anthropology and psychology together constitute the sociobiology of a single primate species*." In the beginning, sociobiology largely focused on behavior, rather than its cognitive and emotional underpinnings (but see Trivers 1971), and also assumed that human behavior largely reflected genetically canalized adaptations rather than culture. Still, it rekindled the debate about biological influences on human nature.

From among the various movements that applied evolutionary thinking to humans (Laland and Brown 2002), one emerged as dominant during the 1990s. Evolutionary Psychology (EP) focused on psychological mechanisms (modules) that underlie behavior and decision-making (Tooby and Cosmides 1992), but largely ignored the fitness consequences of actual behavior in contemporary settings. EP assumed that our cognitive abilities are massively modular. These modules evolved during the Pleistocene, when hominins were hunter-gatherers living in small groups in savanna-like habitats, called the "environment of evolutionary adaptedness." Hence, a certain number of these psychological mechanisms exist as adaptations to problems that we no longer face. They can be identified by the so-called evolutionary functional analysis, which amounts to imagining the problems faced by Pleistocene foragers and suggesting plausible solutions as hypotheses for psychological mechanisms. This line of reasoning elegantly explained the existence of patently maladaptive behaviors in modern humans.

EP generated important insights into sex differences in the criteria for mate choice and clarified the notions of standards of beauty, attractiveness, sexual jealousy (Buss 1994), the incidence of rape (Thornhill and Palmer 2000), patterns of infanticide and aggression, including murder (Daly and Wilson 1988), aggression by male coalitions (Tooby and Cosmides 1988, in Buss 2008), and military history (Diamond 1997; Turchin 2003). It also inspired the functional analysis of disease, both physical and mental (Nesse and Williams 1995), art, music, and dance (Miller 2000), and even literature (Carroll 2007). EP nicely explained why some phobias are much more prevalent than others, in blatant disproportion to current risks (e.g., we are more afraid of snakes than car crashes), and explained our addiction to sweet, fatty, and salty foods. These fears and appetites made perfect sense under hunter-gatherer conditions but are hopelessly maladaptive in our current environment.

This approach has not been without its critics, however (Scher and Rauscher 2003; Buller 2005; Mameli 2007). Among the various points of criticism, two are especially relevant here. First, echoing Wilson's (1978) famous dictum that "genes hold culture on a leash," EP downplays the explanatory power of culture and the force of the autonomous forces that culture creates. Variation in human behavior is due to interactions between a universal set of adaptations shared by all people and the external conditions a child encounters during development. This can produce alternative psychological phenotypes within populations, but can also create variation across populations that may look deceptively like culture (i.e., behavioral innovations that spread and are maintained by often conformity-based social learning), but really is not – EP calls this variation "evoked culture." This shortcoming has been rectified by the successful development of cultural evolution theory (Richerson and Boyd 2005; Henrich and McElreath 2007), which recognizes that much variation in human behavior and the human mind can be due to historically determined processes of innovation and biased social transmission: "*Culture is on a leash, all right, but the dog on the end is big, smart, and independent. On any given walk, it is hard to tell who is leading who*" (Richerson and Boyd 2005).

Second, and most relevant to our argument, EP operates in a historical and phylogenetic vacuum because it ignores the distinction between ancestral and derived human features. EP implicitly assumes that all interesting human traits are derived responses to Pleistocene conditions. Yet, human evolution did not start in the Pleistocene and many of our traits may have been around much longer (Fichtel and Kappeler, this volume; Whiten, this volume). Due to the explosive growth of detailed, long-term primatological studies over the past 50 years, we have now, for the first time in the history of our species, a detailed picture of our closest living relatives.

1.5 Primatology and the Gap

The findings of behavioral primatologists have been spectacular. Monkeys and apes form long-term social relationships that they use to exchange services and favors, including grooming, sex, and coalitionary support in conflicts (Silk and Boyd, this volume); reconcile when they have conflicts; commit infanticide and deploy various social and sexual strategies to reduce that risk (Palombit, this volume); demonstrate sophisticated and surprisingly detailed knowledge about the social goings-on in their groups and, to some extent, in the population at large (Watts, this volume; Dunbar, this volume); grow slowly and acquire numerous social and subsistence skills, in part by social learning, and thus show signs of culture (Whiten, this volume); and so on. Chimpanzees also engage in lethal intergroup aggression, up to the point of eliminating males from neighboring communities (Crofoot and Wrangham, this volume), they make and use tools, and field studies continue to refine our knowledge of how cultural traditions are maintained in wild populations (Matsuzawa 1994; Matsuzawa et al. 2001).

As a result, we now know more about primate behavior than about the behavior of almost any other taxon except temperate diurnal birds. Along with developments in molecular biology, which showed that humans are African great apes, who split off from the rest of the hominoid lineages a mere 6–8 mya (Glazko and Nei 2003), everybody should now be fully aware of our behavioral and genetic similarity to the great apes, and to primates more generally. Thus, we can no longer afford to ignore our primate roots, and must explore the consequences of these hard facts. Indeed, primatologists have gradually begun to use the broad insights into primate behavior to weigh in on topics such as the evolution of human sexuality (Hrdy 1997), language (Savage-Rumbaugh 1999, Cheney and Seyfarth, this volume), parenting (Hrdy 2009), between-group violence (Wrangham and Peterson 1996; (Crofoot and Wrangham, this volume), technology and culture (McGrew 2004; van Schaik 2004; Laland and Galef 2009), and morality (Ridley 1996; de Waal 2006). The development of these ideas has been largely independent of work within the EP movement.

Comparative primatology looks for evidence of both convergence (homoplasy) and common descent (homology) in specific traits. At first, it may seem futile to look for convergences in traits that seem to be unique, but each complex trait can be found to have elements in common with traits in other, sometimes distantly related lineages, which can shed light on their function in each lineage. Culture is a prime example. Although human culture is clearly different than culture in other taxa, it shares some elements with the cultural traditions of other primates, particularly great apes (Whiten, this volume). This approach, thus, enriches our understanding of human evolution. Second, by explicitly reconstructing the ancestral states of human traits, primatology can distinguish between shared and derived human features, something the history-free EP approach is unable to do. Ironically, the examples we quoted above from EP tend to refer to behavioral tendencies we share with many other organisms, whereas primatologists have generally focused on explaining the most clearly derived ones, building on observations on nonhuman primates.

Like any other species, *H. sapiens* is connected to its relatives by descent from common ancestors. No species is fundamentally distinct from its close relatives, certainly not if they shared a common ancestors as recently as humans and the two chimpanzees. The similarities between humans and apes generated by the research of primatologists are numerous, and they do not require any other explanation than that they have been present for a long time, and apparently are not patently maladaptive, allowing them to persist in both taxa. However, in the euphoria of finding numerous fundamental similarities and in defiance of the remaining defenders of human uniqueness, many have proclaimed that humans are not fundamentally different from the other great apes. Fundamentalists can argue that the differences are nonessential features that have been simply layered onto the primate core. We are just a "third chimpanzee" (Diamond 1992), with an "inner ape" inside of us (de Waal 2005). Hence, there are quantitative, not qualitative differences between us and other primates.

Nothing could be further from the truth. Every species is not just connected to others, it is also unique, or else it would not be a separate species. Perhaps the most remarkable thing about humans from a comparative perspective is how different we have become from our fellow great apes in the rather short time that separates us from them. We will briefly summarize these differences below. The real challenge is to explain these derived traits. Which of the myriad aspects of the mind and behavior of humans are unique, and why did they evolve only in our own species?

1.6 Uniquely Human

We noted above that defining humans is a parlor game with roots going back to the Greek philosophers, but during the past decades, this exercise has acquired a more solid foundation based on comparative analyses. But the difficulties of adequately characterizing humans as a species and distinguishing humans from other primate taxa are often overlooked. The overwhelming majority of humans now live in settled societies, surrounded by written texts or even more modern media technology, in very large societies that have multilayered organizations and with numerous institutions. However, all of this is very recent, with the oldest elements not even 12,000 years old. The few remaining hunter-gatherers have life styles that are most similar to those in which humans lived for most of their history, including natural levels of fertility (Hawkes 2006). In addition, we can assume that human universals reflect our most ancient human roots. This relies on the argument that if humans display a common trait across our vast range of social and environmental conditions, this common trait must also have been in the early modern humans that evolved in Africa, and then populated the world. Especially, where the two sets of traits overlap, we can have some confidence in their deep roots. Finally, important insights into human nature have emanated recently from cross-cultural experimental studies of human economic choices (e.g., Fehr and Rockenbach 2004; Henrich et al. 2005).

These developments have made it possible to update and organize the existing lists of the derived traits of humans relative to the reconstructed traits of the last common ancestor. Whole books have been devoted to the subject (e.g., Antweiler 2007), and this is not the place to develop a detailed list. But we want to do two things here: first, to clarify the procedure and, second, to present a selected summary of the major differences between ourselves and other primates.

A summary of derived features requires that we not only know the differences but also their polarity. A difference between two sister taxa can be due to a change in one, a change in the other, or a divergence in both. Polarity in mind and behavior is usually fairly straightforward (the issue is much less obvious at the genomic level), but to make sure, it is useful to compare humans with the genus *Pan* and with the other great apes, as well. Fortunately, in spite of their remarkably variable social organization and subsistence, the great apes, as a group, are rather homogeneous with respect to cognition (Deaner et al. 2006; Burkart et al. in press), brain size (Schoenemann 2006), and life history (Robson et al. 2006). This homogeneity implies that they are rather conservative, making it easier to infer polarity.

A summary of the nonmorphological and nonphysiological features that are derived in humans relative to the great apes and not discussed in subsequent chapters (see also Flinn et al. 2005; Richerson and Boyd 2005; Burkart et al. in press) would include the following unusual features:

- Cumulative material culture and social institutions and rituals, all critically dependent on language. Culture is perhaps our preeminent adaptation.
- Unusual subsistence ecology, involving skill-intensive hunting and gathering and extremely intense cooperation, also in between-group conflict (Ridley 1996; Kaplan et al. 2000; Gurven 2004).
- Slower development, longer lifespan, accompanied by higher female reproductive rates and midlife menopause (Robson et al. 2006) and extensive allomaternal help (Hrdy 2009).
- Unusual cognitive abilities, including language, long-term planning, causal understanding, and episodic memory. These abilities build on *shared intentionality*, i.e., the ability to participate with others in collaborative activities with shared goals and intentions (Tomasello and Moll, this volume), which also involves language-based teaching.

So, there is a gap, and it would be foolish to deny it. A mere extrapolation of any of the great ape phenomena is very unlikely to explain the dramatic differences, and it would seem that what we are looking for is a set of selective pressures not encountered by the other great apes or a confluence of capacities, ecological circumstances, and a certain amount of serendipity that set our ancestors on a different path than other great apes. This could be a completely novel set of pressures encountered by no species before, such as pressures emanating from cultural evolution (see Silk and Boyd, this volume; McElreath, this volume). However, that still begs the question why cultural evolution became so much stronger in humans than in the other apes. Thus, we should also look for selective pressures that are novel for the apes but convergently present among other primates, other mammals, or among birds, and which may have precipitated the operation of the truly unique cultural selection. It seems unlikely that we will ever settle on a single account of how we became such an unusual species. But we now have a much richer body of theory and comparative data that allow us to develop more complete and compelling hypotheses that we can critically evaluate.

References

Antweiler C (2007) Was ist den Menschen gemeinsam? Über Kultur und Kulturen, Wissenschaftliche Buchgesellschaft, Darmstadt

Benedict R [1934] (2005) Patterns of culture. Mariner, Boston

Buller DJ (2005) Adapting minds: evolutionary psychology and the persistent quest for human nature. MIT Press, Cambridge, MA

Burkart JM, Hrdy SB, van Schaik CP. Cooperative breeding and human cognitive evolution. Evol Anthropol (in press)

Buss DM (1994) The evolution of desire: strategies of human mating. HarperCollins, New York

Buss DM (2008) Evolutionary psychology: the new science of the mind, 3rd edn. Pearson, Boston

Carroll J (2007) Evolutionary approaches to literature and drama. In: Dunbar RIM, Barrett L (eds) Oxford handbook of evolutionary psychology. Oxford University Press, Oxford, pp 637–648
Daly M, Wilson M (1988) Homicide. Aldine de Gruyter, New York
Darwin C (1859) On the origin of species. Murray, London
Darwin C (1871) The descent of man and selection in relation to sex. Murray, London
Deaner RO, van Schaik CP, Johnson V (2006) Do some taxa have better domain-general cognition than others? A meta-analysis of nonhuman primate studies. Evol Psychol 4:149–196
de Queiroz K (2005) Ernst Mayr and the modern concept of species. Proc Natl Acad Sci USA 102:6600–6607
Descartes R (1641/1986) Meditationes de prima philosophia. Meditationen über die Erste Philosophie. Reclam, Stuttgart
de Waal FBM (1993) Aggression as a well-integrated part of primate social relationships: a critique of the Seville statement on violence. In: Silverberg J, Gray JP (eds) Aggression and peacefulness in humans and other primates. Oxford University Press, New York, pp 37–56
de Waal FBM (2005) Our inner ape. Riverhead Books, New York
de Waal FBM (2006) Primates and philosophers: how morality evolved. Princeton University Press, Princeton, NJ
Diamond J (1992) The third chimpanzee: the evolution and future of the human animal. HarperCollins, New York
Diamond J (1997) Guns, germs, and steel. Norton, New York
Dobzhansky T (1961) Man and natural selection. Am Sci 49:285–299
Dobzhansky T, Montagu MFA (1947) Natural selection and the mental capacities of mankind. Science 105:587–590
Enard W, Przeworski M, Fisher SE, Lai CSL, Wiebe V, Kitano T, Monaco AP, Pääbo S (2002) Molecular evolution of FOXP2, a gene involved in speech and language. Nature 418:869–872
Evans PD, Gilbert SL, Mekel-Bobrov N, Vallender EJ, Anderson JR, Vaez-Azizi LM, Tishkoff SA, Hudson RR, Lahn BT (2005) *Microcephalin*, a gene regulating brain size, continues to evolve adaptively in humans. Science 309:1717–1720
Fehr E, Rockenbach B (2004) Human altruism: economic, neural and evolutionary perspectives. Curr Opin Neurobiol 14:784–790
Flinn MV, Geary DC, Ward CV (2005) Ecological dominance, social competition, and coalitionary arms races: why humans evolved extraordinary intelligence. Evol Hum Behav 26:10–46
Freud S [1927] (2005) Die Zukunft einer Illusion. Fischer, Frankfurt
Glazko GV, Nei M (2003) Estimation of divergence times for major lineages of primate species. Mol Biol Evol 20:424–434
Griffin DR (1976) The question of animal awareness: evolutionary continuity of mental experience. Rockefeller University Press, New York
Gurven M (2004) To give and to give not: the behavioral ecology of human food transfers. Behav Brain Sci 27:543–583
Hammock EAD, Young LJ (2005) Microsatellite instability generates diversity in brain and sociobehavioral traits. Science 308:1630–1634
Hawkes K (2006) Life history theory and human evolution: a chronicle of ideas and findings. In: Hawkes K, Paine RR (eds) The evolution of human life history. School of American Research Press, Santa Fe, pp 45–93
Henke W, Tattersall I (2007) Handbook of paleoanthropology, 3 Vols. Springer, Berlin
Henrich J, Boyd R, Bowles S, Camerer C, Fehr E, Gintis H, McElreath R, Alvard M, Barr A, Ensminger J, Smith Henrich N, Hill K, Gil-White F, Gurven M, Marlowe FW, Patton JQ, Tracer D (2005) 'Economic man' in cross-cultural perspective: behavioral experiments in 15 small-scale societies. Behav Brain Sci 28:795–855
Henrich J, McElreath R (2007) Dual-inheritance theory: the evolution of human cultural capacities and cultural evolution. In: Dunbar RIM, Barrett L (eds) Oxford handbook of evolutionary psychology. Oxford University Press, Oxford, pp 555–570

Hrdy SB (1997) Raising Darwin's consciousness: female sexuality and the prehominid origins of patriarchy. Hum Nat 8:1–49

Hrdy SB (2009) Mothers and others: the evolutionary origins of mutual understanding. Harvard University Press, Cambridge, MA

Kant I (1797/1998) Metaphysik der Sitten. Reclam, Stuttgart 1990

Kaplan H, Hill K, Lancaster J, Hurtado AM (2000) A theory of human life history evolution: diet, intelligence and longevity. Evol Anthropol 9:156–185

Laland KN, Brown GR (2002) Sense and nonsense: evolutionary perspectives on human behaviour. Oxford University Press, Oxford

Laland KN, Galef BG Jr (2009) The question of animal culture. Harvard University Press, Cambridge, MA

Lewontin RC, Rose S, Kamin LJ (1984) Not in our genes: biology, ideology, and human nature. Random House, New York

Linnaeus C (1758) Tomus I. Systema naturae per regna tria naturae, secundum classes, ordines, genera, species, cum characteribus, differentiis, synonymis, locis. Editio decima, reformata. Laurentii Salvii, Stockholm

Lovejoy CO (1981) The origin of man. Science 211:341–350

Mameli M (2007) Evolution and psychology in philosophical perspective. In: Dunbar RIM, Barrett L (eds) Oxford handbook of evolutionary psychology. Oxford University Press, Oxford, pp 21–34

Matsuzawa T (1994) Field experiments on use of stone tools in the wild. In: Wrangham RW, McGrew WC, de Waal FBM, Heltne PG (eds) Chimpanzee cultures. Harvard University Press, Cambridge, MA, pp 351–370

Matsuzawa T, Biro D, Humle T, Inoue-Nakamura N, Tonooka R, Yamakoshi G (2001) Emergence of culture in wild chimpanzees: education by master-apprenticeship. In: Matsuzawa T (ed) Primate origins of human cognition and behavior. Springer, Tokyo, pp 557–574

McGrew WC (2004) The cultured chimpanzee: reflections on cultural primatology. Cambridge University Press, Cambridge

Mikkelsen TS, Hillier LW, Eichler EE, Zody MC, Jaffe DB, Yang S-P, Enard W, Hellmann I, Lindblad-Toh K, Altheide TK, Archidiacono N, Bork P, Butler J, Chang JL, Cheng Z, Chinwalla AT, de Jong P, Delehaunty KD, Fronick CC, Fulton LL, Gilad Y, Glusman G, Gnerre S, Graves TA, Hayakawa T, Hayden KE, Huang X, Ji H, Kent WJ, King M-C, Kulbokas EJ, Lee MK, Liu G, Lopez-Otin C, Makova KD, Man O, Mardis ER, Mauceli E, Miner TL, Nash WE, Nelson JO, Pääbo S, Patterson NJ, Pohl CS, Pollard KS, Prüfer K, Puente XS, Reich D, Rocchi M, Rosenbloom K, Ruvolo M, Richter DJ, Schaffner SF, Smit AFA, Smith SM, Suyama M, Taylor J, Torrents D, Tuzun E, Varki A, Velasco G, Ventura M, Wallis JW, Wendl MC, Wilson RK, Lander ES, Waterston RH (2005) Initial sequence of the chimpanzee genome and comparison with the human genome. Nature 437:69–87

Miller G (2000) The mating mind: how sexual selection choice shaped the evolution of human nature. Doubleday, New York

Nesse RM, Williams GC (1995) Why we get sick: the new science of Darwinian medicine. Vintage Books, New York

Richerson PJ, Boyd R (2005) Not by genes alone: how culture transformed human evolution. University of Chicago Press, Chicago

Ridley M (1996) The origins of virtue: human instincts and the evolution of cooperation. Viking, New York

Robson SL, van Schaik CP, Hawkes K (2006) The derived features of human life history. In: Hawkes K, Paine RL (eds) The evolution of human life history. School of American Research Press, Santa Fe, pp 17–44

Rosso L, Keller L, Kaessmann H, Hammond RL (2008) Mating system and avpr1a promoter variation in primates. Biol Lett 4:375–378

Savage-Rumbaugh S (1999) Ape language: between a rock and a hard place. In: King BJ (ed) The origins of language: what nonhuman primates can tell us. School of American Research Press, Santa Fe, pp 115–188

Scher SJ, Rauscher F (2003) Evolutionary psychology: alternative approaches. Kluwer, Boston
Schoenemann PT (2006) Evolution of the size and functional areas of the human brain. Annu Rev Anthropol 35:379–406
Stearn WT (1959) The background of Linnaeus's contributions to the nomenclature and methods of systematic biology. Syst Zool 8:4–22
Thornhill R, Palmer C (2000) A natural history of rape: biological bases of sexual coercion. MIT Press, Cambridge, MA
Tooby J, Cosmides L (1992) The psychological foundation of culture. In: Barkow JH, Cosmides L, Tooby J (eds) The adapted mind: evolutionary psychology and the generation of culture. Oxford University Press, Oxford, pp 19–136
Trivers RL (1971) The evolution of reciprocal altruism. Q Rev Biol 46:35–57
Turchin P (2003) Historical dynamics: why states rise and fall. Princeton University Press, Princeton
van Schaik CP (2004) Among orangutans: red apes and the rise of human culture. Harvard University Press, Belknap, Cambridge, MA
Vargha-Khadem F, Gadian DG, Copp A, Mishkin M (2005) FOXP2 and the neuroanatomy of speech and language. Nat Rev Neurosci 6:131–138
Wang Y-Q, Su B (2004) Molecular evolution of *microcephalin*, a gene determining human brain size. Hum Mol Genet 13:1131–1137
White LA (1949) The science of culture: a study of man and civilization. Farrar, Straus and Giroux, New York
Wilson EO (1975) Sociobiology. Harvard University Press, Cambridge, MA
Wilson EO (1978) On human nature. Harvard University Press, Cambridge, MA
Wrangham RW, Peterson D (1996) Demonic males: apes and the origins of human violence. Houghton Mifflin, Boston

Part II
Family & Social Organization

Chapter 2
The Deep Structure of Human Society: Primate Origins and Evolution

Bernard Chapais

> *"...our primate cousins have 'kinship systems' which contain the elements of human kinship systems, but ... no other primate combines elements in the way that we do...The elements are common: the combination is unique. My contention is, therefore, that it is to the combination of elements that we must look for clues to the uniqueness of human systems, not to the elements themselves."*
>
> Robin Fox 1975: 10–11

Abstract On theoretical grounds, one expects all human societies to share a common structural denominator, or *deep social structure*, which would describe both the unity of human society across cultures and its uniqueness in the animal world. Here, I argue that a powerful model of humankind's deep social structure is the concept of reciprocal exogamy described by Claude Lévi-Strauss – a social arrangement in which groups are bound together through the particular linkage of pair-bonds and kinship bonds. The present analysis provides a phylogenetic test of the exogamy model of human social origins. It shows that reciprocal exogamy breaks down into a number of phylogenetically meaningful components and that the evolutionary history of the whole system may be reconstructed parsimoniously in terms of the combination of a *Pan*-like social structure with a new mating system featuring stable breeding bonds. The concept of deep structure points to the following human universals: stable breeding bonds and their correlate, fatherhood; the multifamily community; strong siblingships; bilateral (uterine and agnatic) kin recognition; incest avoidance; out-marriage (exogamy); matrimonial exchange; dual-phase residence (pre/postmarital); lifetime bonds between dispersed kin; bilateral relations between in-laws; kin-biased and affinity-biased marriage rules; and between-group alliances (supragroup levels of social organization).

B. Chapais
Department of Anthropology, University of Montreal, Montreal, Canada
e-mail: bernard.chapais@umontreal.ca

P.M. Kappeler and J.B. Silk (eds.), *Mind the Gap*,
DOI 10.1007/978-3-642-02725-3_2, © Springer-Verlag Berlin Heidelberg 2010

2.1 Introduction

The idea that all chimpanzee societies share a number of structural features that set them apart, collectively, from all other animal societies – that there exists a chimpanzee *deep social structure* – sounds not only reasonable but is also rather obvious. But the same idea applied to human societies is much less evident. Human societies are so variable cross-culturally that the notion that a common structural denominator underlies all of them, past and present, is not easily conceptualized. Yet, all human societies are the product of a unique set of mental constraints and, if only for that reason they must share, at some level of abstraction, a universal deep structure. What that structure is and how it evolved are questions that I addressed at length in a previous book (Chapais 2008), of which the present chapter is a précis. Owing to space constraints, other parts of the book concerned with methodological, theoretical, epistemological, and historical considerations, and with the evolution of descent groups, are largely ignored.

The deep structure of human society can only be abstracted from the comparative analysis of human societies. It is a matter of cross-cultural sociology. Significantly, however, the topic has never been recognized as such by the discipline best positioned to tackle it, namely, sociocultural anthropology – discussions of the nature of human society's deep structure are not found in anthropology textbooks, for example. A number of reasons account for this. Several influential schools of thought in sociocultural anthropology – historical particularism, the Culture and Personality school, cognitive anthropology, symbolic/interpretive anthropology, postmodernist approaches, among others – consistently emphasized the uniqueness of every culture and focused, accordingly, on differences rather than similarities between societies (for general discussions of relativism in anthropological theory, see Harris 1968; Gellner 1985; Barnard 2000; Deliège 2006). These theoretical perspectives stressed the importance of understanding cultures from the inside, kept away from wide-scale cross-cultural comparisons, and consequently from the search of the general principles governing cultural variation, and in some cases, even denied that such principles existed. Other perspectives, such as structural-functionalism (Radcliffe-Brown 1957), Murdock's "statistical comparatism" (Murdock 1949; Goodenough 1970), cultural ecology (Steward 1955), or cultural materialism (Harris 1979) did conceive of anthropology as a comparative science whose main objective was to organize cultural variation and/or cultural change into a set of general principles. But comparativists were harshly criticized for not delivering the generalizations sought for and for producing, instead, principles that were deemed simplistic and reductionist, or self-evident and meaningless. In any case, the comparativists themselves did not address the issue of the deep structure of human society.

For these reasons and many others, sociocultural anthropology came close to skipping the topic altogether, and it did leave it out of its preoccupations as far as an explicit treatment is concerned. Fortunately, however, anthropological research has produced one particularly powerful model of humankind's deep social structure.

In his book "*Les Structures Élémentaires de la Parenté*" (The Elementary Structures of Kinship), published in 1949, Claude Lévi-Strauss, the father of the French structuralist school, implicitly proposed that *reciprocal exogamy* – intermarriage between members of distinct groups – was the defining characteristic of human societies. Lévi-Strauss never discussed reciprocal exogamy in terms of the deep structure of human society and, moreover, he consistently opposed the idea that it should be understood in terms of a chronologically primitive structure. Nonetheless, his description of it fits particularly well with the criteria and attributes one would think of to characterize an entity such as the common denominator of all human societies: namely, a structure (1) that defines the uniqueness of human societies in relation to all other animal societies, (2) whose evolution coincided with the birth of human society, (3) which embodies the unity of human societies, cross-culturally, (4) which is described at such a high level of abstraction that it may be construed as a correlate, in the social sphere, of the human mind, and (5) which reflects the operation of some of the most elementary principles governing human social relationships.

Lévi-Strauss's description of reciprocal exogamy appears to meet all five criteria formally, if not empirically. First, reciprocal exogamy was said to mark "the transition from nature to culture," or the passage from nonhuman to human society. If one assumes that such a transition took place at some point in time, reciprocal exogamy would coincide with the birth of human society and define its uniqueness. Second, the single most important cognitive process involved in reciprocal exogamy – the ability to engage in relationships based on exchange – was described by Lévi-Strauss as a "universal mental structure": in other words, as an integral property of the human mind. At the same time, the identification of women as the "most precious possession" men could exchange was deemed so fundamental a principle that Lévi-Strauss did not even justify this assumption. Taken together, the last two points suggest that the core principle of reciprocal exogamy, matrimonial exchange, would be the outcome of some underlying biological factors. Third, reciprocal exogamy featured what Lévi-Strauss called the "atom of kinship" and which he defined as the most elementary kinship unit and the basic building block of human societies. As we shall see, the atom of kinship is a structure that amalgamates some of the most basic types of human bonds. For all these reasons and from a strictly sociological viewpoint – that is, independently of any evolutionary considerations – Lévi-Strauss's concept of reciprocal exogamy is a strong candidate for the deep structure of human society.

2.2 What is Reciprocal Exogamy?

What follows is a synthetic summary of reciprocal exogamy as far as it relates to my objective of characterizing the deep structure of human society. This summary is written from a primatological and evolutionarily informed perspective. Accordingly, I stress aspects that Lévi-Strauss did not necessarily emphasize, and I use terms that he did not necessarily employ. In particular, I spell out the structural connections between what I consider to be the core features of reciprocal exogamy from a comparative – interspecific – perspective: intermarriage, supragroup kinship networks, alliances between in-laws, the atom of kinship, sister exchange, and marriage between cross-cousins.

Simply stated, reciprocal exogamy is a social arrangement in which groups are bound together through marital unions and kinship. Reciprocal exogamy is best illustrated with the simplest system described by Lévi-Strauss, *restricted exchange* between two exogamous kin groups, A and B (Fig. 2.1). Intermarriage between groups A and B is exemplified here with a single family per group. The two groups are patrilocal and the two families trade their kinswomen to obtain wives in return, building alliances in the process. Upon marriage, wife Ego moves to her husband's group. Because Ego breeds and raises her children there, the A family will have grandchildren, nieces, nephews, and cousins living in group B, with whom they will come into contact when the two families, or the two groups, visit each other. Given that this process is generalized to all marriages and works in both directions, the resulting kinship network encompasses the two intermarrying groups which become quite intricately connected.

Simultaneously, Ego's marriage reinforces bonds between the A and B families because Ego and her husband act as natural intermediaries between their respective families; that is to say, marriage connects and unites in-laws (or affines). Significantly, preferential bonds between in-laws often translate into marriages among them. Two widespread practices are the levirate and the sororate (Murdock 1949: 29). The levirate is the rule according to which a widow must marry the brother of

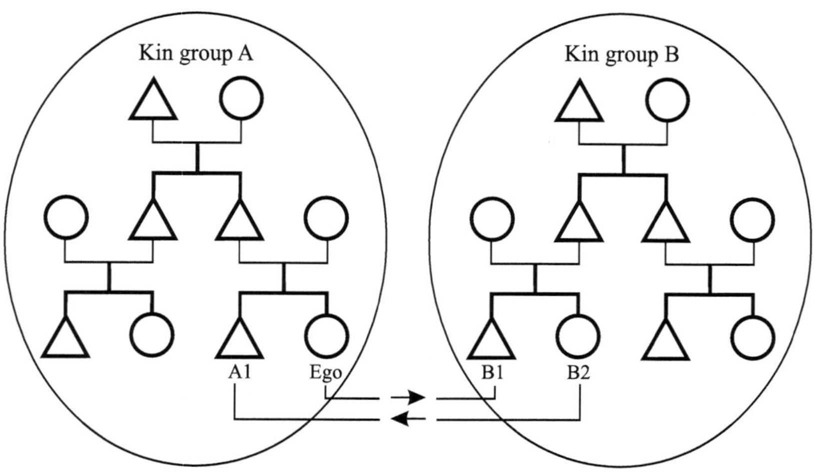

Fig. 2.1 Reciprocal exogamy between two patrilocal kin groups, illustrated by marriage between female Ego and male B1, and marriage between Ego's brother and male B1's sister, the two unions exemplifying sister exchange (or daughter exchange depending on one's viewpoint). *Triangles*: males. *Circles*: females. Thin *U-shaped lines* indicate marriage. *Thick inverted-U lines* indicate siblingships. *Arrows* give the direction of between-group transfer (postmarital residence)

her deceased husband (her brother-in-law). The sororate is the reciprocal rule, a widower marrying the sister of his deceased wife (his sister-in-law). To Lévi-Strauss, the sororate and the levirate were facets of reciprocal obligations between exchanging units.

In sum, reciprocal exogamy binds social groups through two different processes. First, out-marriage distributes close kin across distinct groups, these relatives pursuing their relationship on the long run despite their physical separation, the outcome being further kinship-based bonds between groups. Second, marriage creates or reinforces bonds between more distantly related individuals, the spouses' respective families, generating affinity-based alliances between the groups.

To proceed further with the description of reciprocal exogamy, it is useful to consider how Lévi-Strauss himself accounted for it. His explanation holds in the following assumptions and principles: (1) for some reason, men living in distinct groups needed to ally with each other, (2) reciprocity is a universal mental structure, (3) acts of reciprocity create social partnerships (Mauss 1923), (4) women are the most precious commodities of exchange, (5) men exert some level of control over their kinswomen, and (6) marriage is a means of exchange. From this, it follows that men seeking to form solid alliances with other men would best do so by exchanging their daughters and sisters as spouses for each other. But a major problem crops up at this point. Lévi-Strauss also assumed (7) that men were sexually attracted to their kinswomen – that incest was natural. Therefore, to be in a position to exchange their daughters and sisters, men first had to renounce marrying them, and to achieve this they had to invent the incest prohibition. In that perspective, exogamy and alliance formation are men's ultimate goal, while the incest taboo is the fundamental prerequisite; exogamy and the incest taboo are, thus, two sides of the same coin, and they mark the birth of human society.

With this general picture in mind, we may now introduce what Lévi-Strauss called the *atom of kinship* and which he described as "the most elementary form of kinship that can exist" (1963, p 46) and "the sole building block of more complex systems" (1963, p 48). One might argue that the smallest unit of human kinship is the mother–child bond, but Lévi-Strauss was concerned with social structure, not with dyadic relationships. The atom of kinship rests upon four terms: a brother, his sister, the sister's husband, and their son (Fig. 2.2). A theoretical argument invoked by Lévi-Strauss is that the atom of kinship includes the three types of relations always present in any human kinship structure: a relation of consanguinity (between siblings), a relation of affinity (between spouses), and a relation of descent (between parent and child). A more direct argument, still according to Lévi-Strauss, is that the atom of kinship is the immediate outcome of the incest impediment between brothers and sisters. Owing to the incest taboo, a brother cannot have children with his sister. He, thus, elects to lend her to another male for breeding purposes; so, the sister's children are "the product, indirectly, of the brother's renunciation" as Fox put it (1993: 192). More bluntly, because the brothers cannot reproduce with their sisters, they do so via their sisters' husbands, hence the intimate interconnection between the three basic categories of bonds. It should also be noted that the atom of kinship embodies another chestnut of human kinship

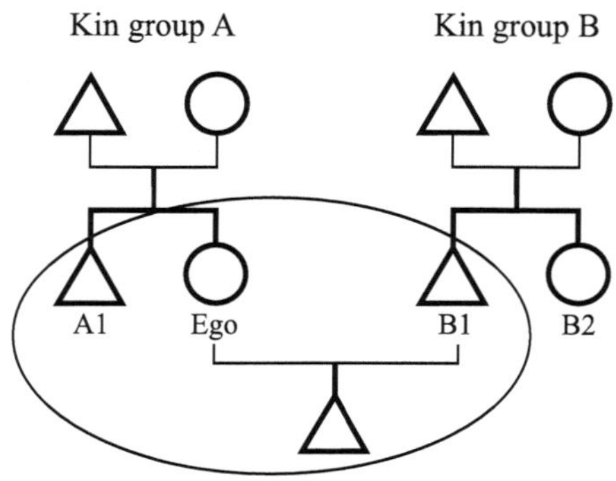

Fig. 2.2 Lévi-Strauss's "atom of kinship" (*circled* individuals). The individuals pictured here are the same as in Fig. 2.1. Definitions of symbols as in Fig. 2.1

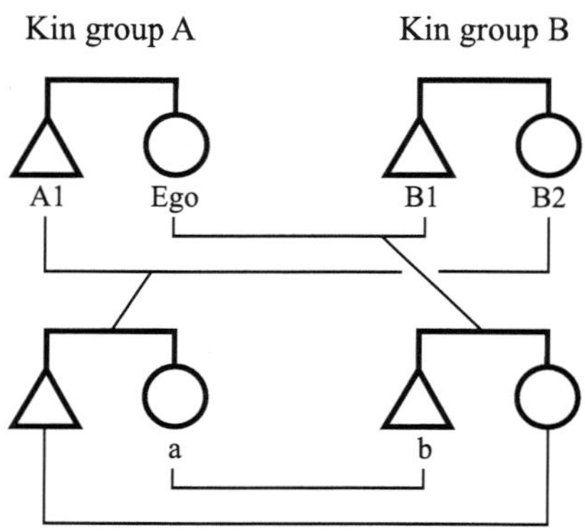

Fig. 2.3 Structural relations between sister exchange (or bilateral marriage between affines) and marriage between cross-cousins. Cross-cousin marriage is the extension of sister exchange to their offspring. Definitions of symbols as in Fig. 2.1. See text for explanations

studies, the widespread phenomenon of avuncular relationships, those special bonds between maternal uncles and their sororal nephews. It is precisely in relation to that problem that Lévi-Strauss (1963) discussed the atom of kinship.

From here, we are in a position to understand still another central aspect of reciprocal exogamy, namely, cross-cousin marriage, which Lévi-Strauss described as the "elementary formula for marriage by exchange" (1969, p 129) and "the simplest conceivable form of reciprocity" (1969, p 48). Cousins are the offspring of siblings and belong to one of two categories: cross-cousins, who are the offspring of opposite-sex siblings, and parallel cousins, the offspring of same-sex siblings. Marriage between cross-cousins is a widespread practice and the object of a prescription in a large number of societies. Let us return to males A1 and B1, in Fig. 2.1, who are married to each other's sisters and thereby, brothers-in-law. If the two males *extend the exchange principle to their own children*, that is, if they exchange their respective daughters as wives for their own sons, this produces marriage between cross-cousins. This is illustrated in Fig. 2.3, extended from Fig. 2.2. If male A1's daughter ("a") marries male B1's son ("b"), this produces a marriage between cross-cousins, female a's father and male b's mother being siblings. In a situation of sister exchange between two groups, it is also the case that female a's mother and male b's father are siblings, so the two individuals are double, or bilateral, cross-cousins.

It may readily be seen that the atom of kinship, with its emphasis on the brother–sister bond, and cross-cousin marriage, which unites the offspring of a brother and a sister, are intimately related structurally. Cross-cousin marriage is one particular manifestation of the brother–sister bond, in which the brother controls the marriage of his sister's children. It is brother A saying to his sister Ego: "Your son will marry my daughter." The brother–sister bond, thus, lies in the very heart of reciprocal exogamy and between-group alliances in Lévi-Strauss's scheme. This is particularly intriguing considering that in nonhuman primates, brothers and sisters are most often separated by natal group dispersal so that brothers are not in a position to exert any influence on their sisters.

In sum, the proposition that reciprocal exogamy embodies the deep structure of human society implies that the distinctiveness of human social organization, compared with all other animal societies, holds in the conjunction and particular linkage of kinship bonds and marriage bonds, a linkage that produces between-group alliances and supragroup levels of social organization; put differently, the essence of human society lies in its "federate" nature. Interestingly, another model concerned with the simplest and earliest human kinship system – the so-called tetradic theory (Allen 2008) – has much in common with reciprocal exogamy. Its central feature is a rule of reciprocal recruitment of spouses between two kin groups (reciprocal exogamy), with pair-bonds uniting affines in the two kin groups. All human kinship systems are supposedly derived from that stem structure. Although Lévi-Strauss's concept of reciprocal exogamy and Allen's tetradic theory differ in a number of respects, they belong to the same family of models and both are compatible with the present evolutionary analysis.

2.3 Phylogenetic Evidence as a Test of the Exogamy Model

Lévi-Strauss described reciprocal exogamy as if it was a cultural creation. Moreover, in accordance with the synchronic perspective of structuralism, he consistently abstained from discussing elementary kinship structures within the evolutionary paradigm. From such an unchronological perspective, reciprocal exogamy appears to be an irreducible entity: a system whose elements owe their origin and existence to the system itself and hence have no evolutionary history of their own. Ever since, Lévi-Strauss has consistently dismissed the relevance of primate studies for understanding the origins of human society (Lévi-Strauss 1985, 2000). But if reciprocal exogamy is, in effect, the deep structure of human society, it might well have an evolutionary past, and its components their own evolutionary histories. This very assertion is not self-evident. Any structure described as the earliest human social system, be it Allen's tetradic model or Lévi-Strauss's reciprocal exogamy, is, by definition, a normative (rule-governed), symbolically mediated structure. It is so because all aspects of human behavior, from greetings to legal systems, are symbolically mediated and have culturally defined meanings. From this, it follows that the very search for the evolutionary precursor of reciprocal exogamy rests on the assumption that it existed under some *presymbolic* and more or less embryonic form prior to the evolution of the symbolic capacity, and that this form had a biological basis and phylogenetic roots. The following analysis may, thus, be construed as a test of the hypothesis that a primitive version of reciprocal exogamy thrived as a set of behavioral regularities well before the evolution of the symbolic capacity generated several more sophisticated and institutionalized versions of it.

As pointed out earlier, Lévi-Strauss's characterization of reciprocal exogamy meets the formal criteria of a deep social structure from a sociological viewpoint. But if that argument is a necessary condition for validating the present claim, it is not a sufficient one. In theory, other candidate structures might satisfy the same conditions, and here lies the relevance of the evolutionary perspective. If reciprocal exogamy is, in effect, the deep structure of human society, it should also accord with the criteria set by the phylogenetic analysis of the phenomenon. I identify three such criteria. First, the candidate structure ought to have been described, or to be actually describable, in terms of the same basic categories used to describe all other primate societies – i.e., group composition, mating system, dispersal patterns, kinship structure, and so on. In other words, the candidate structure should fit within the general framework of primate comparative sociology. Second, the candidate structure ought to break down into phylogenetically sound components. That is to say, its evolutionary roots should be manifest in a number of building blocks observable in other primates; and the building blocks that are uniquely human should also make sense from that perspective (discussion below). Third, the evolutionary history of the candidate structure ought to be parsimoniously reconstructible in view of our knowledge about primate behavior and human evolution. In particular, it should accord with the characteristics of the last common ancestor that we shared with other primates.

To appraise the significance of these criteria, consider the candidate structure of early human society proposed by the cultural evolutionist Lewis Morgan (1877 1974), who hypothesized that early human society featured group marriage among all members of ego's generation, what he called the consanguine family; or the scenario of his contemporary John McLennan (1865) 1970), who proposed a structure featuring wife capture, female infanticide, and generalized polyandry. Unsurprisingly, none of these structures meet even one of the above three criteria. As I shall argue, Lévi-Strauss's concept of reciprocal exogamy does satisfy the three of them. Put differently, the phylogenetic test of the exogamy model of human society's deep structure lends support to it.

Very few authors have looked into the evolutionary origins of human society as a whole. Sociocultural anthropologists White (1959) and Service (1962) briefly explored the topic (Chapais 2008), but the comparative analysis of human kinship and primate kinship was truly pioneered by Fox (1975, 1979, 1980, 1993), also a sociocultural anthropologist. Fox identified exogamy as the cornerstone of human kinship systems and argued that its two basic elements – kin groups and stable breeding bonds (or kinship and marriage) – were present in nonhuman primates but never in the same species as is the case in humans, who form multifamily kin groups. This led him to conclude that the originality of the human system lay not so much in the two components themselves as in their merging in the same species, and that if humans had not invented kinship and pair-bonds, they had invented other elements such as affinal kinship (in-laws), out-marriage, and female exchange (Fox 1975, 1980). Some 10 years later, Rodseth and his colleagues picked up the thread, noting that a few primate species such as the hamadryas baboon do combine kin groups and stable breeding bonds, and that such species practice "exogamy" in that they form stable breeding bonds after transferring into another group. Such bonds, however, do not translate into alliances between groups because members of the dispersing sex lose contact with their natal kin, so in-laws cannot recognize each other. Rodseth et al. (1991) concluded that the distinctiveness of human society lies in the bilateral recognition of in-laws and the exchange dimension of exogamy.

My own comparative analysis builds upon this earlier work, but takes into account all major aspects of reciprocal exogamy described in the previous section. This leads me to break the phenomenon down into the following phylogenetic building blocks: a multimale–multifemale group composition; stable breeding bonds and its correlate, fatherhood; strong siblingships; kin recognition along both the maternal line (uterine kinship) and the paternal line (agnatic kinship); incest avoidance; out-marriage (exogamy) and its correlate, dual-phase residence (pre/postmarital); lifetime bonds between dispersed kin; bilateral relations between in-laws; kin-biased and affinity-biased marriage rules; female exchange; and between-group alliances (supragroup levels of social organization). I call this set of features the *exogamy configuration*, and in the remainder of this chapter, I go on to demonstrate that a phylogenetic analysis of that configuration supports the view that it describes the deep structure of human society.

2.4 Origins of the Multifamily Community

Where should one start when attempting to reconstruct the evolutionary history of the exogamy configuration? Interestingly, the answer to this question is contained in the timing of the evolution of its most basic feature, the modal composition of human groups. That composition is the *multifamily community* (Rodseth et al. 1991) and its evolutionary origin appears to date back to the *Pan–Homo* split, some 6–7 million years ago (Fig. 2.4). The multifamily community is a rare form of group that combines two independent elements: a multimale–multifemale composition and a mating system featuring stable breeding bonds (monogamous or polygynous). On logical grounds, a multifamily system may evolve through two different paths. According to the first possibility, the multimale–multifemale composition came first, followed by the evolution of stable breeding bonds, as illustrated in Fig. 2.5. Humans and their two closest relatives, chimpanzees and bonobos, form multimale–multifemale groups, which suggests that this trait is homologous in the three species and shared with their last common ancestor. Accordingly, early hominids formed

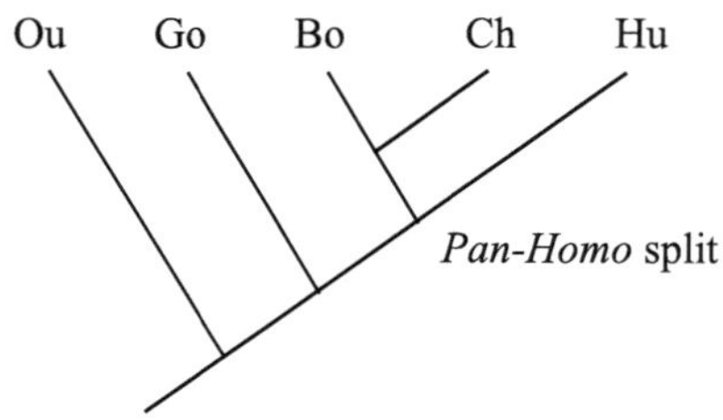

Fig. 2.4 Phylogenetic relationships of humans and apes as assessed by several sets of molecular data (Goodman et al. 1998, 2005; Enard and Pääbo 2004). *Ou* orangutans, *Go* gorillas, *Bo* bonobos, *Ch* chimpanzees, *Hu* humans. The *Pan* genus includes chimpanzees and bonobos

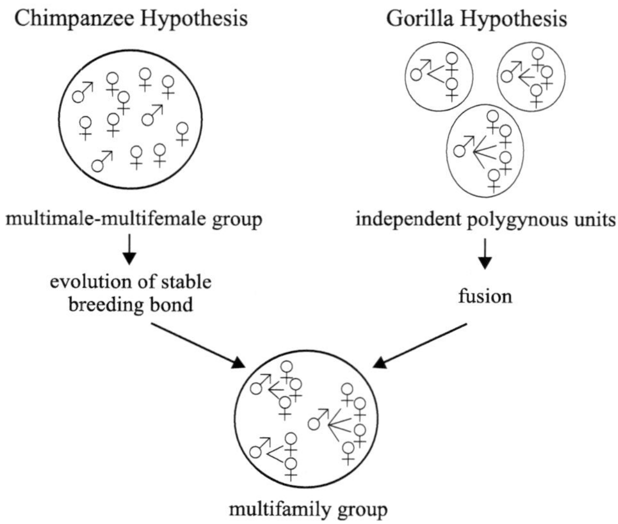

Fig. 2.5 Two evolutionary paths leading to the modal composition of human groups (the multifamily community)

multimale–multifemale groups and had a chimpanzee/bonobo-like promiscuous mating system. It follows that stable breeding bonds and biparental families evolved at some point after the *Pan–Homo* split, producing the multifamily composition.

The second logical possibility is the reverse sequence: families (monogamous or polygynous) appeared first and the multifamily group evolved later through the merging of autonomous families. I call this sequence the "gorilla hypothesis" because it fits with the observation that our third closest relative forms autonomous polygynous social groups. According to one version of this hypothesis, the last common ancestor of the *Pan* and *Homo* lineages had a gorilla-like social structure, which evolved into the multimale-multifemale composition along the *Pan* line and into the multifamily group along the hominid line (Imanishi 1965, Sillén-Tullberg and Møller 1993). According to another, somewhat less parsimonious version, gorilla-like groups had already evolved into the multifamily group prior to the

Pan–Homo split and stable breeding bonds were lost along the *Pan* line (Geary 2005).

As argued at length elsewhere (Chapais 2008), the chimpanzee/bonobo hypothesis is more likely based on a number of arguments. Briefly, it fits better with what we know about the evolutionary history of multifamily groups in other primate species (hamadryas baboons, gelada baboons, drills, and mandrills). A cladistic analysis carried out by Barton (1999) suggests that the multifamily composition did not evolve through the amalgamation of autonomous units, but from an ancestral multimale–multifemale group and the subsequent conversion of the mating system from sexual promiscuity to stable breeding bonds. Second, compared with the gorilla hypothesis, the chimpanzee/bonobo hypothesis requires the smallest number of evolutionary changes to produce the multifamily structure. Third, the gorilla hypothesis implies that independent polygynous units coalesce, presumably through the evolution of reduced levels of competition or higher levels of cooperation between males. But such changes would rather favor the extension of male philopatry in gorilla groups – all males remaining in their natal group – and their transformation into multimale–multifemale groups, not into multifamily groups. Fourth, the chimpanzee hypothesis fits better with the observation that early hominids were anatomically more similar to chimpanzees than to gorillas, and therefore that their behavioral ecology and social structure more closely resembled those of chimpanzees.

2.5 Kinship in Early Hominid Society

The chimpanzee/bonobo hypothesis is an integral part of the larger issue of the suite of traits characterizing the last common ancestor of chimpanzees, bonobos, and humans. The traits common to all three species have been described by Wrangham (1987), Ghiglieri (1987), and Foley (1989), and provide a cladistic model (Moore 1996) of their common ancestor. Besides a multimale–multifemale composition, they include (1) territoriality, (2) male philopatry – male localization coupled with female transfer – and (3) male kin groups. Territoriality characterizes both chimpanzees and bonobos (Newton-Fisher 1999; Wrangham 1999; Boesch and Boesch-Achermann 2000; Watts and Mitani 2001; Fruth and Hohmann 2002; Wilson and Wrangham 2003; Williams et al. 2004; Watts et al. 2006). Territoriality comes along with hostility between males belonging to distinct groups and with the absence of any supragroup social entity. There are no between-group alliances in chimpanzees and bonobos; the local group is the highest level of social organization.

Male philopatry is the rule in chimpanzees and bonobos – all males stay put while a majority of female emigrate (Goodall 1986; Furuichi 1989; Nishida 1990; Kano 1992; Boesch and Boesch-Achermann 2000; Doran et al. 2002; Nishida et al. 2003) – with gorillas displaying a partial pattern of male philopatry (Bradley et al. 2004, discussion in Chapais 2008: 142). In humans, the structural equivalent of male philopatry is the majority residence pattern in humans, with 70% of the 1,153

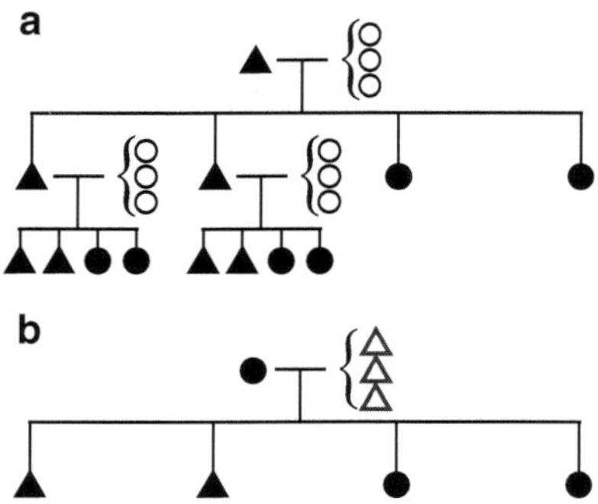

Fig. 2.6 Composition of a single patriline (**a**) and a single matriline (**b**) in a male philopatric primate group. *Circles*: females. *Triangles*: males. *Black symbols*: members of the same matriline or patriline. *Empty symbols*: nonmembers. *Braces* illustrate promiscuous mating

societies in Murdock's (1967) Ethnographical Atlas classified as patrilocal or virilocal. But contrary to previous assertions (Steward 1955; Service 1962; Ember 1978), recent studies indicate that bilocality rather than patrilocality is the majority residence pattern among hunter-gatherers (Alvarez 2004; Marlowe 2004). Notwithstanding this, the prevalence of residential diversity (bilocality, patrilocality, matrilocality) during more recent phases of human evolution is not incompatible with the view that male philopatry and human patrilocality are homologous. As discussed at length elsewhere (Chapais 2008: 142–151, 238–243), ancestral male philopatry in the hominid line remains the most parsimonious hypothesis.

An important correlate of male philopatry in chimpanzee and bonobo groups is their genealogical structure. Male philopatry produces a strong asymmetry in the composition of patrilines and matrilines. Male localization generates kin groups comprised of extensive, multigenerational *patrilines* (Fig. 2.6a), in which a male lives with his sons, brothers, father, uncles, grandfather, and other agnatic kin. Concurrently, female transfer produces small, two-generation *matrilines* (Fig. 2.6b). In a group in which all females emigrate, a male lives with his mother and his maternal siblings, but not with his mother's kin (e.g., his maternal grandfather and uncles) who live in the mother's natal group. Nor does he know his daughters' and sisters' offspring because females emigrate at puberty and lose contact with their natal kin. The next question, then, is: of all the kin types a male lives with, which ones does he recognize as such? To answer this question, it is necessary to consider the domain and basic processes of kin recognition in nonhuman primates in general.

2.5.1 Kin Recognition in NonHuman Primates

Kin recognition is inferred from the preferential treatment of known kin (nepotism). Studies in which nepotism was analyzed according to kin types in groups in which females are the resident sex – female philopatric groups – indicate that females recognize their mother, daughters, sons, brothers, sisters, grandmother, grandoffspring, great-grandmothers, and great-grandoffspring, and, less consistently, their aunts and nieces/nephews. Cousins and more distant relatives do not appear to be part of the domain of kin recognition (Kapsalis and Berman 1996; Chapais et al. 1997, 2001; Bélisle and Chapais 2001; Chapais and Bélisle 2004; Silk et al. 2006). The anthropologist, George Murdock, provided a useful classification of kin types. Ego's primary kin are its mother, father, brothers, sisters, sons, and daughters. Ego's secondary kin are the primary kin of each of its primary kin; they are Ego's grandparents, grandchildren, aunts, uncles, nieces, and nephews. Similarly, Ego's tertiary relatives are the primary kin of one's secondary kin; that is, Ego's first cousins, great-aunts, great-grandparents, among many others (Murdock 1949: 94). Viewed in terms of Murdock's categories, nonhuman primates recognize their primary uterine relatives (mother, sons, daughters, brothers, and sisters) and some of their secondary uterine relatives, namely their grandmother and grandoffspring. Other secondary kin such as aunts and nieces are part of the gray zone of kin recognition. Nonhuman primates also recognize some of their tertiary uterine kin (great-grandrelations), but apparently not others such as cousins.

The cornerstone of uterine kinship is the intimate and enduring bond between mothers and offspring. It is that bond, in all likelihood, that mediates kin recognition between other categories of uterine kin; for example, between sisters. Two different processes are probably at work here. In the first, the mother is merely a passive mediator of familiarity between her daughters. A female would recognize her sister as that particular individual she meets near her mother on a daily basis – because both sisters are independently attracted to the same mother – and with whom she becomes disproportionately familiar over the years. On this basis alone, two sisters are in a privileged position to develop a preferential bond of their own. Similarly, through proximity to her mother, a female is bound to become disproportionately familiar with her maternal grandmother and develop preferential bonds with her (see also Berman and Kapsalis 1999; Berman 2004; Rendall 2004).

This first recognition process focuses on the fact that through proximity to her mother, a female acquires information about her sisters. But in all likelihood, the female is simultaneously acquiring information about her sisters' *relationships* with her mother. The ability to learn about the relationships of others is well documented in nonhuman primates (Cheney and Seyfarth 1980, 1986, 1989, 1999; see also Cheney and Seyfarth 1990, 2004) and provides a distinct cognitive basis for kin recognition. From a female's viewpoint, a sister is not only that particular individual she is disproportionately familiar with, but she may well be, in addition, that close associate of her mother, the one that she protects against certain individuals, grooms at certain rates, tolerates near food sources, and so on. Kin recognition, here, involves Ego classifying others by using her mother as a reference point. In sum, its likely that primates learn the identity of their uterine kin by acquiring information both about their own relationships with their relatives and their mothers' relationships with these same kin. Crucially, the two processes depend on the *lasting character* of the mother–offspring bond. For a newborn sister to be able to recognize her 5 year old sister, the older sister's bond with her mother must last significantly longer than 5 years. Similarly, for a granddaughter to recognize her maternal grandmother, mother–daughter bonds must last significantly longer than the generation length.

2.5.2 Kin Recognition in Early Hominids

With this background information, we may come back to the issue of kin recognition in a chimpanzee/bonobo society and, by way of inference, in early hominid society. In chimpanzees, mother–offspring recognition is well documented, and so is the recognition of maternal siblings (Goodall 1986; Pusey 1990, 2001; Lehmann et al. 2006; Langergraber et al. 2007). But given that males do not normally grow up with their mother's kin, the domain of uterine kin recognition is, in general, limited to these two categories. Interestingly, however, mothers sometimes breed in the group in which they were born, a situation which provides individuals with an opportunity to recognize their mother's kin. For example, in the Gombe colony of wild chimpanzees, 50% of the females stayed and bred in their natal group – whereas in other communities, most females emigrate. In this context Goodall (1990) provided anecdotal accounts of preferential bonds between grandsons and their maternal grandmothers and between maternal uncles and their sororal nephews. Clearly, then, the kin recognition potential of chimpanzees encompasses not only primary maternal kin (mothers, daughters, sons, and maternal siblings), but secondary maternal kin as well, which is not surprising given the cognitive sophistication of chimpanzees.

If mother–offspring recognition is well established in chimpanzees, father–offspring recognition is not. Paternity recognition based on disproportionate levels of familiarity between fathers and offspring is unlikely because females mate with a large number of males (Wrangham 1993, 2002) and do not maintain long-term exclusive relationships with particular ones, including the males with whom they had an offspring (Lehmann et al. 2006). Accordingly, Lehmann et al. (2006) reported remarkably limited and weak effects of paternity on social interactions between adult males and youngsters. Similarly, paternal siblings (half-siblings related through the father) do not appear to recognize each other in chimpanzees (Lehmann et al. 2006; Langergraber et al. 2007). Given that males do not maintain long-term bonds with their fathers, they can hardly recognize their fathers' kin on this basis – e.g., their paternal grandfather and paternal uncles – as females recognize their mother's kin in female philopatric groups. Thus, even positing some degree of paternity recognition, as described by Lehmann et al. (2006) in chimpanzees, or by Buchan et al. (2003) in baboons, such levels of bonding between fathers and sons are unlikely to reveal the agnatic kinship structure in male kin groups as maternity recognition reveals the uterine kinship structure in female kin groups.

Overall, then, the domain of kin recognition in our two closest relatives is normally quite limited (Fig. 2.7). Assuming that early hominids mated promiscuously

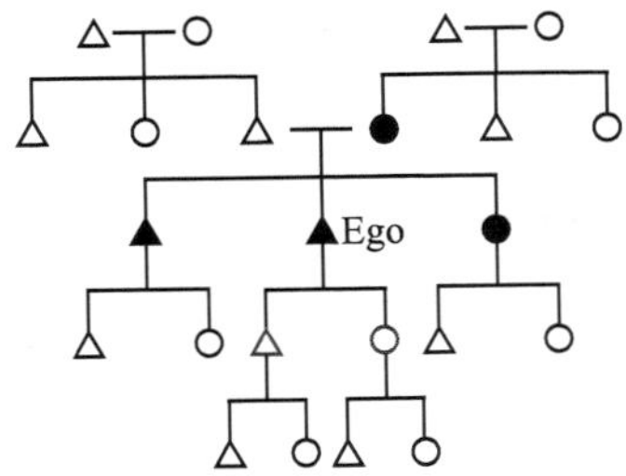

Fig. 2.7 Domain of kin recognition from ego's viewpoint in a male philopatric, chimpanzee-like group, assuming that all females breed outside their natal group. Five generations are pictured but normally only three coexist. *Black symbols*: kin recognized by Ego, assuming that Ego recognizes his uncles, aunts, nieces, and nephews. *Empty symbols*: kin not recognized by Ego

and were male philopatric, male relationships in these groups were minimally differentiated on the basis of kinship. This may appear somewhat intriguing considering that kinship is a central organizing factor in small-scale human societies, which, therefore, would have evolved from a comparatively "kinshipless" type of society. As we shall see, the key to this apparent paradox is the evolution of stable breeding bonds. Table 2.1 summarizes the state of the exogamy configuration prior to the evolution of stable breeding bonds (Phase I).

2.6 The Evolution of Stable Breeding Bonds

The transition from sexual promiscuity to enduring breeding bonds in the course of hominid evolution is the single-most important event that launched the exogamy configuration on its evolutionary path. How did that happen? Answers to this question have traditionally focused on the adaptive aspects of pair-bonding, but they must also take into account the relevant phylogenetic constraints set by the ancestral mating system of hominids. Up to 80% of human societies combine monogamy with polygyny, with the majority of families being monogamous in any given society. Logically, then, hominids went from chimpanzee/bonobo-like sexual promiscuity to the predominantly monogamous multifamily structure. The primate data suggest that this evolution involved two transitions: (1) from sexual promiscuity to generalized polygyny (as in the multiharem structure of hamadryas baboons), and (2) from generalized polygyny to generalized monogamy. A direct passage from sexual promiscuity to generalized monogamy is unlikely for a number of reasons. First, polygyny is the norm in mammals in general. Accordingly, some primate species exhibit the multiharem structure, but none display the multimonogamous pair structure. (Monogamy exists in nonhuman primates but monogamous pairs do not form cohesive groups). Second, the transition from sexual promiscuity

Table 2.1 The cumulative construction of the exogamy configuration in the course of human evolution. Phase I extended from the *Pan–Homo* split to the evolution of stable breeding bonds. Phase II began after the evolution of stable breeding bonds and ended with the evolution of the tribe, which marked the onset of phase III

	Phase I	Phase II	Phase III
Multimale–multifemale groups	Yes	Yes	Yes
Uterine kinship	Yes	Yes	Yes
Incest avoidance	Yes	Yes	Yes
Outbreeding (dual-phase residence)	Yes	Yes	Yes
Stable breeding bonds		Yes	Yes
Paternity recognition (fatherhood)		Yes	Yes
Strong siblingships		Yes	Yes
Multifamily communities		Yes	Yes
Agnatic kinship		Yes	Yes
Out-marriage (exogamy)		Yes	Yes
Pre/postmarital residence		Yes	Yes
Lifetime bonds between dispersed kin			Yes
Bilateral affinity			Yes
Atom of between-group alliances			Yes
Primitive tribe			Yes
Residential diversity			Yes
Marriage between siblings-in-law			Bias
Bilateral marriage between sibling pairs			Bias
Marriage between cross-cousins			Bias
Exchange dimension of exogamy			–

to generalized polygyny finds cladistic support. The few primate species with a multiharem structure belong to taxonomic groups in which the majority of species have a multimale–multifemale composition and a promiscuous mating system. This suggests that the clade's common ancestor had the latter type of mating system (Barton 1999). Third, primate behavioral ecology is compatible with the transition from sexual promiscuity to generalized polygyny, but hardly so with a direct transition to generalized monogamy. The multiharem composition appears to be an adaptation to a low food density that cannot support large aggregations, an hypothesis that finds support in the observation that savanna baboons, which typically form large multimale–multifemale groups, may sometimes subdivide into polygynous units in harsher ecological conditions (Barton 1999). Fourth, as pointed out earlier, a majority of human societies combine monogamy with polygyny, a fact that strongly suggests that the ancestral hominid pattern was generalized polygyny. This leads us to the second step, the transition from generalized polygyny to generalized monogamy.

2.6.1 *Monogamy as Maximally Constrained Polygyny*

The classic view about the origin and function of human pair-bonding is the parental collaboration hypothesis, which conceives of pair-bonds as parental partnerships based on a sexual division of labor (Washburn and Lancaster 1968; Isaac 1978; Alexander and Noonan 1979; Lovejoy 1981; Hill 1982; Fisher 1992, 2006; Kaplan et al. 2000). A competing view holds that pair-bonds in present-day hunter-gatherers do not meet the criteria of cooperative partnerships, from where it is inferred that pair-bonding evolved instead as part of a male's mate guarding strategy (Hawkes 1991, 1993, 2004; Hawkes et al. 2001). The two positions appear mutually exclusive and irreconcilable, but a primatological perspective shows that they are not. It suggests that human pair-bonds *are* parental partnerships, but partnerships that *initially* evolved as mating strategies.

As discussed at length elsewhere (Chapais 2008, pp 162–168), two major correlates of the parental collaboration hypothesis are empirically supported: the costs of raising human children are disproportionately high owing to our larger brain and its correlate, delayed maturation (Kaplan et al. 2000), and the father's economic contribution does alleviate the costs of maternity (Gurven 2004). But one must not confuse the actual working of the human family with its origins. Studies of the mating and parental care systems of mammals in general suggest that pair-bonding did not initially evolve as parental partnerships. Stable breeding bonds in mammals are primarily mating arrangements. In a phylogenetic analysis of mammalian mating and parental care systems, Brotherton and Komers (2003) found that in most monogamously breeding species that exhibit parental collaboration, paternal care had evolved *after* monogamy was already established, and this for reasons other than parental collaboration. This would explain why direct paternal care is absent in several monogamously mating primate species (van Schaik and Kappeler 2003). According to this view, it is monogamy that sets the stage for the evolution of paternal care, rather than parental collaboration driving the evolution of monogamy (Dunbar 1995; Ross and MacLarnon 2000; van Schaik and Kappeler 2003). Applied to the human case, this reasoning suggests that pair-bonding originated as a mating arrangement, which later operated as a *preadaptation* for parental collaboration when delayed maturation evolved and the costs of maternity began to increase. In other words, the father was already present in the family when paternal investment became advantageous and was, presumably, selected for. When viewed sequentially, then, the mating arrangement hypothesis and the parental collaboration hypothesis are basically compatible.

If pair-bonding was not part of a paternal care strategy initially, why did it evolve, and why did hominids forego polygyny for preponderant monogamy? A relatively simple explanation is that monogamy "replaced" polygyny when the costs of polygyny became too high. Consider the following thought experiment. In chimpanzees, male dominance relations are well defined and higher-ranking males have a higher reproductive success (Constable et al. 2001). Now suppose that all males were given similar fighting abilities. In such a situation, conflicts would be extremely costly and males would be better off switching to scramble competition and attempting to copulate with as many females as possible. This would translate into high levels of sexual promiscuity and sperm competition, and a low variance in male reproductive success. At this point, let us carry out the same reasoning, but in a different initial setting: the multiharem structure of hamadryas baboons. If all males were given the same fighting abilities, the monopolization of females by a male would be extremely costly. Any male attempting to defend more than one female would be challenged by equally powerful males. Hamadryas males being harem builders, they would keep doing this, but they would end up forming monogamous bonds. The outcome would be an egalitarian distribution of females among males – generalized monogamy – because it is the arrangement that minimizes conflicts, hence the costs of aggression.

This thought experiment provides an hypothesis for the transition from generalized polygyny to generalized monogamy in the hominid lineage. The pivotal factor is the development of technology, in particular the discovery that tools could be used as weapons. Any tool can be used as a weapon, provided it can inflict injuries. Armed with a deadly weapon, any hominid male was in a position to seriously hurt stronger individuals. In such a context, it should have become extremely costly for a male to monopolize several females. Only males able to monopolize tools, or males forming coalitions, could do so. But then all males could make tools and form coalitions. Thus, a likely side effect of the evolution of tools was a marked increase of the costs of aggression and a corresponding reduction of the variance in male competitive ability. The reasoning rests on the assumption that weapons increased the competitive power of all males in a manner largely independent of their physical prowess. If this assumption is correct, generalized polygyny, with its permanent exclusion of a large fraction of males from the pool of reproductive individuals, would have become unfeasible. It was bound to give way, eventually, to generalized monogamy. Accordingly, the drive for polygyny was merely checked, not eliminated. Polygyny could reemerge whenever some males were able to attract females based on attributes other than physical prowess. Human societies amply testify to this reemergence.

The present explanation is considerably more parsimonious than the parental collaboration hypothesis in so far as the very origin of pair-bonding is concerned. Monogamy is seen not as the outcome of specific selective pressures for paternal investment, but as the mere by-product of other elements merging together over evolutionary time, namely, prior polygyny and the rise of technology.

2.7 Fatherhood and the Expansion of Kinship

Whatever its timing and exact causes, the evolution of stable breeding bonds in the hominid lineage transformed kinship from a low-profile factor in the social organization of early hominids to the prominent role it plays in simple human societies. The key event was the evolution of systematic paternity recognition and fatherhood. From the time the father associated with the mother, children were in a position to recognize their fathers, and the fathers their children, even in the absence of any form of direct paternal care. The processes involved here are similar to those underlying the recognition of uterine sisters, in which the mother acts as the mediator between the two sisters. A child and his father were bound to become disproportionately familiar with each other by virtue of their common preferential bond with the same female, who was a mother to one and a "wife" to the other. The child was also in a position to learn his father's identity by recognizing the characteristics of the relationship between his mother and father, the fact, for example, that his father was his mother's primary male associate, the one who protected her against other males, had sexual interactions with her, and so on.

As soon as young hominids could recognize their father on such a reliable basis, they were in a position to recognize their father's relatives, including their paternal grandfather and grandmothers, their paternal uncles and aunts, and their patrilateral cousins. Again, the processes involved in the recognition of patrilateral kin may be inferred from those underlying the recognition of uterine kin, where the mother is the central reference point from Ego's perspective. Here, it is Ego's father who is the central reference figure and operates as an intermediary between his son and his close kin. Crucially, the recognition of one's patrilateral kin was possible only if fathers and sons engaged in enduring bonds with one another. This raises the question of how such bonds might have evolved. One simple process is suggested by relationships between adult males and adolescent males in chimpanzees. As they start to travel independently of their mothers upon reaching adolescence, male chimpanzees often attempt to form bonds with particular adult males. For example, Pusey (1990) described the relationship between an adolescent male and the group's alpha male who had been a close associate of the adolescent's mother at a time when the latter was still traveling with his mother. The association persisted into adulthood. Jane Goodall used the term "follower" to describe such relationships between a youngster and a particular adult male, stressing that the bond "is almost entirely initiated and maintained by the follower" (Goodall 1986, p 202). Such evidence suggests that nascent father–son bonds in the hominid lineage could have been initiated and maintained by the sons themselves, hence independently of active forms of paternal care, and that they merely required fathers to be selectively tolerant toward their sons, a condition predicted by kin selection theory.

Figure 2.8 illustrates the domain of kin recognition in hominid groups after the evolution of stable breeding bonds. The contrast with Fig. 2.7 is striking. Prior to the evolution of stable breeding bonds, the agnatic kinship structure was present but socially indiscernible. The genealogical structure lay dormant. To reveal it, paternity recognition was needed, and this is precisely what stable breeding bonds

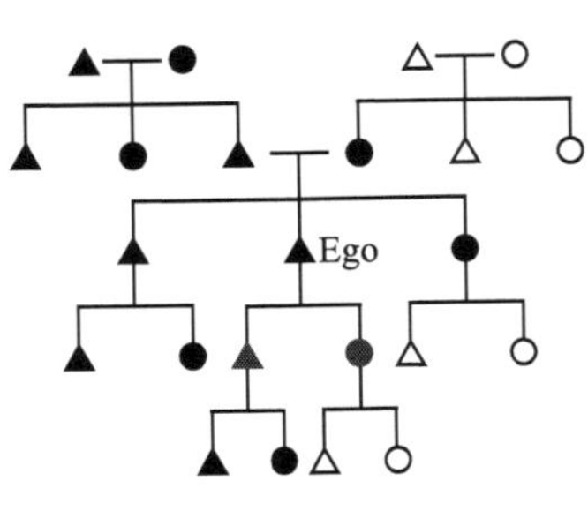

Fig. 2.8 Domain of kin recognition from Ego's viewpoint in a male philopatric, chimpanzee-like group after the evolution of stable breeding bonds. Definitions as in Fig. 2.7

accomplished. With paternity recognition, the role of agnatic kinship in structuring social relationships in male kin groups became comparable to the role of uterine kinship in female kin groups such as macaques and baboons.

Another major consequence of fatherhood on kinship is that it created a whole new type of family. Owing to space constraints, I must skip the reasoning (Chapais 2008, pp 202–215) underlying the following description. From a chimpanzee/bonobo-like bi-generational and monoparental (mother–offspring) unit, the hominid family evolved into a biparental unit integrating three generations of individuals – owing to paternity recognition, grandmothers affiliate with their son's children – and some affines as well, that is, into some sort of extended family. On the basis of the assumption that fathers and sons developed lifetime cooperative partnerships, such families included a well-defined core of *primary agnates* (father and sons) whose cohesiveness stemmed, fundamentally, from the benefits associated with cooperating with a same-sex close kin. Importantly, daughters (or sisters) were an integral part of such units. In chimpanzees, females have loose bonds with their brothers, and with only a fraction of these. Pair-bonding changed that situation drastically. Henceforward, among a young female's most basic bonds in the new hominid family were those that she developed with her primary kin: her mother, father and brothers. This simple fact may be seen as the single most important necessary condition for the evolution of practices such as sister exchange, cross-cousin marriage, and avuncular relationships (discussion below).

2.8 The Origins of Exogamy and Postmarital Residence

"Exogamy lies far back in the history of man" wrote Edward Tylor long ago, "and perhaps no observer has ever seen it come into existence, nor have the precise conditions of its origin yet been inferred" (Tylor 1889, p 267). Tylor could hardly have foreseen that the answer to this enigma lay in our close relatives. From an evolutionary perspective, exogamy (primitive out-marriage) is simply the incidental by-product of the combination of two otherwise typical primate patterns: between-group transfer and pair-bonding. Female dispersal was presumably the rule in the ancestral male kin group. Upon the evolution of stable breeding bonds, females kept emigrating into a new group as before, but instead of mating promiscuously in it, they formed an enduring breeding relationship; they were "marrying-out" so to speak. Significantly, exogamy at that stage was deprived of any exchange dimension: females were transferring between groups on their own initiative, they were not part of transactions between males. That would come later.

We saw that for Lévi-Strauss, female exchange was the primary and most binding form of reciprocity between men, an argument which he based on the empirical observation that throughout the world women are men's "most precious possession." Primatology vindicates this conclusion but through different arguments. Female transfer between groups was the ancestral condition. This helps explains why a female bias in dispersal – not a male one – is widespread cross-culturally. But another, even more basic factor was involved. Throughout the animal kingdom, including primates, females are certainly the most "precious possession" males may compete for, as has been overwhelmingly documented ever since Darwin (1871) 1981) first explained why this was so (for primates, see contributions in Kappeler and van Schaik 2004). In this sense, Lévi-Strauss's assertion fits nicely with sexual selection theory.

Even as they were giving rise to a behavioral form of exogamy, stable breeding bonds were generating a primitive form of what anthropologists call *postmarital residence*. Chimpanzees and bonobos have a dual-phase residence pattern: females spend a prebreeding phase in their natal group, followed by a breeding phase elsewhere. Dual-phase residence is, thus, a phylogenetically primitive pattern. The evolution of pair-bonding transformed that pattern into one comprised of a prepair-bonding phase (or "premarital" phase) spent in the natal group, followed by a postpair-bonding (or "postmarital") phase spent in the new group. Like exogamy, "postmarital" residence emerged from the integration of pair-bonding to male philopatry, a fusion that produced an embryonic form of patrilocality. To many sociocultural anthropologists, in contrast, prior to the invention of the incest taboo, residence had been a single phase spent in one's natal group (e.g., Murdock 1949, p 16).

Table 2.1 summarizes the state of the exogamy configuration immediately after the evolution of stable breeding bonds (Phase II). Compared with the previous stage, several new traits have emerged, but several others are still lacking. At this point in human evolution, hominid groups were independent entities like all other primate groups. Once individuals moved out of their natal group, they ceased to interact with the relatives they left behind. Social life was limited to one's local group. The remaining components of the exogamy configuration had to await the extension of social structure beyond the local group (phase III). They are attributes of between-group alliances, or supragroup social structures. For the sake of simplicity, I use the term *tribe* in a generic sense to refer to such entities.

2.9 The "Atom of Between-Group Alliances"

As pointed out earlier, chimpanzees and bonobos are territorial: they avoid other groups and may even attack strangers. In chimpanzees, intergroup fights are initiated and conducted by adult males, and the targets include other adult males, infants, and, sometimes, mothers. The local community is, thus, the most inclusive level of social organization in our closest relatives. Assuming that early hominids were territorial, it follows that a necessary condition for the evolution of between-group alliances – the tribal level of organization – was the pacification of adult males living in distinct groups. The issue of the origin of the tribe brings Lévi-Strauss's concept of the atom of kinship to the fore and, again, stable breeding bonds appear to hold the key to this major transition in human evolution.

Figure 2.9 pictures two hominid groups after the evolution of stable breeding bonds. The focus is placed on female Ego, born in group A and pair-bonded in

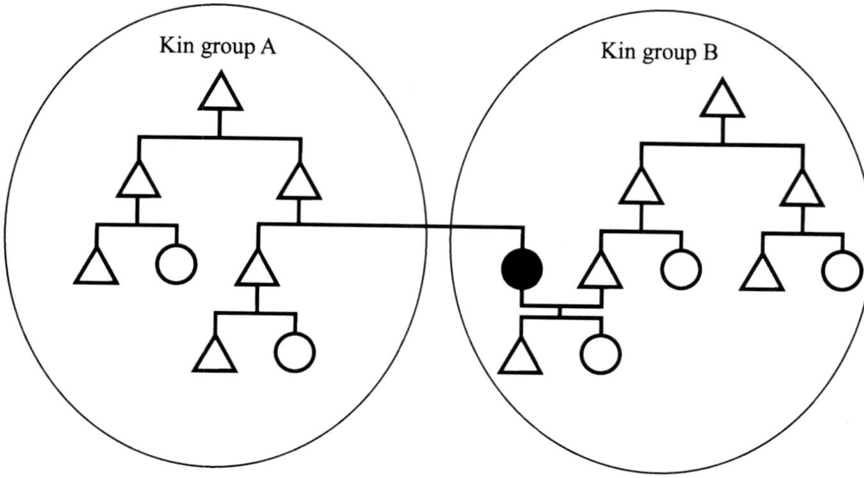

Fig. 2.9 Between-group pacification processes activated during meetings between two male philopatric groups after the evolution of stable breeding bonds. The processes are illustrated by focusing on a single female from group A (*black circle*) after she has transferred to and pair-bonded in group B. The female has children in group B and is recognized by her paternal kin living in A. The female also acts as an intermediary between her kin living in A and her husband and his relatives

group B. Suppose the two groups were to meet at their common border in some nonaggressive way: intergroup meetings have been reported to occur from time to time in bonobos (Idani 1990), but have not been observed in chimpanzees. In the context of such meetings, female Ego would recognize, in addition to her mother and maternal siblings, her father, grandfather, and uncles living in group A, and she would be recognized by them. This could not be the case prior to the evolution of stable breeding bonds, Ego not having experienced a preferential bond with her father. Minimally, a male would be disinclined to attack his daughter, granddaughter, or sister, so Ego would have benefitted from some kind of immunity from her male relatives. The same principle applies to all transferred females and to both directions (group B females transferred in group A), hence to a significant fraction of individuals in both groups. Moreover, a female's immunity against aggression should extend to her own offspring: A male who refrained from attacking his daughter or sister should also refrain from attacking the individual that his daughter or sister carried on her back or belly. Male chimpanzees are known to attack and kill the infants of isolated mothers when they come across them at their common border. From the time males could recognize such individuals as their close kin, male infanticidal attacks should have dropped. In sum, owing to paternity recognition and its impact on agnatic kinship, group A males would be collectively inhibited from attacking several females living in group B and, reciprocally, group B males would be collectively inhibited from attacking several group A females. A state of mutual, though fragmentary, tolerance stemming from the existence of several "kinship bridges" between intermarrying groups would prevail.

Concurrently, another, distinct process would favor between-group pacification, this one involving the mediation of affines. In-laws are the relatives of one's spouse, or the spouses of one's relatives, depending on one's viewpoint. Cognitively speaking, the recognition of in-laws is similar to kin recognition. It requires no more than the ability to recognize preferential bonds between others – e.g., between one's daughter and the latter's husband. When groups A and B came into contact, ego's father could recognize his daughter's husband (his son-in-law) and, reciprocally, ego's husband could recognize his father-in-law. Importantly, from an evolutionary perspective, relationships between in-laws were bound to be, fundamentally, relationships between potential allies. Brothers-in-law, for instance, share a vested interest in the same female, one as a husband, the other as a brother. Both males derive benefits from the female's well-being, the husband through his own reproductive interests with his wife, the brother by virtue of his genetic relatedness with her – inclusive fitness benefits. Crucially, this shared interest is not impeded by sexual competition between the two males: owing to incest avoidance, a brother does not compete with his sister's "husband" for sexual access to his own sister. Minimally, therefore, brothers-in-law should refrain from attacking each other, as should, for that matter, fathers-in-law and sons-in-law and other affines. The importance of the affinity route in the pacification of intergroup relations can hardly be overstated because it is about peaceful relationships between *adult males*, the individuals directly responsible for intergroup conflicts.

The foregoing reasoning presupposes that interbreeding groups met sporadically at their common border in some nonaggressive way, in which case the structure of kinship and affinity bridges just described would be activated. But if interbreeding groups never came into contact in the first place, pacification could not start. In other words, pair-bonding and the expansion of kinship were a necessary condition for pacification, but not a sufficient one. For pacification to get going, some factors had to favor nonaggressive meetings between groups, such as those described for bonobos (Idani 1990). These factors might have operated through a reduction of the levels of feeding competition between groups, an increase in the opportunities for using the same resources simultaneously (food, water, or shelter), and/or an increase in the benefits of between-group cooperation against either other groups or other species. This point needs further investigation.

In the pacification processes envisioned here, the alliance between groups A and B hinges on female Ego who is simultaneously bonded to her male kin in group A and to her husband in group B; or, in the words of Edward Tylor's, on "the peacemaking of the women who hold to one clan as sisters and to another as wives" (Tylor 1889, p 267). The father–daughter/wife–husband triad is a dual-link chain, with female Ego acting as a swivel joint between the two groups. The same applies to the brother–sister/wife–husband triad. Each of these two Ego-centered chains embody simultaneously, the kinship basis and the affinity basis of between-group alliances. Taken together, they may be described as the *atom of between-group alliances*, the smallest social element involved in between-group social structures. This paraphrase of Lévi-Strauss's "atom of kinship" is more than merely analogical. Lévi-Strauss restricted the atom of kinship to the brother–sister–husband triad, neglecting the father–daughter–husband triad. He also erroneously ascribed the atom of kinship to brothers repressing a built-in drive for incest and renouncing marriage with their sisters, and he included the sister's children in it. Notwithstanding these differences, the atom of kinship and the atom of between-group alliances are basically the same thing, structurally speaking: a kinship bond connected to a pair-bond through the intermediary of ego. Lévi-Strauss's atom of kinship, the hub of reciprocal exogamy, does have an evolutionary history.

2.10 The Nascent Tribe

The foregoing discussion points to the nature of the emerging tribe. At this stage in its evolution, the tribe was merely a set of interbreeding local groups exhibiting some levels of tolerance with one another, and the number of local groups forming a tribe was determined by the exact pattern of female transfer between groups. Simple models of female transfer make it possible to specify some of the conditions favoring male pacification through the formation of kinship and affinity bridges between groups (Chapais 2008). These models indicate (1) that male pacification ensues whether female transfer is unidirectional or bidirectional between groups, but that bilateral transfer promotes further congeniality, (2) that the pace of pacification between any two groups depends on the proportion of females in one group moving to the other group: the larger that proportion, the larger the number of kinship bridges between the two groups, and (3) that transfer between groups of substantially different size works to the disadvantage of the smaller group. Combining the three principles, one may infer that intergroup pacification was especially favored in situations where female circulation was bidirectional and concentrated among a small number of groups that were not too dissimilar in size.

If the composition of the primitive tribe reflected the exact pattern of female transfer between its constituent groups, that pattern was itself determined by the geographical distribution of groups. In chimpanzees, for example, female transfer was reportedly frequent between communities whose home ranges overlapped extensively, but infrequent between communities whose home ranges did not overlap (Kawanaka and Nishida 1974; Nishida 1979). Considering that the main factors affecting the geographical distribution of local groups are the presence of physical barriers between them and the distribution of food resources and predators, one may envision the first tribes as *regional entities* whose constituent local communities "exchanged" females and enjoyed various levels of peaceful relations with one another.

Importantly, as between-group hostility markedly decreased *within* the tribe, it remained at its prior level *between* tribes. This aspect of the present model helps resolve the discrepancy between chimpanzees and human foragers with regard to intergroup patterns of violence. Compared with chimpanzees, human hunter-gatherers are much more egalitarian and display substantially lower levels of intergroup competition and violence. So striking is the difference that some authors spoke of a phylogenetic discontinuity between chimpanzees and human foragers as far as patterns of violence were concerned (Knauft 1991; Kelly 2000). But as argued by Rodseth and Wrangham (2004), the local band of hunter-gatherers is not the right social unit for a meaningful comparison with chimpanzees—the tribe is (see also Crofoot and Wrangham, this volume).

2.11 The Evolution of Residential Diversity

Although its boundaries were relatively loose and its structure primitive, the nascent tribe likely brought about some profound changes in the social relationships of hominids. It notably rendered the diversification of postmarital residence feasible. I have hitherto been concerned with human patrilocality and its phylogenetic antecedent, male philopatry. But human groups exhibit several other residence patterns: matrilocality (spouses live with or near the wife's parents), bilocality (spouses live near the husband's parents or the wife's parents), neolocality (both spouses leave their natal home to live elsewhere), and avunculocality (males live with their maternal uncles, wives move to their husbands' location, and their sons return to live with the mother's brothers). Moreover, each broad category constitutes only an ideal or modal type, allowing a fair degree of residential flexibility. For example, several hunter-gatherer societies may be classified as "patrilocal with a matrilocal alternative" (Ember 1975).

In the evolutionary scheme presented here, between-group pacification is a prerequisite for the evolution of novel residence patterns involving the transfer of males between groups, namely, matrilocality, bilocality, and avunculocality. Stated otherwise, the tribal level of organization was a necessary condition for residential diversity. In chimpanzees and bonobos, male territoriality is incompatible with males transferring freely between groups, and this was presumably the case in the ancestral hominid society: a male could hardly move to a group that was collectively defended by males. But after the tribe had evolved, males could move between nonhostile groups of the same tribe – though not between tribes. From then on hominids were in a position to adjust residence patterns to various conditions; for example, to resource distribution, subsistence activities, and patterns of cooperation.

One might object that the occurrence of female philopatry in nonhuman primates – male emigration coupled with female localization – indicates that human matrilocality does not require a tribal level of organization and therefore that matrilocality might well have evolved earlier in the hominid line. But this argument misses the point. First, one must take into consideration the phylogenetic constraints acting on the evolution of residence patterns in the hominid lineage. Ancestral male philopatry and its correlate, male cooperation in territorial defense, had to be circumvented before male transfer and matrilocality became possible. This is precisely what the tribal level of organization achieved. Second, female philopatry is not simply the structural equivalent of human matrilocality. The two patterns differ in some important respects and it is likely that female philopatry is not the evolutionary antecedent of human matrilocality. In female philopatric primates, females cease to interact with their male relatives once the latter have moved out of the group. In marked contrast, human matrilocality consistently comes along with the political control of women by their kinsmen despite their being away (Schneider 1961). Indeed, an important correlate of matrilocality – and avunculocality for that matter – is matrilineal descent, in which the line of authority runs from men to their sister's sons, rather than from women to their daughters. The localization of women is, thus, associated with their maintaining lifetime bonds with their kinsmen married in other groups, hence with the tribal level of organization. This suggests that a prerequisite for the localization of women was that men be in a position to exercise control over their kinswomen despite physical separation from them, a state of affairs that was possible only after the tribal level of organization had evolved.

2.12 The Origins of Exogamy Rules

As described earlier, Lévi-Strauss's theory of reciprocal exogamy features kinship-constrained marriage rules, the most basic of these being sister (or daughter) exchange, the levirate, the sororate, and cross-cousin marriage. Where do these rules come from? The answer proposed here is that they were ultimately derived from the atom of between-group alliances and the ensuing familiarity biases that affected mate selection in the nascent tribe. As illustrated in Fig. 2.10, upon pair-bonding with male B1 and moving into group B permanently, female Ego underwent long-lasting familiarity biases with her husband's close kin, including her brothers-in-law and sisters-in-law. Such biases were likely to translate into amicable relationships between them for reasons already given. In this context, if Ego's husband were to die, Ego might well form a pair-bond with her husband's brother (B3), in which case one obtains the structural equivalent of the levirate – a widow marrying the brother of her deceased husband. Similarly, upon pair-bonding with Ego, male B1 experienced long-lasting familiarity biases with his wife's close kin, including his brothers-in-law and sisters-in-law. If Ego were to die, male B1 could form a pair-bond with his wife's sister (A2), this producing the structural equivalent of the sororate – a widower marrying his deceased wife's sister. If, however, male B1 were to form a pair-bond with Ego's sister A2 while Ego is still alive, this produces a form of sororal polygyny, another widespread practice in human societies.

Interestingly, simple processes akin to those described here have been invoked by cultural anthropologists to explain the levirate and the sororate. Citing figures based on 250 societies, Murdock described the closely related phenomenon of *privileged relationships* between siblings-in-law of opposite sex, "*within which sexual intercourse is permitted before marriage and frequently afterwards as well.*" Murdock argued that both permissive sex and preferred marriage between

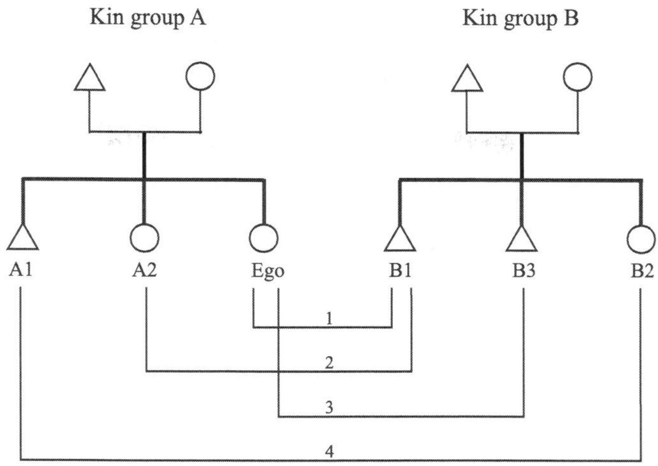

Fig. 2.10 Mating biases stemming from disproportionate levels of familiarity between affines, and their relations with known exogamy rules in humans. The individuals pictured are the same as in Fig. 2.1, except that siblingships include one more individual. (1) Initial pair-bond between female Ego and male B1. (2) Pair-bond between B1 and his wife's sister, the equivalent of sororal polygyny if Ego is still alive, or of the sororate if Ego is dead. (3) Pair-bond between Ego and her husband's brother, the equivalent of the levirate if the husband is dead. (4) Pair-bond between Ego's brother and B1's sister. In conjunction with (1), this produces the structural equivalent of sister exchange

siblings-in-law "*are explicable as extensions of the marital relationship*" and reflect the attraction of people to persons who most closely resemble their spouse. "*The persons who universally reveal the most numerous and detailed resemblances to a spouse*" he wrote, "*are the latter's siblings of the same sex*" ... "*who are likely to have similar physical characteristics*" ... and "*almost identical social statuses since they necessarily belong to the same kin group*" (Murdock 1949, pp 268–269). Murdock's explanation has much in common with the present model. Both conceive of mate selection as being affected by informal regularities, whether the latter stem from familiarity biases, physical similarities, or social compatibilities. To Lévi-Strauss, in contrast, exogamy rules were normative and part of reciprocity agreements. But viewed from an evolutionary perspective, the two types of explanations are compatible: informal regularities paved the way for normative rules; in the same manner, incest avoidances paved the way for incest prohibitions.

These considerations apply to another marriage rule: sister exchange (or daughter exchange, depending on one's viewpoint). Structurally speaking, sister exchange is simply bilateral marriage between two groups of affines. In all likelihood, the *exchange* dimension of the phenomenon is a further and more recent aspect of it, an aspect involving the control of sisters by their brothers (or of daughters by their fathers). As for cross-cousin marriage, it is the extension of sister exchange to the following generation, as described earlier (Fig. 2.3). Thus, from an evolutionary perspective, the relevant question for both sister exchange and cross-cousin marriage concerns the origin of bilateral marriage between affines and, again, the answer proposed here lies in familiarity differentials affecting mate selection. As illustrated in Fig. 2.10, if female Ego is already pair-bonded with male B1, bilateral marriage between affines ensues from B1's sister (B2) pair-bonding with Ego's brother (A1). From male A1's viewpoint, female B2 is a familiar affine, the sister of his brother-in-law. Reciprocally, from female B2's viewpoint, male A1 is also a familiar affine, the brother of her sister-in-law. In short, the most basic of exogamy rules – levirate, sororate, sister exchange, and cross-cousin marriage – might have originated in mate selection biases stemming from disproportionate levels of familiarity between affines.

2.13 Conclusion

If reciprocal exogamy is the deep structure of human society, the configuration of elements listed in Table 2.1 may be seen as the most sophisticated form the structure had reached prior to the evolution of the symbolic capacity. Some crucial elements are still lacking at that stage, notably the actual exchange of kinswomen by men, an aspect which probably required language. Table 2.1 may also be read as the list of human universals that stemmed from the presymbolic and prenormative state of human society.

I began this article by proposing that Lévi-Strauss's concept of reciprocal exogamy was a strong candidate for humankind's deep social structure. I based that hypothesis on the observation that from the outset Lévi-Strauss's characterization of reciprocal exogamy met the formal criteria of such a structure. The present phylogenetic analysis provides further critical evidence for that claim by showing that reciprocal exogamy breaks down into evolutionarily meaningful building blocks. Indeed, a number of components of the exogamy configuration listed in Table 2.1 are observable in nonhuman primates while several others that are not – agnatic kinship, exogamy, postmarital residence, and so on – appear to be the by-products of the combination of building blocks that do exist in nonhuman primates. Hence, it can be said that whether the constituent elements of the exogamy configuration are visible in other primate species or not, they make sense evolutionarily speaking. Had reciprocal exogamy not broken down into phylogenetically meaningful elements, one could not propose that it embodies the deep structure of human society. Correlatively, the phylogenetic reconstruction of the exogamy configuration readily fits with our knowledge about some of the most basic events in the evolutionary sequence that led to human society, namely, an ancestral *Pan*-like society and the subsequent evolution of stable breeding bonds. Had it been problematic to figure out how the exogamy configuration came about in the hominid lineage, there would be more grounds to question its significance.

In sum, Lévi-Strauss's concept of reciprocal exogamy, although issued from an asynchronic theoretical framework, is basically a primate-like, or primate-compatible, structure. The reason it is so is that it centers around two factors of cardinal importance in all primate social structures, sex (mating system) and kinship, and that it hinges on the issue of outbreeding through dispersal from one's group (exogamy). Unknowingly, then, Lévi-Strauss characterized human society in terms of a primate society. It is remarkable that two approaches as distinct as comparative primatology and Lévi-Strauss's structuralism – one avowedly excluding the evolutionary paradigm, the other issuing from it – should converge independently on the issue of the most essential factors that organize human society. This lends further credence to the exogamy model of human origins.

Acknowledgments I am grateful to Robin Fox, Peter Kappeler, and Joan Silk for their helpful comments on the manuscript, and to Julie Cascio for technical assistance with the figures. I also thank several people who provided invaluable comments on my book *Primeval Kinship: How Pair-Bonding Gave Birth to Human Society*, on which the present chapter is based, namely Peg Anderson, Bernard Bernier, Annie Bissonnette, Carol Berman, Robert Crépeau, Michael Fisher, Michel Lecomte, Martin Muller, Jean-Claude Muller, Robert Sussman, Shona Teijeiro, and Richard Wrangham.

References

Alexander RD, Noonan KM (1979) Concealment of ovulation, parental care, and human social evolution. In: Chagnon NA, Irons W (eds) Evolutionary biology and human social behavior: an anthropological perspective. Duxbury, North Scituate, MA, pp 436–453

Allen NJ (2008) Tetradic theory and the origin of human kinship systems. In: Allen NJ, Callan H, Dunbar RIM, James W (eds) Early human kinship: from sex to social reproduction. Blackwell, Oxford, pp 96–112

Alvarez HP (2004) Residence groups among hunter-gatherers: a view of the claims and evidence for patrilocal bands. In: Chapais B, Berman CM (eds) Kinship and behavior in primates. Oxford University Press, New York, pp 420–442

Barnard A (2000) History and theory in anthropology. Cambridge University Press, Cambridge

Barton RA (1999) Socioecology of baboons: the interaction of male and female strategies. In: Kappeler PM (ed) Primate males: causes and consequences of variation in group composition. Cambridge University Press, Cambridge, pp 97–107

Bélisle P, Chapais B (2001) Tolerated co-feeding in relation to degree of kinship in Japanese macaques. Behaviour 138:487–509

Berman CM (2004) Developmental aspects of kin bias in behavior. In: Chapais B, Berman CM (eds) Kinship and behavior in primates. Oxford University Press, New York, pp 317–346

Berman CM, Kapsalis E (1999) Development of kin bias among rhesus monkeys: maternal transmission or individual learning? Anim Behav 58:883–894

Boesch C, Boesch-Achermann H (2000) The chimpanzees of the Taï forest: behavioural ecology and evolution. Oxford University Press, New York

Bradley BJ, Doran-Sheehy DM, Lukas D, Boesch C, Vigilant L (2004) Dispersed male networks in western gorillas. Curr Biol 14:510–513

Brotherton PMN, Komers PE (2003) Mate guarding and the evolution of social monogamy in mammals. In: Reichard UH, Boesch C (eds) Monogamy: mating strategies and partnerships in birds, humans and other mammals. Cambridge University Press, Cambridge, pp 42–58

Buchan JC, Alberts SC, Silk JB, Altmann J (2003) True paternal care in a multi-male primate society. Nature 425:179–181

Chapais B (2008) Primeval kinship: how pair-bonding gave birth to human society. Harvard University Press, Cambridge, MA

Chapais B, Bélisle P (2004) Constraints on kin selection in primate groups. In: Chapais B, Berman CM (eds) Kinship and behavior in primates. Oxford University Press, New York, pp 365–386

Chapais B, Gauthier C, Prud'homme J, Vasey P (1997) Relatedness threshold for nepotism in Japanese macaques. Anim Behav 53:1089–1101

Chapais B, Savard L, Gauthier C (2001) Kin selection and the distribution of altruism in relation to degree of kinship in Japanese macaques (*Macaca fuscata*). Behav Ecol Sociobiol 49:493–502

Cheney DL, Seyfarth RM (1980) Vocal recognition in free-ranging vervet monkeys. Anim Behav 28:362–367

Cheney DL, Seyfarth RM (1986) The recognition of social alliances by vervet monkeys. Anim Behav 34:1722–1731

Cheney DL, Seyfarth RM (1989) Redirected aggression and reconciliation among vervet monkeys, *Cercopithecus aethiops*. Behaviour 110:258–275

Cheney DL, Seyfarth RM (1990) How monkeys see the world: inside the mind of another species. University of Chicago Press, Chicago

Cheney DL, Seyfarth RM (1999) Recognition of other individuals' social relationships by female baboons. Anim Behav 58:67–75

Cheney DL, Seyfarth RM (2004) The recognition of other individuals' kinship relationships. In: Chapais B, Berman CM (eds) Kinship and behavior in primates. Oxford University Press, New York, pp 347–364

Constable JL, Ashley MV, Goodall J, Pusey AE (2001) Noninvasive paternity assignment in Gombe chimpanzees. Mol Ecol 10:1279–1300

Darwin C [1871] (1981) The descent of man and selection in relation to sex. Princeton University Press, Princeton

Deliège R (2006) Une histoire de l'anthropologie. Seuil, Paris

Doran DM, Jungers WL, Sugiyama Y, Fleagle JG, Heesy CP (2002) Multivariate and phylogenetic approaches to understanding chimpanzee and bonobo behavioural diversity. In: Boesch C, Hohmann G, Marchant LF (eds) Behavioral diversity in chimpanzees and bonobos. Cambridge University Press, Cambridge, pp 14–34

Dunbar RIM (1995) The mating system of Callitrichid primates. I. Conditions for the coevolution of pair bonding and twining. Anim Behav 50:1057–1070

Ember CR (1975) Residential variation among hunter-gatherers. Behav Sci Res 10:199–227

Ember CR (1978) Myths about hunter-gatherers. Ethnology 17:439–448

Enard W, Pääbo S (2004) Comparative primate genomics. Annu Rev Genomics Hum Genet 5:351–378

Fisher HE (1992) Anatomy of love: the natural history of monogamy, adultery, and divorce. Norton, New York

Fisher HE (2006) Why we love: the nature and chemistry of romantic love. Henry Holt, New York

Foley RA (1989) The evolution of hominid social behaviour. In: Standen V, Foley RA (eds) Comparative socioecology: the behavioural ecology of humans and other mammals. Blackwell, Oxford, pp 473–494

Fox R (1975) Primate kin and human kinship. In: Fox R (ed) Biosocial anthropology. Halsted Press, New York, pp 9–35

Fox R (1979) Kinship categories as natural categories. In: Chagnon NA, Irons W (eds) Evolutionary biology and human social behavior: an anthropological perspective. Duxbury, North Scituate, MA, pp 132–144

Fox R (1980) The red lamp of incest. EP Dutton, New York

Fox R (1993) Sisters' sons and monkeys' uncles: six theories in search of an avunculate. In: Fox R (ed) Reproduction and succession: studies in anthropology, law and society. Transaction Publishers, New Brunswick, NJ, pp 191–232

Fruth B, Hohmann G (2002) How bonobos handle hunts and harvests: why share food? In: Boesch C, Hohmann G, Marchant LF (eds) Behavioural diversity in chimpanzees and bonobos. Cambridge University Press, Cambridge, pp 231–243

Furuichi T (1989) Social interactions and the life history of female *Pan paniscus* in Wamba, Zaire. Int J Primatol 10:173–197

Geary DC (2005) Evolution of paternal investment. In: Buss DM (ed) The handbook of evolutionary psychology. John Wiley and Sons, Hoboken, NJ, pp 483–505

Gellner E (1985) Relativism and the social sciences. Cambridge University Press, Cambridge

Ghiglieri MP (1987) Sociobiology of the great apes and the hominid ancestor. J Hum Evol 16:319–357

Goodall J (1986) The chimpanzees of Gombe: patterns of behavior. Harvard University Press, Cambridge, MA

Goodall J (1990) Through a window: my thirty years with the chimpanzees of Gombe. Houghton Mifflin, Boston

Goodenough WH (1970) Description and comparison in cultural anthropology. Aldine de Gruyter, Chicago

Goodman M, Porter CA, Czelusniak J, Page SL, Schneider H, Shoshani J, Gunnnell G, Groves CP (1998) Toward a phylogenetic classification of primates based on DNA evidence complemented by fossil evidence. Mol Phylogenet Evol 9:585–598

Goodman M, Grossman LI, Wildman DE (2005) Moving primate genomics beyond the chimpanzee genome. Trends Genet 21:511–517

Gurven M (2004) To give and to give not: the behavioral ecology of human food transfers. Behav Brain Sci 27:543–583

Harris M (1968) The rise of anthropological theory. Thomas Y Crowell, New York

Harris M (1979) Cultural materialism: the struggle for a science of culture. Random House, New York

Hawkes K (1991) Showing off: tests of an hypothesis about men's foraging goals. Ethol Sociobiol 12:29–54

Hawkes K (1993) Why hunter-gatherers work: an ancient version of the problem of public goods. Curr Anthropol 34:341–361

Hawkes K (2004) Mating, parenting, and the evolution of human pair bonds. In: Chapais B, Berman CM (eds) Kinship and behavior in primates. Oxford University Press, New York, pp 443–473

Hawkes K, O'Connell JF, Blurton Jones NG (2001) Hunting and nuclear families: some lessons from the Hadza about men's work. Curr Anthropol 42:681–709

Hill K (1982) Hunting and human evolution. J Hum Evol 11:521–544

Idani G (1990) Relations between unit-groups of bonobos at Wamba, Zaire: encounters and temporary fusions. Afr Study Monogr 11:153–186

Imanishi K (1965) The origin of the human family: a primatological approach. In: Imanishi K, Altmann SA (eds) Japanese monkeys: a collection of translations. SA Altmann, Edmonton, Alberta, pp 113–140. Originally published in Japanese in Jap J Ethnol 25:119–138, 1961

Isaac GL (1978) The food-sharing behaviour of proto-human hominids. Sci Am 238:90–108

Kano T (1992) The last ape: Pygmy chimpanzee behavior and ecology. Stanford University Press, Stanford, CA

Kaplan H, Hill K, Lancaster JB, Hurtado AM (2000) A theory of human life history evolution: diet, intelligence, and longevity. Evol Anthropol 9:156–185

Kappeler PM, van Schaik CP (2004) Sexual selection in primates: new and comparative perspectives. Cambridge University Press, Cambridge

Kapsalis E, Berman CM (1996) Models of affiliative relationships among free-ranging rhesus monkeys (*Macaca mulatta*). I. Criteria for kinship. Behaviour 133:1209–1234

Kawanaka K, Nishida T (1974) Recent advances in the study of inter-unit-group relationships and social structure of wild chimpanzees of the Mahali mountains. In: Kondo S, Kawai M, Ehara A,

Kawamura S (eds) Proceedings from the Symposia of the 5th Congress of the International Primatological Society, Nagoya, Japan, August 1974. Japan Science Press, Tokyo, pp 173–186
Kelly RC (2000) Warless societies and the origin of war. University of Michigan Press, Ann Arbor
Knauft BM (1991) Violence and sociality in human evolution. Curr Anthropol 32:391–428
Langergraber KE, Mitani JC, Vigilant L (2007) The limited impact of kinship on cooperation in wild chimpanzees. Proc Natl Acad Sci USA 104:7786–7790
Lehmann J, Fickenscher G, Boesch C (2006) Kin biased investment in wild chimpanzees. Behaviour 143:931–955
Lévi-Strauss C (1949) Les structures élémentaires de la parenté. Presses Universitaires de France, Paris
Lévi-Strauss C (1963) Structural anthropology. Basic Books, New York. Translation of Anthropologie Structurale, 1958
Lévi-Strauss C (1969) The elementary structures of kinship. Beacon Press, Boston. Translation of Les Structures Élémentaires de la Parenté, 1949
Lévi-Strauss C (1985) The view from Afar. Basil Blackwell, Oxford
Lévi-Strauss C (2000) Apologie des amibes. In: Jamard JL, Terray E, Xanthakou M (eds) En substance: textes pour Françoise Héritier. Fayard, Paris, pp 493–496
Lovejoy CO (1981) The origin of man. Science 211:341–350
Marlowe FW (2004) Marital residence among foragers. Curr Anthropol 45:277–284
Mauss M (1923) Essai sur le don. L'année sociologique, 2ième série
McLennan JF (1970) Primitive marriage: an inquiry into the origin of the form of capture in marriage ceremonies. The University of Chicago Press, Chicago. Reprint of the 1865 edition, A & C Black, Edinburgh
Moore J (1996) Savanna chimpanzees, referential models and the last common ancestor. In: McGrew WC, Marchant LF, Nishida T (eds) Great ape societies. Cambridge University Press, Cambridge, pp 275–292
Morgan LH (1974) Ancient society. Peter Smith, Gloucester, MA Reprint of the 1877 edition, CH Kerr, Chicago
Murdock GP (1949) Social structure. McMillan, New York
Murdock GP (1967) Ethnographic atlas. University of Pittsburgh Press, Pittsburgh
Newton-Fisher NE (1999) Infant killers of Budongo. Folia Primatol 70:167–169
Nishida T (1979) The social structure of chimpanzees of the Mahale mountains. In: Hamburg DA, McCown ER (eds) The great apes: perspectives on human evolution, vol 5. Benjamin/Cummings, Menlo Park, CA, pp 73–121
Nishida T (1990) The chimpanzees of the Mahale mountains: sexual and life history strategies. University of Tokyo Press, Tokyo
Nishida T, Corp N, Hamai M, Hasegawa T, Hiraiwa-Hasegawa M, Hosaka K, Hunt KD, Itoh N, Kawanaka K, Matsumoto-Oda A, Mitani JC, Nakamura M, Norikoshi K, Sakamaki T, Turner L, Uehara S, Zamma K (2003) Demography, female life-history, and reproductive profiles among the chimpanzees of Mahale. Am J Primatol 59:99–121
Pusey AE (1990) Behavioral changes at adolescence in chimpanzees. Behaviour 115:203–246
Pusey AE (2001) Of genes and apes: chimpanzee social organization and reproduction. In: de Waal FBM (ed) Tree of origin: what primate behavior can tell us about human social evolution. Harvard University Press, Cambridge, MA, pp 9–37
Radcliffe-Brown AR (1957) A natural science of society. Free Press, Toronto
Rendall D (2004) Recognizing' kin: mechanisms, media, minds, modules, and muddles. In: Chapais B, Berman CM (eds) Kinship and behavior in primates. Oxford University Press, New York, pp 295–316
Rodseth L, Wrangham RW (2004) Human kinship: a continuation of politics by other means? In: Chapais B, Berman CM (eds) Kinship and behavior in primates. Oxford University Press, New York, pp 389–419
Rodseth L, Wrangham RW, Harrigan AM, Smuts BB (1991) The human community as a primate society. Curr Anthropol 32:221–254

Ross C, MacLarnon A (2000) Evolution of non-maternal care in anthropoid primates: a test of the hypotheses. Folia Primatol 71:93–113

Schneider DM (1961) Introduction: the distinctive features of matrilineal descent groups. In: Schneider DM, Gough K (eds) Matrilineal kinship. University of California Press, Berkeley, pp 1–29

Service ER (1962) Primitive social organization: an evolutionary perspective. Random House, New York

Silk JB, Altmann J, Alberts SC (2006) Social relationships among adult female baboons (*Papio cynocephalus*). I. Variation in the strength of social bonds. Behav Ecol Sociobiol 61:183–195

Sillén-Tullberg B, Møller AP (1993) The relationship between concealed ovulation and mating systems in anthropoid primates: a phylogenetic analysis. Am Nat 141:1–25

Steward JH (1955) Theory of culture change: the methodology of multilinear evolution. University of Illinois Press, Urbana

Tylor EB (1889) On a method of investigating the development of institutions; applied to laws of marriage and descent. J R Anthropol Inst 18:245–269

van Schaik CP, Kappeler PM (2003) The evolution of social monogamy in primates. In: Reichard UH, Boesch C (eds) Monogamy: mating strategies and partnerships in birds, humans and other mammals. Cambridge University Press, Cambridge, pp 59–80

Washburn SL, Lancaster CS (1968) The evolution of hunting. In: Lee RB, DeVore I (eds) Man the hunter. Aldine, Chicago, pp 293–303

Watts DP, Mitani JC (2001) Boundary patrols and intergroup encounters in wild chimpanzees. Behaviour 138:299–327

Watts DP, Muller M, Amsler SJ, Mbabazi G, Mitani JC (2006) Lethal intergroup aggression in chimpanzees in Kibale National Park, Uganda. Am J Primatol 68:161–180

White LA (1959) The evolution of culture: the development of civilization to the fall of Rome. McGraw-Hill, New York

Williams JM, Oehlert GW, Carlis JV, Pusey AE (2004) Why do male chimpanzees defend a group range? Anim Behav 68:523–532

Wilson ML, Wrangham RW (2003) Intergroup relations in chimpanzees. Annu Rev Anthropol 32:363–392

Wrangham RW (1987) The significance of African apes for reconstructing human social evolution. In: Kinzey RW (ed) The evolution of human behavior: primate models. State University of New York Press, Albany, pp 51–71

Wrangham RW (1993) The evolution of sexuality in chimpanzees and Bonobos. Hum Nat 4:47–79

Wrangham RW (1999) Evolution of coalitionary killing. Yearb Phys Anthropol 42:1–30

Wrangham RW (2002) The cost of sexual attraction: is there a trade-off in female *Pan* between sex appeal and received coercion? In: Boesch C, Hohmann G, Marchant LF (eds) Behavioural diversity in chimpanzees and Bonobos. Cambridge University Press, Cambridge, pp 204–215

Chapter 3
Conflict and Bonding Between the Sexes

Ryne A. Palombit

Locked together by their need for partners in sexual reproduction, the sexes undergo an antagonistic dance to the music of time.

Tracy Chapman and Linda Partridge (1996)

Abstract The derivation of human universals from nonhuman data is complicated by the immense diversity of patterns exemplified by both human and nonhuman primates. One approach is to elucidate processes that may operate universally, though the particular phenotypic patterns that result may differ, depending upon the distinctive features of species biology. Below, I argue that sexual conflict and its corollary, sexually antagonistic coevolution, are of central importance for understanding the evolution of reproductive strategies in nonhuman primates. Because sexual conflict is a relatively new area of theory and research, and because primate life histories limit the kinds of data that can be collected, relevant primate data are limited (with the possible exception of one form of conflict: infanticide). Consequently, I review sexual conflict theory, relevant comparative data from nonprimates, and preliminary evidence from select primate studies. Theoretical considerations and empirical evidence suggest significant potential for the widespread action of sexual conflict in nonhuman primates, in both precopulatory and postcopulatory domains of reproduction, and affecting characters ranging from morphology and physiology to sociosexual behavior. Female counterstrategies to male-imposed costs are diverse, but male–female association has been argued to forestall sexual conflict both in the form of precopulatory coercion and of infanticide. In light of evidence for pervasive and diverse effects of sexual conflict in nonhuman primate biology, it is likely that it also constitutes a universal process

R.A. Palombit
Department of Anthropology, Center for Human Evolutionary Studies, Rutgers University, New Brunswick, NJ, USA
e-mail: rpalombit@anthropology.rutgers.edu

P.M. Kappeler and J.B. Silk (eds.), *Mind the Gap*,
DOI 10.1007/978-3-642-02725-3_3, © Springer-Verlag Berlin Heidelberg 2010

underlying human reproduction. I briefly explore several potential sources of human sexual conflict suggested by the nonhuman data.

3.1 An Approach to Universals

Universals are "*mechanisms of human behavior held in common among people all over the world...*" (Boyd and Silk 2006: 590). The variability of human behavior has always bedeviled the search for universals, prompting Fox (1989: 116) to ask how we get beyond the "ethnographic dazzle" to the universals of general, biological importance? The problem is only magnified when we expand the taxonomic context of the analysis to include nonhuman primates, a mammalian order famous for immense diversity in behavior, reproduction, life history, morphology, and physiology. One might say that ethological dazzle threatens to obscure this comparative analysis: how can one discern anything about human universals from this extraordinary variety? There are two solutions to this problem of deriving our family resemblances (*sensu* Fox 1989).

One approach is to search for specific *patterns* of behavior shared between human and nonhuman primates. This orientation towards *substantive* universals necessarily concentrates our attention on a relatively small number of species most closely related to us, notably the chimpanzees (*Pan troglodytes*) and bonobos (*Pan paniscus*), or perhaps the African great apes, or the great apes, generally. To remain useful, however, this approach, focused as it is on elucidating homologous patterns, cannot extend too far beyond this group of primates. This method offers advantages and insights (e.g., Goodall 1971; Wrangham and Peterson 1996; de Waal 2005).

An alternative approach, however, is suggested by Wittgenstein's (1953) theory of universal family resemblances, as captured by the "Churchill face" metaphor (Aaron 1965). Among members of a family, such as the Churchills, there is a distinctive Churchill face, which is recognizable as the same, in some sense, but which cannot be said to have any *one* feature common to *all* faces. In other words, there is *no* shared pattern *per se*. The crucial aspect of this view is its emphasis on a *process* generating predictable patterns not necessarily defined by any one feature or character. The particular patterns will depend upon distinctive features of a species' biology or a population's conditions. It is the *process* that constitutes the universal.

It is this second perspective on behavioral universals that frames this chapter's examination of nonhuman primates. Here, I focus on one process that I believe is paramount for understanding primate reproductive strategies: sexual conflict. Sexual conflict has attracted increasing attention over the last decade, and the studies of this process have now come to outnumber investigations of the conventional forms of sexual selection (intrasexual selection and mate choice) (Pizzari and Snook 2003). Most of this research has focused on invertebrates – particularly insects – although there have also been studies of sexual conflict in some vertebrates, such as fish, birds, and an occasional mammal (e.g., Arnqvist and Rowe 2005). In spite of an early landmark article (Smuts and Smuts 1993), research on sexual conflict in primates has not progressed dramatically.

3.2 What is Sexual Conflict?

As with any relatively new field, there is considerable debate over the definitions, assumptions, and models of sexual conflict (Hosken and Snook 2005; Tregenza et al. 2006). Of course, the notion that male and female reproductive styles do not always coincide perfectly has a long history in evolutionary thinking, beginning with Darwin's (1871) exposition of sexual selection, demonstrated by Bateman's (1948) study of *Drosophila* reproduction, and elaborated by Williams's (1966) "battle of the sexes" metaphor. But it was Trivers (1972) who spotlighted the potential for sexual conflict with an ostensibly simple point: sex differences in parental investment, originating with anisogamy, but amplified in mammals by gestation, lactation, and postnatal care, will generate different reproductive strategies for the males and females, maximizing quantity vs. quality of offspring, respectively. The implication is that reproductive strategies of the sexes not only diverge, but may comprise elements that are *incompatible*. This incompatibility is crucial because different fitness optima for males and females will not generate conflict if they can be achieved simultaneously (Parker 2006). Sexual conflict emerges when strategies among members of one sex impose fitness costs on the other sex. In the resulting evolutionary dialectic, each sex attempts to mitigate these

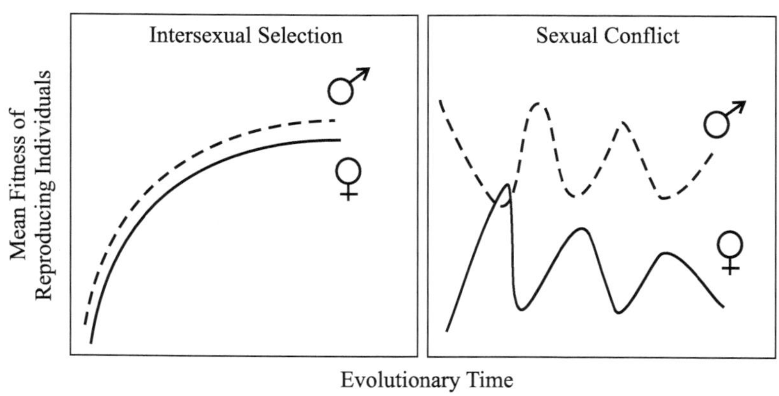

Fig. 3.1 A comparison of average fitness profiles of reproducing males and females over evolutionary time under "conventional" intersexual selection (female choice) (*left*) and sexual conflict (*right*). Under intersexual selection, male fitness (*dashed line*) and female fitness (*line*) often (though not invariably) increase to an asymptote set by natural selection. Under sexual conflict, mutations conferring a net mating benefit to males reduce female fitness, thereby selecting for a female counter-adaptation decreasing male fitness, etc. It is important to note that the figure does not represent the average fitness of males and females in a population, which will coincide with one another (Fisher 1930), but rather the average fitness profiles of reproducing individuals (see Arnqvist 2004; Pizzari and Snook 2004). Figure modified from Pizzari and Snook (2003)

costs and move members of the other sex closer to its own optimum (Gowaty 1997). This coevolutionary dynamic of sexually antagonistic strategies positions sexual conflict as a potential third form of sexual selection, in addition to intrasexual selection and mate choice (Smuts and Smuts 1993; Clutton-Brock and Parker 1995; Chapman et al. 2003; Zeh and Zeh 2003; Tregenza et al. 2006) (Fig. 3.1). It is female avoidance of male-imposed costs that drives sexual conflict, rather than the acquisition of benefits from preferred mating. Parker's (1979) ESS analyses gave rise to the current theoretical framework recognizing two general forms of sexual conflict as sexually divergent optima for either (1) alleles determining a specific trait – *intralocus* conflict – as in the evolution of sexual dimorphism (e.g., Lande 1987; Lindenfors 2002; Cox and Calsbeek 2009); or (2) the outcome of male–female interactions – *interlocus* conflict. This chapter is concerned only with the outcome of male–female interactions.

3.3 Approaches to Studying Sexual Conflict

There are three general approaches to studying sexual conflict. The first method is exemplified by the now classic study of seminal proteins in the fruit fly (*Drosophila melanogaster*) (Rice 1996; Holland and Rice 1999). These proteins originate in accessory glands, are transferred (with sperm) to female mates, and influence females in a number of ways that benefit males, such as: (1) increasing the rate of female egg-laying (Chen 1984); (2) decreasing female receptivity to additional matings (Ravi Ram and Wolfner 2007); and (3) improving sperm competition by displacing the sperm of previous copulators (Harshman and Prout 1994; Clark et al. 1995). Seminal fluids are apparently toxic, such that prolonged exposure to them elevates female mortality (Chapman et al. 1995; Clark et al. 1995; Lung et al. 2002). In order to test the prediction that monogamous mating systems engender less sexual conflict than polygynous systems, Holland and Rice (1999) randomly assigned individual *D. melanogaster* to one of two population treatments: imposed monogamy versus the (control) polygynous ancestral condition. After 47 generations, the monogamous lineage was characterized by lower toxicity of male seminal fluids and lower female resistance to seminal fluids (see also Rice et al. 2005). These data exemplify a key (though not inevitable) corollary of interlocus sexual conflict: *sexually antagonistic coevolution*. This historical approach, tracking changes over evolutionary time, can provide particularly compelling evidence of sexual conflict and sexually antagonistic coevolution, but it is feasible primarily with relatively short-lived animals that can be manipulated in the laboratory.

A second approach, based on quantitative genetics, defines sexual conflict as negative covariance between the sexes in fitness, particularly over generations (Rice and Chippindale 2001; Shuster and Wade 2003; Pizzari and Snook 2003, 2004). For example, red deer (*Cervus elaphus*) stags with greater lifetime reproductive success sired less successful daughters and more successful sons than stags with lower lifetime fitness (Foerster et al. 2007). The negative correlation between the fitness of males and females suggests opposing optimal genotypes for males and females, i.e., sexually antagonistic coevolution. Again, this method is impractical for primates because we know relatively little about lifetime reproductive success, particularly for males, and even less about the selection coefficients and heritability of characters related to fitness.

A third approach considers how certain behavioral, anatomical, or physiological aspects of reproductive strategies among members of one sex impose costs on the other sex, and how phenotypic features of the second sex may function to mitigate those costs (as coevolutionary counterstrategies). The relevant data are collected over relatively short time periods, rarely long enough to demonstrate the effects of sexual conflict on the lifetime reproductive success of individuals. These kinds of analyses can reveal the extent and form of sexual conflict, but they can only indirectly imply the action of sexually antagonistic coevolution. This approach is the only one that is now tractable for studies of nonhuman primates.

3.4 Pre- and PostCopulatory Conflict over Mating: Sexual Coercion

3.4.1 Sexual Coercion: A Conceptual Framework

Aggression between the sexes surrounding mating is termed "sexual coercion" (Smuts and Smuts 1993). Clutton-Brock and Parker (1995) distinguish three forms of sexual coercion: forced copulation, sexual harassment, and sexual intimidation. Although few nonhuman primate studies explicitly differentiate these three categories of sexual coercion, there is evidence that all three may operate in primates.

3.4.2 Forced Copulation

This form of sexual coercion involves the physical restraint and forcible insemination of resistant females. Among primates, forced copulation has been noted occasionally in several species (chimpanzees Tutin 1979; patas monkeys Chism and Rogers 1997; spider monkeys Gibson et al. 2008), but it is regularly observed in only two species, the orangutan (*Pongo pygmaeus*) (van Schaik and van Hooff 1996) and *Homo sapiens* (Smuts 1992; Goetz et al. 2008).

Although forced copulation occurs in a number of different taxa (Table 3.1), it is a less common form of sexual conflict than harassment or intimidation. This may be because forced copulation is only possible under a restricted set of conditions, such as when males are much larger than females (Clutton-Brock and Parker 1995) or when females are isolated and unable to obtain social support. However, neither of these factors provides an entirely satisfactory explanation for the distribution of forced copulation in primates. Although the orangutan is a strongly dimorphic

Table 3.1 Potential forms of (interlocus) sexual conflict

Context	Category	Nature of sexual conflict	Example taxa[a]	Reference[a]
Precopulatory	Forced copulation	Catch and physically restrain female followed by forced insemination; incl. anatomical specializations to grasp and prevent escape of female prior to forced insemination	*Pongo pygmaeus*, *Homo sapiens*, Anseriform birds, some insects	Thornhill and Alcock 1983; Gowaty and Buschhaus 1998; Bertin and Fairbairn 2005; Siva-Jothy 2006; Vahed and Carron 2008, Knott in press
	Harassment, indirect costs of mating or mate guarding	Repeated, persistent courtship or copulation (attempts), by single or especially multiple males; physical aspects of courtship or copulation (e.g., posture, inexperienced males)	Insects, Fish, Anurans, Snakes, Artiodactyls *Microcebus murinus*?	Howard 1980; Clutton-Brock and Parker 1995; Réale et al. 1996; Arnqvist and Nilsson 2000; Shine et al. 2000; Eberle and Kappeler 2004a; Bowcock et al. 2009
	Male dominance displays	Males target females in aggressive displays that function in acquisition and/or maintenance of intragroup dominance status or intergroup spacing	*Papio hamadryas griseipes*, *P. h. ursinus*	Henzi et al. 1998; Kitchen et al. in press
	Sexual intimidation/ punishment	Aggression to (estrus) females that refuse to associate or copulate with male, or that associate or copulate with other male(s)	Primates	Smuts and Smuts 1993; Clutton-Brock and Parker 1995, see text
Postcopulatory (Prezygotic)	Seminal fluid proteins	Proteins beneficially affect outcomes of sperm competition for males, while imposing costs upon female viability and/or reproduction	Insects, Nematodes	Gems and Riddle 1996; Holland and Rice 1999, see text
	Non-fertile sperm	Anucleate sperm reduce female receptivity to subsequent mating	Insects	Cook and Wedell 1999
	Reproductive tract injury	Male-induced changes/injury of female genital tract, typically during copulation, results in decreased sexual interaction	Insects, Rodents, Strepsirrhines? *H. sapiens*?	van der Schoot et al. 1992, Crudgington and Siva-Jothy 2000; Blanckenhorn et al. 2002; Stockley 2002; Low 2005
	Genital plugs, coagulates	Seminal coagulates may: improve sperm transport, reduce sperm loss, physically block intromission by other males, and/or physiologically induce female refractory period	Insects, Rodents, Primates	Matthews and Adler 1978; Simmons and Siva-Jothy 1998; Dixson and Anderson 2002

	Genital lock	Prevention of penis removal from female reproductive tract for prolonged period following ejaculation; due to genital clasping structures or partial enlargement of penis and vaginal adhesion	Insects, *Galago crassicaudatus, Macaca arctoides*	Thornhill and Alcock 1983; Dixson 1998; Werner and Simmons 2008
	Sexual intimidation/ punishment (mate guarding)	Temporary male association with an inseminated female for a prolonged period following ejaculation to aggressively prevent subsequent mating by female	Insects, Primates	Eberle and Kappeler 2004b; Sato and Kohama 2007
	Egg-sperm interaction	Genes of sperm and egg differentially influence processes surrounding capacitation, penetration of egg, and fertilization	Invertebrates, Fish	Rice and Holland 1997; Levitan 2008; Martin-Coello et al. 2009
Postcopulatory (Postzygotic)	Feticide	Male harassment (or forced copulation) of pregnant female promotes or induces spontaneous abortion of implanted zygote or fetus	Equids, Primates	Berger 1983; Pereira 1983; Sommer 1987; Agoramoorthy et al. 1988; Pluháček and Bartoš 2000
	Sexually selected infanticide	Killing of dependent infants to prematurely end lactational amenorrhea and return females to fertilizable (estrus) state	Primates, Fissiped carnivores, Toothed whales	van Schaik 2000a
	Parental investment & genomic imprinting	Activity of genes depends upon sex of parent from which inherited (e.g., paternally derived genes induce disproportionately greater maternal investment in offspring)	Rodents	Keverne 2001; Roulin and Hager 2003
	"Policing"	Male intervenes to curtail female-female aggression, mitigating or eliminating benefits a "winner" could derive via individual or coalitionary competitive superiority	*Gorilla gorilla* subspecies, *Pan troglodytes*	Boehm 1994; Watts 1997; Sicotte 2002; Stokes 2004; Harcourt and Stewart 2007

[a]Taxa and references list are not exhaustive, but rather represent illustrative examples

species, forced copulation is frequently done by small males, who are either subadults or "unflanged" adults with arrested development of secondary sexual characters (Knott 2009). Moreover, in many strongly dimorphic monkeys, males do not exhibit the behavior at all. Social isolation may increase vulnerability to forced copulation. In contrast to the vast majority of highly gregarious anthropoid primates, female orangutans are often alone (Rodman and Mitani 1987). Humans are not solitary, but Emery-Thompson (in press: 361) argues that college-age women experience the highest rate of rape in the United States partly because "*they are the group most likely to be living away from natal kin but not yet with a domestic partner.*" However, social vulnerability does not explain why forced copulations are so rare in chimpanzees (0.2% of the copulations observed by Tutin (1979)) even though females typically disperse from their natal communities and spend much time alone. Possible explanations for the rarity of forced copulation in chimpanzees are female influence on male dominance relations (Stumpf and Boesch 2006) or simply the effectiveness of male sexual coercion in generating mating opportunities (see below), which reduces the benefits of physical restraint and forcible insemination.

Forced copulation in orangutans is commonly considered part of an alternative reproductive strategy of unflanged adult males. The males avoid direct mating competition with large, flanged males by retarding development of secondary sexual traits and relying on force to copulate with uncooperative females that generally prefer flanged males as mates (van Schaik and van Hooff 1996; Atmoko and van Hooff 2004; Maggioncalda et al. 1999). Knott (2009) argues, however, that since forced copulation is not restricted to unflanged males, it is better viewed as a *general* male orangutan strategy to overcome female resistance. Both models are consistent with sexual conflict arguments that forced copulation in nonhuman animals is an alternative mating strategy (Table 3.1).

Thornhill and Palmer (2000) have similarly proposed the controversial hypothesis that human rape reflects an alternative strategy of low-status, socially disadvantaged males to obtain conceptions. Emery-Thompson (2009) rejects this argument on several grounds, including observations that a substantial majority of rapes are perpetrated by men casually or intimately known to their victims (acquaintance rape) and that women often continue their relationships with these attackers. Thus, she contends instead that rape is one of several forms of sexual aggression used by men to maintain *long-term* reproductive access to female mates. Emery-Thompson has shifted the functional focus from immediate copulatory benefits (as in orangutans) to prospective reproductive gains via intimidation and punishment (see below). Again, both hypotheses are based on sexual conflict.

It is important to recognize that forced copulation in humans is an extremely heterogeneous phenomenon (Travis 2003). Some cases of rape may originate in pathological behavior (such as "stranger rape") (Emery-Thompson 2009) or in male tactics of terror and control (e.g., violent rape in the context of warfare; Swiss and Giller 1993). Thus, although a comprehensive understanding of rape in humans will no doubt involve an array of processes and factors, sexual conflict theory seems likely to improve understanding of some forms of the behavior (Emery-Thompson 2009).

3.4.3 Sexual Harassment versus Sexual Intimidation

Sexual harassment refers broadly to the fitness costs of mating to females (*sensu* Daly 1978), particularly costs connected with persistent male courtship, repeated intromission attempts, or the nature of copulation itself. Sexual intimidation refers to situations in which "*males punish females that refuse to associate with them or that associate with other males,*" and is thus designed to reduce female resistance or promiscuity (Clutton-Brock and Parker 1995, p 1353). When males use sexual intimidation tactics, females learn to modify their behavior to minimize the costs of male aggression. This definition is directly similar to Smuts and Smuts's (1993) original definition of sexual coercion. To illustrate the distinction between sexual harassment and sexual intimidation, consider the following examples:

1. During the rut, female sheep (*Ovis* spp.) may be pursued by up to 11 rams at a time, whose repeated attempts to charge, sniff, and mount result in exhaustion and injury to females (Réale et al. 1996) as well as increased mortality, as females evade male suitors on precipitous terrain (Festa-Bianchet 1987).
2. When a female dung fly (*Scatophaga stercoraria*) lands on a dropping occupied by several males, their struggles to copulate and exclude rivals from mating may trample her into the patty, covering her with dung that impairs her ability to fly and sometimes even drowns her (Parker 1970).
3. A male chimpanzee severely attacks an estrous female for "no obvious reason," i.e., in circumstances *unrelated* to ongoing mating, and when the female's sexual swelling is small or flat; many days later, during the period of maximal swelling and mating, she restricts copulations to this male (Goodall 1986: 341).

The various costs imposed on female sheep and dung flies are classified as sexual harassment because they are the indirect by-product of female mate discrimination and male competition, which are particularly relevant when mating attempts are made *repeatedly* or by *multiple* males (or both). The chimpanzee example highlights aggression designed to reduce female resistance or promiscuity, in this case, to promote *future* female mating compliance. Harassment and intimidation can operate in either pre- or postcopulatory contexts. For example, mate guarding is a common manifestation of coercion that can precede or follow copulation. It may comprise threats and attacks on the female herself (*sensu* intimidation) or aggression directed at rival males, thereby imposing indirect mating costs on females (*sensu* harassment).

Harassment and intimidation are behavioral examples of a general distinction in sexual conflict theory between *negative pleiotropic side effects* and *adaptive harm* to females, respectively (Partridge and Hurst, 1998). Many students of sexual conflict maintain that the costs accrued by females are incidental (pleiotropic) by-products of male mating strategies, selected for *not* because of, but *in spite of* the harm to females (Hosken et al. 2003; Morrow et al. 2003; Arnqvist 2004). Conversely, proponents of the adaptive harm hypothesis posit that males benefit from *directly* harming females, if an existing system of phenotypic plasticity promotes female responses that benefit males (e.g., a female injured by a male may increase her resting time, thereby limiting copulation with other males, or she may invest more in her current offspring due to the harm-induced reduction in her reproductive value) (Lessells 1999; Lessells 2005; Johnstone and Keller 2000).

Of crucial importance for understanding these forms of sexual coercion are data addressing not just the magnitude of costs to females, but also the nature of those costs. There are few such data for primates, but studies of other animals reveal costs in the form of reduced foraging efficiency (Rubenstein 1986; Magurran and Seghers 1994; Pilastro et al. 2003; Erez et al. 2005; Heubel and Plath 2008), increased exposure to predation (Magellan and Magurran 2006), injury and associated increased mortality (Hiruki et al. 1993; Miller et al. 1996; Réale et al. 1996; Mühlhäuser and Blanckenhorn 2002), and physiological stress (Moore and Jessop 2003; Shine et al. 2004). These costs are in addition to those associated with suboptimal reproduction due to fertilization by a lower quality male or to lost opportunities for polyandry.

3.4.4 Is Sexual Coercion Beneficial to Females?

It is possible that sexual coercion may actually enhance female fitness by providing a behavioral filter for higher quality males as mates or guaranteeing that females' sons will carry sexually antagonistic traits that enable them to achieve higher reproductive success (Eberhard 1996; Cordero and Eberhard 2003). If the net effect on female fitness is therefore positive, then sexual conflict becomes a mechanism of female choice, which Eberhard (2005) contends explains most male mating aggression to resistant females. This hypothesis has not been supported by some mathematical models (Kirkpatrick and Barton 1997), but there is some related evidence for benefits of coercion to females (Valero et al. 2005).

Most primate researchers assume that sexual coercion reduces the effectiveness of female mate choice and that female preference for *less* aggressive males is a likely counterstrategy to sexual coercion (Smuts and Smuts 1993). This view derives in part from the intensity of both male aggression and toward females and female resistance, which seems to impose high costs on the victims (e.g., chimpanzees: Goodall 1986; Muller et al. in press). Moreover, for most anthropoid primates, group life may provide females with less costly means of evaluating mates than provoking male attacks upon themselves. An arguably more relevant variant of this hypothesis, however, is that females prefer to mate with high-quality males (e.g., dominant males), who also happen to be more aggressive *generally* (which constitutes an indirect cost of mating).

3.4.5 Sexual Harassment and Intimidation in Non-Human Primates

Three conditions promote sexual harassment that occurs when multiple males attempt to mate simultaneously with a single female (Réale et al. 1996; Head and Brooks 2006; Smith and Sargent 2006): (1) a male-biased operational sex ratio;

(2) asynchrony in female estrous; and (3) weak dominance among males (i.-e., reduced or incomplete male ability to control sexual access to females).

All three conditions prevail in nocturnal mouse lemurs (*Microcebus murinus*) studied at Kirindy, western Madagascar: reproducing males tend to outnumber estrous females; females breed on only one night each year, but are receptive on individually different nights of the mating season; male–male competition sometimes involves contests, but scramble competition via extensive roaming behavior is more common (Eberle and Kappeler 2004a, b). On her night of receptivity, a female is typically approached by 2–15 males and mates with almost all of them up to 11 times. Notably, the usual social dominance of females wanes during the mating season, prompting Eberle and Kappeler (2004a: 97) to interpret the high rates of mating with multiple males as "*harassment*" stemming from a temporary female inability to reject suitors. Postcopulatory mate guarding does occur occasionally, raising the possibility of sexual intimidation. But this mate-guarding is based less on aggression directed at the female than on chasing rival males away. Attacks on females occurred in only 4 of the 55 cases of mate guarding and were also largely ineffectual in light of the fact that three of the four females succeeded in deserting the male. These patterns of sexual coercion are generally more consistent with multi-male harassment than with sexual intimidation, as predicted by the demographic, social, and reproductive conditions.

The gregarious (diurnal) strepsirrhines are of comparative interest for distinguishing between harassment and intimidation because intimidation relies particularly on learned cooperation in explicitly gregarious contexts (Clutton-Brock and Parker 1995). Unfortunately, few relevant new data have become available since Smuts and Smuts (1993) to address this question. Brockman's (1999) description of sexual aggression by male sifakas (*Propithecus verreauxi*) suggests harassment rather than intimidation. Multiple males attempt simultaneously to mate with most estrous females during the mating season. Intersexual sexual aggression increases significantly at this time, but the vast majority of it is female aggression to males (not vice versa). Harassment typically takes the form of disrupting an ongoing copulation, and can be perpetrated by either males or females. Although interfering females direct aggression at either copulating partner, males virtually always focus exclusively on the rival male instead of the female. These patterns are collectively inconsistent with the definition of sexual intimidation. Indeed, the data support Smuts and Smuts's (1993) hypothesis that female dominance in some lemurs effectively deters coercion in the form of sexual intimidation. Even so, indirect costs via sexual harassment apparently persist for female sifakas. The nature and magnitude of these costs for female fitness remain unclear, however. Limitation of female choice seems likely, but this possibility needs to be clarified quantitatively (do less harassed females achieve their preferences more often?) as well as tested against the alternative that female resistance functions as mate choice (see below). Moreover, the mating benefits of harassment for the males remain obscure.

A quasi-experimental anecdote concerning ring-tailed lemurs (*Lemur catta*) further supports the notion that female dominance limits sexual intimidation (Parga and Henry 2008). Partly due to the effects of provisioning, a young female reached sexual maturity at an earlier age than usual, but *before* she had developed social dominance over males. This young estrous female subsequently became the target of direct aggression and even forced copulation attempts by a particular adult male.

Data on the diurnal, group-living strepsirrhines also provide a relatively rare primate example of support for the mate choice hypothesis for coercion. In ruffed lemurs (*Varecia variegata*), female conspicuously resist male sexual overtures, even resorting to physical aggression against them. Although males do not typically retaliate with aggression of their own, both Foerg (1982: 119) and Morland (1993) suggest that this sexual antagonism ensures that a female copulates with higher quality ("*strong*") males who "*are more likely to overcome her beating*" long enough to achieve insemination.

Studies of anthropoid primates have made little effort to test between indirect and direct costs to females. Japanese macaques (*Macaca fuscata*) were among the first primates to provide data on sexual coercion, primarily in the form of chases of estrous females or "*possessive following*" (Carpenter 1942; Itani 1982; Enomoto 1981). As Huffman (1987) points out, these patterns were often interpreted as incidental components of male courtship or "*precourtship*" behavior (Itani 1982: 362), thereby implicating sexual harassment. Likewise, a key form of sexual harassment – the costs of mating with *multiple* males – is reflected in the decreased foraging efficiency of females on days they mated polyandrously, compared with days they consorted with the alpha male only (Matsubara and Sprague 2004). Soltis et al. (1997, p 725; 2001, p 486) conclude that male aggression to estrous females is primarily a "*side effect*" of a general mating season increase in overall male aggressiveness and female-maintained proximity to males. Although sexual intimidation does occur, it accounts for a minority of instances of sexual coercion. Subsequent studies of mating-related aggression in this species, however, have emphasized sexual coercion in Clutton-Brock and Parker's (1995) sense of intimidation (Jack and Pavelka 1997; Soltis 1999; Soltis et al. 2001).

Indeed, this interpretation tends to emerge from many recent studies of male aggression over mating in primates (e.g., Kuester et al. 1994; Perry 1997; Reed et al. 1997; Boinski 2000; Colmenares et al. 2002; Arlet et al. 2008; Table 3.1 and references above). This is partly because comparatively few investigations have addressed the Clutton-Brock and Parker (1995) distinction between harassment and intimidation (Soltis et al. (1997) being a notable exception) and have focused on the processes of intimidetion implicit (or explicit) in (Smuts and Smuts 1993). But this emphasis may also reflect the fact that many of the species studied are characterized by gregariousness and male contest competition, which are conditions especially likely to promote sexual intimidation.

One of the more compelling demonstrations of intimidation is provided by the 10-year study of the Kanyawara population of chimpanzees, Kibale, Uganda. It is striking – as well as suggestive of the biological significance of sexual intimidation – that in a species well-known for male–male aggression, male–female aggression occurs at roughly the same rate at Kanyawara (Muller et al. 2009). The majority of this aggression involves male charging displays and chases, but approximately 35% of it entails physical attacks on females (often in coalition with other males). Muller et al. (2007) provide data by directly testing three predictions of the Smuts and Smuts (1993) sexual coercion hypothesis:

Prediction 1: Sexual coercion is costly to females. The intensity of male aggression is difficult to quantify, but assaults on females can involve flailing with branches, pummeling with fists, pulling of hair, and inflicting injuries (Goodall 1986). These attacks are typically assumed to carry costs, such as risk of infection from wounds, but Muller et al. (2007) clarify potential costs with evidence that cycling parous females, who are the primary targets of male coercion, have elevated cortisol levels. The data cannot demonstrate that male coercion directly *causes* hormonally mediated stress in females. A causal connection is suggested, however, by the fact that, compared with parous females, nulliparous females copulated at equivalent rates, spent similar (if not more) time in the company of males, but received relatively less coercion from them (as less preferred sexual partners) *and* had cortisol levels that were not only lower but that did not differ significantly on estrous versus nonestrous days.

Prediction 2: Male mating success is improved by sexual coercion. Previous primate studies had rejected this prediction based on the lack of a positive correlation between overall rates of male aggression to females and male mating success (Bercovitch et al. 1987; Soltis 1999; Stumpf and Boesch 2006). Muller et al. (2009) provide a more direct assay of the selective impact of sexual coercion by demonstrating that male chimpanzees copulated at significantly higher rates with females that they were more aggressive to, than with females that they were less aggressive to (Fig. 3.2).

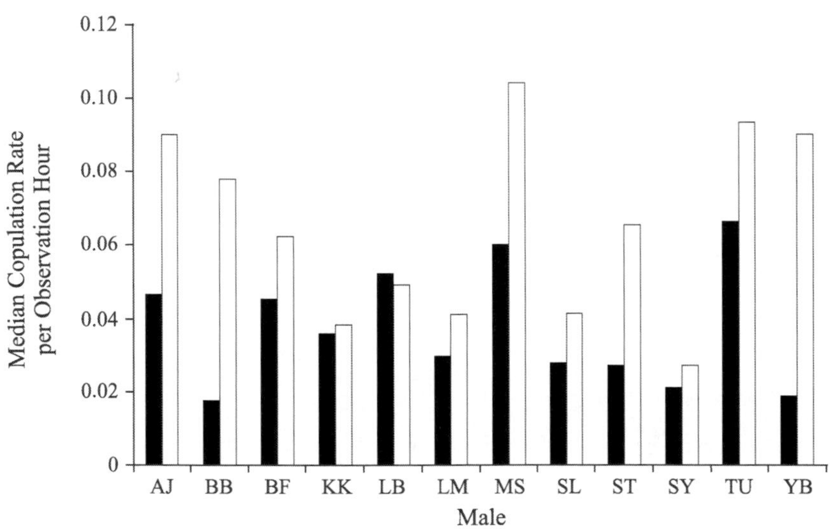

Fig. 3.2 Median dyadic rates of aggression for each of 13 male chimpanzees (*Pan troglodytes*) with 15 parous females. For each male, the median copulation rates were calculated with females who received above (*white*) or below (*black*) the median amount of aggression for that male. The difference was significant (Wilcoxon signed-rank test, $p = 0.002$). Data from Muller et al. (2007)

Prediction 3: Coercion is not simply an indirect cost of female choice. One of the noteworthy aspects of this study is its test of the alternative hypothesis that male aggression reflects female choice. Muller and colleagues marshal several lines of evidence to reject the possibility that intersexual aggression is a by-product of female mating preferences for aggressive males. First, male rank was uncorrelated with aggression directed at females. Second, the relationship between male coercion and mating success with targeted females also held for low-ranking males as well, who are arguably less preferred sexual partners. Finally, a matrix partial correlation analysis revealed a significant association between male aggression directed at individual females and the copulation rate with those females, controlling for time spent together.

3.5 PostCopulatory Sexual Conflict: Prezygotic

Sexual conflict after copulation may involve processes occurring at or prior to fertilization (prezygotic) or thereafter (postzygotic) (Table 3.1). The postcopulatory manifestation of intrasexual selection is sperm competition (along with associated factors such as genital locks, penis morphology, etc.), which has attracted much study (e.g., Birkhead and Møller 1998). Postcopulatory intersexual selection is cryptic female choice (Eberhard 1996), which primarily concerns the differential treatment of sperm in the reproductive tracts of polyandrously mating females (as well as associated phenomena, e.g., abortion). The important question here, however, concerns the potential for conflict between these two postcopulatory processes: how do the benefits to females of cryptic mate choice via multimale mating compare with the costs incurred from male adaptations for sperm competition? Current data are too few to answer this question for primates. Although sperm competition is relatively well investigated (Gomendio et al. 1998), cryptic female choice remains virtually unstudied (Reeder 2003), with the possible exception of *H. sapiens* (Baker and Bellis 1995; Thornhill and Gangestad 1996). Therefore, I highlight below two areas where postcopulatory-prezygotic sexual conflict might occur in primates.

3.5.1 Genital Coagulates

One possible source of conflict concerns enzymes acting on seminal vesicular proteins to congeal ejaculates into structures ranging from a soft coagulum to a more substantial, rubbery plug. Seminal coagulation is more pronounced in primates with multimale mating patterns (compared to unimale systems), suggesting a male strategy to block rival sperm access to the cervical Os (Dixson and Anderson 2002). What is not known, however, is whether these coagulates impose costs on females. Plugs can be dislodged by subsequent male partners or by inseminated females in *L. catta* (Parga 2003) and *P. troglodytes* (Dixson and Mundy 1994), suggesting low potential for sexual conflict (at least over remating) or the existence of an effective female counterstrategy to male manipulation. Intersexual conflict may be more relevant in taxa where females cannot remove plugs, such as *M. murinus* (Eberle and Kappeler 2004a). But even in these cases, conflict cannot be assumed as plugs potentially confer benefits to females, such as facilitating fertilization via sperm retention or transport. This could be valuable in a species like *M. murinus*, in which females are in estrous for only a few hours on a single night each year.

3.5.2 Penis Morphology and Female Injury

In strepsirrhines, keratinized penile spines, plates, or papillae are so conspicuous, widespread, and variable as to have long informed taxonomy (Bearder et al. 1996). Similar, but generally simpler, anatomical features are also found in a few platyrrhines and catarrhines (Dixson 1998). Spines develop upon sexual maturity (Perkin 2007), suggesting testosterone mediation and a mating-related function, but the precise nature of that function remains obscure. Adaptive hypotheses include tactile facilitation of ejaculation, removal of sperm or copulatory plugs, genital locking of partners, stimulation of reproductive readiness in females or of synchrony between partners, and Fisherian female choice (Dixson 1989; Eberhard 1990; Harcourt and Gardiner 1994).

Comparative data from insects suggest an alternative explanation: sexual conflict. In the cowpea weevil (*Callosobruchus maculatus*), the penis is equipped with spines that damage the female genital tract during copulation, reducing her likelihood of subsequent mating, and thereby enhancing sperm competition outcomes for the male (Crudgington and Siva-Jothy 2000; Hotzy and Arnqvist 2009). In primates, the magnitude of spinosity is negatively correlated with the duration of female sexual receptivity during the ovarian cycle (Stockley 2002), suggesting that penile spines similarly improve male sperm competition success by restricting female mating. The precise mechanism underlying this association is unclear, however. Penile spines could stimulate ovulation or associated neuroendocrine reflexes, but they could also cause "*short-term local damage to the female genital tract, making continued sexual activity painful or aversive*" (Stockley 2002, p 130).

Correspondingly, sexual conflict theory may shed light on the function of human practices of genital modification (e.g., Wilson 2008). The patterns and frequency of female genital cutting vary substantially across populations, and the effects on female (and male) sexual behavior and reproduction are strongly debated (Gruenbaum 2001). Reason (2004) argues that in one West African population, the practice enhances female reproductive success because it is a virtual prerequisite for marriage and because men invest significantly more in the offspring of wives who are circumcised. Both patterns are consistent with a sexual conflict interpretation, but clearly more study of human behavioral ecology in the context of relevant cultural influences is needed to test this hypothesis against alternative explanations. As Low (2005, p 76) concludes, although current data on genital modification "*may not prove [sexual] conflicts of interests, they are suggestive*."

3.6 Postcopulatory Sexual Conflict: Postzygotic

Precopulatory intimidation by male chimpanzees can only be fully understood in the context of postcopulatory sexual conflict in the form of infanticide. Muller et al. (2009) argue that sexual coercion, particularly as practiced by high-ranking males, is a counterstrategy to limit female promiscuity, and that promiscuity is itself a counterstrategy to male infanticide (see also Stumpf et al. 2008). This scenario highlights the nature of sexually antagonistic coevolution: male infanticide favors female promiscuity, which favors male sexual coercion, etc.

Infanticide figures prominently in Smuts and Smuts' (1993) original discussion of sexual coercion, but it does not fit easily within Clutton-Brock and Parker's (1995) more general harassment-intimidation dichotomy. It is initially difficult to appreciate that male infanticide might constitute incidental harm to females, since an infant's death seems so directly detrimental to the mother's fitness. But this proposition becomes clearer when we consider that the specific "problem" that lactating females pose to reproducing males is a straightforward consequence of primate biology: a nursing infant is, in the words of Altmann et al. (1978: 1029), a "*perfect contraceptive*." The function of sexually selected infanticide, then, is to disrupt this contraceptive system, not to harm the mother (or reduce her fitness) per se. Thus, following the broader theoretical logic of Partridge and Hurst (1998) and Lessells (2005), if, speculatively, males possessed other means of effectively counteracting the contraceptive – e.g., by manipulating the mother's hormonal state or accelerating weaning – and if the costs of such a strategy did not exceed the costs of infanticide, then males would *not* be selected to kill infants (but could still achieve the same reproductive benefit). Under such conditions, the death of infant, if it occurred, would be an incidental by-product of the male manipulative strategy, not a necessary harmful component of that strategy. This is not to say that male attacks on infants can not, in principle, function as sexual intimidation, if their mothers' mating compliance forestalls further aggression directed at them. As Clutton-Brock and Parker (1995) point out, however, this mechanism of *indirect* sexual intimidation predicts that male threats and attacks will also extend to *juveniles,* which is neither predicted by the sexual selection hypothesis nor is a common correlate of infanticidal behavior (Hrdy 1974; van Schaik and Janson 2000).

Male infanticide is still, however, a drastic form of sexual conflict. It reflects adaptive harm (*sensu* Johnstone and Keller 2000) insofar as infanticidal males exploit a preexisting feature of female reproductive plasticity, such that infant loss often accelerates resumption of ovulatory cycling. Although the adaptive significance of infanticide in primates continues to be debated, the available evidence is more consistent with the sexual selection argument (Borries et al. 1999; Soltis et al. 2000; van Schaik 2000b) than with competing hypotheses, such as the generalized aggression model (Bartlett et al. 1993) and the social pathology argument (Dolhinow 1977). Thus, infanticide appears a likely manifestation of postcopulatory sexual conflict in primates, as well as, arguably, the most studied form of sexual conflict (Hausfater and Hrdy 1984; van Schaik and Janson 2000).

3.7 A Counterstrategy: Male–Female Association

The counterstrategies to sexual conflict are as diverse as the manifestations of conflict itself. They may be morphological, such as the thick skin of female blue sharks (*Prionace glauca*) vulnerable to bites from "courting" males (Pratt 1979), or the large body size of some female lemurs, which is argued to limit sexual coercion during the mating season (Foerg 1982; Taylor and Sussman 1985; Richard 1992; Morland 1993; Brockman 1999).

Female sexual behavior – particularly promiscuity – can limit sexual conflict, both in the form of precopulatory coercion and postcopulatory infanticide. The convenience polyandry hypothesis holds that conceding copulations allows females to avoid the costs of resistance to coercive males (Thornhill and Alcock 1983; Mesnick and le Boeuf 1991; Blyth and Gilburn 2006). This explanation is less often invoked as an anticoercion counterstrategy in primates than in other animals, but one example is Eberle and Kappeler's (2004a, p 97) argument that the multimale mating of female mouse lemurs reflects " '*making the best of a bad job*' *in the face of male harassment*." The counteractive value of convenience polyandry is improved when it is supplemented with postcopulatory mechanisms of cryptic female choice (e.g., spermicides) (Holman and Snook 2006), but this remains unstudied in nonhuman primates. In the postcopulatory domain, both theoretical models and empirical evidence suggest that female promiscuity offers significant potential to limit infanticide by confusing paternity (Hrdy 1979; Ebensperger 1998; van Schaik and Janson 2000; Wolff and MacDonald 2004; Pradhan and van Schaik 2008).

Association with males is a hypothesized female counterstrategy to sexual conflict, again in both the form of sexual coercion and of male infanticide. Sustained proximity to a large, dominant male reduces estrous female exposure to male harassment and intimidation in Japanese macaques (Matsubara and Sprague 2004) and chimpanzees (Wrangham 1986), as well as in many other taxa (insects: Thornhill and Alcock 1983; fish: Pilastro et al. 2003; Dadda et al. 2005; birds: Gowaty and Buschhaus 1998; bighorn sheep: Réale et al. 1996; elephant seals: Mesnick and le Boeuf 1991). This function has also been suggested for the temporary consortships of female orangutans at risk of forced copulation (Mitani 1985; Fox 2002; Setia and van Schaik 2007). Thus, protection from sexual coercion is an alternative functional hypothesis for consortships, independent (though not mutually exclusive) of mate guarding, and female choice hypotheses (Manson

1997). The relevance of this hypothesis for understanding variation in consortships has not been explored thoroughly.

Reducing the costs of precopulatory sexual harassment may similarly underlie sexual swellings. Previous analyzes have suggested that sexual swellings might benefit females because they incite male–male competition, which then facilitates insemination by high-quality males (Clutton-Brock and Harvey 1976) or copulation with many males (Hrdy and Whitten 1987). Alternatively, sexual swellings might serve to reduce the costs of harassment or intimidation by ensuring mate guarding by a dominant male who keeps other males away. The adaptive value of this counterstrategy, however, must be measured against the (coercion) costs of advertising estrous, the benefits of multimale mating, and the benefits of the alternative counterstrategy of reducing coercion via *concealment* of receptivity.

Male–female association is also a proposed counterstrategy to postcopulatory conflict in the form of infanticide (Wrangham 1979; van Schaik and Dunbar 1990; van Schaik and Kappeler 1997). Empirical evidence supports this hypothesis in numerous taxa, including insects, birds, and rodents, and a few primate species (reviewed by Palombit 2000). Mountain gorilla (*Gorilla beringei*) groups have long been viewed as associations of females with a male protector, but whether he deters infanticide or predation is debated. A recent mathematical simulation supports the antiinfanticide hypothesis (Harcourt and Greenberg 2001), but Harcourt and Stewart (2007) argue that rejection of the antipredation hypothesis is premature. Recently, this argument was extended to orangutans with Setia and van Schaik's (2007) suggestion that lactating females use male long calls to stay loosely associated with adult male protectors.

Van Schaik and Dunbar's (1990) hypothesis that social monogamy is an anti-infanticide strategy remains one of the most interesting versions of this hypothesis. Evidence that infanticide has selected for social monogamy is strong in some nonprimate taxa such as burying beetles (*Nicrophorus* spp.) and tropical house wrens (*Troglodytes aedon*), but interpretations of the gibbon data have generated divergent conclusions (Palombit 1999, 2000; Sommer and Reichard 2000; Fuentes 2002; van Schaik and Kappeler 2003). Recent tests of the hypothesis in prosimians, such as fork-marked lemurs (*Phaner furcifer*), avahis (*Avahi occidentalis*), and spectral tarsiers (*Tarsius spectrum*), have not consistently supported the hypothesis (Schülke and Kappeler 2001; Thalmann 2001; Gursky 2002). However, this intriguing hypothesis awaits further direct testing in the taxa it primarily addresses: the gibbons.

One population in which long-term data continue to suggest an antiinfanticide function of male–female bonding is the chacma baboon (*Papio hamadryas griseipes*) of the Okavango Delta, Botswana (see also Weingrill 2000). Like yellow baboons (*P. h. cynocephalus*) and olive baboons (*P. h. anubis*) of east Africa, these baboons live in relatively large, multimale, multifemale groups, with female philopatry and dominance relationships in both sexes. In contrast to its east African cousins, however, the chacma baboon exhibits comparatively high rates of infanticide (Palombit 2003). Infanticide is the primary source of mortality for infants, and accounts for at least 38% of infant mortality, though this rate may be as high as 75%

in some years (Cheney et al. 2004). The patterning of infanticide in this population is more consistent with the sexual selection hypothesis than with alternative hypotheses (Palombit et al. 2000). Infanticide is generally committed by males that have recently immigrated into a group and attained alpha status. The relatively short tenure of alpha males (approximately 7 months, on average) combined with their apparently greater monopolization of matings (Bulger 1993) creates conditions that enhance the potential benefits of infanticide. In other words, a new alpha male is confronted with a short period of relatively exclusive sexual access to females. Conversely, since loss of an infant significantly accelerates resumption of fertile cycling in females, lactating mothers are confronted with a significant threat of infanticide.

Unsurprisingly, lactating females exhibit conspicuous and aroused aversion to newly immigrated alpha males, including continual retrieval of infants, screaming, and tail-up displays (Busse 1984). They almost always establish a "friendship" with an unrelated, adult male shortly after parturition (Busse 1981; Palombit et al. 1997). Friendships can be unambiguously differentiated from a female's relationships with other males in the group on the basis of spatial association, grooming, infant handling, and vocal interaction (reviewed by Palombit 2009). Ad libitum evidence suggests that friendship status increases a male's defense of infants during potentially (or actual) infanticidal attacks. Although several males may rush to the scene of such attacks, it is primarily the male friend of the infant's mother who provides direct, apparently costly forms of defense, such as fighting or threatening the alpha male, or carrying the infant. Experimental playback experiments further showed that male friends were more likely to respond to their female friends' screams than to the screams of other females, and females' screams were more likely to provoke responses form their male friends than from other males (Palombit et al. 1997). These experiments also revealed that the solicitude of male friends was tied closely to the presence of infants: playback of female screams shortly after infants died elicited similarly weak responses from all males, regardless of their friendship status. Alternative benefits of friendships to females, such as protection from harassment from higher-ranking females, lack empirical support (Palombit 2009).

Since these original observations, a series of hormonal studies in this population have further supported the antiinfanticide function of heterosexual friendships. Following the immigration of a new male, glucocorticoid levels rise in females generally, but remain high over subsequent weeks only among anestrous females, not among cycling females (Beehner et al. 2005; Wittig et al. 2008). This is a striking difference because cycling females are the primary targets of the protracted, aggressive chasing that seems to facilitate a new male's rise to alpha status (Kitchen et al. 2009). Thus, hormonal patterns suggest that it is females at risk of infanticide (not simply of aggression) from the new male who experience greater stress upon his arrival in the group. This is further substantiated by *additional* increases in glucocorticoids among lactating females when a new alpha actually commits an infanticide (Engh et al. 2006) or among the (few) lactating females who lack male friends at the time of male immigration (Beehner et al. 2005).

A final indication of the potential importance of friendships is that females compete with one another for them (Palombit et al. 2001). This is reflected partly by the positive correlation between the dominance ranks of male and female friends, and partly by observations of high-ranking females displacing subordinate rivals from friendship with a particular male. Competition among females for males is relatively rare in mammals (Berglund et al. 1993; Andersson 1994), and in this case, it suggests that males provide a service with important fitness consequences for females. It is not immediately obvious why male protection is not shareable among multiple lactating females, but since *friendship* status appears so crucial, females may compete for social access to males in order to develop this relationship.

Sexual conflict hypotheses for male–female bonding are potentially relevant to understanding human pair-bonding, although space precludes a thorough treatment of this question here. Early models argued that a durable pairbond between the sexes was part of an adaptive suite of traits including reproductive monogamy and a division of labor between females and provisioning males (Murdock 1949; Washburn and Lancaster 1968; Lovejoy 1981). An alternative hypothesis emphasizes the importance of male protection of females from sexual conflict in the form of sexual coercion and/or infanticide (Betzig 1992; Smuts 1992; Mesnick 1997; Hrdy 1999; Hawkes 2004). A recent cross-cultural analysis rejected the male protection hypotheses partly because pairbond stability (overall divorce rates in a society) was uncorrelated with general male aggressiveness (overall rates of male homicides and assaults) (Quinlan and Quinlan 2007). However, this conclusion is limited in the same way that the lack of a correlation between overall male aggressiveness and mating success in chimpanzees may overlook the fact that sexual coercion significantly increases a male's mating success with the *particular* females he targets (see above). Thus, the hypothesis must be tested with human data addressing *specifically* how risk of sexual coercion or infanticide to individual women varies with the nature of their pair bonds. Since male partners are themselves sometimes a source of sexual coercion to women (Rodseth and Novak 2009), these analyses must differentiate between the costs of pair-bonding with men and the protective benefits of pair bonds from other men. The variety of current evidence suggests the possibility that the different selective pressures proposed may each promote pair-bonding under different conditions (Quinlan 2008). This proposition merits greater scrutiny.

3.8 Conclusions and Future Directions

Sexual conflict is inevitable and ubiquitous: the question is not whether it occurs, but how and when, and to what degree sexually antagonistic coevolution has acted, compared with other mechanisms of sexual selection (Hosken and Snook 2005: S1, Andersson and Simmons 2006). Sexual conflict theory situates explanations in the "arms race" perspective previously reserved for more conventional coevolutionary adversaries, such as predators and prey (Dawkins and Krebs 1979). The Red Queen hypothesis, that any gain in fitness by one unit of evolution is balanced by equivalent losses in fitness by others (van Valen 1973), may provide the most appropriate framework for analyzing reproductive strategies as a zero-sum game between opposing males and females (Chapman and Partridge 1996; Rice and Holland 1997). This does not mean, however, that conflict universally characterizes the *phenotypic* expression of male and female interaction. Affiliation and intersexual cooperation may be one outcome of this coevolutionary conflict, as suggested for chacma baboon friendships. Indeed, the chacma baboon and chimpanzee together highlight the view of universals as process, rather than as pattern. Current evidence suggests that sexually selected infanticide has generated two distinct modes of female counterstrategy in these species: promiscuity and association with males. The patterns are different, but the underlying process that generates the patterns is the same: sexual conflict. This chapter has focused mostly on sexual conflict over mating, but it may also occur at the level of sex chromosomes, gamete interaction, parental investment, group size and composition, and group dynamics (Table 3.1).

Sexual coercion via intimidation/punishment is likely to be a common, if not universal feature of life among animals that live gregariously and modify their behavior through learning (Clutton-Brock and Parker 1995). The attention following the publication of the Smuts and Smuts (1993) model has enlarged the data base for male mating aggression to females. Somewhat surprisingly, however, relatively few studies have rigorously tested the full set of constituent predictions (but see Muller and Wrangham 2009) or differentiated analytically between sexual harassment and intimidation. Costs to females are often an assumed rather than measured consequence of overt aggression, or are assessed qualitatively (e.g., as an "injury"). A key goal for future studies is quantitative measurement of these costs (as Muller et al. [2009] do). These data will help address some other questions: do females do worse reproductively when mating with more coercive or persistent males, as predicted by theory? The hypothesis that females may derive benefits from coercion also merits greater study. Likewise, the costs of coercion to males are virtually ignored, but may be significant. For example, the seminal fluids of bushcrickets inhibit receptivity of females to further mating in a manner similar to *D. melanogaster*, but males that deliver greater quantities of these fluids *also* experience longer sexual refractory periods themselves (Vahed 2007). Information on costs to males, combined with data addressing covariation in male coercion and fitness, will help to clarify the trade-offs of coercion or manipulation of females versus alternative mating strategies. Most primate studies of sexual conflict have focused on sexual coercion, but male manipulation in the form of antagonistic seduction, and concomitant females resistance (Holland and Rice 1999) merits more attention.

The life history of primates, as well as the practical constraints on an experimental study of them, significantly limit the kinds of data that can be collected. Nevertheless, there are compelling reasons to study sexual conflict in primates. Until fairly recently, much of the research on sexual conflict was conducted on (invertebrate) taxa that conform more or less to the Bateman (1948) principle that males are selected to mate and females not (Partridge and Hurst, 1998, Tregenza et al. 2006). Our understanding of the full significance of sexual conflict will be improved by greater study of systems violating this assumption, i.e., taxa in which remating is potentially beneficial to females. Additionally, as Clutton-Brock and Parker (1995) emphasize, models of sexual conflict have generally focused on relatively simple social contexts. The study of highly social species promises to reveal important and subtle influences of social relationships on the economic trade-offs of sexual coercion and resistance. In spite of the methodological difficulties they pose, primates are excellent subjects to achieve all of these goals.

In summary, conflict among genes is "*a universal feature of life*" (Burt and Trivers 2006, p 3). This is true not only for genes within a genome, but also for genes residing in the genomes of the interacting entities we call "male" and "female."

Acknowledgments I am extremely grateful to Peter Kappeler and Joan Silk for their generous cooperation and advice in the preparation of the manuscript. Field research was funded by NSF (BCS-0117213), the Leakey Foundation, the Wenner-Gren Foundation and the Center for Human Evolutionary Studies (Rutgers University).

References

Aaron RI (1965) Wittgenstein's theory of universals. Mind 74:249–251

Agoramoorthy G, Mohnot SM, Sommer V, Srivastava A (1988) Abortions in free-ranging hanuman langurs (*Presbytis entellus*) - a male induced strategy? Hum Evol 3:297–308

Altmann J, Altmann SA, Hausfater G (1978) Primate infant's effects on mother's future reproduction. Science 201:1028–1029

Andersson M (1994) Sexual selection. Princeton University Press, Princeton, NJ

Andersson M, Simmons LW (2006) Sexual selection and mate choice. Trends Ecol Evol 21:296–302

Arlet ME, Molleman F, Chapman CA (2008) Mating tactics in male grey-cheeked mangabeys (*Lophocebus albigena*). Ethology 114:851–862

Arnqvist G (2004) Sexual conflict and sexual selection: lost in the chase. Evolution 58: 1383–1388

Arnqvist G, Nilsson T (2000) The evolution of polyandry: multiple mating and female fitness in insects. Anim Behav 60:145–164

Arnqvist G, Rowe L (2005) Sexual conflict. Princeton University Press, Princeton, NJ

Atmoko S, van Hooff JARAM (2004) Alternative male reproductive strategies: male bimaturism in orangutans. In: Kappeler PM, van Schaik CP (eds) Sexual selection in primates: new and comparative perspectives. Cambridge University Press, Cambridge, pp 196–207

Baker RR, Bellis MA (1995) Human sperm competition: copulation, masturbation and infidelity. Chapman & Hall, New York

Bartlett TQ, Sussman RW, Cheverud JM (1993) Infant killing in primates: a review of observed cases with specific reference to the sexual selection hypothesis. Am Anthropol 95:958–990

Bateman AJ (1948) Intrasexual selection in *Drosophila*. Heredity 2:349–368

Bearder SK, Honess PE, Bayes M, Ambrose L, Anderson M (1996) Assessing *Galago* diversity – a call for help. Afr Primates 2:11–15

Beehner JC, Bergman TJ, Cheney DL, Seyfarth RM, Whitten PL (2005) The effect of new alpha males on female stress in free-ranging baboons. Anim Behav 69:1211–1221

Bercovitch FB, Sladky KK, Roy MM, Goy RW (1987) Intersexual aggression and male sexual activity in captive rhesus macaques. Aggress Behav 13:347–358

Berger J (1983) Induced abortion and social factors in wild horses. Nature 303:59–61

Berglund A, Magnhagen C, Bisazza A, König B, Huntingford F (1993) Female-female competition over reproduction. Behav Ecol 4:184–187

Bertin A, Fairbairn DJ (2005) One tool, many uses: precopulatory sexual selection on genital morphology in *Aquarius remigis*. J Evol Biol 18:949–961

Betzig LL (1992) Of human bonding: cooperation or exploitation? Soc Sci Inform 31:611–642

Birkhead TR, Møller AP (1998) Sperm competition and sexual selection. Academic Press, New York

Blanckenhorn WU, Hosken DJ, Martin OY, Reim C, Teuschl Y, Ward PI (2002) The costs of copulating in the dung fly *Sepsis cynipsea*. Behav Ecol 13:353–358

Blyth JE, Gilburn AS (2006) Extreme promiscuity in a mating system dominated by sexual conflict. J Insect Behav 19:447–455

Boehm C (1994) Pacifying interventions at Arnhem Zoo and Gombe. In: Wrangham RW, McGrew WC, de Waal FBM, Heltne PG (eds) Chimpanzee cultures. Harvard University Press, Cambridge, MA, pp 211–226

Boinski S (2000) Social manipulation within and between troops mediates primate group movement. In: Boinski S, Garber PA (eds) On the move: how and why animals travel in groups. University of Chicago Press, Chicago, pp 421–468

Borries C, Launhardt K, Epplen C, Epplen JT, Winkler P (1999) DNA analyses support the hypothesis that infanticide is adaptive in langur monkeys. Proc Roy Soc Lond B 266:901–904

Bowcock H, Brown GP, Shine R (2009) Beastly bondage: the costs of amplexus in cane toads (*Bufo marinus*). Copeia 1:29–36

Boyd R, Silk JB (2006) How humans evolved, 4th edn. WW Norton, New York

Brockman DK (1999) Reproductive behavior of female *Propithecus verreauxi* at Beza Mahafaly, Madagascar. Int J Primatol 20:375–398

Bulger JB (1993) Dominance rank and access to estrous females in male savanna baboons. Behaviour 127:67–103

Burt A, Trivers RL (2006) Genes in conflict: the biology of selfish genetic elements. Belknap Press, Cambridge, MA

Busse CD (1981) Infanticide and parental care by male chacma baboons, *Papio ursinus*. PhD Dissertation, Davis, University of California

Busse CD (1984) Tail raising by baboon mothers toward immigrant males. Am J Phys Anthropol 64:255–262

Carpenter CR (1942) Sexual behavior of free-ranging rhesus monkeys (*Macaca mulatta*). I. Specimens, procedures and behavioral characteristics of estrous. J Comp Psychol 33:113–142

Chapman T, Partridge L (1996) Sexual conflict as fuel for evolution. Nature 381:189–190

Chapman T, Liddle LF, Kalb JM, Wolfner MF, Partridge L (1995) Cost of mating in *Drosophila melanogaster* females is mediated by male accessory gland products. Nature 373:241–244

Chapman T, Arnqvist G, Bangham J, Rowe L (2003) Sexual conflict. Trends Ecol Evol 18:41–47

Chen PS (1984) The functional morphology and biochemistry of insect male accessory glands and their secretions. Annu Rev Biochem 29:233–255

Cheney DL, Seyfarth RM, Fischer J, Beehner J, Bergman T, Johnson SE, Kitchen DM, Palombit RA, Rendall D, Silk JB (2004) Factors affecting reproduction and mortality among baboons in the Okavango Delta, Botswana. Int J Primatol 25:401–428

Chism J, Rogers W (1997) Male competition, mating success and female choice in a seasonally breeding primate (*Erythrocebus patas*). Ethology 103:109–126

Clark AG, Aguadé M, Prout T, Harshman LG, Langley CH (1995) Variation in sperm displacement and its association with accessory gland protein loci in *Drosophila melanogaster*. Genetics 139:189–201

Clutton-Brock TH, Harvey PH (1976) Evolutionary rules and primate societies. In: Bateson PPG, Hinde RA (eds) Growing points in ethology. Cambridge University Press, Cambridge, pp 195–237

Clutton-Brock TH, Parker GA (1995) Sexual coercion in animal societies. Anim Behav 49:1345–1365

Colmenares F, Zaragoza F, Hernández-Lloreada MV (2002) Grooming and coercion in one-male units of hamadryas baboons: market forces or relationships constraints? Behaviour 139:1525–1553

Cook PA, Wedell N (1999) Non-fertile sperm delay female remating. Nature 397:486

Cordero C, Eberhard WG (2003) Female choice of sexually antagonistic male adaptations: a critical review of some current research. J Evol Biol 16:1–6

Cox RM, Calsbeek R (2009) Sexually antagonistic selection, sexual dimorphism, and the resolution of intralocus sexual conflict. Am Nat 173:176–187

Crudgington HS, Siva-Jothy MT (2000) Genital damage, kicking and early death. Nature 407:855–856

Dadda M, Pilastro A, Bisazza A (2005) Male sexual harassment and female schooling behaviour in the eastern mosquitofish. Anim Behav 70:463–471

Daly M (1978) The cost of mating. Am Nat 112:771–774

Darwin C (1871) The descent of man, and selection in relation to sex. Murray, London

Dawkins R, Krebs JR (1979) Arms races between and within species. Proc Roy Soc Lond B 205:489–511

de Waal FBM (2005) A century of getting to know the chimpanzee. Nature 437:56–59

Dixson AF (1989) Sexual selection, genital morphology, and copulatory behavior in male galagos. Int J Primatol 10:47–55

Dixson AF (1998) Primate sexuality: comparative studies of the prosimians, monkeys, apes and human beings. Oxford University Press, Oxford

Dixson AF, Anderson MJ (2002) Sexual selection, seminal coagulation and copulatory plug formation in primates. Folia Primatol 73:63–69

Dixson AF, Mundy NI (1994) Sexual behavior, sexual swelling, and penile evolution in chimpanzees (*Pan troglodytes*). Arch Sex Behav 23:267–280

Dolhinow P (1977) Normal monkeys? Am Sci 65:266

Ebensperger LA (1998) Strategies and counterstrategies to infanticide in mammals. Biol Rev Camb Philos Soc 73:321–346

Eberhard WG (1990) Animal genitalia and female choice. Am Sci 78:134–141

Eberhard WG (1996) Female control: sexual selection by cryptic female choice. Princeton University Press, Princeton, NJ

Eberhard WG (2005) Evolutionary conflicts of interest: are female sexual decisions different? Am Nat 165:S19–S25

Eberle M, Kappeler PM (2004a) Selected polyandry: female choice and inter-sexual conflict in a small nocturnal solitary primate (*Microcebus murinus*). Behav Ecol Sociobiol 57:91–100

Eberle M, Kappeler PM (2004b) Sex in the dark: determinants and consequences of mixed male mating tactics in *Microcebus murinus*, a small solitary nocturnal primate. Behav Ecol Sociobiol 57:77–90

Emery Thompson M (2009) Human rape: revising evolutionary perspectives. In: Muller MN, Wrangham RW (eds) Sexual coercion in primates and humans: an evolutionary perspective on male aggression against females. Harvard University Press, Cambridge, Massachnsetts, pp 346–374

Engh AL, Beehner JC, Bergman TJ, Whitten PL, Hoffmeier RR, Seyfarth RM, Cheney DL (2006) Female hierarchy instability, male immigration and infanticide increase glucocorticoid levels in female chacma baboons. Anim Behav 71:1227–1237

Enomoto T (1981) Male aggression and the sexual behavior of Japanese monkeys. Primates 22:15–23

Erez T, Schneider JM, Lubin Y (2005) Is male cohabitation costly for females of the spider *Stegodyphus lineatus* (Eresidae)? Ethology 111:693–704

Festa-Bianchet M (1987) Bighorn sheep, climbing accidents, and implication for mating strategy. Mammalia 51:618–620

Fisher RA (1930) The genetical theory of natural selection. Oxford University Press, Oxford

Foerg R (1982) Reproductive behavior in *Varecia variegata*. Folia Primatol 38:108–121

Foerster K, Coulson T, Sheldon BC, Pemberton JM, Clutton-Brock TH, Kruuk LEB (2007) Sexually antagonistic genetic variation for fitness in red deer. Nature 447:1107–1110

Fox EA (2002) Female tactics to reduce sexual harassment in the Sumatran orangutan (*Pongo pygmaeus abelli*). Behav Ecol Sociobiol 52:93–101

Fox R (1989) The search for society: quest for a biosocial science and morality. Rutgers University Press, New Brunswick, NJ

Fuentes A (2002) Patterns and trends in primate pair bonds. Int J Primatol 23:953–978

Gems D, Riddle DL (1996) Longevity in *Caenorhabditis elegans* reduced by mating but not gamete production. Nature 379:723–725

Gibson KN, Vick LG, Palma AC, Carrasco FM, Taub D, Ramos-Fernández G (2008) Intra-community infanticide and forced copulation in spider monkeys: a multi-site comparison between Cocha Cashu, Peru and Punta Laguna, Mexico. Am J Primatol 70:485–489

Goetz AT, Shackelford TK, Camilleri JA (2008) Proximate and ultimate explanations are required for a comprehensive understanding of partner rape. Aggress Violent Behav 13:119–123

Gomendio M, Harcourt AH, Roldán ERS (1998) Sperm competition in mammals. In: Birkhead TR, Møller AP (eds) Sperm competition and sexual selection. Academic Press, San Diego, pp 667–756

Goodall J (1971) In the shadow of man. Houghton Mifflin, Boston

Goodall J (1986) The chimpanzees of Gombe: patterns of behavior. Belknap Press, Cambridge, MA

Gowaty PA (1997) Sexual dialectics, sexual selection, and variation in reproductive behavior. In: Gowaty PA (ed) Feminism and evolutionary biology: boundaries, intersections, and frontiers. Chapman & Hall, New York, pp 351–384

Gowaty PA, Buschhaus N (1998) Ultimate causation of aggressive and forced copulation in birds: female resistance, the CODE hypothesis, and social monogamy. Am Zool 38:207–225

Gruenbaum E (2001) The female circumcision controversy: an anthropological perspective. University of Pennsylvania Press, Philadelphia

Gursky SL (2002) Determinants of gregariousness in the spectral tarsier (Prosimian: *Tarsius spectrum*). J Zool Lond 256:401–410

Harcourt AH, Gardiner J (1994) Sexual selection and genital anatomy of male primates. Proc Roy Soc Lond B 255:47–53

Harcourt AH, Greenberg J (2001) Do gorilla females join males to avoid infanticide? A quantitative model. Anim Behav 62:905–915

Harcourt AH, Stewart KJ (2007) Gorilla society: conflict, compromise, and cooperation between the sexes. University of Chicago Press, Chicago

Harshman LG, Prout T (1994) Sperm displacement without sperm transfer in *Drosophila melanogaster*. Evolution 48:758–766

Hausfater G, Hrdy SB (1984) Infanticide: comparative and evolutionary perspectives. Aldine, New York

Hawkes K (2004) Mating, parenting, and the evolution of human pair bonds. In: Chapais B, Berman CM (eds) Kinship and behavior in primates. Oxford University Press, Oxford, pp 443–473

Head ML, Brooks R (2006) Sexual coercion and the opportunity for sexual selection in guppies. Anim Behav 71:515–522

Henzi SP, Lycett JE, Weingrill T (1998) Mate guarding and risk assessment by male mountain baboons during inter-troop encounters. Anim Behav 55:1421–1428

Heubel KU, Plath M (2008) Influence of male harassment and female competition on female feeding behaviour in a sexual-asexual mating complex of mollies (*Poecilia mexicana*, *P. formosa*). Behav Ecol Sociobiol 62:1689–1699

Hiruki LM, Stirling I, Gilmartin WG, Johanos TC, Becker BL (1993) Significance of wounding to female reproductive success in Hawaiian monk seals (*Monachus schauinslandi*) at Laysan Island. Can J Zool 71:469–474

Holland B, Rice WR (1999) Experimental removal of sexual selection reverses intersexual antagonistic coevolution and removes a reproductive load. Proc Natl Acad Sci USA 96:5083–5088

Holman L, Snook RR (2006) Spermicide, cryptic female choice and the evolution of sperm form and function. J Evol Biol 19:1660–1670

Hosken DJ, Snook RR (2005) How important is sexual conflict? Am Nat 165:S1–S4

Hosken DJ, Martin OY, Born J, Huber F (2003) Sexual conflict in *Sepsis cynipsea*: female reluctance, fertility and mate choice. J Evol Biol 16:485–490

Hotzy C, Arnqvist G (2009) Sperm competition favors harmful males in seed beetles. Curr Biol 19:404–407

Howard RD (1980) Mating behavior and mating success in woodfrogs, *Rana sylvatica*. Anim Behav 28:705–716

Hrdy SB (1974) Male-male competition and infanticide among the langurs (*Presbytis entellus*) of Abu, Rajasthan. Folia Primatol 22:19–58

Hrdy SB (1979) Infanticide among animals: a review, classification, and examination of the implications for reproductive strategies of females. Ethol Sociobiol 1:13–40

Hrdy SB (1999) Mother nature: a history of mothers, infants, and natural selection. Pantheon Books, New York

Hrdy SB, Whitten PL (1987) Patterning of sexual activity. In: Smuts BB, Cheney DL, Seyfarth RM, Wrangham RW, Struhsaker TT (eds) Primate societies. University of Chicago Press, Chicago, pp 370–384

Huffman MA (1987) Consort intrusion and female mate choice in Japanese macaques, *Macaca fuscata*. Ethology 75:221–234

Itani J (1982) Intraspecific killing among non-human primates. J Soc Biol Struct 5:361–368

Jack KM, Pavelka MSM (1997) The behavior of peripheral males during the mating season in *Macaca fuscata*. Primates 38:369–377

Johnstone RA, Keller L (2000) How males can gain by harming their mates: sexual conflict, seminal toxins, and the cost of mating. Am Nat 156:368–377

Keverne EB (2001) Genomic imprinting, maternal care, and brain evolution. Horm Behav 40:146–155

Kirkpatrick M, Barton NH (1997) The strength of indirect selection on female mating preferences. Proc Natl Acad Sci USA 94:1282–1286

Kitchen DM, Beehner JC, Bergman TJ, Cheney DL, Crockford C, Engh AL, Fischer J, Seyfarth RM, Wittig RM (2009) The causes and consequences of male aggression directed at female chacma baboons. In: Muller MN, Wrangham RW (eds) Sexual coercion in primates and humans: an evolutionary perspective on male aggression against females. Harvard University Press, Cambridge, Massachnsetts, pp 129–156

Knott C (2009) Orangutans: sexual coercion without sexual violence. In: Muller MN, Wrangham RW (eds) Sexual coercion in primates and humans: an evolutionary perspective on male aggression against females. Harvard University Press, Cambridge, Massachnsetts, pp 81–111

Kuester J, Paul A, Arnemann J (1994) Kinship, familiarity and mating avoidance in Barbary macaques, *Macaca sylvanus*. Anim Behav 48:1183–1194

Lande R (1987) Genetic correlations between the sexes in the evolution of sexual dimorphism and mating preferences. In: Bradbury JW, Andersson MB (eds) Sexual selection: testing the alternatives. John Wiley & Sons, London, pp 83–94

Lessells CM (1999) Sexual conflict in animals. In: Keller L (ed) Levels of selection in evolution. Princeton University Press, Princeton, NJ, pp 75–99

Lessells CM (2005) Why are males bad for females? Models for the evolution of damaging male mating behavior. Am Nat 165:S46–S63

Levitan DR (2008) Gamete traits influence the variance in reproductive success, the intensity of sexual selection, and the outcome of sexual conflict among congeneric sea urchins. Evolution 62:1305–1316

Lindenfors P (2002) Sexually antagonistic selection on primate size. J Evol Biol 15:595–607

Lovejoy CO (1981) The origin of man. Science 211:341–350

Low BS (2005) Women's lives there, here, then, now: a review of women's ecological and demographic constraints cross-culturally. Evol Hum Behav 26:64–87

Lung O, Tram U, Finnerty CM, Eipper-Mains MA, Kalb JM, Wolfner MF (2002) The *Drosophila melanogaster* seminal fluid protein Acp62F is a protease inhibitor that is toxic upon ectopic expression. Genetics 160:211–224

Magellan K, Magurran AE (2006) Habitat use mediates the conflict of interest between the sexes. Anim Behav 72:75–81

Maggioncalda AN, Sapolsky RM, Czekala NM (1999) Reproductive hormone profiles in captive male orangutans: implications for understanding developmental arrest. Am J Phys Anthropol 109:19–32

Magurran AE, Seghers BH (1994) A cost of sexual harassment in the guppy, *Poecilia reticulata*. Proc Roy Soc Lond B 258:89–92

Manson JH (1997) Primate consortships: a critical review. Curr Anthropol 38:353–374

Martin-Coello J, Benavent-Corai J, Roldan ERS, Gomendio M (2009) Sperm competition promotes asymmetries in reproductive barriers between closely related species. Evolution 63:613–623

Matsubara M, Sprague DS (2004) Mating tactics in response to costs incurred by mating with multiple males in wild Japanese macaques. Int J Primatol 25:901–917

Matthews MK Jr, Adler NT (1978) Systematic interrelationship of mating, vaginal plug position, and sperm transport in the rat. Physiol Behav 20:303–309

Mesnick SL (1997) Sexual alliances: evidence and evolutionary implications. In: Gowaty PA (ed) Feminism and evolutionary biology: boundaries, intersections, and frontiers. Chapman & Hall, New York, pp 207–260

Mesnick SL, le Boeuf BJ (1991) Sexual behavior of male northern elephant seals. II. Female responses to potentially injurious encounters. Behaviour 117:262–280

Miller EH, Ponce de León A, Delong RL (1996) Violent interspecific sexual behavior by male sea lions (Otariidae): evolutionary and phylogenetic implications. Mar Mamm Sci 12:468–476

Mitani JC (1985) Sexual selection and adult male orangutan long calls. Anim Behav 33:272–283

Moore IT, Jessop TS (2003) Stress, reproduction, and adrenocortical modulation in amphibians and reptiles. Horm Behav 43:39–47

Morland HS (1993) Reproductive activity of ruffed lemurs (*Varecia variegata variegata*) in a Madagascar rain forest. Am J Phys Anthropol 91:71–82

Morrow EH, Arnqvist G, Pitnick S (2003) Adaptation versus pleiotropy: why do males harm their mates? Behav Ecol 14:802–806

Mühlhäuser C, Blanckenhorn WU (2002) The costs of avoiding matings in the dung fly *Sepsis cynipsea*. Behav Ecol 13:359–365

Muller MN, Wrangham RW (2009) Sexual coercion in primates and humans: an evolutionary perspective on male aggression against females. Harvard University Press, Cambridge, Massachnsetts

Muller MN, Kahlenberg SM, Thompson ME, Wrangham RW (2007) Male coercion and the costs of promiscuous mating for female chimpanzees. Proc Roy Soc Lond B 274:1009–1014

Muller MN, Kahlenberg SM, Wrangham RW (2009) Male aggression against females and sexual coercion in chimpanzees. In: Muller MN, Wrangham RW (eds) Sexual coercion in primates and humans: an evolutionary perspective on male aggression against females. Harvard University Press, Cambridge Massachnsetts, pp 184–217

Murdock GP (1949) Social structure. Macmillan, New York

Palombit RA (1999) Infanticide and the evolution of pair bonds in nonhuman primates. Evol Anthropol 7:117–129

Palombit RA (2000) Infanticide and the evolution of male-female bonds in animals. In: van Schaik CP, Janson CH (eds) Infanticide by males and its implications. Cambridge University Press, Cambridge, pp 239–268

Palombit RA (2003) Male infanticide in wild savanna baboons: adaptive significance and intraspecific variation. In: Jones CB (ed) Sexual selection and reproductive competition in primates: new perspectives and directions. American Society of Primatologists, New York, pp 367–412

Palombit RA. 'Friendship' with males: a female counterstrategy to infanticide in the okavango chacma baboons. In: Muller MN, Wrangham RW (eds) Sexual coercion in primates and humans: an evolutionary perspective on male aggression against females. Harvard University Press, Cambridge Massachnsetts, pp 377–409

Palombit RA, Seyfarth RM, Cheney DL (1997) The adaptive value of 'friendships' to female baboons: experimental and observational evidence. Anim Behav 54:599–614

Palombit RA, Cheney DL, Fischer J, Johnson S, Rendall D, Seyfarth RM, Silk JB (2000) Male infanticide and defense of infants in chacma baboons. In: van Schaik CP, Janson CH (eds) Infanticide by males and its implications. Cambridge University Press, Cambridge, pp 123–152

Palombit RA, Cheney DL, Seyfarth RM (2001) Female-female competition for male 'friends' in wild chacma baboons (*Papio cynocephalus ursinus*). Anim Behav 61:1159–1171

Parga JA (2003) Copulatory plug displacement evidences sperm competition in *Lemur catta*. Int J Primatol 24:889–899

Parga JA, Henry AR (2008) Male aggression during mating: evidence for sexual coercion in a female dominant primate? Am J Primatol 70:1187–1190

Parker GA (1970) The reproductive behaviour and the nature of sexual selection in *Scatophaga stercoraria* L. (Diptera: Scatophagidae). II. The fertilization rate and the spatial and temporal relationships of each sex around the site of mating and oviposition. J Anim Ecol 39:205–228

Parker GA (1979) Sexual selection and sexual conflict. In: Blum MS, Blum NA (eds) Sexual selection and reproductive competition in insects. Academic Press, New York, pp 123–166

Parker GA (2006) Sexual conflict over mating and fertilization: an overview. Phil Trans Roy Soc B 361:235–259

Partridge L, Hurst LD (1998) Sex and conflict. Science 281:2003–2008

Pereira ME (1983) Abortion following the immigration of an adult male baboon (*Papio cynocephalus*). Am J Primatol 4:93–98

Perkin AW (2007) Comparative penile morphology of East African galagos of the genus Galagoides (family Galagidae): implications for taxonomy. Am J Primatol 69:16–26

Perry S (1997) Male-female social relationships in wild white-faced capuchins (*Cebus capucinus*). Behaviour 134:477–510

Pilastro A, Benetton S, Bisazza A (2003) Female aggregation and male competition reduce costs of sexual harassment in mosquitofish *Gambusia holbrooki*. Anim Behav 65:1161–1167

Pizzari T, Snook RR (2003) Sexual conflict and sexual selection: chasing away paradigm shifts. Evolution 57:1223–1236

Pizzari T, Snook RR (2004) Sexual conflict and sexual selection: measuring antagonistic coevolution. Evolution 58:1389–1393

Pluháček J, Bartoš L (2000) Male infanticide in captive plains zebra, *Equus burchelli*. Anim Behav 59:689–694

Pradhan GR, van Schaik CP (2008) Infanticide-driven intersexual conflict over matings in primates and its effects on social organization. Behaviour 145:251–275

Pratt HL Jr (1979) Reproduction in the blue shark, *Prionace glauca*. Fish Bull 77:445–470

Quinlan RJ (2008) Human pair-bonds: evolutionary functions, ecological variation, and adaptive development. Evol Anthropol 17:227–238

Quinlan RJ, Quinlan MB (2007) Evolutionary ecology of human pair-bonds: cross-cultural tests of alternative hypotheses. Cross-Cultural Res 41:149–169

Ravi Ram K, Wolfner MF (2007) Seminal influences: *Drosophila* Acps and the molecular interplay between males and females during reproduction. Integr Comp Biol 47:427–445

Réale D, Boussès P, Chapuis J-L (1996) Female-biased mortality induced by male sexual harassment in a feral sheep population. Can J Zool 74:1812–1818

Reason LL (2004) The behavioral ecology of female genital cutting in northern Ghana. Res Econ Anthropol 23:175–202

Reed C, O'Brien TG, Kinnaird MF (1997) Male social behavior and dominance hierarchy in the Sulawesi crested black macaque (*Macaca nigra*). Int J Primatol 18:247–260

Reeder DM (2003) The potential for cryptic female choice in primates: behavioral, anatomical, and physiological considerations. In: Jones CB (ed) Sexual selection and reproductive competition in primates: new perspectives and directions. American Society of Primatologists, New York, pp 255–303

Rice WR (1996) Sexually antagonistic male adaptation triggered by experimental arrest of female evolution. Nature 381:232–234

Rice WR, Chippindale AK (2001) Intersexual ontogenetic conflict. J Evol Biol 14:685–693

Rice WR, Holland B (1997) The enemies within: intergenomic conflict, interlocus contest evolution (ICE), and the intraspecific Red Queen. Behav Ecol Sociobiol 41:1–10

Rice WR, Holland B, Harrison R (2005) Experimentally enforced monogamy: inadvertent selection, inbreeding, or evidence for sexually antagonistic coevolution? Evolution 59:682–685

Richard AF (1992) Aggressive competition between males, female-controlled polygyny and sexual monomorphism in a Malagasy primate, *Propithecus verreauxi*. J Hum Evol 22:395–406

Rodman PS, Mitani JC (1987) Orangutans: sexual dimorphism in a solitary species. In: Smuts BB, Cheney DL, Seyfarth RM, Wrangham RW, Struhsaker TT (eds) Primate societies. University of Chicago Press, Chicago, pp 146–154

Rodseth L, Novak SA (2009) The political significance of gender violence. In: Muller MN, Wrangham RW (eds) Sexual coercion in primates and humans: an evolutionary perspective on male aggression against females. Harvard University Press, Cambridge Massachnsetts, pp 292–321

Roulin A, Hager R (2003) Indiscriminate nursing in communal breeders: a role for genomic imprinting. Ecol Lett 6:165–166

Rubenstein DI (1986) Ecology and sociality in horses and zebras. In: Rubenstein DI, Wrangham RW (eds) Ecological aspects of social evolution. Princeton University Press, Princeton, NJ, pp 282–302

Sato Y, Kohama T (2007) Post-copulatory mounting behavior of the West Indian sweet-potato weevil, *Euscepes postfasciatus* (Fairmaire) (Coleoptera: Curculionidae). Ethology 113:183–189

Schülke O, Kappeler PM (2001) Protection from infanticide, male resource defense, over-dispersed females and the evolution of pair-living in a nocturnal lemur, *Phaner furcifer*. Folia Primatol 72:182–183

Setia TM, van Schaik CP (2007) The response of adult orang-utans to flanged male long calls: inferences about their function. Folia Primatol 78:215–226

Shine R, O'Connor D, Mason RT (2000) Sexual conflict in the snake den. Behav Ecol Sociobiol 48:392–401

Shine R, Phillips B, Langkilde T, Lutterschmidt DI, Waye H, Mason RT (2004) Mechanisms and consequences of sexual conflict in garter snakes (*Thamnophis sirtalis*, Colubridae). Behav Ecol 15:654–660

Shuster SM, Wade MJ (2003) Mating systems and strategies. Princeton University Press, Princeton, NJ

Sicotte P (2002) The function of male aggressive displays towards females in mountain gorillas. Primates 43:277–289

Simmons LW, Siva-Jothy MT (1998) Sperm competition in insects: mechanisms and the potential for selection. In: Birkhead TR, Møller AP (eds) Sperm competition and sexual selection. Academic Press, San Diego, pp 341–434

Siva-Jothy MT (2006) Trauma, disease and collateral damage: conflict in cimicids. Phil Trans Roy Soc B 361:269–275
Smith CC, Sargent RC (2006) Female fitness declines with increasing female density but not male harassment in the western mosquitofish, *Gambusia affinis*. Anim Behav 71:401–407
Smuts BB (1992) Male aggression against women: an evolutionary perspective. Hum Nat 3:1–44
Smuts BB, Smuts RW (1993) Male aggression and sexual coercion of females in nonhuman primates and other mammals: evidence and theoretical implications. Adv Stud Behav 22:1–63
Soltis J (1999) Measuring male-female relationships during the mating season in wild Japanese macaques (*Macaca fuscata yakui*). Primates 40:453–467
Soltis J, Mitsunaga F, Shimizu K, Yanagihara Y, Nozaki M (1997) Sexual selection in Japanese macaques. I. Female mate choice or male sexual coercion? Anim Behav 54:725–736
Soltis J, Thomsen R, Matusabayshi K, Takenaka O (2000) Infanticide by resident males and female counter-strategies in wild Japanese macaques (*Macaca fuscata*). Behav Ecol Sociobiol 48:195–202
Soltis J, Thomsen R, Takenaka O (2001) The interaction of male and female reproductive strategies and paternity in wild Japanese macaques, *Macaca fuscata*. Anim Behav 62:485–494
Sommer V (1987) Infanticide among free-ranging langurs (*Presbytis entellus*) at Jodhpur (Rajasthan/India): recent observations and a reconsideration of hypotheses. Primates 28:163–197
Sommer V, Reichard U (2000) Rethinking monogamy: the gibbon case. In: Kappeler PM (ed) Primate males: causes and consequences of variation in group composition. Cambridge University Press, Cambridge, pp 159–168
Stockley P (2002) Sperm competition risk and male genital anatomy: comparative evidence for reduced duration of female sexual receptivity in primates with penile spines. Evol Ecol 16:123–137
Stokes EJ (2004) Within-group social relationships among females and adult males in wild western lowland gorillas (*Gorilla gorilla gorilla*). Am J Primatol 64:233–246
Stumpf RM, Boesch C (2006) The efficacy of female choice in chimpanzees of the Taï Forest, Côte d'Ivoire. Behav Ecol Sociobiol 60:749–765
Stumpf RM, Thompson ME, Knott CD (2008) A comparison of female mating strategies in *Pan troglodytes* and *Pongo* spp. Int J Primatol 29:865–884
Swiss J, Giller JE (1993) Rape as a crime of war: a medical perspective. J Am Med Assoc 270:612–615
Taylor LL, Sussman RW (1985) A preliminary study of kinship and social organization in a semi-free-ranging group of *Lemur catta*. Int J Primatol 6:601–614
Thalmann U (2001) Food resource characteristics in two nocturnal lemurs with different social behavior: *Avahi occidentalis* and *Lepilemur edwardsi*. Int J Primatol 22:287–324
Thornhill R, Alcock J (1983) The evolution of insect mating systems. Harvard University Press, Cambridge, MA
Thornhill R, Gangestad SW (1996) The evolution of human sexuality. Trends Ecol Evol 11:98–102
Thornhill R, Palmer CT (2000) A natural history of rape: biological bases of sexual coercion. MIT Press, Cambridge, MA
Travis CB (2003) Evolution, gender, and rape. MIT Press, Cambridge, MA
Tregenza T, Wedell N, Chapman T (2006) Sexual conflict: a new paradigm? Phil Trans Roy Soc B 361:229–234
Trivers RL (1972) Parental investment and sexual selection. In: Campbell B (ed) Sexual selection and the descent of man, 1871–1971. Aldine, Chicago, pp 136–179
Tutin CEG (1979) Mating patterns and reproductive strategies in a community of wild chimpanzees (*Pan troglodytes schweinfurthii*). Behav Ecol Sociobiol 6:29–38
Vahed K (2007) Comparative evidence for a cost to males of manipulating females in bushcrickets. Behav Ecol 18:499–506

Vahed K, Carron G (2008) Comparison of forced mating behaviour in four taxa of *Anonconotus*, the Alpine bushcricket. J Zool Lond 276:313–321

Valero A, Hudson R, Luna EÁ, Macías Garcia C (2005) A cost worth paying: energetically expensive interactions with males protect females from intrasexual aggression. Behav Ecol Sociobiol 59:262–269

van Schaik CP (2000a) Infanticide by male primates: the sexual selection hypothesis revisited. In: van Schaik CP, Janson CH (eds) Infanticide by males and its implications. Cambridge University Press, Cambridge, pp 27–60

van Schaik CP (2000b) Vulnerability to infanticide by males: patterns among mammals. In: van Schaik CP, Janson CH (eds) Infanticide by males and its implications. Cambridge University Press, Cambridge, pp 61–71

van Schaik CP, Dunbar RIM (1990) The evolution of monogamy in large primates: a new hypothesis and some crucial tests. Behaviour 115:30–62

van Schaik CP, Janson CH (2000) Infanticide by males and its implications. Cambridge University Press, Cambridge

van Schaik CP, Kappeler PM (1997) Infanticide risk and the evolution of male-female association in primates. Proc Roy Soc Lond B 264:1687–1694

van Schaik CP, Kappeler PM (2003) The evolution of social monogamy in primates. In: Reichard UH, Boesch C (eds) Monogamy: mating strategies and partnerships in birds, humans and other mammals. Cambridge University Press, Cambridge, pp 59–80

van Schaik CP, van Hooff JARAM (1996) Toward an understanding of the orangutan's social system. In: McGrew WC, Marchant LF, Nishida T (eds) Great ape societies. Cambridge University Press, Cambridge, pp 3–15

van der Schoot P, van Ophemert J, Baumgarten R (1992) Copulatory stimuli in rats induce heat abbreviation through effects on genitalia, but not through effects on central nervous mechanisms supporting the steroid hormone-induced sexual responsiveness. Behav Brain Res 49:213–223

van Valen L (1973) A new evolutionary law. Evol Theory 1:1–30

Washburn SL, Lancaster CS (1968) The evolution of hunting. In: Lee RB, DeVore I (eds) Man the hunter. Aldine, Chicago, pp 293–303

Watts DP (1997) Agonistic interventions in wild mountain gorilla groups. Behaviour 134:23–57

Weingrill T (2000) Infanticide and the value of male-female relationships in mountain chacma baboons (*Papio cynocephalus ursinus*). Behaviour 137:337–359

Werner M, Simmons LW (2008) The evolution of male genitalia: functional integration of genital sclerites in the dung beetle *Onthophagus taurus*. Biol J Linn Soc 93:257–266

Williams GC (1966) Adaptation and natural selection: a critique of some current evolutionary thought. Princeton University Press, Princeton, NJ

Wilson CG (2008) Male genital mutilation: an adaptation to sexual conflict. Evol Hum Behav 29:149–164

Wittgenstein L (1953) Philosophical investigations. Blackwell, London

Wittig RM, Crockford C, Lehmann J, Whitten PL, Seyfarth RM, Cheney DL (2008) Focused grooming networks and stress alleviation in wild female baboons. Horm Behav 54:170–177

Wolff JO, MacDonald DW (2004) Promiscuous females protect their offspring. Trends Ecol Evol 19:127–134

Wrangham RW (1979) On the evolution of ape social systems. Soc Sci Info 18:335–368

Wrangham RW (1986) Ecology and social relationships in two species of chimpanzee. In: Rubenstein DI, Wrangham RW (eds) Ecological aspects of social evolution. Princeton University Press, Princeton, NJ, pp 352–378

Wrangham RW, Peterson D (1996) Demonic males: apes and the origins of human violence. Houghton Mifflin, Boston

Zeh JA, Zeh DW (2003) Toward a new sexual selection paradigm: polyandry, conflict and incompatibility. Ethology 109:929–950

Chapter 4
The Unusual Women of Mpimbwe: Why Sex Differences in Humans are not Universal

Monique Borgerhoff Mulder

Abstract Parental investment theory provides a strong basis for generalizations about how male and female mating strategies might vary, and has generated a large number of successful predictions regarding gender differences in human reproductive strategies. There are, however, many situations in which traditional sex roles are not observed, and behavioral ecologists are beginning to determine how and why this might be. In this chapter, I explore the implications of generalizations about universal sex differences for our understanding of gender differences in sexual and reproductive strategies of humans. First, I examine recent work within behavioral ecology on the status of parental investment as a determinant of sex differences in reproductive strategies. Second, I summarize analyses of reproductive strategies in a rural forager-horticultural population in western Tanzania where variance in women's reproductive success is not significantly different from that of men and where women use serial matings rather more effectively than do men to outcompete their competitors, to show that key sex differences predicated on the mammalian pattern of parental investment are not necessarily observed. Third, I broaden this discussion of an obvious ethnographic exception to examine the relationship between human pair bonds and parental investment, to show again that sex differences in parental investment provide only a partial story. The implications of these observations for claims of universal sex differences and the gap between studies of human and nonhuman reproductive strategies are discussed in the conclusion.

M.B. Mulder
Department of Anthropology, University of California at Davis, Davis, CA, USA
e-mail: mborgerhoffmulder@ucdavis.edu

P.M. Kappeler and J.B. Silk (eds.), *Mind the Gap*,
DOI 10.1007/978-3-642-02725-3_4, © Springer-Verlag Berlin Heidelberg 2010

4.1 Introduction

Much of the legitimacy of applying evolutionary approaches to the study of human behavior has been predicated on the existence of universal sex differences. In our species, men are on average taller (Alexander et al. 1979) and stronger than women, and die earlier and from different causes than women (Teriokhin et al. 2004). Additionally, they are generally thought to show higher variance in reproductive success than women (Barrett et al. 2002). These apparently universal sex-varying traits can be attributed to the common mammalian pattern of reproduction, in which gestation and lactation fall exclusively to women, paternity certainty is never assured, and even small amounts of paternal care are provided facultatively (Trivers 1972). As such, male fitness has, for a long time, been seen as limited by competition over access to females, and female fitness limited by access to resources that can often be acquired through males (Emlen and Oring 1977; Wrangham 1980).

Starting in the 1980s, predictions derived from parental investment theory sparked an evolutionary literature addressing human reproductive and mating strategies (reviewed in Cronk et al. 2000; Low 2000; Dunbar and Barrett 2007). Several findings emerge that suggest (or at least are interpreted as) human universals. For example, rich ethnographic and comparative studies demonstrate the prevalence of competition among men over women (Irons 1979; Betzig 1986; Chagnon 1988; Daly and Wilson 1988; Hawkes 1991), although mating competition is mediated through diverse avenues such as political office, murder, wealth accumulation, or the provision of public goods. Similar kinds of work explore how women (or their parents on their behalf) choose and compete for desirable mates (Dickemann 1979; Buss 1989; Gangestad and Simpson 2000), again through a variety of means, including cognitive preferences, dowry payments, and olfactory cues. Sex differences in mating preferences are also evident, with men tending to favor health and fecundity in their mates whereas women look for ambition and resources, as evidenced both by reported preferences (e.g., Buss 1989; Cashdan 1993) and actual behavior (e.g., Borgerhoff Mulder 1989, 1990). The exquisite sensitivity of mechanisms underlying such preferences to ecological and social circumstances (reviewed in Gangestad 2007) have helped to bring the study of human behavior into mainstream evolutionary theory, as well as to promote popular awareness of humans as yet another uniquely evolved species (e.g., Ridley 1994).

With the success of this work, there nevertheless emerged a dangerous tendency to generalize from specific observations to universal sex differences. Such generalizations are problematic for several reasons (e.g., Smith et al. 2001). First, recognizing the importance of culturally transmitted norms Boyd and Silk (2005, using Buss's 1989 data) demonstrate how cultural factors explain a great deal more of the cross-cultural variation in mate choice preferences than does gender. Second, objecting to the stereotypic portrayal of women as at the mercy of their biology and the antics of men, Hrdy (1986), Smuts (1992) and Gowaty (1997) provide cogent qualitative and quantitative support for the view that women can and do operate with agency and employ a wide array of strategies to subvert and counter the strategies of men. Third, anthropologists and others emphasize the importance of the social and ecological environment in generating variability not fixity in sex roles according to principles well established in behavioral and cultural evolutionary theory (e.g., Laland and Brown 2002).

In recent years, studies of sex differences have become more nuanced. In part, this reflects a growing awareness among psychologists and feminist scholars that gender differences have often been inflated, even derived from poor science (Shibley-Hyde 2005). Evolutionary social scientists too are using ethnographic data to emphasize the flexibility in gender roles, and the overlap in gender differences (e.g., Bliege Bird and Bird 2008). In fact, it is now indeed time to ask "are men and women really all that different?" (Borgerhoff Mulder 2004; Brown et al. 2009), and to reevaluate the extent to which sex differences in human reproductive strategies are contingent on sex differences in postzygotic investment.

In this chapter, I scrutinize the notion of universal sex differences in the reproductive strategies of men and women. My goal is cautionary. I do not argue that parental investment theory is wrong, but rather that other factors need to be taken into consideration, factors that may be of particular importance in humans. To demonstrate this point, I present empirical data on a horticultural-hunter-fisher population in Tanzania (Pimbwe) where the variation in fitness among women equals the variance in fitness of men and, quite contrary to the normative pattern, women benefit more from multiple marriages than do men. Finally, I consider how anthropologists think about the relationship between pair bonding and parental care. I finish by considering what my conclusions mean for the gap between human and nonhuman studies, the theme of this volume.

4.2 Parental Investment Theory and Beyond

Models predicated on the differential postzygotic investment of males and females (Trivers 1972) have dominated the study of sexual and reproductive strategies in most mammals, and provided a theoretical context for the classic finding that males benefit more from multiple matings than do females. Key to this discussion has been the regression of reproductive success on mating success, known as "Bateman's gradient (Bateman 1948). Whichever sex has the steepest gradient is the sex that experiences the stronger sexual selection pressure on traits that enhance mating success (Andersson and Iwasa 1996).

In mammals, gestation and lactation fall exclusively to females, paternity certainty is never assured, and paternal care is provided facultatively. Therefore male fitness is seen as limited by competition over mates, and female fitness by access to resources that can in some but not all cases, be acquired through males (Emlen and Oring 1977; Wrangham 1980). Thus, the reproductive strategies of each sex, in particular decisions over mating effort and parenting effort, are analyzed as a product of sex differences in parental investment. Trivers' model (in an expanded form that deals more explicitly with the operational sex ratio and potential reproductive rates, Clutton-Brock and Vincent 1991) can in fact predict much of the variation in sexual selection across taxa and has important implications for sex roles. As noted earlier, its successes in predicting human sex differences in reproductive strategies brought prominence to the new discipline of evolutionary social science (Borgerhoff Mulder et al. 1997).

In the intervening years, theoretical and empirical work in behavioral ecology has taken a richer and more dynamic approach to sex roles (reviewed in Borgerhoff Mulder 2009a). First, there has been a rethinking of the internal logic and consistency of Trivers', and specifically Maynard Smith's (1977) model (Queller 1997; Houston and McNamara 2005; Kokko et al. 2006). These revisions do not change the basic prediction that the caring sex is more likely to be choosier and the object of more competition, but fundamentally alters the evolutionary sequence. In the conventional sequence, differences in prezygotic investment determine potential reproductive rates which then shape payoffs to postzygotic care. In the revised sequence, prezygotic investment generates the conditions for sexual selection as numerically abundant male gametes compete for access to rare female gametes. This lowers the confidence of males in paternity and, given male-male competition for access to females (and/or female choice), creates an elite subset of males that are more eligible to mate (Kokko and Jennions 2003). This revised logic gives more salience to sex differences in competition over mates and less to sex differences in parental care.

Second, and independent of these revisions, both theoretical and empirical work shows that anisogamy does not always produce classic sex roles (Gowaty 2004) and that competition and choice are not mutually exclusive (Kokko et al. 2006), as indeed long recognized in empirical studies of nonhuman primates (Hrdy 1986). In other words, choosiness is not simply a function of operational sex ratios, with the limiting sex enjoying the luxury of choice; it is also dependent on variance in quality among potential mates (Owens and Thompson 1994; Johnstone et al. 1996), the costs of reproduction (Kokko and Monaghan 2001; Maness and Anderson 2007), and extrinsic survival rates (Gowaty and Hubbell 2005).

Third, there is evidence that there are some species in which females are the principal caregivers, but compete more frequently and more intensively with each other than do males. In meerkats (*Suricata suricatta*, Clutton-Brock et al. 2006) and many other cooperatively breeding vertebrates (Holekamp et al. 1996; Hauber and Lacey 2005), females gain greater reproductive benefits from dominance than do males (e.g., Engh et al. 2002, for spotted hyenas, *Crocuta crocuta*), and accordingly are more competitive with one another, thereby demonstrating that sex differences in parental investment are not the only mechanism capable of generating sex differences in reproductive competition. Finally in some species, notably cooperative breeders with single breeding pairs, sex differences in fitness variances are unrelated to differences in mate number, thus providing evidence that counters Bateman's gradient (the idea that males benefit more from multiple mates than do females, Hauber and Lacey 2005). Higher female than male variance in fitness is also observed in sex role-reversed species such as dusky pipefish, Syngnathus floridae (Jones et al. 2000) and wattled jacanas (*Jacana jacana*) (Emlen and

Wrege 2004). Recognition of these additional selective considerations generates a much richer picture of how competition and choice can figure in the strategy of each sex and how these may vary over the life time and across populations.

In short, contemporary perspectives within behavioral ecology provide a broader framework within which to study the great diversity of sex differences in nature than that afforded by the simple parental investment model that guided seminal work in the evolutionary social sciences until the late 1990s.

4.3 The Unusual Women of Mpimbwe

The Pimbwe live in the Rukwa Valley of present day western Tanzania. Impacts from German, Belgian, and British colonial escapades in this central African region were indirect (Tambila 1981), but colonial wildlife policies had more severe impacts, effectively displacing Pimbwe from parts of their traditional chiefdom (Borgerhoff Mulder et al. 2007). In the socialist era (mid 1970s), Pimbwe families were settled in government villages, but many have now returned to ancestral lands that lie outside areas protected for wildlife. Modern Pimbwe rely primarily on a mix of subsistence and cash crops, supplemented by foraged resources and poultry keeping. Small enterprise activities, such as trading, traditional medicine, hunting, fishing, honey production, carpentry, and beer brewing supplement farm income for men and women. Livelihoods are unpredictable because of highly seasonal rainfall that creates critical periods of food shortage and labor demand (Wandel and Holmboe-Ottesen 1992; Hadley et al. 2007), poor infrastructure that makes cash cropping risky, and very poor health services. Between 40 and 50% of households in the district fall below the basic needs poverty line (United Republic of Tanzania 2005), and development initiatives are seriously jeopardized by prevalent beliefs in witchcraft. These and following general observations are based on intermittent fieldwork between July 1995 and February 2008, as well as previous studies in the area.

The traditional marriage pattern, reported as clan controlled, monogamous, and accompanied by bridewealth (Willis 1966), must have been seriously challenged by the high rates of labor outmigration in the colonial period (Tambila 1981). Marriage is now effectively characterized by cohabitation, initiated with a facultative transfer of bridewealth and a celebration (Fig. 4.1). Polygyny appears never to have been common. Nowadays, marriage can be defined as sharing in the production and consumption of food and shelter, with the expectation of exclusive sexual relations. Divorce is permitted and, like marriage, can be defined by the physical movement of one or both partners out of the house, requiring no legal or formal procedures. Divorces occur often when one spouse starts an extramarital relationship, with both sexes tending to claim responsibility for abandoning the relationship. At divorce, children under the age of 8 are supposed to stay with the mother (or the mother's kin), whereas older children should stay with their father. In practice, the fate of children is quite variable. Sometimes fathers "kidnap" very young children from

Fig. 4.1 A longterm monogamously married husband and wife sitting outside their house in Mirumba

their mothers, sometimes mothers leave a recently weaned child with a divorced husband; older children may live with a range of maternal or paternal kin.

Given these residence patterns, parental care is highly facultative. Wives typically take primary responsibility for the direct care of their own small children, with some assistance from older children and/or other kin, including their own mothers or husband's mothers. Regarding indirect care, the bulk of farming is done by husbands and wives, but there is considerable variability within marriages as to how the fruits of joint farm labor are allocated among family subsistence needs, joint family benefits (like health and education), individual cash purchases, or capital for individual economic enterprises (such as using maize for beer brewing). These allocations prompt frequent spousal arguments, and one spouse may even place locks on the family granary to exclude "inappropriate" use of resources by the other spouse. There are no significant heritable resources in this population; men and women get access to land and houses opportunistically from maternal or paternal relatives (or from unrelated individuals) who happen to have unused land or living sites available in the village. Commonly they clear land and build houses anew, such that there is very little to inherit in the way of bequests.

Basic demographic data were collected in all households of a single village in seven different study periods between 1995 and 2006 (for details see Borgerhoff Mulder 2009a) and analyses include only individuals who are assumed to have neared completion of their reproduction (>44 years), yielding 138 men with a mean age of 60.3 years (range 45.3–92.7) and 154 women with a mean age of 59.2 years (range 45.0–86.8) dropping younger men (<55 years), produced statistically equivalent results. Variables used in the analyses presented here are age, sex, number of livebirths, reproductive success (measured as the number of offspring surviving to

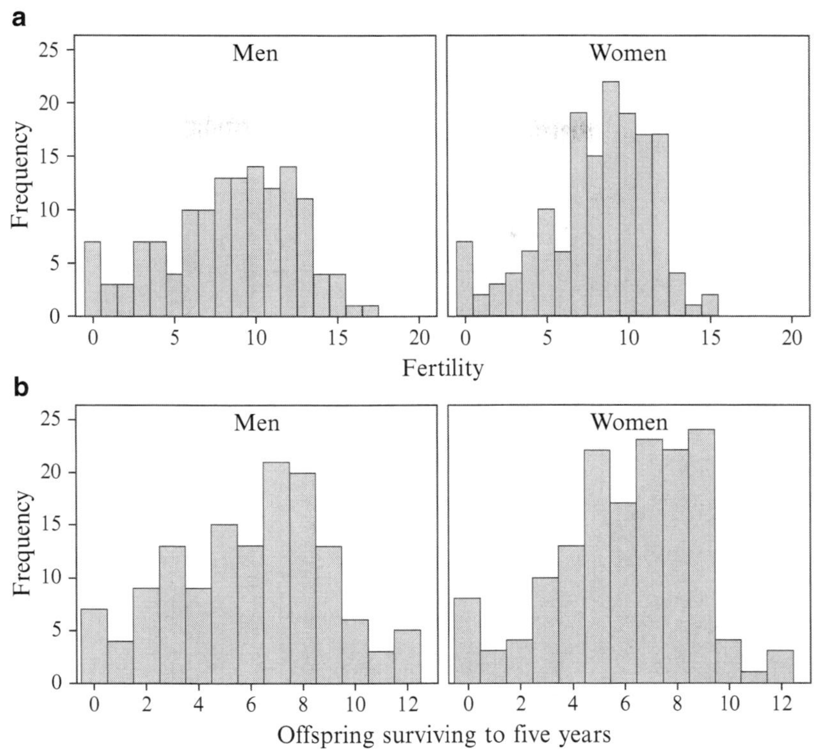

Fig. 4.2 Variance in (**a**) fertility and (**b**) reproductive success for Pimbwe men and women

5 years of age, beyond which mortality is low), and number of spouses over the lifetime, categorized as 1, 2, 3, or more.

Among men and women who had completed their reproductive careers, only 3 (2.2%) men and 2 (1.3%) women had never married, indicating that marriage is virtually universal in this population. The mean values of fertility (men 8.41; women 8.17) and reproductive success (men 5.99; women 6.14) are not statistically different from each other, which suggests that there is no distortional sex bias to the sample. While men show greater variance in fertility (16.16) than women (11.34, Levene's test for equality of variances $F = 5.87$, $p = 0.016$, Fig. 4.2a), there is no significant difference in the variances in completed reproductive success (men 9.00; women 7.27, Levene's test $F = 2.15$, ns, Fig. 4.2b).

When fertility (Fig. 4.3a) and the numbers of offspring reaching 5 years of age (Fig. 4.3b) are shown in relation to number of spouses (1, 2 and 3 or more), an unexpected pattern emerges. Men fail to benefit in terms of fitness from multiple marriages, but women who marry three or more times produce more surviving children than do other women. Fertility and completed reproductive success are regressed on age, sex, and the number of spouses in a number of different models (Table 4.1a, b). Generally, across models, the number of spouses is negatively

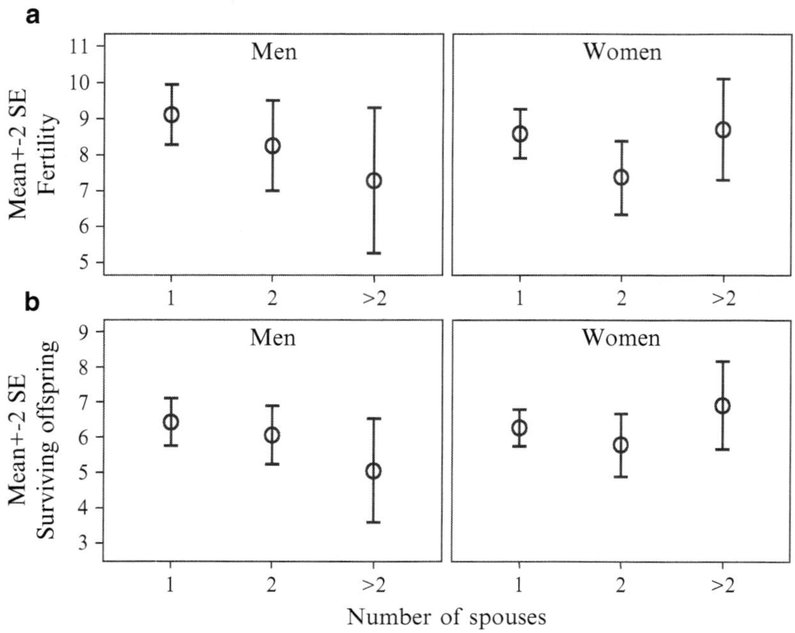

Fig. 4.3 The associations between number of spouses and (**a**) fertility and (**b**) reproductive success for Pimbwe men and women. The mean is shown with a *circle*, and the standard error (*2) with a *bar*. For statistics see Table 4.1

Table 4.1 Regression models for how sex, age, and number of spouses affect fertility and number of surviving offspring (showing beta, standard error, and significance). (a) fertility, (b) number of surviving offspring

(a)	Model 1	Model 2	Model 3	Model 4
Sex	−0.244 (0.433)	−0.172 (0.425)	−0.285 (0.415)	−1.862 (0.985)+
Age		0.067 (0.020)***	0.061 (0.020)**	0.067 (0.020)***
No. of Spouses			−0.516 (0.284)+	−2.045 (0.912)*
Sex* No. of spouses				1.015 (0.575)+

(b)	Model 1	Model 2	Model 3	Model 4
Sex	−0.144 (0.333)	−0.174 (0.332)	−0.117 (0.328)	−1.281 (0.776)
Age		0.028 (0.015)+	0.022 (0.015)	0.028 (0.015)+
No. of spouses			−0.223 (0.224)	−1.578 (0.718)*
Sex*#Spouses				0.899 (0.453)*

***$p<0.001$, **$p<0.01$, *$p<0.05$, + $0.05<p<0.10$

associated with fertility and the number of surviving offspring, and there are interaction effects between spouse number and sex, which reflect the pattern shown in Fig. 4.3, namely that men suffer reproductively from multiple marriages in a way that women do not. The models also show that age is positively associated with fertility and less consistently with reproductive success, suggesting (in this postreproductive sample) that levels of fertility were slightly higher in cohorts that finished reproduction in the 1980s than in the 2000s, which is to be expected in a community where some individuals are beginning to choose smaller family sizes.

In this population, sex differences in fitness variances are not pronounced. While men do show greater variance in the number of live births, reproductive success is equally variable across the sexes, suggesting that men with very high fertility raise few of these "extra" children. A sex difference, or lack thereof, in variance does not, however, shed much light on the operation of sexual selection (Sutherland 1985; Hubbell and Johnson 1987). Much more illuminating is the relationship between breeding success and physiological or behavioral phenotypes (Clutton-Brock 1988). Thus, the finding that men and women benefit differently from multiple marriages is interesting. While the data are very variable (large standard errors), women appear to gain more from multiple mating than do men; furthermore, the same statistically significant effect is observed as a control variable in a different sample of younger women (Borgerhoff Mulder 2009b). Note, however, that it would be analytically more revealing to look at the probability of bearing (and successfully raising) a child as a function of the marital status of the parents – the present analysis shows only that reproductive performance is correlated with the number of spouses married over the lifetime. Note also that these findings take no allowance of the marital status of the spouse. It is tempting to think that women who have married many men should be married to men who have had many wives, but this is not necessarily the case.

Possible reasons why men and women make multiple serial marriages are discussed in more detail elsewhere (Borgerhoff Mulder 2009a). Ethnographic observations suggest that lazy and heavy-drinking men are often divorced and end up marrying postreproductive women. In some cases, these men have dependent children and clearly remarry to find help from a new wife in raising their kids. For these men, multiple marriages therefore, rather counter-intuitively, reflect parenting effort rather than mating effort, although why anyone would want to marry them remains a puzzle.

There are a host of hypotheses to account for multiple mating in females (e.g., Jennions and Petrie 2000; Setchell and Kappeler 2003), entailing both direct and indirect benefits. Women may gain direct benefits by mating with and marrying multiple men if their husbands help them to obtain the resources needed to support reproduction. Pimbwe women benefit from the farming activities of men, as well as from the products of their hunting, fishing, honey production, and other enterprises, but all of these are highly unpredictable, in part because of poor farming conditions and in part because of the current illegality of entering protected areas where fish and meat are plentiful (Borgerhoff Mulder et al. 2007). Given the potentially high inter- and intraindividual variability in male provisioning, it is quite possible that women switch mates to maximize economic income, the "musical chairs" hypothesis reviewed by Choudhury (1995). A similar argument has been made for the instability of marriages among the poor in contemporary USA (Kaplan and Lancaster 2003, see also Maness and Anderson 2007 for Nazca boobies, *Sula granti*). Parallels can also be drawn with baboons (*Papio cynocephalus*) where serial if nonexclusive pair bonds produce temporary male protectors for mothers whose success in raising offspring is heavily influenced by their social networks and matrilineally inherited dominance rank (as reviewed by Silk 2007).

As regards indirect (or genetic) benefits, numerous mechanisms have been proposed (including the maximization of male genetic potential, bet hedging, prevention of inbreeding, and confusion of paternity certainty to avoid infanticide). The most plausible in this context is the idea that a woman can afford to forego the benefits of paternal care (and to risk the dangers of a stepfather in house) for mates with high genetic potential. This argument has been made most forcefully for humans by Gangestad and Simpson (2000) and is particularly plausible in environments with high disease loads where demonstration of heritable fitness is very important (Hamilton and Zuk 1982). In support of this explanation is the fact that Mpimbwe is beset by all of the health problems typical of rural tropical Africa (Hadley and Patil 2006; Hadley et al. 2007) and has a minimal health care infrastructure. If Pimbwe women were choosing genetically superior males and keeping them we would expect once-married women to show highest fitness, but they do not show this. In addition, if Pimbwe women were choosing genetically superior males and then losing them to other women, we would expect multiply married men to show elevated fitness, which is not observed.

In sum, the Pimbwe analysis, while provisional, provides clear evidence that a key sex difference predicated on mammalian patterns of parental investment, the Bateman gradient, is not observed. Whether this results from sex differences in the range of quality in potential mates (Owens and Thompson 1994; Johnstone et al. 1996), costs of reproduction (Kokko and Monaghan 2001; Maness and Anderson 2007), extrinsic mortality rates (Gowaty and Hubbell 2005), or other factors is not yet known. However, the cautionary tale here is that just because humans are typical mammals with all the polygynous tendencies predicated on gestation and lactation, conventional sex-differentiated reproductive strategies are not assured!

Therefore, if Pimbwe women enter into marital bonds to reap direct benefits, does this mean that pair bonds are best thought of as a universal adaptation whereby women trade sex for paternal care? The simplicity of this scenario is alluring, but again the reality more complex.

4.4 Pair Bonds in Humans

Claims regarding the universality of human pair bonds are controversial but this is so because they are often confused with statements about origins. The ethnographic record displays a range of grouping patterns, from small two-adult "family" groups to large multimale/multifemale bands (e.g., Pasternak et al. 1997), but *within these formations*, specialized relationships emerge, between (usually) heterosexual individuals, typically glossed as "marriage." Although it is widely recognized that these bonds do not map precisely onto sexual relationships (Fox 1967), male sexual access to females is key to the definition of marriage, even if it is given quite different salience across cultures (Bell 1997). Furthermore, despite the well-known "double standards" in sexuality (Betzig 1989), Jankowiak et al.'s (2002) survey of detailed ethnographic material shows that in all the 66 societies studied, men *and women* actively mate guard, indicating that sexual propriety is a core component to marital unions, even if much violated. Precise definitions of marriage may be elusive, but in all cases, rights and responsibilities are exchanged (Needham 1962), concerning legitimacy of offspring (Gough 1959), property (Leach 1955) and economics (Murdock 1949). In short, marital bonds are about sexual access, and although additional rights and responsibilities are emphasized in different cultural contexts, these bonds are always identifiable. Societies sanctioning total promiscuity as the principle mating system do not exist (Bell 1997; Kunstadter 1963; Rodseth et al. 1991). Note that this is a claim about the universality of pair bonding in the ethnographic record, not about its more contentious evolutionary origins (as discussed in Knight and Power 2005), to which I now turn.

Several sources of evidence point, at least indirectly, to a long history of pair bonding in our species, for example, relatively limited sperm competition (Birkhead 2000), little sexual dimorphism in size dating back in our lineage to 1.8 mya (McHenry 1996), and a distinctive patterning of testosterone production with pair bond status (e.g., lower levels in undergraduate men in well-established romantic relationships: Gray et al. 2004). Furthermore, a notion of "romantic love" is observed across the vast majority of human cultures (Jankowiak and Fischer 1992), mediated by various neuro-endocrinological systems (e.g., Carter 1998). Much more controversial is the role of male provisioning in the evolution of pair bonded behavior.

Pair bonds evolving from male provisioning were once central to narratives of human origins (Washburn and Lancaster 1968). Bipedal hominins could carry meat (Lovejoy 1981), opening up the possibility for a complementary division of labor in which males provide resources to females encumbered with dependent offspring in return for sexual exclusivity. Modern attempts to unravel the origins of pair bonding attribute very different roles to paternal provisioning. In some formulations, males provision because both parents are assumed to have identical reproductive goals. Thus, Fisher (1989) suggests that divorce rates peak after 4 years, because this is when a typical forager child is safely through the period of dependence, and both parents are free to look for new spouses. Others view the relationship as one in which both cooperation and competition exist. In a carefully argued scenario that links intelligence, longevity, altriciality and diet, as coevolved traits, Kaplan et al. (2000) posit that long lifespan and a cooperative division of labor coevolved as humans moved into a foraging niche where food acquisition (hunting) required great skill and knowledge, and partners could benefit from specialization and exchange (Gurven et al. 2009). In other scenarios, marital bonds are believed to be entirely independent of paternal provisioning. In one version, pair bonds are deemed to have emerged from mate guarding, with males favoring pair bonds to avoid incessant fighting over females (Symons 1979) and females to find refuge from harassment (Blurton Jones et al. 2000) and/or infanticide (Hawkes 2004). A different idea that again relies not on paternal provisioning (rather on *female* provisioning) is that once females discovered how to increase the diversity and density of food value through cooking, they were worth monopolizing (Wrangham et al. 1999). In the latter kinds of scenarios, paternal care is more likely to have evolved after the emergence of pair bonds in our lineage, and not be a necessary condition for the evolution of pair bonding (Chapais 2008).

Phylogenetic analyzes of the relationship between mating systems and paternal care shed light on this origins debate. In mammals, Brotherton and Komers (2003) show that monogamy evolved more often in the absence of paternal care than in its presence, and propose that paternal care most likely arose subsequently (although in birds biparental care may have preceded avian pair bonds, Burley and Johnson 2002). Similar conclusions are being reached for nonhuman primates. Since direct paternal care is present only in some species, it is most likely that monogamy is a preadaptation facilitating the evolution of paternal care rather than a consequence (Palombit 1999; van Schaik and Kappeler 2003). In short, despite the apparent universality of pair bonds in contemporary human populations and cogent models that these evolved to subsidize the high costs of reproduction and encephalization (Kaplan et al. 2000; Gurven and Kaplan 2006), there is little clear comparative evidence that pair bonds evolved in mammals to facilitate paternal provisioning, nor is there much evidence from nonhuman primates to support such a claim. Humans, however, with their exceedingly large brain and unusually long lifespan, may be unique in this respect.

To gain more insight into the nature of the pair bond, anthropologists turn to ethnographic materials, both within and between population analyzes. One question pertinent to this debate is whether men's economic activities are best characterized as mating effort or parental effort. There are cases where men work particularly hard when their mates are lactating (Hadza: Marlowe 2003), allowing the latter to do less work at this energetically demanding time (Hiwi and Ache: Hurtado et al. 1992). These data suggest that men's activities are a form of parenting effort. Similarly, there are cases where men's allocation of effort to food production fits closely with expectations derived from the paternal effort model (Tsimane: Gurven and Kaplan 2006); Although Tsimane men invest in mating effort through extra-marital relations, they do so early in the marriage when there are few, if any, children to care for, rather than later, when paternal contributions are most needed (Winking et al. 2007). In addition, confronted with hypothetical scenarios, Ache and Hadza men show preferences for hunting groups with good hunters that yield high returns for provisioning but low returns for mating effort, rather than groups of poor hunters where provisioning benefits are small and mating benefits large (Wood and Hill 2000; Wood 2006). These patterns support the parental effort hypothesis, as do demographic and economic data that show how brides are hard to find (Borgerhoff Mulder 1990; Cashdan 1993) and quick to leave (Betzig 1989) when men's provisioning resources are not forthcoming, patterns also observed in the modern US when marriage transitions are analyzed in relation to income (Nakosteen and Zimmer 1997; Burgess et al. 2003). However, in other studies, men's economic activities are better characterized as mating effort, as with Hawkes' (1993) analysis of hunting in the Hadza and Smith et al.'s (2003) study of turtle feasts of Mer islanders in the Torres Straits. In both these cases, it is argued that men specialize on risky resources, ones that are better characterized as public goods to be ostentatiously shared (the show off hypothesis: Hawkes 1991) than as reliable streams of paternal provisioning. Despite strong arguments on each side, it seems most likely that men everywhere exhibit elements of both paternal and mating effort in their economic exploits (Anderson et al. 1999a, b; Gurven and Hill 2009), for example in the Marlowe's analyzes of the Hadza (1999).

Cross cultural data can also be used to address, at least indirectly, the question of whether pair bonding is associated with the provision of paternal care or with the defense (or guarding) of mates. Quinlan and Quinlan (2007) find that marital stability (a low divorce rate) is associated with substantial male contributions to subsistence, an absence of alternative caretakers, and late weaning, all indices of the potential value of male assistance. However, marital stability is also associated with high levels of polygyny, a possible proxy for the difficulty men face in finding new mates. The authors therefore conclude that pair bonds likely evolved in response to multiple selection pressures – a need for male care and a strategy for each sex to deal with intense mate competition. In fact, there may even be two different kinds of human pair bonds – one geared to child rearing and one geared toward male reproductive competition (Quinlan 2008).

All in all, studies of contemporary populations do little to resolve whether pair bonds evolved for paternal investment or mate guarding. Why so little progress? First, despite the wealth of empirical studies we have puzzling cases where different data sets from the same population support different models. Second, comparative studies raise multiple problems for interpretation – for example is divorce prevalence really a valid indicator of the importance of pair bonds? A more appropriate variable might be the number of people who actually do marry in the population. Third, contemporary correlates of a trait do not necessarily shed light on its evolutionary origins. While human behavioral ecologists maintain that studies of current behavioral diversity illuminate the flexibility of human nature to different social and ecological triggers (e.g., Smith et al. 2001), extrapolating to evolutionary sequences is much more tenuous. Now that pair bonding is in place, paternal care might be very important in contemporary populations. However, to claim that it was the original selective pressure for pair bonds is an interpretative error, Chapais (2008: 169) dubs the "pitfall of modern family reference" (see also Marlowe 2007). Finally, investigators often talk about men and women as if they had a single sex-specific strategy. In the modern USA, for example, upper strata parents use biparental care to invest in highly profitable education for their children, whereas lower strata women raise children alone and both sexes have serial and simultaneous relationships (Kaplan and Lancaster 2003); multiple strategies characterize many other populations (see, for example, Dickemann 1982). In so far as an individual's optimal mating and reproductive strategy depend on the behavior of both same- and different-sexed conspecifics, multiple strategic spaces emerge. Only recently have theoreticians begun to tackle this, as greater computing power allows model parameters to be generated by individual strategies rather than fixed at *a priori* levels (e.g., Cotar et al. 2008, and for more general discussion see Kokko et al. 2006).

As such, it is perhaps more productive to think about pair bonds in terms of sexual conflict (Borgerhoff Mulder and Rauch in prep). Sexual conflict theory does not inherently avoid the pitfalls of making generalizations about sex differences, but it provides a more dynamic framework for analyzing variable sex roles. For example, it examines the question of how much a wife and husband should work and when they should break their contract, in terms of broader market (supply and demand) conditions. Economists focus on each spouse's bargaining power, the well-being a lady can expect without the cooperation of a gentleman, and *vice versa*; differential bargaining power sets each spouse's "threat point" (Manser and Brown 1980), the resource sharing contractual arrangement at which the lady (or gentleman) would be better leaving than staying. Anything that improves an individual's bargaining power with their spouse, such as relatively larger earnings, gender-biased divorce laws, or greater chances of remarriage, increases that person's share of the marital pot (e.g., Lundberg and Pollak 1996). Accordingly, the benefits of a marriage are not expected to be shared equally, but in accordance with bargaining power. The implications of these considerations for marital stability, marital assortment, and equilibrial states of the marriage market can be explored with game theory, as demonstrated in models of the "better options" hypothesis for divorce in birds (e.g., McNamara et al. 1999) where divorce rates are shown to decrease with individual quality (fewer "better options" assuming a good first pairing), and age (no time to recoup the costs of divorce).

A key factor affecting these conflicts is whether spousal labor is complementary or substitutable (Kaplan and Lancaster 2003). If mum feeds and dad protects the baby, parental roles are complementary, in so far as each activity is valueless without the other. If mum (or dad) uses salary to pay rent and school fees, parental roles are substitutable and marriages become more brittle (and subject to corner solutions). Changes in divorce rates and prevalence cross culturally might be usefully examined from this perspective. Rather intriguingly, Quinlan and Quinlan (2007) find that divorce is least common in populations where both sexes contribute approximately equally to household production, suggesting (albeit indirectly) that marriage is indeed a more stable institution where spouses' work is complementary.

Thinking about the marriage as a complementary division of labor raises the interesting question of when and how the benefits of specialization can be offset by positive assortment of skills (Borgerhoff Mulder 2004). When are the public goods produced by a breadwinner and homemaker eclipsed by two extremely successful breadwinners who might bicker over homemaking? In traditional economies, where most tasks are gender-specialized, the benefits of the sexual division of labor are unlikely to be dwarfed by positive assortment for skills. One might expect, however, that in a modern economy, where most jobs can be done by men or women, positive assortment counterbalances the division of labor (corporate executives intermarry and hire an *au pair*). Positive assortment among mates for various skills is found in many modern economies (e.g., Logan et al. 2008). Indeed a study testing whether US university students show preferences to assort on similar traits (rich men like rich women, and *vice versa*) or on reproductive potential (high earning men prefer beautiful women, and *vice versa*) showed that the first pattern is much stronger than the second (Buston and Emlen 2003); this suggests that the complementary marital relationship is disappearing in some western societies. We see here how far an evolutionary-based discussion of marriage in terms of sexual conflict and bargaining theory has moved us away from the simple constraints of mammalian reproduction.

4.5 Mind the Crack: Concluding Observations

What are the implications of this discussion for universal sex differences and the gap between studies of human and nonhumans? Early evolutionary studies identified predicted sex roles, but failed to consider variation. Nowadays, with more sophisticated theoretical models and richer empirical evidence, we see that the roles of men and women can be highly variable. The Pimbwe study shows how some women, despite common mammalian constraints, can use multiple sequential pair bonds to out-reproduce their monogamous counterparts. Similarly, the broader discussion of the relationship between pair bonds and paternal care reveals limits to the conventional view of marriage as a trade of sex for paternal resources. Men and women have negotiable roles in marriage, for which models from behavioral ecology (beyond conventional parental investment theory) and economics can be brought to bear.

In what sense do we differ from nonhuman primates in this respect? Anthropologists once viewed regularly patterned family behavior (and particularly marriage, e.g., Levi-Strauss 1949) as what divides us from the other primates, contrasting the apparent promiscuity of our apelike cousins to, initially, our nice nineteenth century stable monogamous families, and later the wide range of family types evident in the ethnographic record. However, the gap is nowhere so big as once thought and is better viewed as continuum (e.g., Foley and Lee 1989; Rodseth et al. 1991) or just a crack. This is so in part because over the years so much has been learned about the complexity of nonhuman primate kinship and social behavior. With long-term studies of known individuals, the sophistication of primate kinship behavior is now well appreciated. The narrowing gap also reflects the fact that the human pair bond, with dad provisioning mum in exchange for sex, is now no longer viewed as necessarily the supreme human adaptation that sent us down our distinct evolutionary route. In fact, as I have tried to argue, the contemporary human pair bond is a highly variable trait which, while most likely universal in some form, functions very differently in different social and economic contexts, and has hotly contended evolutionary origins.

Another reason why crack (rather than gap) better characterizes the distinction between sex-differentiated reproductive strategies in human and nonhuman primates is that there are so many parallels between contemporary debates within primatology and human behavioral ecology. I end with a consideration of some of these.

First, nonhuman primates generally exhibit rather low levels of direct paternal care. Males can be important for protection, particularly from infanticide (e.g., Palombit et al. 1997), and males do in some cases provide direct care to their offspring (Buchan et al. 2003), but with a few exceptions (e.g., Goldizen 1987), paternal care is not extensive among primates. Furthermore, as we have seen in humans, there is lengthy debate over whether male activities are designed for improving offspring survival or enhancing access to mates (e.g., Smuts and Gubernick 1992; van Schaik and Paul 1996). In short, as in mammals more generally, male care is not extensive, and where it occurs, its relationship to paternity, and its impact on offspring survival, is debatable (Woodroffe and Vincent 1994). The situation is rather similar in the study of humans, as reviewed in this chapter. The traditional view that paternal care is central to our evolutionary trajectory is now in question, and suggestions that pair bonds originated for mate defense and avoidance of harassment are gaining attention. Humans do, however, look different with respect to the *current function* of pair bonds: the division of labor over the production and consumption of food among spouses is particularly developed in our species, hence the foray into economics for new theoretical tools (and see Noë et al. 2001, for similar applications of economic theory to nonhuman primates).

Secondly, related to low paternal care is the significance of female-bonded kin groups in nonhuman primates. If males are not helping, who is? In many species, amongst them cercopithecine monkeys (Silk 2007), the fertility and reproductive success of females is heavily influenced by social networks and matrilineally derived access to resources. Such social systems are increasingly being used as a model with which to think about human evolutionary origins and the importance of cooperation among kin in subsidizing the costs of childbearing (Hrdy 2005a, b). This marks a radical departure from seeing our origins in the social organization of the male-bonded apes (e.g., Rodseth et al. 1991).

A final parallel trend lies in the prevalence of multiple mating by females. Female primates commonly mate with multiple males to avoid infanticide, and rather rarely mate for resources or good genes; indeed, preferences for dominant individuals dissolve once the male loses dominance, suggesting that protection is more prominent in this preference than genetic quality (Setchell and Kappeler 2003). Interestingly, human behavioral ecologists turned initially to birds for models for mating systems (Flinn and Low 1986; Borgerhoff Mulder 1990) because of the clear importance of paternal investment in many extent societies. This, however, may have distracted us away from very different (and non paternal care-based) arguments for the *origins*, if not the *current function*, of pair bonds in our species.

In sum, there are several parallels between the arguments advanced in this chapter and current debates within primatology that support the claim that we should perhaps be minding a crack rather than a gap.

Acknowledgments Fieldwork was funded by the LSB Leakey Foundation, the Packard Foundation and UC Davis. Thanks to the editors and two reviewers for the helpful comments on the manuscript.

References

Alexander RD, Hoogland JL, Howard RD, Noonan KM, Sherman PW (1979) Sexual dimorphisms and breeding systems in pinnipeds, ungulates, primates, and humans. In: Chagnon NA, Irons W (eds) Evolutionary biology and human social behavior: an anthropological perspective. Duxbury, North Scituate, MA, pp 402–435

Anderson KG, Kaplan H, Lam D, Lancaster J (1999a) Paternal care by genetic fathers and stepfathers. II. Reports by Xhosa high school students. Evol Hum Behav 20:433–451

Anderson KG, Kaplan H, Lancaster J (1999b) Paternal care by genetic fathers and stepfathers. I. Reports from Albuquerque men. Evol Hum Behav 20:405–431

Andersson M, Iwasa Y (1996) Sexual selection. Trends Ecol Evol 11:53–58

Barrett L, Dunbar RIM, Lycett J (2002) Human evolutionary psychology. Palgrove, Basingstoke, UK

Bateman AJ (1948) Intrasexual selection in *Drosophila*. Heredity 2:349–368

Bell D (1997) Defining marriage and legitimacy. Curr Anthropol 38:237–253

Betzig LL (1986) Despotism and differential reproduction: a darwinian view of history. Aldine de Gruyter, New York, NY

Betzig LL (1989) Causes of conjugal dissolution: a cross-cultural study. Curr Anthropol 30:654–676

Birkhead TR (2000) Promiscuity: an evolutionary history of sperm competition. Harvard University Press, Cambridge, MA

Bliege Bird R, Bird DW (2008) Why women hunt: risk and contemporary foraging in a Western Desert Aboriginal community. Curr Anthropol 49:655–693

Blurton Jones NG, Marlowe FW, Hawkes K, O'Connell JF (2000) Paternal investment and hunter-gatherer divorce rates. In: Cronk L, Chagnon N, Irons W (eds) Adaptation and human behavior: an anthropological perspective. Aldine de Gruyter, New York, NY, pp 69–90

Borgerhoff Mulder M (1989) Early maturing Kipsigis women have higher reproductive success than later maturing women, and cost more to marry. Behav Ecol Sociobiol 24:145–153

Borgerhoff Mulder M (1990) Kipsigis women's preferences for wealthy men: evidence for female choice in mammals. Behav Ecol Sociobiol 27:255–264

Borgerhoff Mulder M (2004) Are men and women really so different? Trends Ecol Evol 19:3–6

Borgerhoff Mulder M (2009a) Serial monogamy as polygyny or polyandry? Marriage in the Tanzanian Pimbwe. Hum Nat 20(2):130–150

Borgerhoff Mulder M (2009b) Tradeoffs and sexual conflict over women's fertility preferences in Mpimbwe. Am J Hum Biol 21(4):478–487

Borgerhoff Mulder M, Rauch KL (in press) The battle of the sexes. Evol Anthropol

Borgerhoff Mulder M, Richerson PJ, Thornhill NW, Voland E (1997) The place of behavioral ecological anthropology in evolutionary social science. In: Weingart P, Mitchel SD, Richerson PJ, Maasen S (eds) Human by nature: between biology and the social sciences. Lawrence Erlbaum Associates, Majwaj, NJ, pp 253–282

Borgerhoff Mulder M, Caro TM, Msago AO (2007) The role of research in evaluating conservation strategies in Tanzania: the case of the Katavi-Rukwa ecosystem. Conserv Biol 21:647–658

Boyd R, Silk JB (2005) How humans evolved. Norton, New York, NY

Brotherton PNM, Komers PE (2003) Mate guarding and the evolution of social monogamy in mammals. In: Reichard UH, Boesch C (eds) Monogamy: mating strategies and partnerships in birds, humans and other mammals. Cambridge University Press, Cambridge, pp 42–58

Brown GR, Laland KN, Borgerhoff Mulder M (2009) Bateman's principles and human sex roles. Trends Ecol Evol 24:297–304

Buchan JC, Alberts SC, Silk JB, Altmann J (2003) True paternal care in a multi-male primate society. Nature 425:179–181

Burgess S, Propper C, Aassve A (2003) The role of income in marriage and divorce transitions among young Americans. J Popul Econ 16:455–475

Burley NT, Johnson K (2002) The evolution of avian parental care. Phil Trans Roy Soc B 357:241–250

Buss DM (1989) Sex differences in human mate preferences: evolutionary hypotheses tested in 37 cultures. Behav Brain Sci 12:1–49

Buston PM, Emlen ST (2003) Cognitive processes underlying human mate choice: the relationship between self-perception and mate preference in Western society. Proc Natl Acad Sci USA 100:8805–8810

Carter CS (1998) Neuroendocrine perspectives on social attachment and love. Psychoneuroendocrinology 23:779–818

Cashdan E (1993) Attracting mates: effects of paternal investment on mate attraction strategies. Ethol Sociobiol 14:1–23

Chagnon NA (1988) Life histories, blood revenge, and warfare in a tribal population. Science 239:985–992

Chapais B (2008) Primeval kinship: how pair-bonding gave birth to human society. Harvard University Press, Cambridge, MA

Choudhury S (1995) Divorce in birds: a review of the hypotheses. Anim Behav 50:413–429

Clutton-Brock TH (1988) Reproductive success: studies of individual variation in contrasting breeding systems. University of Chicago Press, Chicago, IL

Clutton-Brock TH, Vincent ACJ (1991) Sexual selection and the potential reproductive rates of males and females. Nature 351:58–60

Clutton-Brock TH, Hodge SJ, Spong G, Russell AF, Jordan NR, Bennett NC, Sharpe LL, Manser MB (2006) Intrasexual competition and sexual selection in cooperative mammals. Nature 444:1065–1068

Cotar C, McNamara JM, Collins EJ, Houston AI (2008) Should females prefer to mate with low-quality males? J Theor Biol 254:561–567

Cronk L, Chagnon NA, Irons W (2000) Adaptation and human behavior: an anthropological perspective. Evolutionary foundations of human behavior. Aldine de Gruyter, New York, NY

Daly M, Wilson M (1988) Evolutionary social psychology and family homicide. Science 242:519–524

Dickemann M (1979) The ecology of mating systems in hypergynous dowry societies. Soc Sci Inf 18:163–195

Dickemann M (1982) Commentary on John Hartung's 'polygyny and inheritance of wealth'. Curr Anthropol 23:8

Dunbar RIM, Barrett L (2007) Oxford handbook of evolutionary psychology. Oxford University Press, Oxford

Emlen ST, Oring LW (1977) Ecology, sexual selection, and the evolution of mating systems. Science 197:215–223

Emlen ST, Wrege PH (2004) Size dimorphism, intrasexual competition, and sexual selection in Wattled Jacana (*Jacana jacana*), a sex-role reversed shorebird in Panama. Auk 121:391–403

Engh AL, Funk SM, Van Horn RC, Scribner KT, Bruford MW, Libants S, Szykman M, Smale L, Holekamp KE (2002) Reproductive skew among males in a a female-dominated mammalian society. Behav Ecol 13:193–200

Fisher HE (1989) Evolution of human serial pairbonding. Am J Phys Anthropol 78:331–354

Flinn MV, Low BS (1986) Resource distribution, social competition, and mating patterns in human societies. In: Rubenstein DI, Wrangham RW (eds) Ecological aspects of social evolution: birds and mammals. Princeton University Press, Princeton, NJ, pp 217–243

Foley RA, Lee PC (1989) Finite social space, evolutionary pathways and reconstructing hominid behavor. Science 243:901–906

Fox R (1967) Kinship and marriage: an anthropological perspective. Penguin, Baltimore, MD

Gangestad SW (2007) Reproductive strategies and tactics. In: Dunbar RIM, Barrett L (eds) Oxford handbook of evolutionary psychology. Oxford University Press, Oxford, pp 321–332

Gangestad SW, Simpson JA (2000) The evolution of human mating: trade-offs and strategic pluralism. Behav Brain Sci 23:573–644

Goldizen AW (1987) Tamarins and marmosets: communal care of offspring. In: Smuts BB, Cheney DL, Seyfarth RM, Wrangham RW, Struhsaker TT (eds) Primate societies. University of Chicago Press, Chicago, IL, pp 34–43

Gough KE (1959) The nayars and the definition of marriage. J R Anthropol Inst 89:23–34

Gowaty PA (1997) Sexual dialectics, sexual selection, and variation in reproductive behavior. In: Gowaty PA (ed) Feminism and evolutionary biology: boundaries, intersections and frontiers. Chapman & Hall, New York, NY, pp 351–384

Gowaty PA (2004) Sex roles, contests for the control of reproduction, and sexual selection. In: Kappeler PM, van Schaik CP (eds) Sexual selection in primates: new and comparative perspectives. Cambridge University Press, Cambridge, pp 37–54

Gowaty PA, Hubbell SP (2005) Chance, time allocation, and the evolution of adaptively flexible sex role behavior. Integr Comp Biol 45:931–944

Gray PB, Chapman JF, Burnham TC, McIntyre MH, Lipson SF, Ellison PT (2004) Human male pair bonding and testosterone. Hum Nat 15:119–131

Gurven M, Hill K (2009) Why do men hunt? A reevaluation of ‘man the hunter’ and the sexual division of labor. Curr Anthropol 50:51–74

Gurven M, Kaplan H (2006) Determinants of time allocation across the lifespan: a theoretical model and an application to the Machiguenga and Piro of Peru. Hum Nat 17:1–49

Gurven M, Winking J, Kaplan H, von Rueden C, McAllister L (2009) A bioeconomic approach to marriage and the sexual division of labor. Hum Nat 20:151–183

Hadley C, Patil CL (2006) Food insecurity in rural Tanzania is associated with maternal anxiety and depression. Am J Hum Biol 18:359–368

Hadley C, Borgerhoff Mulder M, Fitzherbert E (2007) Seasonal food insecurity and perceived social support in rural Tanzania. Public Health Nutr 10:544–551

Hamilton WD, Zuk M (1982) Heritable true fitness and bright birds: a role for parasites? Science 218:384–387

Hauber ME, Lacey EA (2005) Bateman’s principle in cooperatively breeding vertebrates: the effects of non-breeding alloparents on variability in female and male reproductive success. Integr Comp Biol 45:903–914

Hawkes K (1991) Showing off: tests of an hypothesis about men’s foraging goals. Ethol Sociobiol 12:29–54

Hawkes K (1993) Why hunter-gatherers work: an ancient version of the problem of public goods. Curr Anthropol 34:341–361

Hawkes K (2004) Mating, parenting, and the evolution of human pair bonds. In: Chapais B, Berman CM (eds) Kinship and behavior in primates. Oxford University Press, Oxford, pp 443–473

Holekamp KE, Smale L, Szykman M (1996) Rank and reproduction in the female spotted hyena. J Reprod Fertil 108:229–237

Houston AI, McNamara JM (2005) John Maynard Smith and the importance of consistency in evolutionary game theory. Biol Phil 20:933–950

Hrdy SB (1986) Empathy, polyandry, and the myth of the coy female. In: Bleier R (ed) Feminist approaches to science. The ATHENE series. Pergamon Press, New York, NY, pp 119–146

Hrdy SB (2005a) Comes the child before the man: how cooperative breeding and prolonged postweaning dependence shaped human potentials. In: Hewlett BS, Lamb ME (eds) Hunter-gatherer childhoods: evolutionary, developmental and cultural perspectives. Transaction, New Brunswick, NJ, pp 65–91

Hrdy SB (2005b) Cooperative breeders with an ace in the hole. In: Voland E, Chasiotis A, Schiefenhoevel W (eds) Grandmotherhood: the evolutionary significance of the second half of female life. Rutgers University Press, New Brunswick, NJ, pp 295–317

Hubbell SP, Johnson LK (1987) Environmental variance in lifetime mating success, mate choice, and sexual selection. Am Nat 130:91–112

Hurtado AM, Hill K, Kaplan H, Hurtado I (1992) Trade-offs between female food acquisition and child care among hiwi and ache foragers. Hum Nat 3:185–216

Irons W (1979) Cultural and biological success. In: Chagnon NA, Irons W (eds) Evolutionary biology and human social behavior: an anthropological perspective. Duxbury, North Scituate, MA, pp 257–272

Jankowiak WR, Fischer EF (1992) A cross-cultural perspective on romantic love. Ethnology 31:149–155

Jankowiak WR, Nell MD, Buckmaster A (2002) Managing infidelity: a cross-cultural perspective. Ethnology 41:85–101

Jennions MD, Petrie M (2000) Why do females mate multiply? A review of the genetic benefits. Biol Rev Camb Philos Soc 75:21–64

Johnstone RA, Reynolds JD, Deutsch JC (1996) Mutual mate choice and sex differences in choosiness. Evolution 50:1382–1391

Jones AG, Rosenqvist G, Berglund A, Arnold SJ, Avise JC (2000) The Bateman gradient and the cause of sexual selection in a sex-role-reversed pipefish. Proc Roy Soc Lond B 267:677–680

Kaplan H, Lancaster JB (2003) An evolutionary and ecological analysis of human fertility, mating patterns, and parental investment. In: Wachter KW, Bulatao RA (eds) Offspring: human fertility behavior in biodemographic perspective. National Academies Press, Washington DC, pp 170–223

Kaplan H, Hill K, Lancaster JB, Hurtado AM (2000) A theory of human life history evolution: diet, intelligence and longevity. Evol Anthropol 9:156–185

Knight C, Power C (2005) Grandmothers, politics, and getting back to science. In: Voland E, Chasiotis A, Schiefenhoevel W (eds) Grandmotherhood: the evolutionary significance of the second half of female life. Rutgers University Press, New Brunswick, NJ, pp 81–98

Kokko H, Jennions MD (2003) It takes two to tango. Trends Ecol Evol 18:103–104

Kokko H, Monaghan P (2001) Predicting the direction of sexual selection. Ecol Lett 4:159–165

Kokko H, Jennions MD, Brooks R (2006) Unifying and testing models of sexual selection. Annu Rev Ecol Evol Syst 37:43–66

Kunstadter P (1963) A survey of consanguine or matrifocal family. Am Anthropol 65:56–66

Laland KN, Brown GR (2002) Sense and nonsense: evolutionary perspectives on human behaviour. Oxford University Press, Oxford

Leach ER (1955) Polyandry, inheritance and the definition of marriage. Man 55:182–186

Levi-Strauss C (1949) The elementary structures of kinship. Beacon Press, Boston, MA

Logan JA, Hoff PD, Newton MA (2008) Two-sided estimation of mate preferences for similarities in age, education, and religion. J Am Stat Assoc 103:559–569

Lovejoy CO (1981) The origin of man. Science 211:341–350

Low BS (2000) Why sex matters: a Darwinian look at human behavior. Princeton University Press, Princeton, NJ

Lundberg S, Pollak RA (1996) Bargaining and distribution in marriage. J Econ Perspect 10:139–158

Maness TJ, Anderson DJ (2007) Serial monogamy and sex ratio bias in Nazca boobies. Proc Roy Soc Lond B 274:2047–2054

Manser M, Brown M (1980) Marriage and household decisionmaking: a bargaining analysis. Int Econ Rev 21:31–44

Marlowe FW (1999) Showoffs or providers? The parenting effort of Hadza men. Evol Hum Behav 20:391–404

Marlowe FW (2003) A critical period for provisioning by Hadza men: implications for pair bonding. Evol Hum Behav 24:217–229

Marlowe FW (2007) Hunting and gathering: the human sexual division of foraging labor. Cross Cult Res 41:170–195

Maynard Smith J (1977) Parental investment: a prospective analysis. Anim Behav 25:1–9

McHenry HM (1996) Sexual dimorphism in fossil hominids and its socioecological implications. In: Steele J, Shennan S (eds) The archaeology of human ancestry: power, sex and tradition. Routledge, New York, NY, pp 91–109

McNamara JM, Forslund P, Lang A (1999) An ESS model for divorce strategies in birds. Phil Trans Roy Soc B 354:223–236

Murdock GP (1949) Social structure. Free Press, New York, NY

Nakosteen RA, Zimmer MA (1997) Men, money and marriage: are high earners more prone than low earners to marry? Soc Sci Q 78:66–82

Needham R (1962) Notes on the comparative method and prescriptive alliance. Bijdragen tot de Taal-, Land- en Volkenkunde 118:160–182

Noë R, van Hooff JARAM, Hammerstein P (2001) Economics in nature: social dilemmas, mate choice and biological markets. Cambridge University Press, Cambridge

Owens IPF, Thompson DBA (1994) Sex differences, sex ratios and sex roles. Proc Roy Soc Lond B 258:93–99

Palombit RA (1999) Infanticide and the evolution of pair bonds in nonhuman primates. Evol Anthropol 7:117–129

Palombit RA, Seyfarth RM, Cheney DL (1997) The adaptive value of ‘friendships’ to female baboons: experimental and observational evidence. Anim Behav 54:599–614

Pasternak B, Ember CR, Ember M (1997) Sex, gender and kinship: a cross-cultural perspective. Prentice Hall, Upper Saddle River, NJ

Queller DC (1997) Why do females care more than males. Proc Roy Soc Lond B 264:1555–1557

Quinlan RJ (2008) Human pair-bonds: evolutionary functions, ecological variation, and adaptive development. Evol Anthropol 17:227–238

Quinlan RJ, Quinlan MB (2007) Evolutionary ecology of human pair-bonds: cross-cultural tests of alternative hypotheses. Cross Cult Res 41:149–169

Ridley M (1994) The red queen: sex and the evolution of human nature. Penguin Books, London

Rodseth L, Wrangham RW, Harrigan AM, Smuts BB (1991) The human community as a primate society. Curr Anthropol 32:221–254

Setchell JM, Kappeler PM (2003) Selection in relation to sex in primates. Adv Study Behav 33:87–173

Shibley-Hyde J (2005) The gender similarities hypothesis. Am Psychol 60:581–592

Silk JB (2007) The adaptive value of sociality in mammalian groups. Phil Trans Roy Soc B 362:539–559

Smith EA, Borgerhoff Mulder M, Hill K (2001) Controversies in the evolutionary social sciences: a guide for the perplexed. Trends Ecol Evol 16:128–135

Smith EA, Bliege Bird RB, Bird DW (2003) The benefits of costly signaling: meriam turtle hunters. Behav Ecol 14:116–126

Smuts BB (1992) Male aggression against women: an evolutionary perspective. Hum Nat 3:1–44

Smuts BB, Gubernick DJ (1992) Male infant relationships in nonhuman primates: paternal investment of mating effort. In: Hewlett BS (ed) Father-child relations: cultural and biosocial contexts. Aldine de Gruyter, New York, NY, pp 1–30

Sutherland WJ (1985) Chance can produce a sex difference in variance in mating success and account for Bateman’s data. Anim Behav 33:1349–1352

Symons D (1979) The evolution of human sexuality. Oxford University Press, New York, NY

Tambila A (1981) A history of Rukwa Region (Tanzania) ca. 1870–1940: aspects of economic and social change from precolonial to colonial times. PhD thesis, University of Hamburg

Teriokhin AT, Budilova EV, Thomas F, Guegan J (2004) Worldwide variation in life-span sexual dimorphism and sex-specific environmental mortality rates. Hum Biol 76:623–641

Trivers RL (1972) Parental investment and sexual selection. In: Campbell B (ed) Sexual selection and the descent of man 1871–1971. Aldine de Gruyter, New York, NY, pp 136–179

United Republic of Tanzania (2005) Tanzania poverty and human development report. Mkuki na Nyota, Tanzania

van Schaik CP, Kappeler PM (2003) The evolution of social monogamy in primates. In: Reichard UH, Boesch C (eds) Monogamy: mating strategies and partnerships in birds, humans and other mammals. Cambridge University Press, Cambridge, pp 59–80 van Schaik CP, Paul A (1996) Male care in primates: does it ever reflect paternity? Evol Anthropol 5:152–156

Wandel M, Holmboe-Ottesen G (1992) Food availability and nutrition in a seasonal perspective: a study from the Rukwa region in Tanzania. Hum Ecol 20:89–107

Washburn SL, Lancaster CS (1968) The evolution of hunting. In: Lee RD, DeVore I (eds) Man the hunter. Aldine, Chicago, IL, pp 293–303

Willis RG (1966) The Fipa and related peoples of South-West Tanzania and North-East Zambia. Ethnographic survey of Africa: East Central Africa pt. 15. International African Institute, London

Winking J, Kaplan H, Gurven M, Rucas S (2007) Why do men marry and why do they stray? Proc Roy Soc Lond B 274:1643–1649

Wood BM (2006) Prestige or provisioning? A test of foraging goals among the Hadza. Curr Anthropol 47:383–387

Wood B, Hill K (2000) A test of the 'showing-off' hypothesis with Ache hunters. Curr Anthropol 41:124–125

Woodroffe R, Vincent A (1994) Mother's little helpers: patterns of male care in mammals. Trends Ecol Evol 9:294–297

Wrangham RW (1980) An ecological model of female-bonded primate groups. Behaviour 75:262–300

Wrangham RW, Jones JH, Laden GD, Pilbeam D, Conklin-Brittain NL (1999) The raw and the stolen: cooking and the ecology of human origins. Curr Anthropol 40:567–594

Part III
Politics & Power

Chapter 5
Dominance, Power, and Politics in Nonhuman and Human Primates

David P. Watts

As long as politics is the shadow cast on society by big business, attenuation of the shadow will not change the substance . . . Power today resides in control of the means of production exchange, publicity, transportation, and communication. Whoever owns them rules the life of the country. The machinery of power . . . is business for private profit . . . reinforced by command of the . . . means of publicity and propaganda.

(John Dewey, quoted in Westbrook 1991: 440, 442)

Abstract Dominance is a common, although not universal, characteristic of social relationships in nonhuman primates. One individual is dominant to another when it consistently wins the agonistic interactions between them. Attainment of high dominance rank can bring reproductive payoffs, mostly because it confers priority of access to monopolizable food sources (for females) or to mating opportunities (for males). For females in particular, a wide variation exists in the frequency of intense aggression, the directionality of aggression within dyads, the tolerance of high-ranking individuals, and other aspects of "dominance style." This variation reflects variation in ecology and is also influenced by phylogenetic history. Variation also exists in male dominance style, although it has not received as much attention. Dominance is one component of power, which also encompasses other sources of asymmetry in relationships that affect the relative ability of individuals to carry out their goals against the interests of others. Leverage is an important source of power in many nonhuman primates; an animal has leverage over another when it controls a resource or service that cannot be appropriated by force, such as agonistic support. Individuals behave politically when they try to increase or maintain their power relative to that of others by manipulating social relationships, both their own and those of others. The concept of power applies universally to gregarious primates, although asymmetries do not occur between adults of all

D.P. Watts
Department of Anthropology, Yale University, New Haven, CT, USA
e-mail: david.watts@yale.edu

P.M. Kappeler and J.B. Silk (eds.), *Mind the Gap*,
DOI 10.1007/978-3-642-02725-3_5, © Springer-Verlag Berlin Heidelberg 2010

species, but we should only ascribe politics to species in which the actors have knowledge about the relationships between others in their groups. Some parts of the literatures in political science and political anthropology not only provide useful frameworks for a comparative investigation of power and politics in nonhuman primates, but also highlight qualitative differences between humans and other primates, especially regarding the importance of ideologies and political rhetoric.

5.1 Introduction

Dewey's remarks seem incongruous in a chapter about dominance, power, and politics in nonhuman primates. The influential philosopher of pragmatism was writing about humans in the context of twentieth century industrial capitalism; his ideas about politics, power, and the proper social role of "the means of publicity" were grounded in – and contentious in – that context. They do not describe human universals, and "politics" and "power" have many definitions in political anthropology, a subdiscipline that includes multiple, often contrasting and sometimes complementary, explanatory paradigms that cover the entire range of current and historically known human social arrangements (reviewed in Kurtz 2001, and Lewellen 2003, among others). However, Dewey makes an implicit point about nonhuman primates: without the capacity for language and for symbolically mediated systems of meaning, they cannot engage in "publicity and propaganda" or seek profit, nor can they invent, pursue, or argue about ideologies, whether capitalism or any other: these are qualitative differences from humans. Can we then justifiably ascribe politics and power dynamics to them and seek commonalities with humans?

Any answer starts with the fact that individuals of most primate species maintain long-term social relationships with conspecifics. These often have affiliative dimensions, but they also involve competition over food, mates, and other resources, the outcome of which can have major effects on fitness. Competition may or may not involve social interaction and may or may not lead to dominance relationships. Competitive interactions, or contests, are crucial to the concept of dominance, which becomes part of a social relationship when one individual can monopolize resources at the other's expense or usurp them from the others by using force or threatening to do so, even though not all aggression concerns immediate access to resources or contests over status, and some contests are decided by unprovoked, unilateral submission rather than by aggression.

However, not all resources accessible via social interaction can be appropriated by force, and dominance can be subsumed within a broader category of power (Lewis 2002). Variation in power among individuals potentially allows for social maneuvering that sometimes warrants the label of "politics." Alliance formation strategies illustrate these points well. Most contests are dyadic, but many primate species stand out when compared with most mammals because of the frequency with which they form coalitions, in which two or more individuals collaboratively direct aggression at joint targets (Harcourt 1992). Alliances develop when particular dyads repeatedly and consistently form coalitions; they feature prominently in the competitive strategies of macaques, chimpanzees, and many other species. One effect of coalitions is to help individuals to win contests they would otherwise lose. Male baboons can sometimes forcibly take over consortships with estrous females from higher-ranking males, for example, but require coalition partners to do so. Not all potential partners are equally effective, and males may compete for access to the best partners (Seyfarth 1977; Noë 1990, 1992). Males cannot coerce others to join them; the ability to grant or withhold support gives potential partners power over their would-be allies, and males whose value is higher than that of their partners are likely to take disproportionate shares of the benefits of successful coalition formation (Noë 1990, 1992). Strategic pursuit of alliances and negotiation over the distribution of their costs and benefits easily bring to mind human political maneuvering, as de Waal (1982) argued, for the ways in which male chimpanzees (*Pan troglodytes*) use coalitions in their complex strategies of competition for status. Other nonhuman primates that also engage in complex social maneuvering for status by using alliances, exchanging social services like grooming, and testing the strength of social bonds arguably also have politics (e.g., white-faced capuchins, *Cebus capucinus*: Perry and Manson 2008).

However, we should be wary of using "politics" too loosely (e.g., Boehm 1997) and of anthropomorphizing chimpanzees, baboons, capuchins and other nonhumans in the service of superficial and misleading extrapolations to humans (e.g., Fukuyama 1998). In the following section, I define dominance and briefly review concepts of power. Differentials in power among individuals are nearly universal in primates; dominance relationships and dominance hierarchies are not, although they exist in the majority of taxa. I review competing explanations for variation in dominance "style" and briefly summarize evidence concerning the relationship of dominance rank to reproductive success. I also consider the aspects of social relationships in nonhuman primates that involve politics; I argue that politics is far from universal and only characterizes species capable of triadic awareness (i.e., knowledge about the social relationships between others in one's social group). Chimpanzees necessarily figure in large measure in my discussion of politics, because they have been the main subject of relevant research and speculation. Finally, I offer some comparisons between human and nonhuman primates.

5.2 What is Dominance?

Following Hinde (1976; cf. Dunbar 1988), I regard dominance as a property of social relationships, not a characteristic of individuals, who are "dominant" or "subordinate" only in the context of social relationships. This perspective differs from the one common in social psychology, in which dominance is seen as a human personality trait that varies quantitatively among individuals (e.g., Maner et al. 2008). By convention, we can determine whether dominance is part of a social relationship by quantifying the direction and outcomes of agonistic interactions, i.e., those that involve aggressive and/or submissive acts and/or signals. For example, supplants are agonistic acts in which a stationary individual moves away at the approach of another, who takes its place, while "pant grunts" are formal vocal signals of subordinate status in chimpanzees (Marler 1976) and "silent bared teeth faces" are formal visual signals of subordinate status in and some (but not all) macaque species (Thierry 2000). If one of two members of a dyad consistently wins the agonistic interactions between them, it is dominant to the consistent loser, which in turn is subordinate to the consistent winner. Since the same individual can be dominant to some members of his or her social group and subordinate to others, terms like "dominant individuals" or "dominance interactions," while convenient shorthand, are best avoided because they risk reifying dominance as an inherent aspect of individual phenotypes.

Linear dominance hierarchies, of varying degrees of stability, form when most or all group members or all members of one class of individuals (e.g., adult females) have dominance relationships. These occur in many, but not all, group-living primate species. For example, both macaques (*Macaca* spp.) and mountain gorillas (*Gorilla beringei beringei*) live in multifemale groups; macaque females typically form stable, linear dominance hierarchies (de Waal 1986, 1989; de Waal and Luttrell 1988; Chapais 1992; Thierry et al. 2000; Thierry 2007), but mountain gorilla females do not necessarily do so and many female dyads lack decided agonistic relationships (Watts 1994).

5.3 Functions of Dominance

Group-living primates have many means to manage the tensions that arise from conflicts of interest among individual group members (de Waal 1986, 1989, 2000). The argument that establishing decided dyadic dominance relationships and (when these occur) linear dominance hierarchies limits aggression, and thus mitigates such tensions and reduces the costs of aggression – notably injury risk – has a long history (e.g., Bernstein 1976; Chapais 1991). De Waal (1986) argued that formal signals of subordinate status and "conditional reassurance" from higher to lower-ranking individuals limit the use of force to express competitive tendencies, and stated (*ibid.*: 475) "a well recognized hierarchy promotes social bonds and reduces violence," as evidenced by data from macaques, chimpanzees, and other species.

But this begs the question of why status striving should so often be all that prominent. The standard answer is that high rank, by conferring priority of access to resources, can lead to high reproductive success. A sex difference in the key resource over which individuals compete exists, with females competing mostly for food (although safety can also be important; van Schaik and van Noordwijk 1986) and males mostly for mating opportunities. The literature on the association between rank and reproductive success is too vast to summarize here. In brief, the weight of the evidence generally supports the hypothesis that rank is positively correlated with reproductive success for both sexes. Data on some species show the expected positive relationship for females (e.g., yellow baboons, *Papio cynocephalus*: Altmann and Alberts 2005; long-tailed macaques, *Macaca fascicularis*: van Noordwijk and van Schaik 1999). Exceptions exist (e.g., chacma baboons, *Papio hamadryas ursinus*: Cheney and Seyfarth 2007), and the strength of social bonds can influence female reproductive success independently of rank (Silk et al. 2003), but rank effects may only become discernable during times of prolonged food shortfalls and thus only become evident in very long-term studies (Cheney and Seyfarth 2007). Harcourt (1987) noted that no cases were known in which female reproductive success was inversely related to rank; this is still true. Accumulating genetic evidence also supports the "priority of access" model for males in many species (reviewed in Di Fiore 2003), including yellow baboons (Alberts et al. 2006), long-tailed macaques (van Noordwijk and van Schaik 2004; Engelhardt et al. 2006), mandrills (*Mandrillus sphinx*: Setchell et al. 2005), and chimpanzees (Boesch et al. 2006), although the number of males per group and the degree of estrus synchrony among females also influence reproductive skew among males. Behavioral variation linked to differences in personality (or "behavioral syndromes") also can influence male reproductive success independently of dominance rank (Bergman et al. 2008).

5.4 Sources of Variation in Female Dominance Style: Ecology, Phylogeny, and Self-Structuring

Relationships between females vary widely among primate species that form multifemale groups, as do those between males and females and, in multimale groups, those between males. Females in some species may not form dominance hierarchies (e.g., mountain gorillas: Watts 1994; blue monkeys, *Cercopithecus mitis*: Cords 2000). Aggression directionality and intensity varies among other species (e.g., aggression goes unidirectionally down the hierarchy in rhesus macaques, *Macaca mulatta*, but is common up the hierarchy in stump-tailed macaques, *M. arctoides*: de Waal 1986, 1989; de Waal and Luttrell 1988). Dyadic agonistic asymmetries can vary, sometimes consistently, across contexts, and in some species, they are susceptible to the influence of third parties. Kawai (1958) used the term "dependent rank" to refer to the attainment of dominance by young female Japanese macaques (*Macaca fuscata*) over larger adolescent and adult females subordinate to their mothers, with the help of coalitionary support from kin and, sometimes, nonkin. White (1996) described bonobos (*Pan paniscus*) as having female "feeding priority," because females usually win contests with males over food, and males sometimes avoid patches where females are feeding, but males are not necessarily submissive to females in other contexts. Similar variation occurs in tolerance of higher-ranking individuals for their subordinates, dynamics of conflict management and resolution, and other behavior related to preserving social bonds in the face of potentially serious competition. De Waal (1986, 1989; cf. de Waal and Luttrell 1988) introduced the concepts of "formal dominance" (based on the directionality of formal signals of relative status) and "real dominance" (based on the actual outcomes of agonistic interactions) to clarify understanding of this variation and referred to particular patterns of variation as the "dominance style" of a species or group of species.

Feeding efficiency can crucially influence female reproductive success; thus, variation in the relative strength and intensity of scramble and contest feeding competition may lead to predictable variation in female dominance styles (van Schaik 1989). The socioecological model (Sterck et al. 1997; cf. Koenig 2002) accounts well for much of this variation by considering the effects of competitive regimes; by incorporating sexual conflict, it accounts for much variation in male–female relationships and helps to explain why in almost all species that form stable social groups, males associate permanently with those females. Briefly, this model holds that when within-group scramble competition predominates and contest competition is inconsequential, females in multifemale groups should either form weak and unstable dominance hierarchies or not form them at all, and females can transfer between groups to minimize scramble competition and/or to choose mates. In contrast, when feeding on clumped resources monopolizable by single individuals or by coalitions that include only some group members is important – i.e., when contest competition for food is important – and monopolization confers nutritional advantages, linear dominance hierarchies occur. Given associated female philopatry, nepotism is the main basis for coalition formation, and related females help each other to acquire and maintain ranks. High within-group contest competition should result in strong ("despotic") hierarchies, stabilized by nepotism. When between-group contest competition also strongly influences female fitness, however, high-ranking females – those able to win within-group contests against most or all others – should be tolerant of lower-ranking females, and hence hierarchies should be weaker, so that high-ranking females can retain their subordinates' support in contests with other groups. Finally, high between-group contest competition combined with low within-group contest competition should lead to egalitarian relationships (no dominance hierarchies or weak ones) combined with female philopatry.

Much evidence supports the model, although accounting for intertaxon diversity in female social relationships minimally also requires other consideration of overall food abundance and variation in food nutritional quality (Isbell 1991; Pruetz 2009). For example, within-group contest competition is relatively weak, female agonistic relationships are egalitarian, and female transfer is common in Thomas langurs (*Presbytis thomasi*), and within-group contest competition is relatively high in "despotic" long-tailed macaques (Sterck and Steenbeek 1997). Likewise, variation in the strength of within-group contest competition accords with variation in the strength of dominance hierarchies among populations of hanuman langurs (*Semnopithecus entellus*: Koenig et al. 1998) and among three species of squirrel monkeys (*Saimiri* spp.: Boinski et al. 2002; cf. Mitchell et al. 1991). Blue monkeys at Kakamega face high between-group contest competition, but within-group contest competition is weak; correspondingly, females are philopatric, but do not form linear dominance hierarchies (Cords 2000). Such egalitarianism in species with high between-group contest competition (e.g., some guenons) may be a general solution to a collective action problem that female superior competitors face, but whether this is actually so depends on how much they could gain by winning within-group contests. If potential gains rarely outweigh the costs of aggression perhaps because most food patches can accommodate all group members (but not more than one group), egalitarianism or tolerance requires some other explanation. Lions provide a valuable comparative example of female egalitarianism: females in the same pride cooperatively defend carcasses against females of neighboring prides and defend cubs against infanticidal males, and they engage in communal care of cubs; they do not form dominance hierarchies despite high potential for within-pride contests over food (Packer et al. 1990; Heinsohn and Packer 1995).

Despite the considerable success of the socioecological model, its generality and functional logic have been strongly challenged, notably with respect to variation in female dominance style among macaque species. Macaques fall into semi-discrete grades along a spectrum in which the directionality of aggression, the relative frequency of high-intensity aggression, the extent of kin biases in social behavior, conciliatory tendencies, and tolerance co-vary (Thierry 2007). This co-variation seems to have a strong phylogenetic signal (*ibid.*; Thierry et al. 2000). In despotic species like rhesus and Japanese macaques, rigid hierarchies co-occur with relatively high rates of intense aggression, unidirectional aggression down the hierarchy, strong nepotism and consistent operation of the "youngest ascendancy rule" in rank acquisition (maturing females assume ranks immediately below those of their mothers and above any older sisters), and low conciliatory tendencies. At the other end of the spectrum, high tolerance is associated with weaker nepotism and less kin-bias in social relationships, low rates of intense aggression, less consistent operation of the youngest ascendancy rule, and high conciliatory tendencies. The extreme despots belong to the *fascicularis* lineage; the most tolerant species belong to the *silenus-sylvanus* lineage; and other members of these lineages and species in the *sinica-arctoides* lineage occupy various intermediate positions (Thierry 2007), although not all variation sorts neatly by phylogeny (e.g., Tibetan macaques (*M. thibetana*) belong to the relatively "tolerant" *sinica-arctoides* lineage, but have despotic female dominance; Berman et al. 2004). Moreover, limited field data indicate that variation in hierarchy strength may not correspond to variation in the strength of within-group contest competition and that high between-group contest competition does not obviously characterize the more tolerant species (Cheney 1992; Berman et al. 2004; Ménard 2004; Thierry 2007).

Thierry (2007; cf. Matsumara 1999; Matsumara and Kobayashi 1998) has proposed that these grades are different evolutionarily stable outcomes of selection on a collection of traits that are linked because all are mediated by the same underlying neurobiological and hormonal mechanisms (e.g., the effects of serotonin on anxiety and aggression intensity). Contrary to the "collective action" explanation for tolerance among female macaques, high-ranking females are not forced to restrain their competitive tendencies to induce cooperation from low-ranking females, because their competitive tendencies are already low. Shallower dominance gradients in tolerant species than in more despotic ones also could mean that for a female of a tolerant species, proportionately more of the other females in her group are potentially valuable allies, although alliances would have less importance for her if the co-evolved system involves low competitive tendencies. Alternatively, Preuschoft and van Schaik (2000) proposed that low asymmetries in fighting ability and less consistent or predictable agonistic support in tolerant than in despotic species means that females of tolerant species need to probe others repeatedly to assess their current willingness to act as allies. This leads to high tolerance for spatial proximity regardless of rank difference, relatively infrequent retaliation against females that direct mild aggression at higher-ranking partners, more need for reconciliation, and more frequent grooming between nonkin. In their view, tolerance is really "calculated generosity."

Problems with the sociological model are not restricted to macaques. For example, female hanuman langurs at Ramnagar engage in high within-group contest competition for food and, as expected, form despotic dominance hierarchies, but are not tolerant despite also facing high between-group feeding competition (Lu et al. 2008). Between-group contest competition seems to be generally a poor predictor of female dominance style.

Hemelrijk has used a series of agent-based simulation models to argue that self-structuring could account for much of the observed variation in dominance styles (e.g., Hemelrijk 1996, 1999a,b, 2000a,b, 2002; Hemelrijk et al. 2003). Model entities are assigned initial "dominance" values for which the gradients vary across simulations. They then follow one of various alternative sets of rules that determine how they move and how they behave on encounters with others moving in the same space. They may estimate their capacity to win agonistic interactions with others based on their past histories of interaction; alternatively, they may assess the risk of conflicts by directly evaluating their own fighting ability relative to that of the entities they have encountered. Group cohesion also varies, along with the probability that more than two entities meet simultaneously and "coalitions" form. Entities are sometimes divided into species in which aggression intensity is typically high and others in which it is low, and sex differences in attack intensity can vary. Crucially, contest outcomes are probabilistically determined, and both winning and losing are self-reinforcing (winning reinforces an agent's dominance value, whereas losing decreases it), with effects of winning greater for the subordinate of two interactors and the losing effect stronger on the dominant. The self-structuring effects of these simulations produce dominance hierarchies that vary along the axes of "despotism-egalitarianism" and "tolerance-intolerance", in the degree of nepotism, and in the extent to which male and female dominance hierarchies overlap. Inclusion of variation in food clumping and in sexual attraction between males and females also influences variation in male–female hierarchy overlap (Hemelrijk et al. 2003).

Hemelrijk states that these models are "caricatures" that do not reflect the complete behavior of real monkeys and apes. For example, dyads with established social relationships almost certainly use memory-based assessment (Silk 2002), coalition formation is nonrandom, and triadic awareness, which depends on learning and memory, leads to strategic decisions about intervening in conflicts and soliciting coalition partners. As Preuschoft and van Schaik (2000) state, nonhuman primates often have accurate information about relationships in their groups. Clear evidence that monkeys classify others on the basis of kinship (e.g., Bergman et al. 2003) argues against the possibility that apparent nepotism is simply a byproduct of variation in the intensity of aggression (Hemelrijk 1999b).

It remains an open question whether we should jettison the socioecological model in favor of game theoretic analyses of alternative equilibria that result from selection on linked traits (Thierry 2007) or can improve it by incorporating data on other aspects of food distribution and quality (Pruetz 2009), better data on the extent to which low-ranking females participate in contests between groups, and consideration of other potential sources of leverage for low-ranking females (e.g., willingness to engage in cooperative defense of infants against infanticidal males; Lu et al. 2008). Ultimately, alternative social equilibria can only be stable within bounds set by species' ecology, and we need much better quantitative data on feeding competition in many species, notably macaques. However, in looking for bivariate associations between competitive regimes and categorical distinctions like despotic versus egalitarian or tolerant versus intolerant, the model's advocates fail to acknowledge that these probably are parts of coevolving complexes that cannot be atomized (Thierry 2007; cf. Thierry 2000; Thierry et al. 2000). Hemelrijk's nonadaptationist models can help to provide proximate explanations for dominance style variation. Superficially, their outcomes strongly resemble the spectrum of dominance styles evident in macaques and some other nonhuman primates. However, they may reproduce the surface structure of dominance styles precisely because in the real world, the variables that lead to self-structuring – winning and losing effects, aggression intensity, group cohesion, sex differences in agonistic power – help to determine developmental outcomes within ranges of possible phenotypic space set by varying histories of evolutionary response to competitive regimes, and whatever explanatory power they have is not independent of phylogeny and ecology.

Occasional references to winner and loser effects in the real world occur; for example, Preuschoft and van Schaik (2000) note that individuals that repeatedly lose contests to various opponents can become "trained losers," who defer to most or all opponents, whereas those that repeatedly win may show the opposite pattern. Any such effects could be hormonally mediated, perhaps by testosterone. An extensive literature on humans (reviewed in Archer 2006) indicates that the relationship between testosterone and competition in both sexes is complex. One apparent generalization based on a meta-analysis of relevant studies (*ibid.*) is that male testosterone levels rise slightly in anticipation of sports competition; they also increase from before to after the competition, with the increase greater in winners than in losers, although variation in personality, in attribution of causality for the outcome, and other factors can influence the magnitude of change. Additionally, losing sometimes decreases testosterone, although this effect may occur only in individuals that attribute their losses to intrinsic factors (e.g., Mehta and Josephs

2006), and drops in testosterone in association with high social anxiety can make losers less willing to compete again (Maner et al. 2008). Differential effects of winning and losing like these, if persistent, could help to produce the kind of self-reinforcement that Hemelrijk envisions.

Attempts to monitor short-term fluctuations in testosterone induced by competitive interactions in wild nonhuman primates face formidable logistical hurdles, but longer-term increases or decreases in testosterone respectively associated with winning or losing contests for alpha male status in mandrills (Setchell et al. 2008) and for control of one-male units in geladas (Beehner and Bergman 2009) suggest that such effects occur. Further circumstantial support comes from the evidence that testosterone influences decisions by male chacma baboons in Botswana (Beehner et al. 2006; Bergman et al. 2006) whether to engage in contests. Testosterone was highest in males rising in rank, and current levels predicted male rank and mating success over the subsequent eleven and seven months, respectively (Beehner et al. 2006). Males avoided others with high testosterone more often than those with low testosterone; relative rank had little effect, and adjacently ranked males (probably each other's most serious current competitors) with high combined testosterone were most likely to avoid each other.

5.5 Variation in Male Dominance Style

Sex differences in dominance style are common, as expected given that variation in female fitness depends mostly on the outcome of competition for food, whereas variation in male reproductive success depends mostly on the outcome of mating competition, and given that dispersal is often male-biased. For example, in some chimpanzee communities, many female dyads lack dominance relationships (e.g., Ngogo: Wakefield 2008), and females usually do not form linear hierarchies. Instead, they reduce feeding competition by adjusting gregariousness to food availability. In contrast, males typically form steep linear hierarchies (e.g., Ngogo: Watts and de Vries 2009), although male–male social relationships are also highly affiliative. Male dominance ranks in nonhuman primates usually depend on individual fighting ability and show an inverse-U shaped relationship to age, even in cercopithecine species in which females form stable hierarchies in which ranks depend on predictable nepotistic and mutualistic support, not on fighting ability (Chapais 2001). Nepotistic effects on male ranks prior to dispersal are more likely in those cercopithecines in which adult size dimorphism is relatively low than in those in which adult males are much larger than females (Pereira 1988). Post-dispersal nepotistic effects are unlikely, and nepotism may not be consistently important even in species with male philopatry. For example, while some maternal brothers form alliances in chimpanzees, most allies are not maternally related (Langergraber et al. 2007). Bonobos (*P. paniscus*) may be exceptional, in that high-ranking females may form alliances with their sons that enable the sons also to attain high rank (Kano 1992), although how pervasive such effects are is unknown.

In general, the extent of intersexual overlap in dominance ranks varies inversely with sexual dimorphism in body size in cercopithecines, presumably because the risk of female aggression to males increases as dimorphism in body size and canine size increases (Packer and Pusey 1979; Thierry et al. 2000); these relationships help to explain the inverse relationship between the sex difference in initial "dominance" (i.e., fighting ability) and male–female rank overlap in Hemelrijk et al.'s (2003) self-structuring model. But in macaques, rank overlap among males and females seems to depend on the extent of kin-biases in female behavior and the strength of female alliances against males; thus, it varies along phylogenetic lines rather like variation in female dominance style and is somewhat independent of variation in sexual dimorphism (Thierry et al. 2000). Variation in male dominance style in macaques also partly mirrors that among females. For example, tolerance between males is lower, serious aggression is more common, and the directional consistency of aggression higher in rhesus macaques than in more tolerant species (Thierry 2007). However, the influence of male rank on mating tactics and reproductive success depends on variation in female mating synchrony in a manner independent of female dominance style variation (reviewed in Thierry 2007). Rank strongly affects reproductive success in nonseasonally breeding species, including highly tolerant ones (e.g., Tonkean macaques, *Macaca tonkeana*), in which low-ranking males must win aggressive challenges against high-ranking males to have good prospects for siring offspring (cf. van Noordwijk and van Schaik 1985). In contrast, reproductive skew is low in seasonally breeding species, including despotic ones (e.g., rhesus macaques); severe aggression is correspondingly uncommon except in small groups, and rank tends to increase with age and length of group residence.

Variation in population density and demography can strongly influence the costs and benefits of dispersal and of alternative mating tactics for males (e.g., yellow baboons: Alberts and Altmann 1995; macaques: Soltis 2004). Such variation may not fundamentally affect male dominance style; for example, male yellow baboons in low-density populations may delay dispersal, but their ranks, and thus to important extents, their reproductive success, still depend on their fighting ability (Alberts and Altmann 1995). Nevertheless, the potential for such effects bears further investigation, as does variation in the steepness of male hierarchies.

5.6 Power and Politics

Variation in dominance style reflects variation in power and raises questions about political maneuvering. The term "power" is used widely, but often without explicit definition. For example, Datta (1983a,b) implied that power in rhesus macaques (*M. mulatta*) comprises individual fighting ability plus any competitive advantages gained from coalitionary support. Likewise, "politics" is often not explicitly defined. De Waal (1982) characterized the tactics of alliance formation (including competition for partners and disruption of potential alliances between rivals), strategic and sometimes conditional use of reconciliation, and other ways in which chimpanzees manipulate their social partners and manipulate relationships between other individuals, as political. Subsequently, de Waal (1989) specified that the need for allies in within-community competition for status and the need for all males in a community to cooperate in aggression against neighboring communities are "internal and external 'political reasons'" why male chimpanzees require effective mechanisms to cope with within-community competition. In describing challenges for the alpha position among male chimpanzees at Mahale, Nishida and Hosaka (1996) also stressed the complexity and flexibility of alliance formation strategies and the varied ways in which males use social resources like grooming and meat sharing to maintain alliances and referred to such behavior as political (cf. Mitani and Watts 2001; Watts 2002; Mitani 2006). Duffy et al. (2007) characterized the Kanyawara alpha male's selective tolerance of his allies mating behavior in exchange for coalitionary support as a political tactic from which all parties benefited, despite potential tradeoffs (less than maximum mating monopolization vs. prolonged tenure as alpha; additional current mating opportunities vs. foregone attempt to challenge for the alpha position).

A review of the many ways in which political anthropologists have characterized power is beyond the scope of this chapter, but brief attention to some of this literature provides a useful context for considering power and politics in nonhumans. Some explanatory paradigms (e.g., postmodernist ones) provide little or no basis for comparative analysis (sometimes deliberately), but others, such as "processualism," are more amenable. Processualists see politics as the processes involved in determining and implementing public goals and the differential use of power by group members concerned with those goals (Swartz et al. 1966; Adams 1977; Kurtz 2001; Lewellen 2003; Swartz et al. 1966; Box 5.1). Power has many dimensions, not all political; for example, it is embedded in healing rituals (Kurtz 2001). Weber's (1965 [1947]) definition of power as the relative ability of one actor to carry out his or her own will despite resistance underlies the relevant notion of political power, and many anthropologists extend politics to the pursuit of individual goals as well as group goals. For example, Kurtz (2001: 21) writes:

> "Politics is all about power: about how political agents create, compete for, and use power to attain public goals that, at least on the surface, are perceived to be for the common good of the political community. Yet just as open and more covertly, political power is used to attain private goals for the good of the individuals involved."

Processualists divide power into several broad categories (Box 5.1). Independent power is based on individual capabilities. Dependent power is granted, delegated, or allocated to others by someone who has independent power, but that individual in turn is subject to consensual power (i.e., the assent of the people).

Anything that contributes to or maintains power counts as support; two basic supports are coercion and legitimacy (Box 5.1).

Box 5.1. Selected examples of how "power" and "politics" have been conceptualized in the literature on political anthropology.

Politics: *"The processes involved and determining public goals ... and the differential use of power by the members of the group concerned with those goals."* (Swartz et al. 1966: 7)

Power: *"... a quality ascribed to people, and often also to things, that concerns their relative abilities to cope with the real world or their potential effects on it ... [power cannot exist alone,] but must be recognized by others and by the individual possessing it."* (Adams 1977: 389)

"... the probability that one actor in a social relationship can carry out his will despite resistance." (Weber 1965 [1947])

An ability to control the behavior of others and/or to gain controlling influence over others; the ability of one individual to bend another to his or her will. (Lewellen 2003)

Components of Power in the "Processualist Paradigm" (Adams 1977; Lewellen 2003):

- **Independent power:** *"A relation of dominance based on the direct capabilities of an individual, such as knowledge, skills, or personal charisma."* (Lewellen 2003: 231).
- **Dependent power**: Power that is granted, allocated, or delegated, either by someone who has independent power over its recipient or to a more powerful individual by his or her supporters.
- **Granted power**: Decision-making rights that one individual gives to another.
- **Allocated power**: Power given by a group of people to a certain individual (e.g., a "Big Man," a shaman).
- **Delegated power**: Decision-making rights given to a number of different people by a single individual who has a concentration of individual power.
- **Consensual power**: Leadership that arises from the assent of the people (based on tradition, faith in the personal qualities of the leader, etc.) rather than from force alone.
- **Support**: A broad concept that includes virtually anything that contributes to or maintains political power. Two basic supports are coercion and legitimacy.
- **Legitimacy**: A primary basis for power that derives from the people's expectations about the nature of power and how it should be attained (e.g., by election, by holding redistributive feasts) and used.

Box 5.1. continued
Power resources (Kurtz 2001):

- Material resources:

  1. **Tangible resources**: Culturally defined goods (e.g., money) that individuals compete for and use to attain their goals.
  2. **Human resources**: allies and supporters that any political agent requires to be a leader.

- Ideational resources:

  1. **Symbolic resources**: material objects, mental projections, actions, ideas, or words that humans infuse with ambiguous, multiple, and disparate meanings; political symbols may be anything in the social or physical environment that leaders can use to convince people to support them; fluid, changeable, respond to shifting social, cultural, and political conditions.
  2. **Ideological resources**: a political ideology is "a system of hypotheses, principles, and postulates that justify the exercise of authority and power, asserts social values and moral and ethical principles, sets forth the causal connections between leaders and the people they govern, and furnish guides for action...around a set of beliefs and ideas." (Kurtz 2001: 35).
  3. **Informational resources**: Information both includes and produces knowledge; to the extent that leaders can produce and harness the flow of information, it becomes a source of political power.

Kurtz (2001) presents a rather different perspective, in which political power depends on the ability of agents to acquire and maintain control over resources; this resonates with the emphasis on resource competition in the primatological literature, but redistribution of resources often accompanies control in humans. In political struggles, those who control more resources tend to win against those who control less, but agents who control less can win if they use resources more wisely and skillfully. Some resources are material; these include tangible objects and goods as well as human supporters (Box 5.1). Other crucial resources are ideational. These include ideological, symbolic, and informational resources (Box 5.1). Political rhetoric – the deliberate exploitation of eloquence in public speaking or writing by leaders to persuade others – is the most common source of information as political power and is a pervasive and extremely important alternative to coercion. Independent power thus depends on the ability to control culturally constructed ideational resources as well as material resources; a leader must be good at using ideational resources to attract supporters, but maintaining their support ultimately depends on providing them with sufficient tangible resources.

Chapais (1991) and Lewis (2002) explicitly have characterized power in nonhuman primates by drawing on literature in political science. This literature mostly concerns modern nation states and lacks the comparative ethnographic focus of political anthropology, but shares certain themes with it. Chapais (1991) adapted an explanatory framework developed by Bacharach and Lawler (1980). Echoing Weber, power in this framework is defined as the capacity of an actor to carry out its own will despite resistance. It can be "aggression-based" – that is, based on aggression or threat and thus coercive, although this category also includes power based on control of resources, services, or knowledge – or "dependence-based," with the power of actors derived from support that others give them. Acquisition and maintenance of dominance rank in nonhuman primates provide good examples of both categories (Chapais 1991). It generally involves aggression-based power, but sometimes depends largely on agonistic support from third parties (notably among females in macaques, baboons, and vervets; Chapais 1992) and is thus dependence-based. Dependence-based power in this sense is in keeping with Kawai's (1958) "dependent rank." It contrasts with the processualists' dependent-power subcategories of granted and delegated power (Adams 1977; Box 5.1), in which the arrow of dependency points in the opposite direction and those who have granted or delegated power to others can withdraw it; it is closer to allocated power, although without the collectively agreed conferral of status that this term implies. But dependence by leaders is a major theme in political anthropology, and recurring emphasis on the need to attract and retain supporters (*"the single biggest problem that any leader confronts"*; Kurtz 2001: 34) indicates that dependency is bidirectional in humans. When one individual, human or nonhuman, has dependence-based power over another and has some ability to manipulate outcomes for it, their relationship can take on a political dimension. Chapais (1991) noted that many of the sources of power (whatever allows one to control its basis) that Bacharach and Lawler (1980) identified in humans do not apply to nonhuman primates. For example, "normative power," a form of dependence-based power in which one individual or group can bestow symbolic rewards on another, is exclusively human, as is most, and perhaps all, power based on possession of knowledge that cannot be acquired simply by observational learning. Control of information beneficial to others is an important source of power in humans (Bacharach and Lawler 1980; Chapais 1991) and a potential source in nonhuman primates, but nonhuman primates do not appear to bargain over information (Chapais 1991), and manipulating others by withholding information is uncommon and may be limited to a few taxa (e.g., chimpanzees: Hare et al. 2006). More generally, nonhuman primates make little, if any, use of ideational resources.

Lewis (2002) applied "power" to asymmetries in social relationships that can originate in individual differences in resource holding potential (i.e., fighting ability) or in differences in the strength of alliances, but also in the possession by some individuals of inalienable resources, broadly defined, that can influence the fitness of others. Thus, it comprises both dominance and leverage. Dominance is the agonistic component of power; it is based on force or threat of force. "Intrinsic" dominance depends solely on interindividual differences in fighting ability and is thus roughly equivalent to Chapais' (1991) "aggression-based power." "Derived" dominance (cf. Datta 1983a, b) depends also, or largely, on the relative strength of coalitionary support and resembles Chapais' (1991) "dependence-based power." When individuals could benefit from resources that others control or services that they could provide, but cannot take these by force, the resource holders and service providers have leverage; implicitly, this idea is also included in dependence-based power.

If politics is all about creation of, competition over, and use of power, it necessarily involves social maneuvering, and not all potential sources of leverage are necessarily available for political use. For example, Lewis (2002) considers possession of fertilizable eggs as a source of leverage that can allow estrous females to gain temporary social advantages (e.g., increased receipt of grooming from males) without changes in dominance, but females may gain such advantages simply because others respond to signals of fertility or receptivity or to proceptive behavior, not because they use these signals to manipulate others socially. In contrast, when individuals have alternative options for distributing allogrooming or coalitionary support, strategic deployment of these options may qualify as political, especially given variation in partner quality. One individual has leverage over another when the second depends on it in some way, and such dependence can be mutual; their interactions can become political when one or both partners can manipulate socially determined outcomes that the other values (Chapais 1991).

Leverage can either increase or decrease power asymmetries, as illustrated by classic examples of male alliances in baboons and chimpanzees. Noë (1990) documented the formation of coalitions by two or three mid-ranking male yellow baboons against higher-ranking males in several contexts, notably in attempts to separate high-ranking males from estrous females with whom those males were consorting. In Lewis' (2002) terms, coalitions temporarily increased the allies' power relative to that of their opponents without changing their dominance ranks (they still behaved submissively to their opponents in dyadic encounters; Noë 1990), but the highest-ranking of the three allies had leverage over the other two (and thus increased his relative power over them) because they had little chance of succeeding without his participation. This presumably explained why he took over the consorts in all observed cases in which the coalitions succeeded. Noë characterized coalition formation by these males as a "veto game" with the highest-ranking ally acting as the "veto player," and pointed out that variation in the value of potential partners should lead to shopping for, and bargaining over, their services (cf. Noë 1992; Noë and Hammerstein 1994). Variation in partner value means that dependence, while bidirectional, is not always symmetrical, but bargaining can also occur in established alliances, and asymmetries can be constrained, when weaker partners have enough leverage (Noë 1990, 1992). Baboon males obliged to negotiate with weaker allies bear some resemblance to "weak leaders" among humans, who "have allies whose commitments to them are transactional...based on what they can get for their support, which therefore is tenuous" (Kurtz 2001: 45).

Derived power is important in status competition among male chimpanzees, and alpha males usually depend on allies to attain and maintain their positions. The third-ranking of three males involved in a status struggle in the Arnhem Zoo community mated more often than expected, based on his rank, while another male depended on his support to consolidate his newly attained status as alpha (de Waal 1982). However, his mating frequency declined when the new alpha's position was secure. Nishida (1983) described a similar case in a wild community that had only three adult males. Duffy et al. (2007) found broader leverage effects in a larger chimpanzee community at Kanyawara, in which mating success for males other than the alpha was positively correlated with the frequency with which they gave the alpha coalitionary support, independently of their own dominance ranks. The alpha male exerted some control over others' mating success by disrupting many copulation attempts; the frequency with which he disrupted those of individual males was inversely related to the amount of support they gave him.

Lewis (2002; see Chapais 1991, for a somewhat different categorization) defined four proximate characteristics of power: its base, or source (e.g., fighting ability); its means, or the way in which individuals negotiate it in relationships (e.g., by using force); its amount, which can vary with context and can be expressed as a probability of winning contests; and its scope, or the range of behavior that an individuals can invoke in others by using dominance or leverage. This framework can help to resolve several longstanding debates and to clarify some terminological confusion. Lewis (*ibid.*) gave de Waal's (1986) "formal" and "real" dominance as one example: formal dominance involves the use of formalized status signals that are consistent across contexts; thus, its scope differs from that of real dominance, which can vary contextually and which also involves derived dominance and leverage.

The issue of "female dominance" in lemurs also benefits from reframing as a question about power. Male–female asymmetries are pronounced in some lemur species; notably, female ring-tailed lemurs (*Lemur catta*) win all contests against males and evoke formalized submissive signals from them, and aggression is unidirectional from females to males (Pereira et al. 1990; Pereira and Kappeler 1997). However, females are less powerful in other species, and the scope of power differs; for example, aggression is bidirectional in brown lemurs (*Eulemur fulvus*), males win many contests, and formal dominance does not exist (*ibid.*). Researchers have sometimes used "co-dominance" to refer to species in which neither sex consistently wins contests against the other (e.g., gibbons: Leighton 1987) or "female feeding priority" to refer to those in which females typically win contests over food, but not always in other contexts and in which aggression is bidirectional and formal dominance between males and females absent (e.g., bonobos, *P. paniscus*: White 1996). Specifying whether sex differences in the amount of power exist would be more productive than arguing about whether a given species has "female dominance" or just "female feeding priority" and about whether these are different phenomena. As Lewis (2002: 154) notes, *"female dominance to males occurs only when female fighting ability is superior to that of males in intersexual dyadic interactions."* Likewise, if no such dyadic asymmetry exists between males and females, no dominance exists, even if a leverage asymmetry means that the sexes differ in relative power; "co-dominance" is a meaningless term in this context.

Flack and de Waal (2004) have made the most elaborate attempt to define power in nonhuman primates and to link it to politics, on the one hand, and to dominance style, on the other. In their scheme, dominance style operates on the level of social relationships and refers strictly to the discrepancy between the inherent agonistic asymmetry between individuals (roughly, fighting ability) and the degree to which they express this asymmetry, as indicated by the directional consistency of aggression, the typical intensity of aggression, and the types of signals used to indicate dominance or subordination. When inherent asymmetries are high and are routinely expressed in social interactions, for example, dominance relationships are despotic; when they are high to moderate, but only weakly expressed, relationships are tolerant. They did not explicitly distinguish intrinsic from derived dominance, but included alliance formation as a *"contextually and temporally stable factor"* (p. 169) that makes agonistic outcomes more predictable and thus influences agonistic relationships.

Flack and de Waal (*ibid.*) linked dominance style to politics via "social power," which they defined (p. 167) as "the degree of implicit agreement among group members that an individual is capable of using force in polyadic social situations." In turn, force "leads to the reduction or elimination of the choices of others" (p. 168). They restricted social power to species that use formal signals of dominance or subordination (e.g., macaques, chimpanzees), arguing that "implicit agreement" can only occur if individuals consistently acknowledge status differentiation in nonaggressive contexts. They proposed that we can operationalize social power by comparing, for each individual, the number of others from whom it receives and to whom it gives signals of subordinance or dominance, the frequency of these signals, and the way in which they are distributed among other group members. Such operationalization provides a basis for interspecific comparison, but, as Flack and de Waal acknowledge, their concept of social power is far narrower than those of Chapais and Lewis, and it would exclude many species (e.g., ring-tailed lemurs and gorillas have social power, but brown lemurs and chimpanzees do not). When social power in their sense exists, groups have power structures, which lead to "political systems" at the societal level that reflect "the interplay between the power structure and conflict management" (p. 157). They classified political systems (Table 5.1) based on four main factors: how much social power is concentrated in single individuals versus distributed among all group members, who intervenes in conflicts and what intervention strategies they follow, the extent to which interventions in conflicts reinforce or reduce social power differentials, and how equally resources are distributed.

5.7 Politics and Cognition

The claim that some nonhuman animals engage in politics has been criticized. For example, Schubert (1991) argued that nonhuman primate "politics" is mostly a metaphor and particularly criticized Hrdy (1977) for referring to "regimes" and "usurpation of power" by extra-group males that lead to "regime changes" in hanuman langurs. Boehm (1997) was more sympathetic when comparing power

Table 5.1 Dominance style, social power, and political systems (after Flack and de Waal 2004)

Dominance style	Distribution of social power	Political system
Despotic	Uniform; increases as *social power rank* (SPR) increases	*Hierarchy*: resource allocation determined mostly by SPR; conflict interventions reinforce system
Tolerant	Concentrated in a few individuals and distributed uniformly among others so as to increase with SPR	*Informal oligarchy*: some resource allocation by SPR; powerful 3rd parties intervene in conflicts impartially or to favor least powerful participants, others intervene to reinforce hierarchy
	Concentrated in a few individuals; others have approximately equal power	*Constrained*: Leveling coalitions, policing by powerful individuals, and mediation maintain system
Relaxed	Temporally stable, but small, differences in social power	*Equal outcome system*: maintained by coalitions against individuals intolerant of subordinates and mediation by powerful individuals; policing can lead to equal resource distribution; may be institutional roles
Egalitarian	No temporally stable differences in social power, but some individuals may temporarily be more powerful than others	*Equal opportunity system*: maintained by punishment of norm breakers and of nonpunishers; can lead to division of labor among arbiters and impartial policing to mediate conflicts between coalitions

asymmetries in chimpanzees and human hunter-gatherers. He defined *"political intelligence"* as *"the decision making capacity that enables social animals to further their self-interests in situations that involve rivalry and quests for power and leadership*." He ascribed political intelligence to chimpanzees and other nonhumans, but both too broadly and too narrowly, regarding all forms of agonistic behavior involved in such decisions (e.g., bluffing, appeasement, aggression, deference) as its manifestations. By implication, leverage that involves social manipulation also represents political intelligence. This perspective risks conflating politics with power. Attempts to manipulate relationships between others (e.g., separating interventions by male chimpanzees; de Waal 1982) may well be political, but invoking political intelligence adds little to the understanding of, for example, how opponents assess each others' fighting abilities or why targets of aggression might sometimes appease their attackers by directing nonaggressive acts or signals to them that reduce the probability of further attack. Conversely, restricting political intelligence to decisions about *"whether to try to dominate or submit to"* others (Boehm 1997: 354) neglects how coalition formation can influence power dynamics by reinforcing dominance relationships, by providing successful partners with temporary advantages (e.g., takeovers of consorts with estrous females by male baboons), or by conferring derived dominance in some relationships (e.g., alliances that allow male chimpanzees to attain and maintain alpha status). Correspondingly, it omits the importance of leverage and of market effects in competition for allies (Noë 1990, 1992; Noë and Hammerstein 1994).

Boehm (1997) argued that political decisions do not necessarily require complex cognition. However, we should turn this around to argue that decisions that enable individuals "to further their self-interests in situations of rivalry and quests for power" are only political if they involve cognitively complex social manipulation. This begs the question of what qualifies as complex cognition, but at least any species capable of triadic awareness can behave politically in this sense. Individuals who can acquire knowledge about the social relationships between others in their groups can use this knowledge in calculated, sometimes opportunistic, ways to gain extrinsic dominance and to exert leverage over allies and potential supporters (Preuschoft and van Schaik 2000) and to avoid or mitigate the costs of conflicts. Flack and de Waal (2004) seem to imply that "political systems" require such knowledge. It is unclear whether their restriction of social power to "polyadic situations" means that it operates only in polyadic interactions; I take it to mean that it can only operate in groups of three or more individuals, in which two or more can "agree" that another can use force against them. But they state that conflict mediation by socially powerful individuals, which occurs in "constrained" and "equal outcome" political systems (Table 5.1), requires cognitive empathy (the ability to take others' perspectives), and list several macaque species and chimpanzees as possible exemplars of these systems. Cognitive empathy would certainly allow for triadic awareness. Additionally, their scheme applies specifically to social variation in macaques, all of which are presumably capable of such awareness.

The list of species in which triadic awareness has been formally demonstrated either experimentally or through statistical modeling is short, but further formal investigation would undoubtedly lengthen it. Chacma baboons show triadic awareness in many ways (reviewed in Cheney and Seyfarth 2007), including avoiding close maternal kin of higher-ranking females from whom they have just received threats and responding more strongly to playbacks that simulate dominance rank reversals between adult females belonging to different matrilines (which are rare and threaten to disrupt the entire female dominance hierarchy) than to those simulating within-matriline rank reversals (also rare, but with much less potential to disrupt other dominance relationships; Bergman et al. 2003). Playback studies also showed triadic awareness in vervets (reviewed in Cheney and Seyfarth 1990). Perry et al. (2004) showed that white-faced capuchins at Lomas Barbudal solicited coalition partners who had better-quality social relationships with themselves than with their opponents more often than expected by chance, and also solicited partners who outranked their opponents more often than expected by chance, although they might have done this simply by preferentially soliciting partners higher-ranking than themselves (cf. Range and Noë 2004, for mangabeys, *Cercocebus torquatus*). Nonrandom solicitation of potential coalition partners on the basis of relative relationship quality, relative rank, and/or relatedness to opponents has also been demonstrated in bonnet macaques (*Macaca radiate*: Silk 1999) and Japanese macaques (*M. fuscata*: Schino et al. 2006). Conflict management and resolution tactics apparently based on the recognition of relatedness between opponents and third parties (e.g., kin-redirected reconciliation and aggression; reviewed in Das 2000 and Watts et al. 2000) provide indirect evidence that triadic awareness is widespread among cercopithecines and perhaps other primates, and use of separating interventions is one of many forms of circumstantial evidence for triadic awareness in chimpanzees.

Triadic awareness combined with comparative knowledge of partner value allows for politics, which is partly a strategic use of such knowledge. A female chacma baboon who opportunistically forms a bridging alliance or a revolutionary alliance (Chapais 1991, 1992) to challenge another female to whom she is subordinate, but who has just lost a challenge to a third individual (Engh et al. 2006; Cheney and Seyfarth 2007) is behaving politically. Likewise, while he was alpha male in the Mahale M group of chimpanzees, Ntologi behaved politically by regularly directing separating interventions at other males, mostly at his main rivals, Nsaba and Kalunde (who nevertheless formed an alliance that allowed Kalunde to defeat Ntologi, although his tenure as alpha was then short; Nishida and Hosaka 1996). Politics also encompasses the strategic use of knowledge about variation in partner value and the corresponding strategic use of leverage, such as the decision by Yeroen, the third ranking male in the Arnhem Zoo chimpanzee colony, to ally himself with Nikkie, then a weak alpha, rather than Luit; Nikkie needed the derived dominance provided by a strong alliance more (de Waal 1982). Several primate species tested in lab settings can solve tasks that require coordinated efforts by two partners (e.g., brown capuchins, *Cebus apella*: Mendres and de Waal 2000). This suggests that they recognize the necessity of acting jointly and the value of partners and the services they can provide; his would facilitate political exploitation of variation in partner value and political negotiation over services. An alternative explanation is that they simply learn contingencies between their actions and obtaining rewards (*ibid.*). However, this seems inadequate to explain differential recruitment of partners on the basis of their task-solving skills by chimpanzees (Melis et al. 2006).

Politics may require triadic awareness, but it can occur at the level of dyads. One possible political tactic would be to induce a loser effect in a potential rival by targeting him or her sufficiently to prevent, or at least forestall, any competitive challenge. Unpredictable attacks independent of direct contests over resources might be particularly effective (Silk 2002). Rank changes were common in the baboon group studied by Bergman et al. (2006) and males in that population did not form alliances. Induced loser effects, if they occurred, might thus have often been short lived, but this does not preclude the possibility that they form part of individual competitive strategies. Male chimpanzees are good candidates for such effects because male philopatry means that adolescent males will become rivals of those in older age cohorts, who could benefit by delaying challenges from the adolescents and who could increase the delay by forming coalitions against them. This might help to explain why adult males direct aggression at adolescents at high rates – often higher than those for adult dyads – and sometimes single out particular individuals for persistent attacks (Pusey 1990, Watts unpubl. data). Aggression sometimes produces direct benefits – e.g., adults sometimes steal meat from adolescents – but much of it may be punishment, in that it has an immediate energetic cost, but changes the target's future behavior in favor of the aggressor (Clutton-Brock and Parker 1995), essentially via negative conditioning. Thinking of punishment as a manifestation of political intelligence in chimpanzees seems reasonable, but the common occurrence of age-related loser effects in ungulates suggests that we should not always assume that political intelligence is involved. Female dominance rank in many ungulates increases with age (reviewed in Côté 2000), and age is the main influence on rank among female mountain goats (*Oreamnos americanus*), which form stable hierarchies in which neither the body size nor the horn length influences rank and in which attacks by adult females lead to persistent subordination younger females even after the younger females attain adult size (*ibid.*)

5.8 Politics in Human versus Non-Human Primates

Power and politics in human and nonhuman primate share some similarities and contrast in many ways, only a few of which I will briefly consider. First, politics and power in nonhuman primates revolve around social relationships (or, in Flack and de Waal's (2004) view, politics arises from social relationships via the mediating effects of social power), and political anthropologists sometimes also stress that power is a component of human social relationships (e.g., Adams 1977). Face-to-face interactions in which individuals or small coalitions either directly assert their own interests, with at least an implicit threat of force, or use social persuasion to do so form part of human politics and would have characterized most human political behavior during our evolutionary history in small-scale societies (Archer 2006). In some respects, the forms and outcomes of such interaction resemble chimpanzee political interactions and other aspects of power in nonhuman primates (reviewed in Chapais 1991). Humans use visual and vocal threats, engage in physical aggression and contest access to resources as individuals or members of small coalitions, and compete over social partners. However, even at the interpersonal level, humans have sources of power unavailable to other primates. An obvious example is the use of weapons in intraspecific aggression, a uniquely human source of aggressive coercion that distinguishes power relations in our species from those of other nonhuman primates (Gat this volume). But another obvious example, one that highlights human cognitive uniqueness and is probably more important than physical coercion, is the use of political rhetoric (Kurtz 2001).

Aggression-based dominance hierarchies like those in nonhuman primates, whether they result solely from intrinsic dominance or also involve derived dominance, are uncommon in humans, at least among adults (Chapais 1991). Nor should the association of outstanding hunting skill with high reproductive success, common among human foragers, be equated with the positive relationship between dominance rank and male reproductive success in many nonhuman primates.

Hunting skill may provide leverage, but without conferring coercive power or any formal political authority. Thus, Ache men known to be good hunters can get preferential treatment from others in their bands regarding decisions about group movements, but because they can threaten to transfer to other bands and thereby depriving others of their hunting returns, not because they have any formal political power (Hill and Kaplan 1988). Leaders in hunter-gatherer societies and other small-scale human groups are "episodic" or "weak" (Kurtz 2001) and must lead by example and by persuading others (e.g., Tsimane village chiefs: Gurven and Winking 2008). Regardless of the extent to which power ultimately rests on coercive ability, humans have many more sources of dependence-based power than do any nonhuman primates, thus many more ways to control and manipulate the needs of others (Chapais 1991) and many ways to create needs. As Chapais (*ibid.*: 216) notes, *"the most powerful individuals (the ones most able to control the behavior of others) are those with the greatest number of individuals depending on them for the satisfaction of needs."* The obverse of manifold sources of dependence-based power is the existence of far more sources of leverage than are available to any nonhumans, such as possession of specialized skills or knowledge and even influence over others' prestige and self-esteem.

Coalitions in nonhuman primates are often conservative or polarizing, in that they accentuate dyadic power asymmetries (Chapais 1992, 2001; Preuschoft and van Schaik 2000). When female macaques follow the "support the high born" rule by forming coalitions with other females whose mothers outrank their opponents, they help the females they support to attain or maintain dominance over the opponents; given that females intervene mostly against targets that they also outrank, such support reinforces the existing dominance hierarchy and is best regarded as mutualism (Chapais 1992, 2001). Likewise, male chimpanzee coalitions at Ngogo mostly include partners who both outrank their targets (Watts and de Vries 2009). But leveling coalitions also occur in some primates. In these, either coalition partners compensate for dyadic power asymmetries by collaborating against higher-ranking opponents, or high-ranking individuals intervene in conflicts on behalf of subordinates against opponents that they outrank and thereby suppress within-group competition, (e.g., Barbary macaques, *Macaca sylvana*: Preuschoft et al. 1998; Preuschoft and Paul 2000; yellow baboons: Noë 1992; Noë and Hammerstein 1994; chimpanzees: Nishida and Hosaka 1996). At an extreme, individuals can reverse dyadic power asymmetries by forming revolutionary alliances (Chapais 1992); in these, partners collaborate to reverse dominance ranks with targets to which they are subordinate (e.g., male chimpanzees: de Waal 1982; Nishida and Hosaka 1996; female chacma baboons: Engh et al. 2006).

But leveling coalitions in cercopithecines and among male chimpanzees co-occur with linear dominance hierarchies and, although they may limit the ability of high-ranking individuals to monopolize resources (e.g., they may reduce reproductive skew among males), they do not lead to the "egalitarian" politics of human hunter-gatherer societies known from the ethnographic record (Boehm 1997). Recognition of individual merit and accordance of limited authority to certain individuals occurs alongside tracking of people's behavior and linguistic communication about whether this stays within bounds acceptable to culturally variable moral communities. Humans also have broad means to constrain powerful individuals; some of these, like imposition of moral sanctions, are unavailable even to chimpanzees (Boehm 1997; Kurtz 2001). Likewise, a growing body of evidence indicates that people often engage in third-party punishment at some cost to themselves, although considerable cross-cultural variation in willingness to do so exists (Henrich et al. 2006; Gurven and Winking 2008) and people's behavior in economic games may not match that in the real world (Gurven and Winking 2008, Plourde this volume).

In experiments, captive chimpanzees retaliate against others who have stolen food from them (Jensen et al. 2007a) and respond negatively to situations in which they have received unfairly small rewards (Brosnan et al. 2005). De Waal and Luttrell (1988) documented negative reciprocity (or "bidirectionality") in agonistic interventions against others among the Arnhem Zoo chimpanzees – that is, individuals often supported others against third parties who in turn often intervened against them; male chimpanzees at Ngogo show similar bidirectionality (Watts unpubl. data). De Waal and Luttrell (1988) labeled this bidirectionality as a "revenge system," and referred to aggression by food possessors toward others who were trying to obtain food from them in a food sharing experiment, but were reluctant to share when they possessed the food, as "moralistic aggression." De Waal (1989) subsequently argued that revenge and moralistic aggression "introduce powerful sanctions to a social system." However, such behavior appears to be entirely egocentric. Chimpanzees seem to be unconcerned with whether others have been treated fairly (Jensen et al. 2007b), and although retaliation could be considered a form of punishment, currently available evidence indicates that they do not engage in costly third party punishment (Jensen et al. 2007a). Such findings seriously question Flack and de Waal's (2004) categorization of political systems in which, for example, they consider chimpanzees as possible representatives of the "equal outcome" category (Table 5.1), in which policing by powerful individuals leads to an equal distribution of resources; how this could occur when powerful individuals are unconcerned with whether others achieve fair outcomes and are perhaps incapable of such concerns (but see de Waal 2008, for a contrary view), is unclear.

Absent the ability to form moral communities, chimpanzees could also not control the cost and efficiency of punishment by delegating authority to enforce social norms to individuals and institutionalized subgroups at different levels of a hierarchical society: an ability essential for the formation of large scale, hierarchical societies in the first place (Dubreuil 2008). As Boehm (1997) argued, politics among male chimpanzees remain individualistic, even when males try to achieve social goals as members of small coalitions. Although Flack and de Waal (2004) note that "equal opportunity" political systems (Table 5.1) occur only in humans – and, implicitly, only humans construct institutional roles of the sort that can characterize "equal outcome systems" (Table 5.1) – they present their classification of political systems as if it is a continuum, when in fact it incorporates these and other qualitative disjuncture between humans and nonhuman primates. More notably, not even chimpanzees can engage in symbolically mediated leverage, competition, and manipulation, and their lack of language stringently limits their ability to use information as a source of power. In his comparison of power in humans and nonhuman primates, Chapais (1991) listed many similarities; virtually all of the contrasts involve semiosis or otherwise derive from cognitive differences. Aggressive coercion in nonhuman primates never involves moral justification or attribution of blame to victims, for example, nor does "normative power," based on symbolically based group norms, occur in nonhumans. Primatologists who write about politics in apes and monkeys need to recognize the crucially important semiotic and ideational dimensions of human politics – struggles over meaning and ways in which language, including political rhetoric, mediates these – and to pay attention to corresponding human cognitive uniqueness. We should acknowledge that – to paraphrase Plotkin's (2003) summary comment on calling socially learned behavioral traditions in chimpanzees "culture" – it is not politics as we know it in humans, but it is politics of a kind.

References

Adams RN (1977) Power in human societies: a synthesis. In: Fogelson RD, Adams RN (eds) The anthropology of power: ethnographic studies from Asia, Oceania and the New World. Academic, New York, NY, pp 387–410

Alberts SC, Altmann J (1995) Balancing costs and opportunities: dispersal in male baboons. Am Nat 145:279–306

Alberts SC, Buchan JC, Altmann J (2006) Sexual selection in wild baboons: from mating opportunities to paternity success. Anim Behav 72:1177–1196

Altmann J, Alberts SC (2005) Growth rates in a wild primate population: ecological influences and maternal effects. Behav Ecol Sociobiol 57:490–501

Archer J (2006) Testosterone and human aggression: an evaluation of the challenge hypothesis. Neurosci Behav Rev 30:319–345

Bacharach SB, Lawler EJ (1980) Power and politics in organizations. Jossey-Bass, San Francisco, CA

Beehner JC, Bergman TJ (2009) The 'bleeding heart baboon': does chest patch color signal male quality in geladas? Am J Phys Anthropol 138(Suppl. 48):37

Beehner JC, Bergman TJ, Cheney DL, Seyfarth RM, Whitten PL (2006) Testosterone predicts future dominance rank and mating activity among male chacma baboons. Behav Ecol Sociobiol 59:469–479

Bergman TJ, Beehner JC, Cheney DL, Seyfarth RM (2003) Hierarchical classification by rank and kinship in baboons. Science 302:1234–1236

Bergman TJ, Beehner JC, Cheney DL, Seyfarth RM, Whitten PL (2006) Interactions in male baboons: the importance of both males' testosterone. Behav Ecol Sociobiol 59:480–489

Bergman TJ, Phillips-Conroy JE, Jolly CJ (2008) Behavioral variation and reproductive success of male baboons (*Papio anubis* x *Papio hamadryas*) in a hybrid social group. Am J Primatol 70:136–147

Berman CM, Ionica CS, Li J (2004) Dominance style among *Macaca thibetana* on Mt. Huangshan, China. Int J Primatol 25:1283–1312

Bernstein IS (1976) Dominance, aggression and reproduction in primate societies. J Theor Biol 60:459–472

Boehm C (1997) Egalitarian behavior and the evolution of political intelligence. In: Byrne R, Whiten A (eds) Machiavellian intelligence II: extensions and evaluations. Cambridge University Press, Cambridge, pp 341–364

Boesch C, Kohou G, Néné H, Vigilant L (2006) Male competition and paternity in wild chimpanzeez of the Taï forest. Am J Phys Anthropol 130:103–115

Boinski S, Sughrue K, Selvaggi L, Quatrone R, Henry M, Cropp S (2002) An expanded test of the ecological model of primate social evolution: competitive regimes and female bonding in three species of squirrel monkeys (*Saimiri oerstedii*, *S. boliviensis*, and *S. sciureus*). Behaviour 139:227–261

Brosnan SF, Schiff HC, de Waal FBM (2005) Tolerance for inequity may increase with social closeness in chimpanzees. Proc Roy Soc Lond B 272:253–258

Chapais B (1991) Primates and the origins of aggression, politics, and power among humans. In: Loy JD, Peters CB (eds) Understanding behavior: what primate studies tell us about human behavior. Oxford University Press, Oxford, pp 190–218

Chapais B (1992) The role of alliances in social inheritance of rank among female primates. In: Harcourt AH, de Waal FBM (eds) Coalitions and alliances in humans and other animals. Oxford University Press, Oxford, pp 29–59

Chapais B (2001) Primate nepotism: what is the explanatory value of kin selection? Int J Primatol 22:203–229

Cheney DL (1992) Intragroup cohesion and intergroup hostility: the relation between grooming distribution and intergroup competition among female primates. Behav Ecol 3:334–345

Cheney DL, Seyfarth RM (1990) How monkeys see the world: inside the mind of another species. University of Chicago Press, Chicago, IL

Cheney DL, Seyfarth RM (2007) Baboon metaphysics: the evolution of a social mind. University of Chicago Press, Chicago, IL

Clutton-Brock TH, Parker GA (1995) Punishment in animal societies. Nature 373:209–216

Cords M (2000) The agonistic and affiliative relationships of adult females in a blue monkey group. In: Whitehead PF, Jolly CJ (eds) Old world monkeys. Cambridge University Press, Cambridge, pp 453–479

Côté S (2000) Dominance hierarchies in female mountain goats: stability, aggressiveness and determinants of rank. Behaviour 137:1541–1566

Das M (2000) Conflict management via third parties: post-conflict affiliation of the aggressor. In: Aureli F, de Waal FBM (eds) Natural conflict resolution. University of California Press, Berkeley, pp 263–280

Datta SB (1983a) Relative power and acquisition of rank. In: Hinde RA (ed) Primate social relationships: an integrated approach. Blackwell, London, pp 93–102

Datta SB (1983b) Relative power and the maintenance of dominance. In: Hinde RA (ed) Primate social relationships: an integrated approach. Blackwell, London, pp 103–112

de Waal FBM (1982) Chimpanzee politics: power and sex among apes. Johns Hopkins University Press, Baltimore, MD

de Waal FBM (1986) The integration of dominance and social bonding in primates. Q Rev Biol 61:459–479

de Waal FBM (1989) Dominance 'style' and primate social organization. In: Standen V, Foley RA (eds) Comparative socioecology: the behavioral ecology of humans and other mammals. Blackwell, Oxford, pp 243–263

de Waal FBM (2000) The first kiss: foundations of conflict resolution in animals. In: Aureli F, de Waal FBM (eds) Natural conflict resolution. University of California Press, Berkeley, pp 15–33

de Waal FBM (2008) Putting the altruism back into altruism: the evolution of empathy. Annu Rev Psychol 59:279–300

de Waal FBM, Luttrell L (1988) Mechanisms of social reciprocity in three primate species: symmetrical relationship characteristics or cognition? Ethol Sociobiol 9:101–118

Di Fiore A (2003) Molecular genetic approaches to the study of primate behavior, social organization, and reproduction. Yrbk Phys Anthropol 46:62–99

Dubreuil B (2008) Strong reciprocity and the emergence of large-scale societies. Philos Soc Sci 38:192–210

Duffy KG, Wrangham RW, Silk JB (2007) Male chimpanzees exchange political support for mating opportunities. Curr Biol 17:R586–R587

Dunbar RIM (1988) Primate social systems. Cornell University Press, Ithaca, NY

Engelhardt A, Heistermann M, Hodges JK, Nürnberg P, Niemitz C (2006) Determinants of male reproductive success in wild long-tailed macaques (*Macaca fascicularis*) – male monopolization, female mate choice or post-copulatory mechanisms? Behav Ecol Sociobiol 59:740–752

Engh AL, Beehner JC, Bergman TJ, Whitten PL, Hoffmeier RR, Seyfarth RM, Cheney DL (2006) Female hierarchy instability, male immigration and infanticide increase glucocorticoid levels in female chacma baboons. Anim Behav 71:1227–1237

Flack JC, de Waal FBM (2004) Dominance style, social power, and conflict management. In: Thierry B, Singh M, Kaumanns W (eds) Macaque societies: a model for the study of social organization. Cambridge University Press, Cambridge, pp 157–181

Fukuyama F (1998) Women and the evolution of world politics. Foreign Aff 77:24–40

Gurven M, Winking J (2008) Collective action in action: prosocial behavior in and out of the laboratory. Am Anthropol 110:179–190

Harcourt AH (1987) Dominance and fertility among female primates. J Zool 213:471–487

Harcourt AH (1992) Coalitions and alliances: are primates more complex than non-primates? In: Harcourt AH, de Waal FBM (eds) Coalitions and alliances in humans and other animals. Oxford University Press, Oxford, pp 445–471

Hare B, Call J, Tomasello M (2006) Chimpanzees deceive a human competitor by hiding. Cognition 101:495–514

Heinsohn R, Packer C (1995) Complex cooperative strategies in group-territorial African lions. Science 269:1260–1262

Hemelrijk CK (1996) Dominance interactions, spatial dynamics and emergent reciprocity in a virtual world. In: Maes P, Mataric MJ, Meyer J-A, Pollack J, Wilson SW (eds) Proceedings of the fourth international conference on simulation of adaptive behavior, vol 4. MIT, Cambridge, MA, pp 545–552

Hemelrijk CK (1999a) Effects of cohesiveness on intersexual dominance relationships and spatial structure among group-living virtual entities. In: Floreano D, Nicoud J-D, Mondada F (eds) Advances in artificial life. Fifth european conference on artificial life, vol 1674. Springer Verlag, Berlin, pp 524–534

Hemelrijk CK (1999b) An individual oriented model of the emergence of despotic and egalitarian societies. Proc Roy Soc Lond B 266:361–369

Hemelrijk CK (2000a) Sexual attraction and inter-sexual dominance. In: Moss S, Davidsson P (eds) Multi agent based simulation. Second international workshop on multi-agent based simulation. Springer, Berlin, pp 167–180

Hemelrijk CK (2000b) Towards the integration of social dominance and spatial structure. Anim Behav 59:1035–1048

Hemelrijk CK (2002) Despotic societies, sexual attraction, and the emergence of male 'tolerance': an agent based model. Behaviour 139:729–747

Hemelrijk CK, Wantia J, Dätwyler M (2003) Female co-dominance in a virtual world: ecological, cognitive, social, and sexual causes. Behaviour 140:1247–1273

Henrich J, McElreath R, Barr A, Ensminger J, Barrett C, Bolyanatz A, Cardenas JC, Gurven M, Gwako E, Henrich N, Lesorogol C, Marlowe F, Tracer D, Ziker J (2006) Costly punishment across human societies. Science 312:1767–1770

Hill K, Kaplan H (1988) Tradeoffs in male and female reproductive strategies among the Ache. Part 1. In: Betzig L, Borgerhoff Mulder M, Turke P (eds) Human reproductive behavior: a Darwinian perspective. Cambridge University Press, Cambridge, pp 277–289

Hinde RA (1976) Interactions, relationships, and social structure. Man 11:1–17

Hrdy SB (1977) The langurs of Abu: female and male strategies of reproduction. Harvard University Press, Cambridge

Isbell LA (1991) Contest and scramble competition: patterns of female aggression and ranging behavior among female primates. Behav Ecol 2:143–155

Jensen K, Call J, Tomasello M (2007a) Chimpanzees are vengeful, but not spiteful. Proc Natl Acad Sci USA 104:13046–13050

Jensen K, Call J, Tomasello M (2007b) Chimpanzees are rational maximizers in an ultimatum game. Science 318:107–109

Kano T (1992) The last ape. Pygmy chimpanzee behavior and ecology. Stanford University Press, Palo Alto, CA

Kawai M (1958) On the rank system in a natural group of Japanese monkeys, I and II. Primates 1:111–148

Koenig A (2002) Competition for resources and its behavioral consequences among female primates. Int J Primatol 23:759–783

Koenig A, Beise J, Chalise MK, Ganzhorn JU (1998) When females should contest for food – testing hypotheses about resource density, distribution, size, and quality with hanuman langurs (*Presbytis entellus*). Behav Ecol Sociobiol 42:225–237

Kurtz DV (2001) Political anthropology: paradigms and power. Westview Press, Boulder, CO

Langergraber KE, Mitani JC, Vigilant L (2007) The limited impact of kinship on cooperation in wild chimpanzees. Proc Natl Acad Sci USA 104:7786–7790

Leighton DR (1987) Gibbons: territoriality and monogamy. In: Smuts BB, Cheney DL, Seyfarth RM, Wrangham RW, Struhsaker TT (eds) Primate societies. University of Chicago Press, Chicago, IL, pp 135–145

Lewellen TC (2003) Political anthropology: an introduction. Praeger, Westport, CT

Lewis RJ (2002) Beyond dominance: the importance of leverage. Q Rev Biol 77:149–164

Lu A, Koenig A, Borries C (2008) Formal submission, tolerance, and socioecological models: a test with female hanuman langurs. Anim Behav 76:415–428

Maner JK, Miller SL, Schmidt NB, Eckel LA (2008) Submitting to defeat: social anxiety, dominance threat, and decrements in testosterone. Psychol Sci 19:764–768

Marler P (1976) Social organization, communication, and graded signals: the chimpanzee and the gorilla. In: Bateson PPG, Hinde RA (eds) Growing points in ethology. Cambridge University Press, Cambridge, pp 239–280

Matsumara S (1999) The evolution of 'egalitarian' and 'despotic' social systems among macaques. Primates 40:23–31

Matsumara S, Kobayashi T (1998) A game theory model for dominance relations among group living animals. Behav Ecol Sociobiol 42:77–84

Mehta P, Josephs R (2006) Testosterone change after losing predicts the decision to compete again. Horm Behav 50:684–692

Melis AP, Hare B, Tomasello M (2006) Chimpanzees recruit the best collaborators. Science 311:1297–1300

Ménard N (2004) Do ecological factors explain variation in social organization? In: Thierry B, Singh M, Kaumanns W (eds) Macaque societies: a model for the study of social organization. Cambridge University Press, Cambridge, pp 237–262

Mendres KA, de Waal FBM (2000) Capuchins do cooperate: the advantages of an intuitive task. Anim Behav 60:523–529

Mitani JC (2006) Demographic influences on the behavior of chimpanzees. Primates 47:6–13

Mitani JC, Watts DP (2001) Why do chimpanzees hunt and share meat? Anim Behav 61:915–924

Mitchell C, Boinski S, van Schaik CP (1991) Competitive regimes and female bonding in two species of squirrel monkeys (*Saimiri oerstedi* and *S. sciureus*). Behav Ecol Sociobiol 28:55–60

Nishida T (1983) Alpha status and agonistic alliance in wild chimpanzees (*Pan troglodytes schweinfurthii*). Primates 24:318–336

Nishida T, Hosaka K (1996) Coalition strategies among adult male chimpanzees of the Mahale mountains, Tanzania. In: McGrew WC, Marchant LF, Nishida T (eds) Great ape societies. Cambridge University Press, Cambridge, pp 114–134

Noë R (1990) A veto game played by baboons: a challenge to the Prisoner's Dilemma as a paradigm for reciprocity and cooperation. Anim Behav 39:78–90

Noë R (1992) Alliance formation among male baboons: shopping for profitable partners. In: Harcourt AH, de Waal FBM (eds) Coalitions and alliances in humans and other animals. Oxford University Press, Oxford, pp 285–321

Noë R, Hammerstein P (1994) Biological markets: supply and demand determine the effect of partner choice in cooperation, mutualism, and mating. Behav Ecol Sociobiol 35:1–11

Packer C, Pusey AE (1979) Female aggression and male membership in troops of Japanese macaques and olive baboons. Folia Primatol 31:212–218

Packer C, Scheel D, Pusey AE (1990) Why lions form groups: food is not enough. Am Nat 136:1–19

Pereira ME (1988) Agonistic interactions of juvenile savanna baboons I. Fundamental features. Ethology 79:195–217

Pereira ME, Kappeler PM (1997) Divergent systems of agonistic behaviour in lemurid primates. Behaviour 134:225–274

Pereira ME, Kaufman R, Kappeler PM, Overdorff D (1990) Female dominance does not characterize all of the lemuridae. Folia Primatol 55:96–103

Perry S, Manson JH (2008) Manipulative monkeys: the capuchins of Lomas Barbudal. Harvard University Press, Cambridge

Perry S, Barrett HC, Manson JH (2004) White-faced capuchin monkeys show triadic awareness. Anim Behav 67:165–170

Plotkin H (2003) The imagined world made real: towards a natural science of culture. Rutgers University Press, New Brunswick, NJ

Preuschoft S, Paul A (2000) Dominance, egalitarianism, and stalemate: an experimental approach to male-male competition in Barbary macaques. In: Kappeler PM, van Schaik CP (eds) Primate males: causes and consequences of variation in group composition. Cambridge University Press, Cambridge, pp 205–216

Preuschoft S, van Schaik CP (2000) Dominance and communication: conflict management in various social settings. In: Aureli F, de Waal FBM (eds) Natural conflict resolution. University of California Press, Berkeley, pp 77–105

Preuschoft S, Paul A, Kuester J (1998) Dominance style of female and male Barbary macaques (*Macaca sylvanus*). Behaviour 135:731–755

Pruetz J (2009) The socioecology of adult female patas monkeys and vervets in Kenya. Pearson Prentice Hall, Upple Saddle River, NJ

Pusey AE (1990) Behavioural changes at adolescence in chimpanzees. Behaviour 115:203–246

Range F, Noë R (2004) Can simple rules account for the pattern of triadic interactions in juvenile and adult female sooty mangabeys? Anim Behav 69:445–452

Schino G, Tiddi B, Polizzi di Sorrentino E (2006) Simultaneous classification by rank and kinship in Japanese macaques. Anim Behav 71:1069–1074

Schubert G (1991) Primate politics. In: Schubert G, Masters RD (eds) Primate politics. Southern Illinois University Press, Carbondale, IL, pp 37–56

Setchell JM, Charpentier M, Wickings EJ (2005) Mate guarding and paternity in mandrills (*Mandrillus sphinx*): factors influencing alpha male monopoly. Anim Behav 70:1105–1120

Setchell JM, Smith T, Wickings EJ, Knapp LA (2008) Social correlates of testosterone and ornamentation in male mandrills. Horm Behav 54:365–372

Seyfarth RM (1977) A model of social grooming among adult female monkeys. J Theor Biol 65:671–698

Silk JB (1999) Male bonnet macaques use information about third-party rank relationships to recruit allies. Anim Behav 58:45–51

Silk JB (2002) Practice random acts of aggression and senseless acts of intimidation: the logic of status contests in social groups. Evol Anthropol 11:221–225

Silk JB, Alberts SC, Altmann J (2003) Social bonds of female baboons enhance infant survival. Science 302:1231–1234

Soltis J (2004) Mating systems. In: Thierry B, Singh M, Kaumanns W (eds) Macaque societies: a model for the study of social organization. Cambridge University Press, Cambridge, pp 135–151

Sterck EHM, Steenbeek R (1997) Female dominance relationships and food competition in the sympatric Thomas langur and long-tailed macaque. Behaviour 134:739–774

Sterck EHM, Watts DP, van Schaik CP (1997) The evolution of female social relationships in nonhuman primates. Behav Ecol Sociobiol 41:291–309

Swartz MJ, Turner VW, Tuden A (1966) Political anthropology. Aldine, Chicago, IL

Thierry B (2000) Covariation of conflict management patterns across macaque species. In: Aureli F, de Waal FBM (eds) Natural conflict resolution. University of California Press, Berkeley, pp 106–128

Thierry B (2007) Unity in diversity: lessons from macaque societies. Evol Anthropol 16:224–238

Thierry B, Iwaniukn AN, Pellis SM (2000) The influence of phylogeny on the social behaviour of macaques (Primates: Cercopithecidae, genus *Macaca*). Ethology 106:713–728

van Noordwijk MA, van Schaik CP (1985) Male migration and rank acquisition in wild long-tailed macaques (*Macaca fascicularis*). Anim Behav 33:849–861

van Noordwijk MA, van Schaik CP (1999) The effects of dominance rank and group size on female lifetime reproductive success in wild long-tailed macaques, *Macaca fascicularis*. Primates 40:105–130

van Noordwijk MA, van Schaik CP (2004) Sexual selection and the careers of primate males: paternity concentration, dominance-acquisition tactics and transfer decisions. In: Kappeler PM, van Schaik CP (eds) Sexual selection in primates: a comparative perspective. Cambridge University Press, Cambridge, pp 208–229

van Schaik CP (1989) The ecology of social relationships amongst female primates. In: Standen V, Foley RA (eds) Comparative socioecology: the behavioral ecology of humans and other mammals. Blackwell, Oxford, pp 195–218

van Schaik CP, van Noordwijk MA (1986) The hidden costs of sociality: intra-group variation in feeding strategies in Sumatran long-tailed macaques (*Macaca fascicularis*). Behaviour 99:296–315

Wakefield ML (2008) Grouping patterns and competition among female *Pan troglodytes schweinfurthii* at Ngogo, Kibale National Park, Uganda. Int J Primatol 29:907–929

Watts DP (1994) Agonistic relationships between female mountain gorillas. Behav Ecol Sociobiol 34:347–358

Watts DP (2002) Reciprocity and interchange in the social relationships of wild male chimpanzees. Behaviour 139:343–370

Watts DP, de Vries H (2009) Linearity and strength of male dominance hierarchies at Ngogo, Kibale National Park, Uganda. Am J Phys Anthropol 138(Suppl. 48):267

Watts DP, Colmenares F, Arnold K (2000) Redirection, consolation, and male policing: how targets of aggression interact with bystanders. In: Aureli F, de Waal FBM (eds) Natural conflict resolution. University of California Press, Berkeley, pp 281–301

Weber M ([1947] 1965) The theory of social and economic organization. In: Parsons T (ed), (trans: Henderson AM, Parsons T). The Free Press, New York, NY

Westbrook R (1991) John Dewey and American democracy. Cornell University Press, Ithaca, NY

White FJ (1996) Comparative socio-ecology of *Pan paniscus*. In: McGrew WC, Marchant LA, Nishida T (eds) Great ape societies. Cambridge University Press, Cambridge, pp 29–41

Chapter 6
Human Power and Prestige Systems

Aimée M. Plourde

Abstract Prestige is a major source of social power in human societies, and one that is often much more important than physical dominance. This observation, while obvious, gains significance in light of the fact that prestige does not exist in other primates and so, must be a derived property of our species. Understanding prestige systems will, therefore, further our understanding of social power and inequality in human societies, and contribute to anthropological and archeological theories of the evolution of social structures and institutions. Here, an evolutionary theory for the emergence of prestige in the human lineage is discussed, and a model to explain the reason why material items may have begun to be used in competition for prestige presented, in which I propose that they initially functioned as costly signals of an individual's skill and expertise, aimed at learning individuals. It is then hypothesized how and why prestige became involved in the emergence of socio-political and economic ranking through the increasing importance of leadership and collective action, with the result that the signal content of prestige goods became linked not simply to prestige but also to coercion and dominance through the possession of wealth, elevated social class, and positions of authority.

6.1 Introduction

Like many other primates, humans live in social groups with complex status hierarchies, are deeply motivated to strive for higher status, and spend a great deal of time and energy in doing so. But in this, we see a significant difference between humans and other primates; while social power in nonhuman species is generally based on dominance, in human societies both dominance and prestige are important

A.M. Plourde
AHRC Centre for the Evolution of Cultural Diversity, Institute of Archaeology, University College London, London, England
e-mail: aimee.plourde@ucl.ac.uk

P.M. Kappeler and J.B. Silk (eds.), *Mind the Gap*,
DOI 10.1007/978-3-642-02725-3_6, © Springer-Verlag Berlin Heidelberg 2010

determinants of social power. Prestige, from an evolutionary perspective, is a funny thing; it is a source of social power that is freely conferred on an individual by others as a result of the respect and admiration in which they hold that person.

The quest for prestige has led humans to do many things that, at first pass, seem quite puzzling. For instance, people spend enormous amounts of money on items, from Gucci handbags to Mont Blanc pens, which, from a functional stance, are hardly (if at all) better than the ones available at a fraction of the price. Some prestige items, including some designer sunglasses and jewel-encrusted wristwatches are, indeed, known to perform worse than economy models. Alternatively, many prestige items may be of higher quality or functionality than nonprestige equivalents, but it is often unclear whether the difference is "worth" the extra cost; for example, a Ferrari may, in fact, outperform an economy car, but its owner is unlikely to be able to take advantage of that fact in the course of daily life. Instead, displaying them to others appears to be at least as important a reason for the purchase of these kinds of objects as their purported function. And the importance of prestige display is not limited to Western, or contemporary, societies. The Potlatch ceremony performed in Native American groups located along the Pacific Northwest Coast, in which not only the display but the destruction of both prestige and utilitarian items occurred, is a classic case demonstrating the importance of prestige goods in social dynamics, while some "Big Man" societies of Highland New Guinea have actually been termed "prestige economies" because of the critical role played by prestige goods in socioeconomic dynamics there. Further, archeological data from cultures as diverse as the Han empire of China, Phaeronic Egypt, and the Inca empire of South America suggest that the display of luxury goods was as important, if not more so, in these societies as in modern ones.

This chapter presents an evolutionary model to explain how such seemingly wasteful and perplexing behavior may have emerged in humans. I will begin by exploring the unique aspects of the operation of prestige in human power systems, particularly in comparison with the power systems in other primate species, and theories for the emergence of prestige in the human lineage. I will then present an evolutionary model to explain why material items may have begun to be used in competition for prestige, in which I propose that they initially functioned as honest advertisements of an individual's skill and expertise, aimed at learning or unsuccessful group members. Following from this model, I hypothesize how and why prestige became involved in the emergence of sociopolitical and economic ranking through the increasing importance of leadership and collective action, with the result that the signal content of prestige goods became linked not simply to prestige but also to coercion and dominance through the possession of wealth, elevated social class, and positions of authority.

6.2 The Basis of Social Power in Primates

Many nonhuman primate species form societies characterized by social ranking, where an individual's place within the dominance hierarchy is determined by their ability to defeat others in agonistic conflicts (Ellis 1995; Henrich and Gil-White

2001; Watts, this volume). Size, strength, kin relations, and political alliances all play parts in how dominance relationships are determined in nonhuman primate groups.

Humans' status hierarchies differ from those of other primates in two important ways. First, the repertoire of coercive behaviors is greatly expanded. In addition to personal qualities (e.g., strength, fighting ability), social qualities (e.g., affiliation with kin, political alliances), humans can also coerce others through economic strength, and through the possession of an office of authority. Individuals who control resources or hold offices of authority can coerce others by offering positive incentives for preferred behavior or threatening negative consequences for deviating from preferred courses of action. This expansion of the ways in which an individual can dominate others via coercion is the product of the rise of complex social organization and institutions, which emerged over the course of the last 10,000 years and is a relatively recent addition to human power relations. Perhaps an even more striking difference between humans and other primates, and other social species more generally, is that humans also possess a noncoercive means to increase social status that is based on prestige.

6.3 Prestige: A Unique Source of Social Power

"Prestige," as used in everyday speech generally refers to having the respect of others, even deference from them, which is freely conferred, rather than being compelled through violence, threat, or coercion. Prestigious people are honored by their peers and wield influence over them. The source of this respect and influence is the possession of skill and/or knowledge in some domain of activity, particularly one that is valued (Henrich and Gil-White 2001, p 167). An important ramification of this is that a person may be highly skilled in one domain, and thus have a large amount of prestige in that arena, but have little prestige in other areas. The number of independent, simultaneous prestige hierarchies of this kind could, in theory, be as numerous as the number of recognized skills (Henrich and Gil-White 2001, p 170), which in contemporary societies is quite a large number, but in more traditional societies, including most societies in the past, was probably much more restricted.

Prestige is likely to be even more important than dominance in human status hierarchies (Barkow 1975, p 554). Ethnographic and historical evidence suggests that prestige systems constitute a "human universal" (sensu Brown 1991), and there are differences in the extent of prestige conferred on individuals within groups of all sizes. In contrast, skews in power based on dominance do not exist in all societies. Given its ubiquity, any complete explanation of prestige will necessarily include an evolutionary account for its appearance in humans, addressing the questions of what adaptive function the psychologies and behaviors associated with prestige dynamics could serve, and what in the evolutionary past of humans led to strong selection for prestige that was not present in all other species, including those most closely related to us.

6.4 The Evolutionary Origins of Prestige

Henrich and Gil-White (2001) outlined a theory for the origins of prestige in human social dynamics as a result of changes in the nature of cultural transmission. They suggest that selection favored new strategies to augment social learning, as culturally transmitted knowledge became increasingly complex and important for success. Young or migrant individuals, who have a need to learn new skills, would benefit more from learning from successful individuals than from other possible types of models. Selection should, therefore, favor psychological mechanisms that (1) assist the learning individual to determine which members of her group are successful, (2) generate admiration for and attraction to successful individuals as motivation to maintain proximity to them and to copy their behaviors, and (3) cause the learning individual to treat successful individuals with respect and deference, in order to gain proximity. Ethnographic observations suggest that prestigious individuals may receive other kinds of positive benefits in social relationships, such as leniency after a transgression of group norms or a failure to reciprocate in dyadic relationships, and more support for self and family members following injury (see Bateson 1958, p 91, Hawkes 1991; Henrich and Gil-White 2001, pp 182–183). Henrich and Gil-White further suggest that paying attention to the distribution of respect and deference by group members could provide an efficient means for determining which group members are most successful, reducing the amount of time spent by learners in personal observation of others' success rates and reducing error.

6.4.1 Competition for Prestige

Henrich and Gil-White's theory provides an explanation for why learning individuals and successful models would benefit from prestige dynamics. The model assumes that there is some cost associated with gaining access to successful models; that is, learners literally "pay deference" to skilled models. Such deference is advantageous to learners as long as the net benefits that they gain from what they learn provides a level of success that is higher than they would have gained from learning from parents or other close relatives without cost. As Shennan and Steele (1999) have observed in a survey of ethnographic data, when the transmission of skills and knowledge is strictly vertical (from parents or other close relatives), as might be the case when variance in skill levels is low, the opportunity to acquire prestige benefits from learning individuals should be limited.

Successful individuals should welcome the social perks and favors that they obtain from learning individuals, provided that these benefits outweigh the costs associated with the presence of admirers. Moreover, if the benefits derived from prestige are sufficiently high, then, over time, psychologies that motivate individuals to desire prestige and to strive for it should be favored. These motivations may be built upon the more ancient motivational systems that encourage individuals to strive for high dominance status, and competition for prestige should be expected to emerge as a result.

How then might a successful individual act, given that prestige is freely conferred by others, to attract learning individuals and the social benefits that they might confer? Henrich and Gil-White suggest that when competition for prestige is high, successful individuals should modify their behavior to attract and retain the admiration and benefits from learning individuals, such as being less arrogant and more approachable (Henrich and Gil-White 2001, pp 171, 178–179). Another strategy would be to signal her degree of skill in order to convince prospective admirers that her skills are better than those of other potential models. The performance of the skill in question is, perhaps, the most obvious way to accomplish this, but I have argued elsewhere (Plourde 2006, 2008, in press) that the production of a signal with physical items would work particularly well because goods are more durable than actions.

The idea that material goods (as well as behavioral displays) could function as signals of personal qualities, including social standing, is an old one; Veblen ((1899) 1994) put forward the idea that people used "conspicuous consumption" and display of costly luxuries to advertise their degree of wealth over a century ago. Signaling theory has a substantial literature in economics and evolutionary biology. Applications of signaling theory to human behavior from evolutionary psychology and human behavioral ecology were initially focused on sexual selection (for example, see Miller 2000) but interest in other kinds of signals has been increasing. For instance, several scholars have proposed that contributions to public goods may be costly signals made in an attempt to gain prestige and related personal benefits. The contributions to public goods may include hunting large game animals for public consumption (Bliege Bird et al. 2002; Smith et al. 2003) and the punishment of transgressors for violations of group norms (Gintis et al. 2001). Archeologists have also begun to draw on costly signaling theory in relation to status striving by political elites (Neiman 1997; Boone 2000).

6.4.2 Prestige Goods as Costly Signals of Skill and Expertise

Physical objects that accurately reflect some aspect of an individual's skills or knowledge could act as advertisements in prestige competition. These prestige goods could signal the kind of technical expertise involved in hunting and craft production, levels of knowledge about environmental information, or the quality of an individual's relationships with people outside of his/her local group. Of course, goods could also be used to signal success at a general level, providing a measure of the amount of "extra" time and labor that an individual has. This kind of signal would be of interest to many different kinds of audiences, including potential mates and rivals, as well as individuals looking for a skilled tutor. The specific reference to a skill is what distinguishes prestige goods from other kinds of signals of wealth and success that probably did (and do) exist; mates' and rivals' primary interest would be in the level of success held by the signaler, not the details of the skills involved in producing that success.

Signals produced by skilled individuals must be costly to produce in order to maintain their value as honest signals (Zahavi 1975; Zahavi and Zahavi 1997) and prevent signalers from deceiving their audience about their skills. This reflects a potential conflict of interest between the signalers and the recipients. Signalers always benefit from receiving deference and other social perquisites from admirers, regardless of their level of skill, whereas learners only benefit if they limit deference to skilled or knowledgeable tutors.

A formal model of this idea (Plourde 2008) supports the logic behind it and demonstrates that when (1) the cost of making the signaling is higher for unskilled than for skilled signalers, (2) the benefits gained by the signaler from deference outweigh the cost of producing the signal, and (3) the benefits to learners (signal responders) from increasing their skills outweigh the cost of deference, a signaling/responding strategy can both invade a nonsignaling equilibrium, and can resist invasion by a nonsignaling strategy. Although the model is an extreme simplification of the complexities that must be involved in prestige dynamics, it nonetheless suggests that this kind of behavioral strategy and the attendant psychologies supporting it provide a viable explanation for why prestige goods came into existence.

This account is partially supported by data from ethnographic and historical studies of traditional societies. These accounts suggest people who are considered to be highly skilled within their community are preferentially copied, have higher status, receive deference and privileges, and are excused from some social obligations (Henrich and Gil-White 2001, pp 180–187). However, in these accounts, it is often unclear how competence and prestige were determined by community members, or whether the display of prestige goods influenced prestige. It is not known whether any of the goods displayed actually reflected skills or knowledge that deferential admirers were interested in learning. In contrast, the use of material goods as signals of wealth and social class in contemporary stratified societies is ubiquitous (Clark 1986, p 82; Johnson and Earle 2000). In many cases, these displays do not provide evidence of expertise in specific domains that have particular importance for learners. This leaves us with the following puzzle: if prestige goods came into existence in order to be used in signaling skills and expertise, how is it that the most eye-catching examples of conspicuous displays in contemporary settings signal wealth and social class, qualities which may have little to do with specific skills or expertise? Answering this question requires us to reconsider why a signaling strategy might have evolved.

6.4.3 Selection for Signaling in Prestige Competition

The time frame during which the psychological mechanisms underlying prestige-based exchange of knowledge for deference came into existence is unknown.

However, it was likely to be part of the transition to fully modern *Homo sapiens*, because prestige seems to be a universal feature of contemporary human societies, and archeological evidence also attests to the increasing importance of cultural knowledge during this transitional period. If signaling via prestige goods existed as a strategy before this time, then the signal's content could not have been to signify wealth and social class, assuming that wealth and class distinctions did not yet exist. This implies that signaling of wealth and social status probably tapped into phylogenetically older signaling psychologies and behaviors. This brings forward the question of how and when the use of novel competitive strategies, including the costly production of signals, might have been favored. Previously, I have suggested two factors that could have changed the social context in which prestige dynamics were operating in ways which may have favored an expertise-signaling strategy: (1) increases in group size and (2) increases in the complexity and/or the number of skills relevant to success (Plourde 2006, in press).

Increasing the size of groups could change the dynamics of prestige to favor a signaling strategy because, all other things being equal, the extent of variation in skill and expertise should increase with the number of individuals in the group. From the perspective of learning individuals, in large groups, it should be more difficult to identify the most skilled individual within the pool of potential models. Thus, as group size increases, it would be more beneficial for skilled individuals to advertise their expertise. Further, with more skilled individuals in the group, the learning individual has more potential models from which to choose, thus increasing the competition between skilled individuals for followers.

Alternatively, if the complexity or number of skills and realms of expertise increase, so should the difficulty associated with mastering all the skills necessary for success, making it all the more important for followers to choose the most skilled models. This should cause followers to invest more in deference to secure access to the most skilled models. As the benefits derived from deference increase, it could be worthwhile for skilled individuals to incur more costs in competition with other potential models, and these costs may include the production of prestige goods.

It is worth noting at this point that these theories for the origins of prestige and for the development of prestige signaling do not rely on the presence of a sociopolitical hierarchy, competition for offices of leadership, or the existence of social inequality. However, the factors that promote increases in levels of competition for prestige could very well result from the same kinds of environmental and social factors that are often implicated in the emergence of social inequality: increases in population size, increasing group circumscription, and technological innovation.

6.5 Prestige, Signaling and the Origins of Inequality

If the same overarching forces drove both competition for prestige and the emergence of socio-politico-economic differentiation, this might explain why evidence for both of these often overlaps in the archeological record. An example that demonstrates this pattern comes from the Lake Titicaca Basin region in the South-Central Andean Highlands of South America, where the first unambiguous evidence of prestige goods (in fact, the earliest known beaten gold artifacts in the Americas to date) – beads made from nonlocally derived gold (Aldenderfer et al. 2008) – appear in the archeological record before other kinds of indicators of social ranking, including differences in house size, settlement sizes, or diet, intensified agricultural production, or the creation of civic/ceremonial architecture (for a review, see Stanish 2003). The relatively early appearance of prestige goods here suggests that the production of such items predated the existence of heritable rank and wealth differences, and could have been involved in their creation (Plourde 2006; Plourde and Stanish 2006). I hypothesize that prestige and competition for it, including the use of material goods as signals of skill and expertise, often played a critical role in "transegalitarian" societies (Owens and Hayden 1997; Hayden 2001) – namely those in which the development of social and economic institutions of inequality occurs – primarily as a means by which leaders were chosen.

6.5.1 Prestige Competition and Selection for Leadership Roles

Ethnographic observations of how leaders are chosen in contemporary "egalitarian" societies – ones without formalized or permanent rank differences – suggest that prestige plays an important part in who is selected, or accepted, by the group to act as a leader in collective activities, such as combat (Chacon 2004), intergroup alliance, and trade (Johnson and Earle 2000). Leaders generally come from among those who are recognized as having the most experience and/or skill relevant to the task at hand: in other words, they are respected, and in all likelihood have prestige in that particular domain of activity or knowledge. In such societies, a leader's authority is ephemeral and generally restricted in scope to the activity at hand. But in groups experiencing increasing selection for the ability to act collectively, due to increasing circumscription and competition with neighboring groups or some other combination of factors, the need for leadership would occur more frequently and would be more important to group success, thus selecting for more elaborate and permanent roles of authority.

Many ethnographic and historic examples support the idea that groups with the ability to organize themselves in efforts for the common good, such as military action and cooperative tasks for production, often do better than neighboring groups (Richerson and Boyd 1999). For instance, if intergroup conflict over resources or land increases in frequency and intensity, groups with an individual recognized to lead and organize group defense and raiding might be more successful than groups without such a leader. The same logic could apply to leading and organizing group labor for public goods projects, such as the construction and maintenance of irrigation systems, fish weirs, and the like. If efficient collective action becomes more relevant to a group's success, then the presence of a leader to coordinate such activity may become more common over time. In this way, competition for personal prestige, and signaling skill via the display of prestige goods, could become linked to competition for positions of authority.

6.5.2 Prestige Competition and Status Striving

Individuals who occupy positions of authority gain personal benefits and opportunities to augment the wellbeing of themselves and their families. For example, Johnson and Earle (2000, p 126) observe that headmen and Big Men are often polygynous, and thus likely have elevated reproductive success. Clark and Blake (1994) coined the term "aggrandizer" to refer to ambitious men who are described in ethnographic and historical accounts of tribal and chiefly societies. Aggrandizers desire prestige and use material goods to gain prestige and attract followers, and are thought to be largely responsible for the beginnings of social hierarchy.

Hayden's (1998) model of the evolution of prestige technology posits that the major changes in human societies that have led to the emergence and elaboration of sociopolitical ranking result from the actions of aggrandizing individuals who attempt to promote their personal self-interest through various means, including the use of prestige goods. Hayden posits that "aggrandizing personalities" exist in all human populations, but need only constitute a small percentage of a group in order to provoke change in social structure. He defines a person with this personality as ambitious, socially, politically and economically aggressive, acquisitive, risk-taking, as someone who manipulates other individuals in order to promote his or her own self-interest, and who often acts selfishly rather than in the interest of the community. He concludes that "*...aggrandizers have, in effect, an inner motor, an inner drive to increase their own standard of living and their own reproductive success*" (1998: 18–19).

The model of the function of prestige goods function that I have presented here contributes to our understanding of the evolution of aggrandizing behaviors. An evolutionary perspective suggests that all people will have an "inner motor" that motivates them to improve their well-being and their own reproductive success, and to enhance the welfare of their kin. I propose that the desire to attend to, admire, own, and display prestige goods exists in all individuals, and is not limited to a certain personality "type." Instead, as an individual's skill level changes over the course of her lifetime, so too should her interest in attending to and obtaining and displaying prestige goods. Which behavioral strategy a person uses at a given time could be considered state-dependent, as the decision to obtain and display prestige goods will depend on the level of skill and success currently possessed by the individual, and will fluctuate over the course of the individual's life as her competence changes or the availability of superior models changes.

The development of elevated social positions, and eventually of ranked social classes, would have had considerable impact on the expression of these psychological dispositions. Both competition for prestige and the desire to possess and display prestige goods will be augmented when the benefits derived from holding a leadership position increase. More importantly, once social ranking in wealth and status classes is established, this would constitute a new kind of information that would be beneficial to convey to potential mates, allies, and competitors. If prestige goods are used as signals to relay status information, this could intensify the desire to own and display prestige goods and influence leaders' interactions with followers. For example, leaders might distribute prestige goods to their followers. This would be an effective strategy because it would take advantage of the psychology already in place that encourages ownership of prestige items as a means of increasing prestige and deriving the associated benefits, and as a form of imitation of the behaviors of successful individuals. Aggrandizing individuals use the display and distribution of prestige items to group members to tap into these deep-seated desires for their own benefit, and this constitutes one of the mechanisms by which hierarchical power relations within the group develop. In fact, it would seem likely that prestige competition could create a positive feedback system in which, as the family and allies of would-be leaders contribute to the production of prestige goods, the goods themselves would inevitably come to embody status. This sort of Veblenian process has been modeled by Boone (1992, 1998, 2000); Boone and Kessler 1999).

6.5.3 Signaling and Group Competition

Another significant transformation of human power relations is due to the increasing impact of interaction between politically defined and hierarchically organized groups. The competitive interaction between communities that would favor the existence of collective action and leadership, and the benefits to be gained from occupying a leadership role for an individual and his/her kin and corporate group, could lead to the formation of political factions within communities and hierarchical power relations between communities and their leaders. As politically and economically stratified societies emerged through processes of conflict and alliance, the quality of group strength would become much more important than in egalitarian societies. In terms of competition for prestige, this may also be viewed to some degree as a result of leaders' ability to control and direct the labor of others. The amount of labor possessed by a group and the social power embodied in a leader's ability to direct that labor constitute an emergent quality that describes both a leader and his group. As a consequence, the signal content of prestige goods was further expanded to include this aspect of social power. Such signals would function in competition between would-be leaders within a political group as well as between groups.

This idea is not a new one; archeologists have often argued that prestige goods act as signals of group and leader strength in models of the development of factional and peer polity competition. Several scholars have drawn on evolutionary frameworks and methods to examine the costs and benefits of prestige goods and other signals of competitive strength in terms of energy and ultimately, reproductive and cultural fitness. For instance, Neiman (1997) proposed that elaborate carved stone monuments built by members of the Mayan ruling elite functioned as costly signals of ability to win political contests (1997: 270). Success in such contests, he argues, would depend on a variety of factors, including "*physical size, physiological condition, fighting skill, and psychological cleverness, which contribute in varying proportions to an individual's ability to build and maintain coalitions comprised of kin and non-kin.*" Individuals who win political contests increase their reproductive success by gaining greater access to mates and resources. In this way, Neiman links social power as a currency to the currency used in all other evolutionary analyzes of animal behavior. This model is one of selection at the level of the individual but it may also draw on kin selection to the extent that the performance of leaders in political competition relies heavily on the actions of their immediate family and kin networks (Neiman 1997, p 271). This logic can be extended to non-kin corporate groups as well, and thus the problem could also be approached from a group selection perspective.

Pletka (2001) adopts an explicitly group-selection approach in examining the possible function of Neolithic and Bronze Age Danish earthwork monuments (mounded graves and enclosures) as costly signals of a group's ability to defend its rights to land and resources. Earthen monuments, like carved stelae, are just one of many possible ways to signal defensive capability, and I would argue that Pletka's model of honest signaling of group strength via the construction of earthen monuments could be expanded and adapted to apply to all such signals. Like Neiman and Pletka, Boone (2000) uses costly signaling theory to answer the question of why conspicuous displays of wealth expenditure are universally associated with social inequality. His model addresses all material forms of display, including elaborate monumental architecture, exotic and nonfunctional objects, and also elaborate feasts and gift-giving performances (2000: 84–85).

All of these models share strong similarities with the theoretical literature on contests in evolutionary biology. The signaling models developed by evolutionary biologists demonstrate that signals between competitive rivals can evolve when they convey information about unobservable qualities relevant to fighting. Honest signals about fighting ability enable contestants to assess who is more likely to win a fight, and avoid costly contests. However, the winners of such contests attain dominance over their rival, it does not generate prestige. Thus, these models do not provide much insight about how prestige goods first emerged or how such goods were linked to leadership and ranking. I suggest that once leadership and social ranking existed, the use of prestige goods was transformed through involvement in competition between rivals, and took on an expanded signal content having to do with coercion and dominance through their links with authority and social power.

The emergence of these two phenomena – the formation of family and lineage social rank differences and the increasing interaction between factions via their leaders – transformed the social landscape in which prestige goods were being used in dramatic ways. Correspondingly, the content of the signal being made with prestige goods was significantly expanded in these new social contexts. One result was that self-promotion and aggrandizement using prestige goods, once rebuffed by social equals, became a tactic that was used by would-be leaders to advertise their own social power and the success of their group, and to induce followers to accepting their authority.

6.6 Conclusions

The focus of this chapter has been on the role of prestige in human systems of social power. The recognition of the importance of prestige systems in human societies is important for at least two reasons. First, prestige systems are a derived property of human societies, and are not found in any other social species. Second, prestige is arguably more important than dominance in determining interpersonal status relationships in human societies. Thus, an understanding of prestige systems will further our understanding of social power and inequality in human societies. Evolutionary models of prestige systems enhance anthropological and archeological theories of the evolution of social structures and institutions. For instance, admiration of and desire for prestige goods are often cited as the reasons why the display of prestige goods, and their distribution, were important in the emergence of socio-political hierarchy, but the source of these feelings is also often left unexplained or underspecified. Evolutionary models of prestige systems can provide this much needed explanation; here, I have hypothesized that admiration of and desire for prestige goods emerged out of a more general desire for prestige as a means to enhance status, specifically, because such goods acted as a means of advertising skill and expertise. Changes in environmental and social contexts favoring the development of leadership would make the advertisement of skills and expertise pertinent to all group members, not just to the learning individuals, and I suggest that it is because of this reason that prestige goods could play a critical role in the emergence of the social structures and institutions of hierarchy and inequality. In turn, this forged the link between the possession and display of prestige goods and the occupation of positions of authority and wealth that not only predominates their use in contemporary Western society but also appears to have played a powerful part in the development of political hierarchy and social and economic inequality in human societies around the world and through time.

References

Aldenderfer MS, Craig NM, Speakman RJ, Popelka-Filcoff R (2008) Four-thousand-year-old gold artifacts from the lake Titicaca basin, southern Peru. Proc Natl Acad Sci USA 105:5002–5005

Barkow JH (1975) Prestige and culture: a biosocial interpretation. Curr Anthropol 16:553–572

Bateson G (1958) Naven, a survey of the problems suggested by a composite picture of the culture of a new Guinea tribe drawn from three points of view. Stanford University Press, Stanford, CA

Bliege Bird R, Bird DW, Kushnick G, Smith EA (2002) Risk and reciprocity in Merriam food sharing. Evol Hum Behav 23:297–321

Boone JL (1992) Competition, conflict, and development of social hierarchies. In: Smith EA, Winterhalder B (eds) Evolutionary ecology and human behavior. Aldine de Gruyter, New York, pp 301–338

Boone JL (1998) The evolution of magnanimity: when is it better to give than to receive? Hum Nat 9:1–21

Boone JL (2000) Status signaling, social power, and lineage survival. In: Diehl MW (ed) Hierarchies in action: Cui Bono? Occasional Paper No. 27. Center for Archaeological Investigations, Southern Illinois University, Carbondale, IL

Boone JL, Kessler KL (1999) More status or more children? Social status, fertility reduction, and long-term fitness. Evol Hum Behav 20:257–277

Brown DE (1991) Human universals. McGraw-Hill, New York

Chacon RJ (2004) Seeking the headhunter's power: the quest for Arutam among the Achuar Indians of the Ecuadorian Amazon. Paper represented at the 69th annual meeting of the society for American Archaeology, Montreal, Quebec, Canada

Clark G (1986) Symbols of excellence: precious materials as expressions of status. Cambridge University Press, Cambridge

Clark JE, Blake M (1994) The power of prestige: competitive generosity and the emergence of rank societies in Lowland Mesoamerica. In: Brumfiel EM, Fox JW (eds) Factional competition and political development in the new world. Cambridge University Press, Cambridge, pp 17–30

Ellis L (1995) Dominance and reproductive success among nonhuman animals: a cross-species comparison. Ethol Sociobiol 16:257–333

Gintis H, Smith EA, Bowles S (2001) Costly signaling and cooperation. J Theor Biol 213:103–119

Hawkes K (1991) Showing off: tests of an hypothesis about men's foraging goals. Ethol Sociobiol 12:29–54

Hayden B (1998) Practical and prestige technologies: the evolution of material systems. J Archaeol Method Theory 5:1–55

Hayden B (2001) The dynamics of wealth and poverty in the transegalitarian societies of Southeast Asia. Antiquity 75:571–581

Henrich J, Gil-White FJ (2001) The evolution of prestige: freely conferred deference as a mechanism for enhancing the benefits of cultural transmission. Evol Hum Behav 22:165–196

Johnson AW, Earle TK (2000) The evolution of human societies: from foraging group to agrarian state. Stanford University Press, Stanford, CA

Miller G (2000) The mating mind: how sexual selection choice shaped the evolution of human nature. Doubleday, New York

Neiman FD (1997) Conspicuous consumption as wasteful advertising: a Darwinian perspective on spatial patterns in classic Maya terminal monument dates. In: Barton CM, Clark GA (eds) Rediscovering Darwin: evolutionary theory and archaeological explanation. Archaeological papers of the American Anthropological Association, Number 7. American Anthropological Association, Arlington, VA, pp 267–290

Owens DA, Hayden B (1997) Prehistoric rites of passage: a comparative study of transegalitarian hunter-gatherers. J Anthropol Archaeol 16:121–161

Pletka SM (2001) Technology, labor organization, and institutions during the adoption of agriculture in mesolithic and neolithic Denmark. PhD thesis, University of California, Los Angeles

Plourde AM (2006) Prestige goods and their role in the evolution of social ranking: a costly signaling model with data from the late formative period of the northern Lake Titicaca Basin, Peru. PhD thesis, University of California, Los Angeles

Plourde AM (2008) The origins of prestige goods as honest signals of skill and knowledge. Hum Nat 19:374–388

Plourde AM. Prestige goods and the formation of political hierarchy – a costly signaling model. In: Shennan S (ed) Pattern and process in cultural evolution. University of California Press, Berkeley (in press)

Plourde AM, Stanish C (2006) The emergence of complex society in the Titicaca Basin: the view from the north. In: Isbell WH, Silverman H (eds) Andean archaeology III: north and south. Springer, New York, pp 237–257

Richerson PJ, Boyd R (1999) Complex societies: the evolutionary origins of a crude superorganism. Hum Nat 10:253–289

Shennan SJ, Steele J (1999) Cultural learning in hominids: a behavioral ecological approach. In: Box HO, Gibson KR (eds) Mammalian social learning: comparative and ecological approaches. Cambridge University Press, Cambridge, pp 367–388

Smith EA, Bliege Bird R, Bird DW (2003) The benefits of costly signaling: meriam turtle hunters. Behav Ecol 14:116–126

Stanish C (2003) Ancient Titicaca: the evolution of complex society in southern Peru and northern Bolivia. University of California Press, Berkeley

Veblen T [1899] (1994) The theory of the leisure class. Dover Publications, New York

Zahavi A (1975) Mate selection – a selection for a handicap. J Theor Biol 53:205–214

Zahavi A, Zahavi A (1997) The handicap principle: a missing piece of Darwin's puzzle. Oxford University Press, New York

Chapter 7
The End of the Republic

Laura Betzig

Abstract Virgil first read his Georgics, or "farm poem," to the first Roman emperor, Augustus, on his way back from the battle at Actium – where Augustus had put an end to the Roman republic, and established one-man rule. Virgil's natural history was bad: he thought that queen bees were kings, among other things. But the point he was making was good. *Communis natos*, he wrote: in Rome, as in *Apis mellifera* hives, sterile workers and soldiers would help raise the young of their emperor, or queen. Like dominant members of eusocial species – bees, ants, wasps, gall thrips, termites, aphids, beetles, sponge-dwelling shrimp, and naked mole-rats – Roman emperors had enormous reproductive success. They had sexual access to hundreds or thousands of women, who may have borne hundreds or thousands of children. And they got help defending their territories, and provisioning their families, from millions of facultatively sterile workers and soldiers, and from thousands of eunuchs – who made up an obligately sterile caste. This example from human history illustrates the unusual flexibility of human reproductive strategies.

7.1 Introduction

Over 2,000 years ago, in September of 31 BC, Gaius Octavius beat Marc Antony in a naval battle at Actium off the coast of Greece. A year later, on his way back to Rome, Octavian stopped to visit his friend Virgil in the south of Italy – where he recited his *Georgics,* or farm poem, for 4 days straight. Virgil's natural history was bad: he thought queen bees were kings, among other things. But the point he was making was good. Equals would fight to the death, till just one was left on the nest.

L. Betzig
The Adaptationist Program, Ann Arbor, MI, USA
e-mail: lbetzig@gmail.com

P.M. Kappeler and J.B. Silk (eds.), *Mind the Gap*,
DOI 10.1007/978-3-642-02725-3_7, © Springer-Verlag Berlin Heidelberg 2010

That, Virgil thought, was for the best. In his words: "Slay the weak rebel! bid the usurper bleed! Slay, 'ere he waste the hive.' Defend the throne, and let the rightful monarch reign alone." He put it even better in the *Aeneid* he left on his deathbed: "Spare the submissive, and war down the proud" (Virgil, *Georgics*, 4.89–91, *Aeneid*, 6.852–853).

Virgil grew up on a farm in the north of Italy, the son of a man who greatly increased his little property by buying up woodlands and raising bees; and he devoted the fourth part of his *Georgics* to beekeeping, or apiculture (Suetonius, *Life of Virgil*). After the honeybee fight to the death, Virgil knew something about honeybee peace. As long as their king was alive, the hive was all "of one mind." Colony members would work for him; they'd fight for him; and they'd raise his young. "*Communis natos*" is how Virgil put it: everybody helped the emperor breed (Virgil, *Georgics*, 4.154–191, Whitfield 1956).

That made them truly social. Almost half a century ago, the term *eusocial* was first used to describe insect societies, in which workers cooperatively care for a monarch's brood, as members of an obligately sterile caste (Batra 1966; Wilson 1971). Over the last few years, that definition has been expanded to include animals in a variety of taxa – from insects, including bees, ants, wasps, thrips, termites, aphids and beetles, to a sponge-dwelling shrimp, to the East African naked mole-rat. Continuous definitions of eusociality include the wide range of species in which some individuals help care for others' young, and measure an index of reproductive variance, or "skew" (Gadagkar 1994; Sherman et al. 1995; Lacey and Sherman 2005). Discrete definitions restrict eusociality to the small handful of species with obligately sterile castes (Crespi and Yanega 1995; Costa and Fitzgerald 2005; Crespi 2005). But in practice, these definitions overlap. In societies where some individuals are obligately sterile, reproductive skew will usually be high; in societies without sterile castes, reproductive skew will usually be lower.

In either case, Virgil's honeybees and Rome after Actium war exhibited some striking parallels. Both *Apis mellifera* hives and the Roman Empire were provisioned and defended with help from obligately sterile workers, or eunuchs – who cared, directly or indirectly, for breeders' young. And in both honeybee societies and Imperial Rome, reproductive skew was at the high end of the continuum. In *Apis mellifera,* as in most eusocial societies, direct reproduction is restricted to a single queen, and other group members do not breed. In the Roman Empire, direct reproduction was shared, but the emperors had sexual access to hundreds or thousands of women, and thousands of obligately sterile eunuchs never bred. That contributed to a reproductive division of labor, and moved Imperial Rome down the eusocial continuum.

Virgil's honeybees, *Apis mellifera*, live in colonies of tens of thousands of workers, but produce only about ten queens. The first of those queens to emerge searches the others out and kills them, or is killed herself. Afterwards, the survivor's tens of thousands of sterile worker daughters clean cells, feed brood, store nectar, forage for pollen, and defend the hive – sacrificing their viscera, and their lives, with their barbed stings. But the queen – which grows up to twice their length, and lives up to 50 times as long – specializes as an egg-laying machine (Michener 1974; Seeley 1985).

In a similar way, Roman emperors systematically had their competitors wiped out; and their empires, like honeybee hives, were provisioned and protected with help from thousands of eunuchs – who made up a sterile caste. The Roman Empire was defended by millions of celibate soldiers; it was supplied by millions of celibate slaves; and it was administered by a *praepositus sacri cubiculi*, a eunuch "set over the emperor's sacred bedchamber," and by other eunuchs "more in number than flies around the flocks in spring" by the time the emperors left Rome. But, like *Apis mellifera* queens, Roman emperors specialized as breeders. They had sex with the freeborn women procured by family members and friends, senators, and their praetorian guard; and they had sexual access to hundreds or thousands of slaves.

7.2 Senators, Soldiers, Slaves and a Sterile Caste

Over the 357 years the empire remained in Rome, the distance between subject and emperor increased. Some members of senatorial families who had administered the *res publica*, or republic, were executed under the law of *maiestas*, or treason. And others were punished for being promiscuous, under the "moral laws." Millions of soldiers were legally barred from marriage; millions of slave men had little access to women. And by the time Roman emperors moved to Constantinople, thousands of eunuchs worked in the civil service, fighting for and providing for the empire as members of a sterile caste.

7.2.1 Senators

Augustus was the first to investigate a libel under the *maiestas* law in AD 12 – being "provoked" by the senator Cassius Severus, who'd made sarcastic remarks about the burning of republican books (Tacitus, *Annals*, 1.72). Severus paid for his criticism with an exile to Crete, ending his days on the rock of Seriphos. Others were punished for treasonous acts. As early as 22 BC, Lucius Murena and Fannius Caepio, who'd been suspected of a plot to assassinate Augustus, were "seized by state authority and suffered by law what they'd worked to accomplish by violence" (Velleius, *Compendium*, 2.91). Even under benign emperors, hundreds of subjects were put to death. Occasionally, whole lineages were wiped out (Bauman 1974, 1996; Fig. 7.1).

Overall, the number of dead was nontrivial – though hard to pin down. Even before he became an emperor, in December of 43 BC, Octavian put bounties on the heads of 300 senators and 2,000 knights. Cicero was among the proscribed: hunted down in a thicket on his way to the coast, his head and right hand were cut off (Appian, *Civil Wars*, 4.5, Pliny, *Natural History*, 7.148). Augustus' step-son, Rome's second emperor, Tiberius, had bodies – up to 20 a day – thrown on the Stairs of Mourning leading down into the Forum (Suetonius, *Tiberius*, 61).

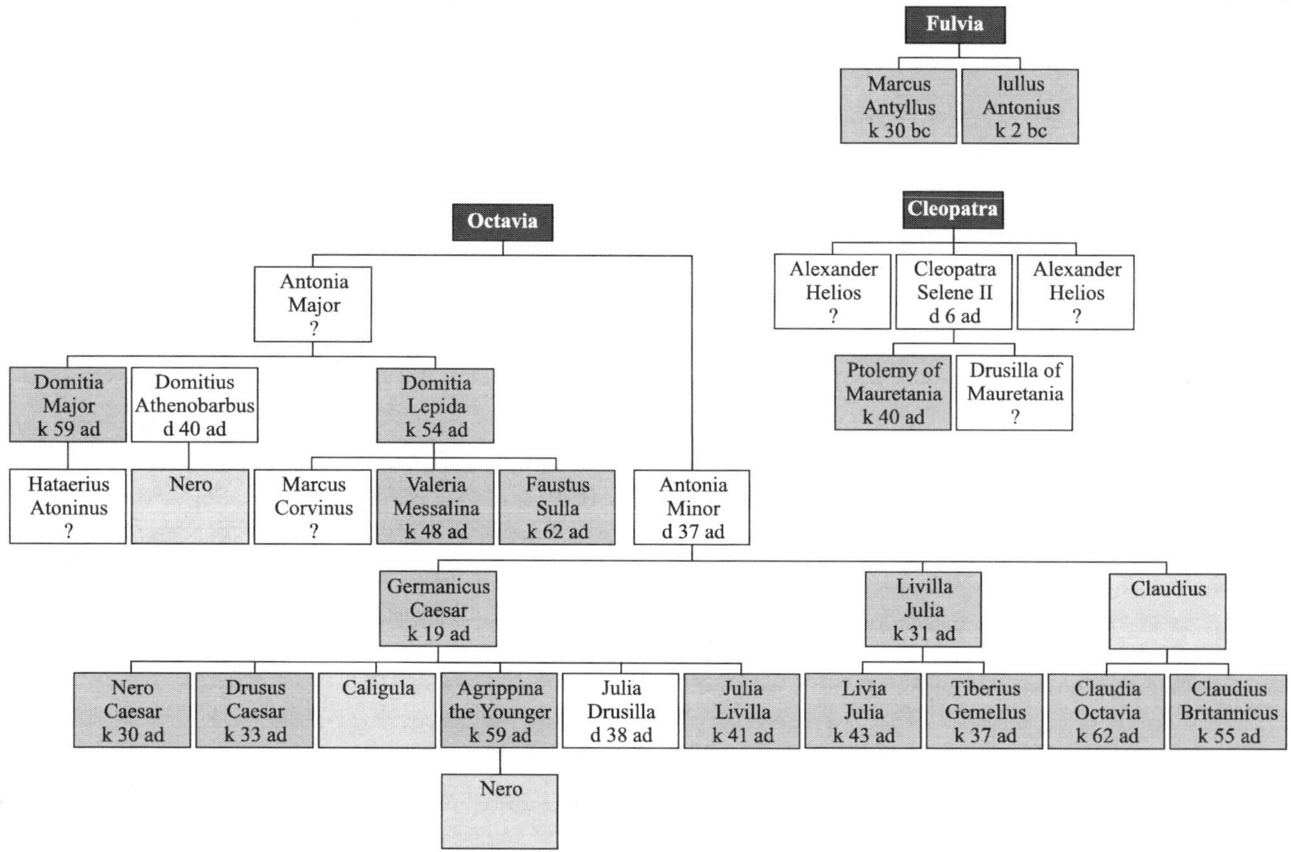

Fig. 7.1 Victims of despotism in Marc Antony's family. Antony had legitimate children by five legitimate wives; many of their descendants were eliminated by Julio-Claudian emperors. See text

And the third emperor, Augustus' great-grandson Caligula, kept lists (*The Dagger* and *The Sword*) of marked subjects, being credited with the remark: "I wish all you Romans had just one neck" (Suetonius, *Gaius*, 50, Seneca, *On Anger*, 3.19). Caligula's uncle Claudius, picked by his soldiers to be Rome's fourth emperor, had up to 300 knights and 35 senators put to death – along with lesser subjects "to the number of the grains of sand and the specks of dust" (Seneca, *Apocolocyntosis*, 13–14, Suetonius, *Claudius*, 29). And the last man in Augustus' dynasty, his great-great-grandson Nero, was arguably the worst: Seneca (the Stoic philosopher, his teacher), Lucan (the poet, and Seneca's nephew), Petronius (the novelist, and Nero's "Arbiter of Taste"), and Thrasea Paetus (another Stoic philosopher) were put to death; after a senatorial conspiracy in AD 65, the flow of blood "fatigued the mind" – though the number of casualties is unknown (Tacitus, *Annals*, 16.16). Victims were accused along with their children, and their relatives were forbidden to mourn (Suetonius, *Tiberius*, 61, *Digest*, 3.2.11.3). As a philosophical emperor would later say, "what is no good for the hive is no good for the bee" (Marcus Aurelius, *Meditations*, 6.54).

The survivors were asked to limit their direct reproduction. One day in 28 BC, wearing a sword and steel corset under his tunic, with ten well-built bodyguards around him, the first emperor persuaded 50 men to withdraw from the senate voluntarily. Then, he "compelled" another 140 to follow their example. Eleven years later, Augustus took another 200 senators off the lists; others were asked to leave, in 11 BC and AD 4 (Syme 1939; Talbert 1984). Many had been politically unfriendly; but others were censored for having too much sex. "On the strength of their own knowledge of their families and their lives, he urged senators to become their own judges."

Augustus did the same to the *equites*, or knights – who were often the richest men in Rome. He "cross-examined" them on their personal affairs: "some, whose lives proved to have been scandalous, were punished; others were only degraded" (Suetonius, *Augustus*, 35, 39, Dio, *History*, 52.42.2).

Augustus' contemporaries were appalled at censors' powers. They invaded "the privacy of our homes"; or, more explicitly, "throwing open every house and extending the authority of the censors even to the bedchamber, they made that office the overseer and guardian of everything that took place in their subjects" houses' (Pliny, *Natural History*, 29.8.18, Dionysius of Halicarnassus, *Roman Antiquities*, 2.15). Other emperors conducted similar purges. Claudius brought back the censorship in AD 47–48, revised the senate and struck knights off the lists – warning a notorious seducer of wives and virgins, "restrain your passions" (Suetonius, *Claudius*, 16). Then, in AD 73–74, Vespasian, with his son Titus beside him, reformed the senate and knights again, removing "undesirable members" and "tightening discipline" (Suetonius, *Vespasian*, 8–9). Domitian made himself perpetual censor (*censor perpetuus*) in AD 84–85, and started another campaign for "improving public morals": men were sentenced for "unnatural" practices, and women for not being chaste (Suetonius, *Domitian*, 8, Dio, *History*, 67.12.2).

On another day in 28 BC, the first emperor probably tried to pass his first set of "moral laws" – though he seems to have had them repealed. Propertius, the poet, remembered how he had "rejoiced when that law was lifted"; and Livy, who wrote histories, was not sure whether "corruption or its corrective" was worse (Propertius, *Poems*, 2.7, Livy, *History*, pr.9). At any rate, in around 18 BC, Augustus tried again: his *lex Julia de maritandis ordinibus* encouraged Roman bachelors to get married and raise legitimate children; but his *lex Julia de adulteriis* had convicted adulterers and adulteresses banished to remote islands – with up to half of their property confiscated by the state. Any man who "knowingly" made his house available for adultery was exiled, as if he'd committed adultery himself; and any man who made a profit out of adultery was punished, even if the adulterer was his spouse, "for it is no small crime to have pimped for one's wife" (Paulus, *Sententiae*, 2.26, *Digest*, 48.5.9, 30, Treggiari 1991).

7.2.2 Soldiers

Constraints on direct reproduction were not limited to the senatorial class: for hundreds of years, soldiers' marriages were banned. The law that soldiers could not "legally have wives" may go back to Augustus: Ovid, the poet, found it odd that the emperor's legislation should encourage civilian breeders, in order to supply his empire with unmarried soldiers; and Cassius Dio, the senator, dated the law against soldiers' marriages to Augustus' reign (Ovid, *Art of Love*, i.113–114, Dio, *History*, 60.24.3).

The Roman Empire, like most empires, was built on conquest. Roman emperors supported millions of soldiers, and many were killed in battle. Some of the survivors had relationships with women, and fathered children. But for centuries, they were legally unable to marry, and their children were considered bastards (Wells 1999).

7.2.3 Slaves

There may have been 10 million slaves in a population of 60 million in the whole Roman Empire in Augustus' time, and none of those slaves had wives (Scheidel in press). As Ulpian, a third-century jurist, summed up: "*Conubium* is the capacity to marry a wife in Roman law"; and "there is no *conubium* with slaves" (*Titului Ulpiani*, v.3–5, also Gaius, *Institutes*, i.57). Most slave men lived and worked on the farms, or in the mines. In the mines, thousands "never saw the light of day" for months; and they constantly risked being crushed to death. And on the farms, thousands worked in chained gangs – where they were driven by "whips rather than words" in the fields, and slept in "an underground prison, as wholesome as possible," with light coming in through window slits high enough up that they couldn't be reached (Pliny, *Natural History*, 33.70, Columella, *On Agriculture*, 1.6.3, 1.8.5). Conditions were not conducive to family life.

But a minority of slaves filled the imperial civil service: the *Familia Caesaris*. They worked from the lowest level to the highest – in menial positions, as footmen (*pedisequi*), watchmen (*custodes*), name callers (*nomenclatores*), or post officers (*tabellarii*); in middle management, as assistants (*adiutores*), archivists (*commentarii*), aides (*vicarii*), or accountants (*tabularii*); and in upper-level cabinet posts, as secretaries of letters (*ab epistulis*), secretaries of documents (*a studiis*), secretaries of petitions (*a libellis*), and secretaries of finance (*a rationibus*). The names of over 4,000 imperial slaves and freed slaves survive on Roman tombs (Weaver 1972). By the end of Augustus' dynasty, the emperor was so close to a slave that Rome became a "slave to two emperors" at once; and by the end of the next dynasty, "most emperors, though masters of their subjects, were the slaves of their freed slaves" (Dio, *History*, 62.12.2, Pliny, *Panegyric*, 88). Some of those slaves had relationships with women, and fathered illegitimate children – who might, like their fathers, have ended up in the *Familia Caesaris* (Rawson 1966; Weaver 1972).

7.2.4 A Sterile Caste

Early Roman emperors filled their bureaucracies with "outcasts" – with foreigners and the poor, or people who lacked important ancestors. Later emperors promoted "dry trees" – or people who lacked sons, or a sterile caste (Isaiah 56: 3–8, Betzig 2005, 2008).

By the end of three centuries , the imperial civil service was filled with eunuchs. Even Maecenas – who patronized Virgil, and worked as a regent for Augustus – was "attended in public" by a pair of *eunuchi*; and even Livia, who was Augustus' last wife, buried a *rarius eunuchus*, along with another half dozen *cubicularii* – bedchamber attendants, customarily castrated – in her family tomb (Seneca, *Moral Letters*, 114, Dunlap 1924). Some *cubicularii* specialized as "workers," and others as "soldiers": they advised emperors on foreign affairs; or were honored at triumphs after foreign wars (Philo of Alexandria, *Embassy to Gaius*, 27.175, Suetonius, *Claudius*, 28).

But there were more of them, and they were more important, as time went on. By the end of the first century, there were "troops of eunuchs" at the imperial court; and by the end of the second century, a hundred castrated Roman citizens waited on an emperor's wife (Suetonius, *Titus*, 7, Dio, *History*, 76.14). By the end of the third century, the empire was administered by a *praepositus sacri cubiculi* – the eunuch "set over an emperor's sacred bedchamber"; and by the fourth century, after Constantine moved the capital to Constantinople, the emperor was surrounded by a thousand cooks, as many barbers and more butlers, a swarm of waiters, and "eunuchs more in number than flies around the flocks in spring" (Malalas, *Chronicle*, 339, Libanius, *Orations*, 18.130). *Castrensi* managed the imperial bodyguard, properties, movables, and treasury (Hopkins 1963; Jones 1964). Eventually, 8 out of 18 administrative ranks were reserved for eunuchs, who consistently outranked the "bearded" civil service (Constantine VII Porphyrogenitus, *On Ceremonies*, 2.52, Tougher 2002, 2008).

Remnants of the senatorial class disapproved. Eunuchs were all "extortionate or despicable" lizards and toads: subservient to their masters, insolent to everybody else (Ammianus Marcellinus, *Roman History*, 16.7.8, Basil of Caesarea, *Letter* 115).

7.3 Emperors

Emperors, on the other hand, bred. Roman emperors married just one, legitimate wife at a time; but they had sex with as many women as they could afford (Betzig 1986, 1992a, b, 1993). Some of those women were freeborn Roman citizens (Syme 1960); but many of those women were slaves (Scheidel 2009). Friends, family members, senators, and his praetorian guard brought freeborn wives and daughters to the imperial bed. And on the side, Roman emperors had sexual access to hundreds or thousands of slaves, whose daughters and sons – *vernae*, or homeborn slaves – filled Latin law, literature, and the imperial civil service: the *Familia Caesaris*.

7.3.1 Free Women

Even his friends admitted that Augustus was an adulterer. But they justified it: "he attracted many women by his comeliness and high lineage." Besides, he had sex "for reasons of state": he was trying to find out what his enemies were up to by getting intimate with their daughters or wives – who were stripped of their clothes, "and inspected as though they were for sale" (Nicolaus of Damascus, *Life of Augustus*, 5, Suetonius, *Augustus*, 69).

Other emperors were provided for by senators, bodyguards, family members, and slaves. Tiberius got help from his slaves. Rummaging through the streets of Rome for freeborn daughters and sons, they "rewarded compliance, overbore reluctance with menaces, and – if resisted by parents or relations – kidnapped their victims, and violated them on their own account" (Tacitus, *Annals*, 6.10). Caligula abused his senate, asking legislators to dinner, and taking their wives to bed, then coming back to announce in a loud voice "how they'd behaved in sexual intercourse" (Seneca, *On Constancy*, 17.2). Claudius, who was sickly, depended on his family – like his mother and his grandmother, who put girls in his bed to satisfy his "healthy" appetite for sex (Dio, *History*, 60.2.4–6). And Nero relied on his praetorian guard. His praetorian prefect, Tigellinus, had a raft towed by gold and ivory boats floated in the Campus Martius on Marcus Agrippa's lake: brothels were set up for high-ranking ladies ("the most beautiful and distinguished in the city"), and whores ("naked prostitutes, indecently posturing and gesturing") were lined up on the quays (Suetonius, *Nero*, 27).

7.3.2 Slave Women

Mostly though, emperors had sex with their slaves (see Tables 7.1–7.3). There were an estimated 6 million slaves in a population of 60 million in the whole Roman Empire by the Augustus' time, and most of them were owned by rich women and men (Scheidel 2009). Great, late republican families saved places for 204, 634, and 652 household slaves in their family tombs, and roughly a third had female names (Treggiari 1975; Saller and Shaw 1984). Many were good looking or young: ages of female slaves ranged from 4 to 35 at time of sale – with a ripe median of around 19; and slaves of "good appearance" cost more (Bradley 1978, Suetonius, *Caesar* 47).

They were expected to be fertile. In legal sources, buyers were due a refund if a slave menstruated twice a month, or never menstruated at all; if she regularly produced "stillborn issue," or was "so doctored that she cannot fulfill the function

Table 7.1 Sex ratios of slaves on Roman tombs. After Treggiari 1975 (Volusii, Statilii, Liviae), Weaver 1972 (Familia Caesaris), and Rawson 1986 (Alumnae, Vernae)

Sample	Ratio of ♀ = Slaves (% ♂)
Volusii family	129:75 (63.2%)
Statilii family	421:213 (66.4%)
Monumentum Liviae	440:212 (67.5%)
Familia Caesaris	3,325:291 (92.0%)
Alumnae	276:139 (66.5%)
Vernae	381:183 (67.6%)

Table 7.2 Ages of *ancillae*, or female slaves, at time of sale, as listed on Egyptian papyri. After Bradley 1978

Range of ages	Number (%) of Ancillae
<5	1 (3.4%)
5–9	3 (10.3%)
10–14	5 (17.2%)
15–19	5 (17.2%)
20–24	7 (24.1%)
25–29	4 (13.8%)
30–35	4 (13.8%)

Table 7.3 Ages of *vernae*, or homeborn slaves, on inscriptions from Roman tombs. Because *vernae* who were commemorated on tombs died young, the implication is that the status of those who survived changed early to liberty, or freed slaves. After Rawson 1986

Range of ages	Number (%) of *Vernae*
<5	103 (32.0%)
5–9	100 (31.1%)
10–14	48 (14.9%)
15–19	34 (10.6%)
20–24	17 (5.3%)
25–29	9 (2.8%)
30–100	11 (3.3%)

of a woman" – as it was "the highest and particular lot of woman to conceive and conserve what she conceives." On the other hand, a slave woman could be emancipated for giving birth to three children, or "if the first child she bears is male"; and it was considered "a good reason for manumission where, for instance, anyone offers for manumission before the council a natural son or daughter" (*Digest*, 1.5.15, 21.1.14–15, 35.5.10, Gaius, *Institutes*, i.19). Three out of four Roman epitaphs belong to a freed slave, and *vernae* – the "homeborn" children born on their masters' estates, to their masters' slave women – were most likely to be freed (Taylor 1961; Hopkins 1978). *Vernae* were brought up in *paedagogia* along with their masters' legitimate children, attended by the same hairdressers (*ornatrici*), anointers (*unctores*), teachers (*praeceptori*), and doctors (*iatroliptae*) (Rawson 1986; Bradley 1991). They often grew up to become knights, or sat in the senate; and hundreds of *Augusta vernae* worked in the civil service (Tacitus, *Annals*, 13.27, Weaver 1972).

7.3.3 Genius

Early in 44 BC, just months before the Ides of March, Julius Caesar was voted "Father of his Country" by his senate, and the inscription *pater patriae* was inscribed onto Roman coins (Suetonius, *Caesar*, 76, Dio, *History*, 44.4.4). Rumors circulated about Caesar after the Ides of March, to the effect that members of the senate had "actually ventured to suggest permitting him to have intercourse with as many women as he pleased, because even at this time, though 50 years old, he still had numerous mistresses." Helvius Cinna, the people's representative, or tribune, was supposed to have drawn up a bill for the commons to pass while Caesar was out of town, "legitimizing his marriage with any woman, or women, he pleased, 'for the procreation of children'" (Dio, *History*, 44.7.3, Suetonius, *Caesar*, 52). And within days after Caesar was butchered in the senate, Cinna was supposed to have been torn, limb from limb, by a crowd of angry men (Plutarch, *Caesar*, 68). But others paid tribute to Caesar. They set up an altar in his honor, and raised a 6-m column of Numidian marble in the Forum, with "To the Father of His Country" written on the bottom (Appian, *Civil Wars*, 1.4, Suetonius, *Caesar*, 85). Cicero's son-in-law, Dolabella, had that column leveled, and made sure the masses that gathered to make sacrifices were slaughtered. "Debauched and wicked free men" were thrown down from the Tarpeian Rock; and "audacious and rascally slaves" were hung up on crosses. As Cicero bragged to his friend Atticus, on May 1st: "Away with the pillar! Contract for paving the site!" (Cicero, *Ad Atticus*, 14.15, *Philippics*, 1.5, Gelzer 1968).

On a pair of bronze pillars outside his enormous, 90-m diameter, mausoleum in Rome, Augustus inscribed his *Res Gestae*, or "History of His Reign." The first emperor was proud to be remembered for having served as a *triumvir*, a senator

Fig. 7.2 A coin issued under emperor Maximinus II, a contemporary of Constantine the Great. The naked emperor holds a cornucopia in his left hand, and the head of Serapis – an Egyptian fertility god – in his right. The inscription reads genio augusti, in honor of the emperor's genius

and a censor, but he may have been proudest of the last accomplishment on his list – having been voted *pater patriae*, or "Father of his Country," by his subjects (*Res Gestae*, 35). Tiberius was offered that honor in the first year of his reign, but in spite of "repeated popular pressure," he turned it down (Tacitus, *Annals*, 1.72).

Caligula, Claudius, and Nero, all capitulated in their first 12 months – though they all declined the epithet at first, in Nero's case, "because of his youth." In later dynasties, *pater patriae* was a commonplace honor – in literature, on coins, and in architecture (Suetonius, *Nero*, 8, Gradel 2004).

In the months before March 15th, when the senate was voting excessive honors for Julius Caesar, they decided that public prayers should be offered every year on his behalf, and that people should "swear by" Caesar's *genius* (Dio, *History*, 44.6.1). They had to do the same for Augustus. Romans had always made sacrifices to the *genius*, or generative power, of the heads of their families, or *gens*; now they would make the same sacrifices to their head of state (Fig. 7.2). When a family's household deities, or Lares, were put out to "feed from the dish," the first emperor was offered the same wine cup; and when a farmer came home from his field to food and wine, he invoked Augustus "as a god," along with his Lares, and offered the emperor prayers (Ovid, *Fasti*, 2, February 22, Horace, *Odes*, 4.5). Tiberius vetoed bills for the dedication of priests and temples to his divinity, and decided not to allow subjects to swear by his generative power – though "if anybody after swearing by it incurred the charge of perjury, he would not prosecute him"; but Caligula had subjects boxed up in small cages or sawn in half, for "failing to swear by his *genius*" (Suetonius, *Tiberius*, 26, *Gaius*, 27, Dio, *History*, 57.8.3, 58.2.8, 59.4.4). Bulls were sacrificed on public altars to the *genius* of Nero, and other emperors, in order to ensure the harvests; and coins, covered with cornucopia, were issued in honor of *genio augusti,* or the emperors' *genius* (Fishwick 1987; Gradel 2004). But people were thrown to wild beasts in arenas, or consumed by fire, for failing to "swear by the *genius* of Caesar," for nearly 300 years (*Acts of Polycarp*, 9 and *Acts of Perpetua*, 6 by Musurillo 1972).

7.4 Reproductive Skew

For more than 300 years after Augustus became the first emperor of the West, proud Roman subjects were put in their place – with respect to politics, and with respect to sex. They were relegated to remote islands, or killed, for offending the majesty of the emperor, under the law of *maiestas*. And they were thrown out of Rome, or removed from the senatorial and equestrian orders, for being promiscuous, under the censorship or *lex Iulia de adulteriis*. They were expected to worship the emperors as *divi Iulii*, or gods; and they were turned into human torches, or thrown to wild animals in the circus, for failing to make sacrifices to the emperor's generative power, or *genius*.

The Roman Empire, like an *Apis mellifera* colony, exhibited many parallels with eusocial animals. There was a reproductive division of labor, with eunuchs helping to care for the emperor's freeborn and slave born children. And as a result, reproductive skew was high. Like social insects – from bees, to ants, to wasps (Wilson 1971), to gall thrips (Crespi 1992), to termites (Thorne 1997), to aphids (Aoki 1977), to beetles (Kent and Simpson 1992), like at least one crustacean – *Synalpheus regalis*, the sponge-dwelling shrimp (Duffy 1996), and like at least one other social mammal – *Heterocephalus glaber*, the naked mole-rat (Jarvis 1981), Roman emperors specialized as breeders. They had sexual access to hundreds or thousands of women, who may have borne hundreds or thousands of children. And they got help defending their territories, and provisioning their families, from the hundreds or thousands of workers and soldiers who made up an obligately sterile caste.

For the more than 100,000 years of *H. sapiens* prehistory, reproductive skew was low. In most cases, successful foragers raised roughly twice as many children as average foragers; but most foragers managed to become parents (Hill and Hurtado 1996; Smith 2004). Like many insects, some birds and a handful of mammals including primates, hunters, and gatherers have always been cooperative breeders: adults, especially, closely related adults, have helped one other feed and protect their broods. But – with the exception of postmenopausal women, who have helped raise their own grandchildren – they have lacked an obligately sterile caste (Foster and Ratnieks 2005; Hrdy 2005).

This changed around 10,000 years ago, with the origin of farming. By the time the first historical records were being kept in the Ancient Near East, "beardless" attendants were waiting on "bearded" kings – who collected large numbers of *lukurs* (or "king's fallow," or virgins) and *nins* (or "queens"), and left surviving records of dozens of sons and daughters (Postgate 1994; Grayson 1995). In Old Kingdom Egypt, the desert god Seth, whose testicles have become "impotent," helps administer the empire for pharaoh; and in the New Kingdom, the names of 49 *sons* – generals, hereditary counts, chiefs of secrets, scribes – survive from Rameses II reign, along with another 111 unnamed *sons* (Fisher 2001; Allen 2005). In the Sanskrit of India's early imperial dynasties, the Maurya and Gupta, emperors are waited on by "third genders"; and harem women are taught to hold an emperor's interest by speaking multilingually and talking to parrots, "even though he may have thousands of other women" (Shamasastry 1951; Burton 1979). In China, in the 2nd millennium BC, there are *huan guan* – court officers, customarily castrated – on Shang dynasty oracle bones; by the 2nd millennium AD, there were a record 100,000 Ming Dynasty eunuchs, and a Sui Dynasty emperor kept a record 100,000 women (Tsai 1996; Ebrey 2003). Across Old World empires including Rome, eunuchs oversaw administration and commanded armies. But emperors specialized as breeding machines.

What accounts for the change? In any society, reproductive skew is expected to increase: (1) as the genetic relatedness of group members goes up; (2) as the social benefits of group membership go up; and (3) as ecological constraints on dispersal increase (Vehrencamp 1983a, b; Emlen 1995).

In eusocial species, relatedness lowers the costs of helping (Hamilton 1964, 1972). And as expected, average relatedness in most skewed societies is high. In the *Hymenoptera*, including *Apis mellifera*, haplodiploidy makes full sister bees, ants, and wasps more closely related than mothers and daughters; so helping is favored (Hughes et al. 2008). But in haplodiploid colonies where queens are inseminated more than once, or in colonies with more than one queen, relatedness is not enough to explain the existence of sterile castes (Keller 1993). And kinship is an inadequate explanation for eusociality in diploid species – from termites, to aphids, to beetles, to shrimp, to naked mole-rats.

Social benefits – including cooperative foraging, and cooperative defense – can also raise reproductive skew. Among other things, group members may benefit each other as sentinels or fighters, groomers or hunters; and some may be willing to limit direct reproduction, and to help others reproduce, as a result (Clutton-Brock 2006). But in many social species, altruism is not voluntary, but enforced: breeders often punish or evict nonbreeders who fail to help (Ratnieks and Wenseleers 2008). Again, that has often been the case in human groups.

Strong evidence suggests that reproductive skew is often a response to ecological constraints. Ecological benefits – including habitats safe from predation, and with plenty of food – may compensate nonbreeders for becoming nepotists (Emlen 1982, 1997). Eusocial species take advantage of resource patches across taxa – from insects, to crustaceans, to mammals (Jarvis et al. 1994; Duffy 1996; Thorne 1997). Societies tend to colonize discrete nesting sites in sharply delineated habitats – from decaying logs, to sponge cavities on coral reefs, to 30 kg tubers dispersed in arid ground. The fact that eusocial species "dominate the central, more stable areas of habitats," while solitary species "flourish in the peripheral, more ephemeral areas," (Wilson and Hölldobler 2005) may indicate that high skew is often an *effect*, rather than a *cause*, of finding a good food source. For *H. sapiens*, that seems to have been the case.

Most civilizations have probably risen up as an effect of ecological constraints. Before sedentary societies spread with agriculture, reproductive skew in most foraging societies was low. But around the Old World – from Sumerian and later civilizations on the Tigris and Euphrates, to the Egyptian civilization that lasted for millennia on the Nile, to Harappan civilization on the Indus, to the Shang and later dynasties on the Yellow River and its tributaries, to Rome – every ancient empire began on "*areas of circumscribed agricultural land*" (Carneiro 1970, 1986). A small minority of men collected up to 100,000 women, and up to 100,000 eunuchs filled sterile castes.

Acknowledgments I thank my old friend, Bernie Crespi, for help with evolutionary biology; and I thank my old friend, Walter Scheidel, for help with Roman history.

References

Allen JP (2005) The ancient Egyptian pyramid texts. Leiden, Brill
Aoki S (1977) *Colophina clematis* (Homoptera, Pemphigidae), an aphid species with 'soldiers'. Kontyû 45:276–282
Batra SWT (1966) Nests and social behavior of halictine bees of India. Indian J Entomol 28:375–393
Bauman RA (1974) Impietas in principepem. Beck, München
Bauman RA (1996) Crime and punishment in ancient Rome. Routledge, New York
Betzig LL (1986) Despotism and differential reproduction: a Darwinian view of history. Aldine de Gruyter, New York
Betzig LL (1992a) Roman polygyny. Ethol Sociobiol 13:309–349
Betzig LL (1992b) Roman monogamy. Ethol Sociobiol 13:351–383
Betzig LL (1993) Sex, succession, and stratification in the first six civilizations. In: Ellis L (ed) Social stratification and socioeconomic inequality. Praeger, New York, pp 37–74
Betzig LL (2005) Politics as sex: the Old Testament case. Evol Psychol 3:326–346
Betzig LL (2008) Fertility in the bible and ancient Middle East. In: Goldberg R (ed) Judaism in biological perspective. Paradigm, Boulder, pp 42–61
Bradley KR (1978) The age at time of sale of female slaves. Arethusa 11:243–252
Bradley KR (1991) Discovering the Roman family: studies in Roman social history. Oxford University Press, New York
Burton RF (1979) The kamasutra of Vatsyayana. Heinemann, London
Carneiro RL (1970) A theory of the origin of the state. Science 169:733–738
Carneiro RL (1986) Further reflections on resource concentration and its role in the rise of the state. In: Manzanilla L (ed) Studies in the neolithic and urban revolutions. BAR, Oxford, pp 245–260
Clutton-Brock TH (2006) Cooperative breeding in mammals. In: Kappeler PM, van Schaik CP (eds) Cooperation in primates and humans: mechanisms and evolution. Springer, Berlin, pp 173–190
Costa JT, Fitzgerald TD (2005) Social terminology revisited: where are we ten years later? Ann Zool Fennici 42:559–564
Crespi BJ (1992) Eusociality in Australian gall thrips. Nature 359:724–726
Crespi BJ (2005) Social sophistry: logos and mythos in the forms of cooperation. Ann Zool Fennici 42:569–571
Crespi BJ, Yanega D (1995) The definition of eusociality. Behav Ecol 6:109–115
Duffy JE (1996) Eusociality in a coral-reef shrimp. Nature 381:512–514
Dunlap JE (1924) The office of the grand chamberlain in the later roman and byzantine empires. In: Boak AER, Dunlap JE (eds) Two studies in later Roman and Byzantine Administration. Macmillan, New York, pp 130–308
Ebrey PB (2003) Women and the family in Chinese history. Routledge, London
Emlen ST (1982) The evolution of helping. I. An ecological constraints model. Am Nat 119:29–53

Emlen ST (1995) An evolutionary theory of the family. Proc Natl Acad Sci USA 92:8092–8099
Emlen ST (1997) Predicting family dynamics in social vertebrates. In: Krebs JR, Davies NB (eds) Behavioral ecology: an evolutionary approach, 4th edn. Blackwell, Oxford, pp 228–253
Fisher MM (2001) The sons of Ramesses II. Harrassowitz, Wiesbaden
Fishwick D (1987) The imperial cult in the Latin West. Brill, Leiden
Foster KR, Ratnieks FLW (2005) A new eusocial vertebrate? Trends Ecol Evol 20:363–364
Gadagkar R (1994) Why the definition of eusociality is not helpful to understand its evolution and what should we do about it. Oikos 70:485–488
Gelzer M (1968) Caesar: politician and statesman. Harvard University Press, Cambridge, MA
Gradel I (2004) Emperor worship and Roman religion. Oxford University Press, New York
Grayson AK (1995) Eunuchs in power: their role in the Assyrian bureaucracy. In: Dietrich M, Loretz O (eds) Vom Alten Orient zum Alten Testament. Festschrift für Wolfram Freiherrn, Neukirchen, pp 85–98
Hamilton WD (1964) The genetical evolution of social behavior. I & II. J Theor Biol 7:1–52
Hamilton WD (1972) Altruism and related phenomena, mainly in social insects. Annu Rev Ecol Syst 3:193–232
Hill K, Hurtado AM (1996) Ache life history: the ecology and demography of a foraging people. Aldine de Gruyter, New York
Hopkins K (1963) Eunuchs in politics in the later Roman Empire. Proc Camb Philol Soc 9: 62–80
Hopkins K (1978) Conquerors and slaves: sociological studies in Roman history I. Cambridge University Press, Cambridge
Hrdy SB (2005) The evolutionary context of human development: the cooperative breeding model. In: Carter CS, Ahnert L, Grossmann KE, Hrdy SB, Lamb ME, Porges SW, Sachser N (eds) Attachment and bonding: a new synthesis. MIT Press, Cambridge, MA, pp 9–32
Hughes WOH, Oldroyd BP, Beekman M, Ratnieks FLW (2008) Ancestral monogamy shows kin selection is key to the evolution of eusociality. Science 320:1213–1216
Jarvis JUM (1981) Eusociality in a mammal: cooperative breeding in naked mole-rat colonies. Science 212:571–573
Jarvis JUM, O'Riain MJ, Bennett NC, Sherman PW (1994) Mammalian eusociality: a family affair. Trends Ecol Evol 9:47–51
Jones AHM (1964) The later Roman Empire 284–602: a social, economic, and administrative survey. Johns Hopkins University Press, Baltimore
Keller L (1993) Queen number and sociality in insects. Oxford University Press, Oxford
Kent DS, Simpson JA (1992) Eusociality in the beetle *Austroplatypus incompertus* (Coleoptera: Curculionidae). Naturwissenschaften 79:86–87
Lacey EA, Sherman PW (2005) Redefining eusociality: concepts, goals, and the levels of analysis. Ann Zool Fennici 42:573–577
Michener CD (1974) The social behavior of bees: a comparative study. Harvard University Press, Cambridge, MA
Musurillo H (1972) Acts of the Christian martyrs. Clarendon, Oxford
Postgate JN (1994) Early Mesopotamia. Routledge, London
Ratnieks FLW, Wenseleers T (2008) Altruism in insect societies and beyond: voluntary or enforced. Trends Ecol Evol 23:45–52
Rawson B (1966) Family life among the lower classes at Rome in the first two centuries of the empire. Class Philol 61:71–83
Rawson B (1986) Children in the Roman familia. In: Rawson B (ed) The family in ancient Rome. Cornell University Press, Ithaca, pp 170–200
Saller RP, Shaw BD (1984) Tombstones and Roman family relations in the principate: civilians, soldiers and slaves. J Roman Stud 74:124–156
Scheidel W (2009) Sex and empire: a Darwinian perspective. In: Morris I, Scheidel W (eds) The dynamics of ancient empires: state power from Assyria to Byzantium. Oxford University Press, Oxford, pp 255–324

Scheidel W (2009) The Roman slave supply. In: Bradley K, Cartledge P (eds) Cambridge world history of slavery. I. The ancient mediterranean world. Cambridge University Press, Cambridge, pp 230–250
Seeley TD (1985) Honeybee ecology. Princeton University Press, Princeton
Shamasastry R (1951) Kautilya's arthashastra. University Bombay Press, Bombay
Sherman PW, Lacey EA, Reeve HK, Keller L (1995) The eusociality continuum. Behav Ecol 6:102–108
Smith EA (2004) Why do good hunters have higher reproductive success? Hum Nat 15:343–364
Syme R (1939) The Roman revolution. Oxford University Press, Oxford
Syme R (1960) Bastards in the Roman aristocracy. Proc Am Philos Soc 104:323–327
Talbert RJA (1984) The senate of imperial Rome. Princeton University Press, Princeton
Taylor LR (1961) Freedmen and freeborn in the epitaphs of imperial Rome. Am J Philol 82:113–132
Thorne BL (1997) Evolution of eusociality in termites. Annu Rev Ecol Syst 28:27–54
Tougher S (2002) Eunuchs in antiquity and beyond. Duckworth, London
Tougher S (2008) Eunuchs in Byzantine history and society. Routlege, London
Treggiari S (1975) Family life among the staff of the Volusii. Trans Proc Am Philol Assoc 105:393–401
Treggiari S (1991) Roman marriage. Iusti coniuges from the time of cicero to the time of ulpian. Oxford University Press, Oxford
Tsai SH (1996) The eunuchs in the Ming Dynasty. SUNY Press, Albany
Vehrencamp SL (1983a) A model for the evolution of despotic versus egalitarian societies. Anim Behav 31:667–682
Vehrencamp SL (1983b) Optimal degree of skew in cooperative societies. Am Zool 23:327–335
Weaver PRC (1972) Familia caesaris. Cambridge University Press, Cambridge
Wells CM (1999) Celibate soldiers: Augustus and the army. Am J Ancient Hist 14:180–190
Whitfield BG (1956) Virgil and the bees: a study in ancient apicultural lore. Greece & Rome 3:99–117
Wilson EO (1971) The insect societies. Harvard University Press, Cambridge, MA
Wilson EO, Hölldobler B (2005) Eusociality: origins and consequences. Proc Natl Acad Sci USA 102:13367–13371

Part IV
Intergroup Relationships

Chapter 8
Intergroup Aggression in Primates and Humans: The Case for a Unified Theory

Margaret C. Crofoot and Richard W. Wrangham

Abstract Human warfare and intergroup aggression among primates have traditionally been considered to be largely unrelated phenomena. Recently, however, chimpanzee intergroup violence has been proposed to show evolutionary continuities with war among small-scale societies because both systems involve interactions among temporary subgroups, deliberate attempts to hunt and maim, and demographically significant death rates. Here, we ask whether the functional similarities between intergroup aggression among humans and chimpanzees can be extended to troop-living primates. In most primates, patterns of intergroup aggression involve brief encounters among stable troops, rare violence, and almost no killing. Although they, therefore, show little behavioral resemblance to warfare, growing evidence indicates that intergroup dominance is adaptively important in primates because it predicts long-term fitness. This suggests that in all primates, including humans, individuals use coalitions to maintain or expand access to resources by dominating their neighbors. Thus, while the style of coalitionary aggression depends on each species' evolutionary ecology, we propose that the essential functional reasons for intergroup competition are consistent across group-living primates and humans: strength in numbers predicts long-term access to resources.

8.1 Introduction

Although societies can sometimes spend decades without practicing war, the capacity for warfare is clearly a human universal. However, the question of why humans readily engage in war is unresolved from an evolutionary perspective. In this chapter, we review evidence suggesting that war between groups of

M.C. Crofoot (✉) and R.W. Wrangham
Department of Human Evolutionary Biology, Harvard University, Cambridge, MA, USA
email: mccrofoot@gmail.com, wranghamrichard@gmail.com

P.M. Kappeler and J.B. Silk (eds.), *Mind the Gap*,
DOI 10.1007/978-3-642-02725-3_8, © Springer-Verlag Berlin Heidelberg 2010

humans tends to serve the same essential functions as aggression between nonhuman primates. This might seem unsurprising given that there are obvious and extensive behavioral similarities between human and nonhuman primate intergroup aggression (van der Dennen 1995). Until recently, however, two major barriers have inhibited the development of a unified theoretical explanation for these two phenomena.

First, there has been considerable resistance among war scholars towards using evolutionary theory to explain why war occurs. Thus, military organization has been widely viewed as a socially constructed device that challenges rather than conforms to evolutionary principles. For instance, Collins (2008) argues that war systems are designed to overcome an instinctive tendency to avoid conflict. Furthermore, because warfare is not archeologically visible until about 10,000 years ago, it is sometimes claimed to be a recent invention which, therefore, requires explanation in terms of culture rather than biology (Ferguson 2000). A similar argument notes that warfare is rare or unknown in some hunter-gatherers and must therefore be unnatural (Fry 2006). The fact that aggression is oftentimes not the main motivation of soldiers in battle also suggests an important discontinuity from intergroup aggression among animals (Hinde 1993). For these and many other reasons, such as the complexity of human military and political organization and the novelty of weapon technology, as well as the fear that an evolutionary explanation of warfare will encourage more war (Sponsel 1996; Fry 2006), the problem of war has often been considered to be social or cultural rather than biological.

A second obstacle to conceptual unification has been the lack of a coherent theory for intergroup aggression among primates. Aggression between primate social groups is highly variable, whether in terms of the frequency and intensity of encounters, the resources being contested, or the sex of the participants (Cheney 1987). This variation is observed not only between species, but also between populations of the same species (e.g., *Macaca fuscata*: Saito et al. 1998; Sugiura et al. 2000) and between seasons within the same population (e.g., *Cercopithecus sabaeus*: Harrison 1983; *Cercocebus galeritus*: Kinnaird 1992). It has been difficult to discern unifying patterns amid such variation, especially since studies of intergroup interactions (IGIs) tend to be opportunistic and have relatively small sample sizes.

Other factors have also hampered the efforts to understand the broader significance of intergroup aggression in primates. First, a tendency to treat feeding and mating competition between social groups as unrelated and, oftentimes, mutually exclusive phenomena, has created an artificial division between species where males compete over mates and species where females compete over food. Recent studies have demonstrated that males can defend food resources either directly, or as a by-product of their mate defense (Fashing 2001; Harris 2005, 2006a), highlighting the flaws of this dichotomy and indicating that closer attention must be paid to the functional implications of intergroup aggression (Harris 2007). Second, the role that intergroup resource competition plays in determining individual fitness remains disputed. While Wrangham (1980) proposed that success in intergroup competition provides reproductive advantages for individuals living in large social groups, empirical data (Janson 1985), meta-analyses (Majolo et al. 2008), and literature reviews (van Schaik 1983) have suggested. If, as van Schaik claims, "*intergroup feeding competition ... [is not an] important determinant of an individual's fitness*," this calls into question the need for an adaptive theory for intergroup aggression among primates.

Thus, traditionally neither students of war nor primatologists had much reason to develop a common theory uniting human and primate intergroup aggression. However, in recent decades, the discovery of human-like patterns of killing between neighboring communities of chimpanzees (*Pan troglodytes*) has provoked evolutionary explanations of chimpanzee violence (Manson and Wrangham 1991; Wrangham and Peterson 1996; Wrangham 1999; Wilson and Wrangham 2003; Williams et al. 2004; Wilson et al. 2004; Watts et al. 2006; Boesch et al. 2008), and has inspired parallel development of an evolutionary biology of human warfare (van der Dennen 1995; Thayer 2004; Kelly 2005; Gat 2006; Roscoe 2007; Smith 2007). These efforts, which we review below, suggest that important elements of intergroup violence among humans and chimpanzees can be explained by the hypothesis that groups use aggression to achieve dominance over their neighbors. According to this idea, intergroup dominance promotes fitness by a variety of mechanisms, including access to more land and more females. We call this the intergroup dominance hypothesis.

In this chapter, we consider whether the intergroup dominance hypothesis can account for patterns of IGI among troop-living primates, chimpanzees, and humans.

8.2 Nonhuman Primates Living in Troops

Intergroup relationships have long been of interest to primatologists, as dominance relationships are known to mediate competition for resources at the individual level, and could therefore also do so at the group level (Huntingford and Turner 1987; Dunbar 1988). Yet, few studies have quantitatively investigated the relationships among neighboring primate social groups or explored how group-level dominance influences access to resources. For example, of the 60 studies cited in Cheney's 1987 review of primate IGIs that included information about intergroup dominance relationships, approximately half (25) concluded that such relationships existed. However, most of these studies provided only verbal descriptions of the relationships between social groups and only one-third (20/60) reported the number of interactions on which their conclusions were based. In addition, the majority (28/35) of studies that failed to find evidence for intergroup dominance were of species in which groups defended home ranges as territories, and in which dominance relations are therefore hard to detect (Cheney and Seyfarth 1987). Single interactions between territorial groups rarely result in noticeable boundary changes, but large groups have been shown to have higher quality ranges (Cheney and Seyfarth 1987), make more incursions into their neighbors' ranges (Struhsaker 1967), and expand their range at the expense of smaller neighboring groups (Cheney and Seyfarth 1987). Thus, it is not clear if these studies failed to detect intergroup dominance relationships because such relationships did not exist or because the studies were too short or too narrowly focused to adequately address the question.

To test the intergroup dominance hypothesis, three kinds of data are required. First, numerous observations of encounters between neighboring, habituated primate groups are needed to determine if intergroup dominance relationships exist. Investigating the relationships among several habituated social groups, rather than a single habituated group and its unhabituated neighbors, is critical because the presence of human observers may alter the behavior of unhabituated primates and potentially decrease their competitive ability (Zinner et al. 2001). Large sample sizes are essential because many factors can affect the outcome of intergroup encounters; multivariate statistics may be needed to elucidate how these factors interact to shape the relationships between neighboring groups (e.g., Kitchen et al. 2004a; Pride et al. 2006; Crofoot et al. 2008). Relatively few studies meet both these criteria, and several that do have not yet published relevant analyses (e.g., *Presbytis thomasi*: Steenbeek 1999; Steenbeek and van Schaik 2001, *Cercopithecus mitis*: Cords 2002, 2007). Nevertheless, studies meeting these criteria demonstrate the presence of consistent intergroup relationships in a range of phylogenetically, socially, and ecologically disparate species (see Table 8.1), suggesting that the pattern may well be widespread among primates.

Second, data on how group dominance affects access to resources is required to understand the functional implications of these relationships. Typically, high-ranking groups are predicted to have priority of access to food resources and consume a higher quality diet than their lower-ranking neighbors. However, few studies have collected the detailed behavioral and ecological data required to demonstrate such patterns (but see Table 8.1).

Finally, demographic data are needed to investigate whether the foraging advantages attained through intergroup dominance lead to increased fitness. Members of high-ranking groups are expected to have higher reproductive rates, higher offspring survival rates, and/or lower mortality rates than their counterparts in low-ranking groups.

In short, the combination of detailed behavioral, ecological, and demographic data required to test the intergroup dominance hypothesis is found in only a small number of primates. For this reason, we focus on three species where the data are particularly complete.

8.2.1 Wedge-Capped Capuchins (*Cebus olivaceus*)

While studies of competition between primate social groups sometimes treat numerical superiority as a sufficient proxy for group dominance (e.g., Koenig 2000; Cooper et al. 2004), the balance of power between opposing primate social groups can be influenced by a range of additional factors, including the behavior,

Table 8.1 Intergroup interactions and dominance relationships in nonhuman primates: a selective review

Species	Groups studied	Interactions observed	Length of study	IG dominance?	Description of IG relationship	Effects of IG dominance on foraging success or behavior	Effects of IG dominance on demographics	Reference
Verreaux's sifaka *(Propithecus verreauxi)*	5	19	3 months	Yes	In 10 of the 11 interactions with a clear winner, the group with more individuals (or, if groups were equal in size, with more males) won	Subordinate groups slept in the core area of their range	No data available	Benadi et al. (2008)
White-faced capuchins *(Cebus capucinus)*	6	58	6 months	Yes	Large group size and proximity to home range center increased the likelihood of winning interactions	Groups traveled further, faster, and stopped less frequently after losing interactions. Small groups spent more time feeding and foraging and less time socializing than large groups	No data available	Crofoot et al. (2008), Crofoot (2008)
Tibetan macaques *(Macaca thibetana)*	5	58	7 months	Seasonal	An IG dominance hierarchy based on group size existed in the birth season, but not in the mating season	Dominant groups were able to monopolize tourist feeding sites	No data available	Zhao (1997, 1999)
Savannah baboons (1) *(Papio cynocephalus ursinus)*	4	10	12 months	No	Most encounters did not involve agonism and mingling of troops was observed. Displacements occurred in two of ten encounters	The two groups with more males had larger, higher quality home ranges, used overlap areas more frequently, and had shorter day ranges than their neighbors	The two groups with more males also had higher birth rates and a higher proportion of juveniles	Anderson (1981)
Savannah baboons (2)	4	110	23 months	Yes	Both location and rival group identity influenced the outcome of intergroup encounters, and group-dominance seemed to depend on the relative number of males	No data available	No data available	Kitchen et al. (2004a)

(*continued*)

Table 8.1 (continued)

Species	Groups studied	Interactions observed	Length of study	IG dominance?	Description of IG relationship	Effects of IG dominance on foraging success or behavior	Effects of IG dominance on demographics	Reference
Black and white colobus (*Colobus guereza*)	6	115	19 months	Yes	Group dominance relationships were strong and linear and depended on characteristics of each group's male	High ranking groups had higher quality home ranges than low ranking groups	No data available	Harris (2006a,b)
Japanese macaques (*Macaca fuscata* (Yakushima))	7	151	14 years intermittently	Yes	Dominance relationships were consistent and stable over time. Larger groups were generally dominant to smaller groups, especially when the difference in group size was large	Large social groups frequently displaced their smaller neighbors from food trees	The ratio of infants to adult females was higher in large groups, suggesting that members of large groups had higher reproductive rates	Suzuki et al. (1998), Takahata et al. (1998), Sugiura et al. (2000)
Japanese macaques (*Macaca fuscata* (Kinkazan))	3	63	14 years intermittently	?	Groups rarely displaced one another during intergroup interactions (IGIs) (7/63 interactions), but in cases with clear outcomes, larger social groups tended to dominate their smaller neighbors (5/7 interactions)	Large social groups rarely excluded their smaller neighbors from food resources	Group size was not related to reproductive rate	see Yakushima
Wedge-capped capuchins (*Cebus olivaceus*)	12	169	9 years	Yes	Strong, stable dominance relationships existed among groups. Large groups were dominant to smaller groups, but dominance may have been more closely tied to the number and identity of adult males than group size per se	High-ranking groups had priority of access to high quality areas and consumed fruits with higher sugar contents. They spent less time traveling and foraging than their lower ranking neighbors	Both male and female members of large social groups had higher reproductive rates than their counterparts in small social groups	Robinson (1988), Srikosamatara (1987)

Ring-tailed lemurs (1) *(Lemur catta)*	6	188	12 months	Territorial	Encounter location, rather than group size determined the outcome of IGIs	Large social groups had higher quality home ranges, and their members maintained higher food intake rates than members of small groups except during a period of atypically low food availability. The costs of range defense were lower for individuals living in large social groups	Long-term data from this study site show that reproductive rate decreases with group size, but that females in large social groups may experience lower mortality	Pride et al. (2006), Jolly et al. (2002), Pride (2005a,b)
Ring-tailed lemurs (2)	10		13 years		No data available	No data available	Females in intermediate sized groups had higher reproductive rates than females in either large or small groups	Takahata et al. (2008)

temperament or size of the alpha male (e.g., Harris 2006b), the strength of relationships between group-mates (e.g., Starin 1991) and the location of the interaction (e.g., Pride et al. 2006; Crofoot et al. 2008). For example, using 9 years of data on interactions among 12 wedge-capped capuchin social groups, Robinson (1988) demonstrated a linear dominance hierarchy among groups. Rather than depending solely on group size, this hierarchy was ordered by the number and identity of the adult and subadult males in each group (Robinson 1988). Groups with many males tended to be high-ranking, but in some group dyads, the identity of the interacting males also influenced intergroup relationships, such that a group with a smaller but more potent male cohort outranked a group with a larger number of males.

Group dominance afforded several foraging benefits in this capuchin population. High-ranking groups had greater access to fruit species that were clumped and relatively uncommon, and were able to concentrate their foraging effort in areas with high fruit tree density (Srikosamatara 1987). The fruit species consumed by members of large groups also tended to have higher sugar content than the fruits that made up the majority of small groups' diets. Perhaps to compensate for the costs of subordinacy, low-ranking groups spent more time moving and foraging than high-ranking groups. They traveled further, particularly on days when they encountered one of their neighbors (Srikosamatara 1987), presumably in an effort to make up for decreased foraging efficiency. Such attempts, however, appeared to be ineffective because females belonging to low-ranking groups had lower reproductive rates than their counterparts in high-ranking groups (Robinson 1988).

The demographic ramifications of the relationship between group size and reproductive success in this capuchin population were striking. Because high-ranking groups grew faster than small groups, over time the percentage of the population living in high-ranking groups is expected to increase. However, past a certain size, resource competition within groups is expected to promote group fissioning. The interaction between these opposing pressures structured population growth in Robinson's study population (1988). Low-ranking groups tended to be small and to go extinct, while high-ranking groups grew and eventually "budded off" new small groups. Resource competition between social groups also shaped the genetic structure of this population, as both the female and male members of high-ranking groups contributed disproportionately to population growth, and thus to future generations (Robinson 1988; Valderrama Aramayo 2002).

8.2.2 *Japanese Macaques (Macaca fuscata)*

Relationships between neighboring social groups, and the effect that these relationships have on individual fitness, are expected to depend not only on the physical and social characteristics of the species in question, but also on the distribution and abundance of food and the density of conspecifics in their habitat (Horiuchi 2008). The Japanese macaque populations on Yakushima and Kinkazan Islands illustrate the strong effects that environmental variables can have on intergroup relationships.

These macaque populations have been studied intensively for almost three decades, and numerous comparative studies of their social structure, behavior, ecology, and demography have been undertaken (Yamagiwa 2008). These studies show that intergroup relationships have much larger consequences in the high density Yakushima population than in the low density Kinkazan population.

In Yakushima, relationships among social groups were determined by relative group size (the difference in the number of adults belonging to each group), and interactions between neighbors were found to influence both immediate foraging opportunities and long-term resource access. Dominance relationships among groups were consistent and stable over time (Saito et al. 1998). Larger groups generally displaced smaller groups (74% of interactions: Sugiura et al. 2000) especially when they had a large numeric advantage: when the larger group had at least ten more members than their opponent, they won 94% of interactions (Sugiura et al. 2000). This competitive advantage likely increased the short-term foraging efficiency of females living in large social groups because the majority of IGIs (100/151, i.e., 66.2%) ended with one group displacing the other (Sugiura et al. 2000), and 17% of interactions occurred when two groups simultaneously approached a fruit or nut tree. Thus, members of small groups lost feeding opportunities as a direct consequence of encountering their neighbors. Intergroup dominance also appeared to have long-term consequences because, on average, larger groups had higher reproductive rates, with the number of infants per female increasing linearly with group size (Takahata et al. 1998). This effect emerged, however, only during periods of resource scarcity; birth rates of large (high-ranking) groups were higher than those of their smaller (lower-ranking) neighbors only during years with poor fruit production (Suzuki et al. 1998).

On the island of Kinkazan, approximately 1,300 km northeast of Yakushima, a very different picture of Japanese macaque ecology and behavior emerges. In the Kinkazan population, large group size did not confer competitive or reproductive advantages. Groups encountered one another about one-third as often (0.012 encounters/hour on Kinkazan compared to 0.039 encounters/hour on Yakushima), and fewer interactions were agonistic (11% vs. 49% in Kinkazan and Yakushima, respectively) or involved one group displacing the other (11% on Kinkazan compared to 66% in Yakushima: Sugiura et al. 2000). Although bigger social groups were dominant in five of the seven interactions with a clear winner (Sugiura et al. 2000), these rare displacements did not appear to affect the resource access of smaller groups. As expected, therefore, larger groups did not have higher reproductive rates than their smaller neighbors. Group size was unrelated to reproductive rate both in years with good fruit crops and in years with poor fruit crops (Suzuki et al. 1998).

Why were intergroup relationships different in the Yakushima and Kinkazan populations? In theory, Kinkazan population density could have been low relative to food resources, thanks to hunting, predation, or disease. However, in groups of all sizes on Kinkazan, birth rates were higher during years with large fruit crops than in years with small fruit crops (Suzuki et al. 1998), indicating that the population was food limited. Furthermore, the Kinkazan macaques had no

Table 8.2 Comparison of the Kinkazan and Yakushima population of Japanese macaques

	Yakushima	Kinkazan
Intergroup interaction (IGI) rate[a]	0.039/h	0.012/h
Aggressive IGIs[a]	70/151 encounters	7/63 encounters
Displacements[a]	100/151 encounters	7/63 encounters
Troop density[b]	4.7/km^2	0.2/km^2
Average distance between home range centers[b]	361 m	1,232 m
Home range size[b]	90 ha	221 ha
Average home range overlap[b]	58.70%	55.10%
Food tree density[b]	1,802/ha	94/ha
Average distance between feeding bouts[b]	70 m	151 m
Average travel speed[b]	2.19 m/min	3.08 m/min

[a]Sugiura et al. (2000)
[b]Maruhashi et al. (1998)

predators (Takahata et al. 1998). The Kinkazan population, therefore, did not appear to be living below carrying capacity.

Alternatively, the distribution and/or abundance of food resources in Kinkazan might have made resource defense less economical than in Yakushima. If so, feeding competition between groups is expected to be less intense (Wrangham 1980). Certainly, differences in resource distribution were clear (see Table 8.2). Yakushima had a higher density of food patches than Kinkazan, i.e., 19 times more food trees per hectare (Maruhashi et al. 1998), and a higher overall food abundance (the basal area of food trees per hectare was 2.2 times greater than in Kinkazan). On the other hand, the average size of food trees in Kinkazan (those providing fruits and nuts) was larger (Maruhashi et al. 1998). These differences were correlated with differences in foraging behavior. In Kinkazan, macaques had larger, more evenly used home ranges, and traveled faster and further between feeding bouts, indicating that they worked harder to meet their metabolic requirements (Maruhashi et al. 1998). The more dispersed food sources and lower food abundance in Kinkazan were thus associated with greater foraging effort, suggesting groups competed primarily via scramble rather than contest competition. By contrast, the higher resource density in Yakushima could have increased the profitability of active resource defense, leading to a fitness advantage for individuals living in larger and higher-ranking groups. Further research is thus needed to assess the importance of differences in food distribution between Yakushima and Kinkazan, and to distinguish effects of resource distribution and abundance from those due to differences in population density and encounter rates (Horiuchi 2008).

8.2.3 Ring-Tailed Lemurs (*Lemur catta*)

Most, if not all, primate groups share some portion of their home range with their neighbors. In the overlap zone, intergroup aggression can either occur over specific food patches or over space. Dominance in these interactions can be mediated by the characteristics of the groups, as in wedge-capped capuchins where home ranges overlapped completely and dominant social groups defeated their subordinate neighbors throughout the entire area (Robinson 1988) (i.e., absolute dominance *sensu* Kaufmann 1983). In other species, dominance can be context-dependent because it depends on the location of the interaction (e.g., *Lemur catta*: Pride et al. 2006, *C. mitis*: Cords 2002) (i.e., relative dominance *sensu* Kaufmann 1983). Ring-tailed lemurs in Berenty provide the best-studied example of the latter system. In this species a group's fitness appears to depend on its ability to maintain the ownership of a high-quality area (Pride et al. 2006).

In Berenty, groups of ring-tailed lemurs win interactions in their "typical" ranges (their 85% minimum convex polygon home range), and tend to lose outside these areas, regardless of the strength of the opposing group (Pride et al. 2006). Why large social groups are not able to overpower their smaller neighbors is not understood. This problem presents a challenge for the intergroup dominance hypothesis by calling into question whether, in territorial species, groups are able to translate competitive superiority into increased resource access or higher fitness. Although large social groups do not seem to have a competitive advantage in any single IGI, they might achieve high foraging success (and therefore high fitness) by using their power advantage to defend territories of superior quality. The simplest way to test this is by assessing the long-term effect of group size on reproductive rate.

Two studies have yielded conflicting data on this point. Jolly et al. (2002) found that reproductive rate of lemurs at Berenty decreased with group size, thus indicating no benefits for larger groups. In contrast, Takahata et al. (2006) reported that groups with an intermediate number of adult females had higher reproductive rates than those with either few or many females. Since elevated within-group competition is expected in large groups and was demonstrated in their study, Takahata et al. (2008) concluded that their data, based on ten groups over 13 years, supported the hypothesis that large groups use social dominance over smaller groups to achieve higher fitness.

The discrepancy between the results of Jolly et al. (2002) and Takahata et al. (2006, 2008) has not been fully explained. Takahata et al. (2006) suggest that differences in population density may be responsible, as their study groups were in a high-density area of Berenty (542.3 individuals per km^2), whereas Jolly's study included groups from a range of habitats with a broad range of densities (100–580 individuals per km^2 in the scrub forest and near the tourist station, respectively). However, this explanation is not fully supported because Jolly et al. (2002) found a negative relationship between group size and reproductive rate even in the groups near the tourist station where population density was highest.

Ring-tailed lemurs in Jolly et al.'s (2002) study thus show that large group size does not necessarily confer a reproductive advantage. Nevertheless, two additional lines of evidence from this population provide support for the intergroup dominance hypothesis. First, Pride et al. (2006) demonstrated that large groups defend higher quality home ranges than small groups, and are able to do so at a lower cost to individual members. Females in large social groups are able to share the burden of territorial defense with more group mates, and thus can maintain home ranges in the most productive and stable areas at lower *per capita* cost of defense (Pride et al. 2006). Second, members of large social groups may have reduced mortality compared to members of smaller groups. Pride (2005b) demonstrated that glucocorticoid levels in females at Berenty predicted individual survival, and that females in large social groups had lower cortisol levels than females in small groups (Pride 2005a). This pattern may be explained by the fact that individual participation in intergroup contests declined with increasing group size (Pride et al. 2006). The costs of territorial defense and intergroup resource competition thus seem to be lower for females in large groups because they are shared among more individuals, and this appears to have long term consequences for both survival and individual fitness. In addition, competition for space is intense among ring-tailed lemurs, and members of groups that lose control over their "typical" areas experience high mortality (Koyama 1991; Hood and Jolly 1995; Jolly and Pride 1999; Koyama et al. 2002; Gould et al. 2003). For example, Jolly and Pride (1999) recorded a group of ring-tailed lemurs expanding and fissioning over a 6-year period. In this case, the group expanded as a result of increased resources coming from human food (a tourist project). It reached 19 individuals, compared with group sizes of 3–12 in 11 other groups in the same area, and then fissioned. The two daughter groups were both successful, one using aggression to extend its range at the expense of a neighboring group and the other entirely taking over a neighboring range (Jolly and Pride 1999). Thus, it is possible that even if smaller (and less dominant) groups have higher reproductive rates, their long-term fitness is reduced by the mortality risks associated with range loss and group extinction.

8.2.4 Troop-Living Primates: Discussion

Wedge-capped capuchins, Japanese macaques, and ring-tailed lemurs provide rare examples of relatively complete studies of the long-term consequences of intergroup dominance in troop-living primates. In wedge-capped capuchins, the intergroup dominance hypothesis was clearly supported, because groups had predictable dominance relationships that depended on fighting power, and members of higher-ranking groups had access to better resources and achieved higher reproductive rates. In Japanese macaques, a high-density population experienced a similar dynamic, whereas a low-density population did not. The effects of intergroup dominance were more complicated in ring-tailed lemurs. Although it is not clear whether members of large social groups have higher reproductive rates than members of small groups, the data suggest that they experienced reduced mortality.

Similar evidence of the importance of long-term survival comes from toque macaques (*Macaca sinica*, Dittus 1986). For 7 years, a group of 8–15 females consistently dominated a neighboring group of 7–11 females in conflicts over feeding sites, yet during this period, the reproductive rates of the two groups were not significantly different. However, the dominant group then took over the range of its neighbor, and females in the subordinate group became members of the dominant group. Over the next 8 years, differences in reproduction and survival led to the lineages of the dominant group having 20 females, compared to one descendant from those in the subordinate group. This case suggests that over the long term, the ability to control and defend a home range may be more important for fitness maximization than short-term measures of reproductive rates. Where dominant groups do not achieve high reproductive rates, they may alternatively have higher rates of individual survival and/or superior physical condition.

Although data on the intergroup relations of troop-living primates remain too sparse to provide a definitive test of the intergroup dominance hypothesis, intergroup dominance relationships have been shown to be important in an ecologically and socially diverse set of species. These include both frugivores (*Cebus olivaceus*: Robinson 1988) and folivores (*Colobus guereza*: Harris 2006a,b); territorial species (*L. catta*: Jolly et al. 2002) and species with completely overlapping home ranges (*Cebus olivaceus*: Robinson 1988); and in primates with one-male social systems (*C. guereza*: Harris 2006a, b) and multi-male social systems (*M. fuscata*: Sugiura et al. 2000). In each case, the results challenge van Schaik's (1983) contention that intergroup feeding competition is not important in determining an individual's fitness, and they indicate that when long-term data are available, the intergroup dominance hypothesis is at least partly supported, i.e., that troop-living primates can achieve long-term benefits from success in coalitionary aggression against neighbors. Since hostile IGIs often have no obvious immediate effect in determining access to a particular food patch, these case studies suggest that the long-term implications merit further research.

The behavioral implication of the intergroup dominance hypothesis is that conflicts between groups are not necessarily over immediate access to resources, but instead can represent fights over dominance status. Despite indications from wedge-capped capuchins, Japanese macaques, and ring-tailed lemurs that the benefits of winning such contests are high, escalated aggression rarely occurs during intergroup conflicts in troop-living primates. One possible explanation is that the costs of aggression are high. When the outcome of a conflict can be predicted based either on previous interactions or on physical characteristics of the participants, and when the cost of interacting is high, weaker opponents are expected to withdraw rather than engage in a risky fight they are likely to lose (Maynard Smith and Parker 1976). Neighboring primate social groups interact with one another regularly and thus levels of intergroup aggression may be relatively low because the outcome is a foregone conclusion. In addition, the social groups of troop-living primates are, by definition, cohesive and thus intergroup aggression in these species rarely involves the imbalances of power that are implicated in lethal aggression of chimpanzees and human foragers (below). Observations of intergroup killings in capuchin monkeys lend support to this hypothesis because they suggest that troop-living primates will participate in escalated aggression if the costs are sufficiently low (Gros-Louis et al. 2003). In this instance, the coalitionary nature of the attacks meant that the aggressors could inflict serious wounds on their victim without risking substantial injury themselves (Gros-Louis et al. 2003).

An alternative explanation for the low intensity of intergroup aggression observed in troop-living primates is that the collective action problem inherent in group-level resource competition presents an obstacle to high individual investment (Nunn and Lewis 2001; Nunn and Deaner 2004; Kitchen and Beehner 2007). Why should any individual risk injury by participating in intergroup fights when the benefits gained through such confrontations will be enjoyed by all group members, including individuals that did not take part in securing them? Participation in aggressive intergroup encounters is highly variable (Wilson et al. 2001; Wich et al. 2002a, b; Kitchen 2004, 2006; Kitchen et al. 2004b), and why some individuals rush boldly towards an opposing group, risking injury, while others hang back and watch the excitement from a safe distance remains poorly understood (Kitchen and Beehner 2007). The fact that intergroup dominance relationships exist in a number of troop-living species clearly suggests that primates are able to overcome this collective action problem, but further study is required to demonstrate how this is accomplished.

8.3 Chimpanzees

Chimpanzees form social communities that occupy a stable home range. Within communities, individuals forage in parties (sub-groups) of variable size and composition, including sometimes being alone. Intercommunity interactions often occur only at long distance, mostly through auditory contact. In three populations (Taï, Mahale, Ngogo), they occurred at similar rates, 1–1.5 times per month. Aggression is the principal form of interaction between communities. It occurs mostly when parties meet by chance, but also when one party deliberately approaches another, sometimes by stealth.

The principal actors are adult males and there are two main types of interaction. Battles involve mostly bluff, including numerous calls and aggressive charges made alone or jointly towards opponents. Physical contact is occasional and generally mild, though it can lead to one individual being separated and attacked, and in Taï, it includes herding and temporary forced consortships of females (Boesch et al. 2008). Battles may continue in the same location for up to 45 min (Wrangham pers. observ.), and normally end with one or both sides retreating. Attacks, by contrast, involve a coalition of at least two and generally four or more males violently attacking a member of the neighboring community. Attacks occur both when parties meet by chance and when one party searches for potential victims during boundary patrols or after detecting them at long distance. Attacks are much less common than battles (Wilson and Wrangham 2003; Watts et al. 2006; Boesch et al. 2008).

While chimpanzee intercommunity relations have not been studied in depth, they appear to conform to the three components of the intergroup dominance hypothesis. First, relationships between communities are generally predictable. For example, in Gombe parties from the larger Kasekela community consistently defeated those from the smaller Kahama community (Goodall 1986). Nishida et al. (1985) found a similar relationship at Mahale for M-group and K-group. However, the outcome of specific interactions depends on the local context, such as the relative numbers of males in each party, so parties from small communities can sometimes win intergroup encounters (Wrangham 1999; Boesch et al. 2008).

Second, communities that win interactions improve their access to resources. Thus, two dominant communities in Gombe and Mahale permanently extended their territories at the expense of their neighbors (Nishida et al. 1985; Goodall 1986; Williams et al. 2004). The M-group community in Mahale also exploited its dominance seasonally by taking control of an area normally occupied by the neighboring K-group, whenever the principal food-plant species in the shared area came into fruit.

Third, success in intercommunity aggression had fitness pay-offs. In particular, the dominant Gombe community experienced variation in territory size, which was suspected to result from varying success in competition with neighboring communities. Larger territory size was associated with several indications of greater access to resources, including higher individual body weights and larger parties, and fitness gains are indicated by shorter interbirth intervals and higher infant survival (Williams et al. 2004). Additionally, subordinate communities have twice been observed to go extinct, apparently as a result of aggression from dominant neighbors (Kahama at Gombe, K-group at Mahale). While some individuals from these subordinate communities survived the dissolution of their groups, almost all males died and the females who were known to survive experienced high rates of infanticide (Nishida et al. 1985).

Intercommunity dominance accordingly appears to be beneficial for chimpanzees because it gives both sexes increased access to resources, while males can also gain increased access to females. The question that links intergroup dominance in chimpanzees to human warfare is why intergroup contests are so much more aggressive among chimpanzees than among troop-living primates. In particular, why do chimpanzees sometimes violently attack and kill members of neighboring communities?

According to the imbalance-of-power hypothesis, the fission–fusion social organization of chimpanzees facilitates lethal aggression against members of neighboring groups. Chimpanzees form temporary subgroups that vary in size, so parties with several males sometimes encounter lone males or isolated mothers from neighboring groups. When loners meet large parties, aggressive power is distributed so asymmetrically that the dominant party can afford to express intense violence while experiencing a very low risk of being hurt themselves. The proposed advantage of damaging or killing an opponent is that by reducing the number of coalitionary aggressors in the neighboring community, the attackers increase the relative power of their community. As a result, they become more likely to win future interactions, and therefore to achieve the fitness gains accruing from elevated intercommunity dominance (Manson and Wrangham 1991; Wrangham 1999; Williams et al. 2004; Wilson et al. 2004; Watts et al. 2006; Sherrow and Amsler 2007). The imbalance-of-power hypothesis predicts that the aggressors will be members of the philopatric sex, whether females (as in spotted hyenas, *Crocuta crocuta*) or males (as in chimpanzees) (Wrangham 1999).

The proposal that an asymmetry of power tends to induce attack is supported by data from Gombe, Mahale, Kibale and Taï on the contexts of aggression. For example, in 20 cases recorded by Watts et al. (2006) involving the Ngogo community in Kibale, attacks were conducted by at least three individuals on a victim that was either alone when encountered, or was rapidly isolated from the rest of his/her party. A victim who has members of his/her own community nearby has sometimes been supported and rescued (Boesch et al. 2008). Thus, where power is more evenly balanced, attacks are less likely or can be stopped. The importance of power asymmetry is also indicated experimentally by playbacks showing that the probability of males approaching the location of a male stranger's call, or the speed at which they do so, is predictably increased by the number of males in the listening party (Wilson et al. 2001). As expected, border zones tend to be avoided in general, and males in small parties are particularly unlikely to visit them (Wilson et al. 2001; Wrangham et al. 2007). In sum, the power asymmetries made possible by fission–fusion grouping make lethal violence cheap, provided that aggressors can assess the relative fighting ability of parties correctly.

If escalated aggression is cheap and serves to increase the future dominance of the aggressors' community, it should be directed towards the most effective fighters among the neighbors. Females are not active aggressors in intercommunity interactions in most sites. However, in Taï, females can take part, perhaps because parties there are more stable than elsewhere, power asymmetries are reduced, and intercommunity attacks are rare (Boesch et al. 2008). In other sites attacks are more common and are indeed directed mostly at males. For instance, the probability of attacks on strangers at Gombe was 100% for males ($n = 6$ single males, 16 in parties), <60% for females without sexual swellings ($n = 51$) and <20% for females with sexual swellings ($n = 23$) (Williams et al. 2004). The sex difference is particularly pronounced for lethal aggression. In a review of data from five populations including 16 known and 16 suspected cases of adult deaths from intercommunity aggression, Wrangham et al. (2006) found that 94% of the victims were adult males ($n = 30$ deaths). Intercommunity aggression also involves attacks on infants. Unfortunately, observers can rarely detect the sex of infant victims, but of eight cases where the sex of the victim was known, six were male (75%) (Wrangham et al. 2006).

The imbalance-of-power hypothesis is thus supported by evidence that chimpanzees are sensitive to power imbalances, tend to reduce the number of males in neighboring communities, and gain fitness advantages by doing so. Competing hypotheses, to explain why chimpanzees make deliberate attacks on victims who are outnumbered and over-powered, have mostly focused on the proximate stimuli eliciting violence, and receive little support (Williams et al. 2004; Wilson et al. 2004). First, chimpanzees could, in theory, have a generalized tendency to attack unfamiliar individuals. However, as we have seen, the likelihood of an attack depends on context. Second, specific individuals might be particularly prone to violence. However, although individual variation has been shown for predatory aggression by chimpanzees (Gilby et al. 2008) and for rank-related frequencies of intracommunity aggression (Muller and Wrangham 2004), Wilson et al. (2001) found equally strong responses to playbacks of strangers among all seven adult males in their study. Third, attacks could be provoked by immediate competition over resources. Relevant stimuli could include the presence of sexually active females, the presence of preferred food patches, a season of ecological stress, or a long-term shortage of land or females. None of these has yet been demonstrated to be important, however (Wilson et al. 2004).

The imbalance-of-power hypothesis predicts that chimpanzees will rarely take risks as aggressors in intercommunity interactions. Against this, Boesch et al. (2008: 531) suggest that "*chimpanzees can take large risks when potential benefits are large or when failure to do so could inflict larger costs*." Aggressors have rarely been wounded to date, but further data will enable these alternatives to be more finely discriminated. In particular, evidence that aggressors expose themselves to risk will suggest that competition over detectable resources is more important than current data indicate. At present, the propensity for chimpanzees to violently attack neighbors appears to be best explained by the intergroup dominance hypothesis, including a tendency to use attacks to weaken the neighbors whenever possible.

Chimpanzees are the best-studied primate living in fission–fusion communities, but spider monkeys (*Ateles* spp.) have similar patterns of grouping and territoriality. According to the imbalance-of-power hypothesis, therefore, spider monkeys should show parallel forms of intergroup violence. No intergroup killing has yet been seen in spider monkeys, but recent observations suggest that important elements of their patterns of aggression are similar to those in chimpanzees. In particular, spider monkeys show active defense of territories, larger parties tend to win interactions, and small parties avoid the border zone (Aureli et al. 2006; Wallace 2008). Males have been seen on intergroup raids making incursion into neighboring ranges and attacking lone individuals (Aureli et al. 2006). Cooperative killing has been seen within groups (Campbell 2006; Valero et al. 2006). On the basis of these observations, the imbalance-of-power hypothesis predicts that lethal attacks will eventually be found also in spider monkeys.

The implication of the imbalance-of-power hypothesis for chimpanzees is that selection has favored the propensity to attack male neighbors whenever the costs are perceived to be sufficiently low. Roscoe (2007) presents an alternative idea. He proposes that the reason why unprovoked attacks on strangers occur in chimpanzees more than other nonhuman primates is that chimpanzees are exceptionally intelligent. As a result, he argues, the attackers are so skilled at assessing the long-term benefits that they can evaluate the merits of a risky attack. The cognitive demands implied by Roscoe's proposal are high. According to Roscoe's idea, a chimpanzee is expected to perceive that a violent attack will lead to a reduction in the fighting power of the neighboring group, and hence to an increased likelihood of the aggressors' community winning intercommunity interactions. The chimpanzees should then be able to realize from this that they will obtain increased access to resources. The cognitive challenges seem to us too great for this scenario to be realistic, and we believe that a more parsimonious explanation is that, faced with an uncertain long-term pay-off, chimpanzees are motivated by a psychological reward system that has been favored evolutionarily by the benefits that tend to accrue to judicious killers.

8.4 Humans

War is sometimes defined as being a more exclusive activity than intergroup aggression. For instance, Kelly (2000) and Fry (2006) defined warfare to exclude feuding. Such a definition means that warfare is not considered to have occurred among the Andamanese, for example, even though members of neighboring tribes killed each other whenever they met vulnerable opponents (Fry 2006). Similarly, Fry (2006) considered that among the Murngin, an Australian aborigine group, there was no war even though they practiced six types of warfare according to Warner (1958), their principal ethnographer. For example, "maringo" was defined by Warner (1958: 166) as "*Surprise attack by group, in revenge. Always woundings or death.*"

To avoid confusion and allow easy comparisons with primates, here we define warfare inclusively to mean IGIs among humans, in which coalitions attempt to aggressively dominate or kill members of other groups. Using this definition, warfare is characteristic of most human societies. The few in which it has been recently absent tend to be societies that were politically dominated by their neighbors (Fry 2006).

While cultural and socio-political diversity makes generalization difficult, two broad styles of warfare can be recognized, below and above the military horizon (Turney-High 1949). Below the military horizon, warfare is conducted anarchically in the sense that individuals cannot be ordered to participate. Most interactions involve asymmetric attacks, made either opportunistically or as a result of a deliberate plan. In the cases of planned attacks, the typical motivation is revenge for prior killings. Attacks can continue into a massacre if power is sufficiently imbalanced. Males are the chief targets, but children and women can also be killed. Battles involving deliberate confrontation of opposing sides are rare, though not unknown. When battles occur, they tend to stop after a few deaths. This style of warfare is characteristic of hunter-gatherers and small-scale farming societies (Gat this volume). Hostility is often unrelenting between tribes with different language-groups ("external war"). Within tribes, groups tend to oscillate between conditions of war ("internal war") and peace, often brought about by explicit peace-making ceremonies (Wright 1942; Turney-High 1949; Keegan 1993).

Above the military horizon, warfare is practiced by armies, i.e., institutions in which leaders devise plans and have the power to order soldiers into battle. While asymmetric attacks remain common, battles are especially prominent in warfare above the military horizon. Battles are rarely opportunistic and often require the leaders of opposing forces to agree where and when to fight. The leaders' motivation for fighting includes complex political considerations, and tends to be aimed at destroying or subjugating the opposing army (Keegan 1993; Collins 2008). The soldiers' motivation for fighting varies widely. Individuals may fight from a sense of duty; they may wish to fight out of patriotism or opportunities for loot; or they may fear the consequences of not fighting, such as being killed by the enemy, being killed by their leaders, or letting down their immediate comrades (Keegan 1993).

Human warfare clearly conforms to the intergroup dominance hypothesis, because intergroup dominance relationships are routinely stable for years at a time, and they predict access to resources such as valuable locations or trade routes. Dominant groups also commonly flourish by expanding their territorial ranges or by restricting the access to resources of individuals belonging to subordinate groups. Gat (2006) gives numerous examples.

Warfare also appears broadly to fit the imbalance-of-power hypothesis. The fit is particularly clear below the military horizon, though in humans, there are more sources of power asymmetry than in chimpanzees. As in chimpanzees, power asymmetry between opposing sides comes both from differences in party size and from one side having the element of surprise. In addition, humans routinely use night-time attacks (often initiating attacks just before dawn), and devise special tactics and weapons to give themselves a military advantage. Practitioners of internal war also use deceit (ensnaring familiars by guile) and treachery (betrayal of a trust) to establish a power advantage (Turney-High 1949; Zegwaard 1959; Wadley 2003; Gat 2006). The use of such tactics and the tendency to avoid battles suggest that most killing below the military horizon occurs during asymmetric interactions in which the killers experience low risk of being injured (e.g., Chagnon 1997). These generalizations suggest that the pattern of warfare among foragers and small-scale farming tribes largely conforms to the imbalance-of-power hypothesis.

Above the military horizon, the calculus is more complex because of the distinction between leaders (who are motivated to fight or direct others to fight) and soldiers (who may be reluctant to fight). The existence of hierarchical relationships between leaders and soldiers means that leaders can take substantial risks, deliberately allowing their armies to sustain large casualties. The lack of leaders in chimpanzees or hunter-gatherers, therefore, contributes to explaining why they rarely have lethal battles. Nevertheless, although a steep military hierarchy means that warfare above the military horizon does not necessarily conform to the imbalance-of-power hypothesis, we conjecture that within battles, and in numerous engagements during prolonged wars, aggressive interactions tend to be conducted according to the imbalance-of-power hypothesis. For example, military analyses tend to find that most deaths occur not from direct confrontation, but as a result of killing by the winning side, typically of soldiers who are helpless because they are in retreat or have been captured (Collins 2008).

As for chimpanzees, coalitions of humans with a large power imbalance in their favor could kill opponents either as a result of rational calculation or from emotional satisfaction (Roscoe 2007). Both factors seem likely to apply.

8.5 Discussion

Our central question is whether intergroup aggression can be explained by the same principles among troop-living primates, chimpanzees, and humans. Our review suggests that in each case, the intergroup dominance hypothesis has substantial explanatory power. Our findings differ from a number of recent reviews, which have supported van Schaik's claim that large group size does not provide functionally significant benefits in terms of resource competition in primates (Silk 2007; Majolo et al. 2008). A critical component of our analysis, which may explain this discrepancy, is that we focus on long-term rather than short-term reproductive consequences of intergroup competition.

We note two ramifications. First, the intergroup dominance hypothesis suggests that due to the social structuring of primate populations, individuals from dominant social groups are expected to contribute disproportionately to future generations. This indicates that source-sink dynamics will influence primate and human evolution with respect to intergroup aggression. Thus, in preferred habitats, groups are expected to occur at high density and to act as genetic sources, exporting genes to subpopulations in more marginal habitats. Groups within these successful subpopulations should compete aggressively, and success in competition will, therefore, lead to high fitness for individuals that have evolved to fight well against neighboring groups. Dominant groups are thus expected to export genes which promote success in intergroup aggression. This may contribute to explaining why aggressive intergroup relations sometimes prevail in populations where intergroup aggression provides no obvious benefits to dominant social groups, as discussed for Japanese macaques.

Second, the evidence that dominant groups tend to have a fitness advantage in nonhuman primates implies that many of the psychological mechanisms underlying success in intergroup competition may be similar in humans and other primates. Such mechanisms have hardly been studied. Wrangham (1999) suggested that for chimpanzees, they might include the experience of a victory thrill, an enjoyment of the chase, a tendency for easy dehumanization (or its equivalent for nonhuman primates) and deindividuation, ready coalition formation, and sophisticated assessment of power differentials in the context of intergroup conflict. Depending on the species (e.g., how important coalitions are within groups, or how often each sex participates in aggression between groups), such mechanisms may be differentiated by sex. The evidence that intergroup dominance is often critical in group-living primates thus provokes a series of questions about the degree of similarity and difference in the psychological mechanisms underlying coalitionary aggression between humans and other species. (see Gat, this volume).

In sum, there are notable behavioral and functional similarities between human warfare and intergroup aggression among nonhuman primates. They suggest that coalitionary aggression in both systems is explicable by promoting intergroup dominance and therefore tending to promote the aggressors' fitness. There are also important differences between human warfare and primate intergroup aggression, particularly above the military horizon where the interests of leaders and followers are often in conflict and where lethal battles are a prominent feature. The conceptual framework provided by the relatively simple case of nonhuman primates is, therefore, merely a starting-point for understanding the behavioral ecology and evolutionary psychology of warfare. Some basic outstanding evolutionary problems in the study of warfare include a fuller accounting of individual costs and benefits (such as the extent to which warriors are altruistic), understanding the nature and importance of the emotional rewards experienced by fighters, and understanding the role of social rewards conferred on warriors as a way to increase aggressive motivations. Studies of the evolution of war are promising, but they are at a very early stage.

Acknowledgments We are grateful to Robert Hinde, Joan Silk, and two anonymous reviewers for comments.

References

Anderson CM (1981) Intertroop relations of chacma baboon (*Papio ursinus*). Int J Primatol 2:285–310

Aureli F, Schaffner CM, Verpooten J, Slater K, Ramos-Fernandez G (2006) Raiding parties of male spider monkeys: insights into human warfare? Am J Phys Anthropol 131:486–497

Benadi G, Fichtel C, Kappeler PM (2008) Intergroup relations and home range use in Verreaux's sifaka (*Propithecus verreauxi*). Am J Primatol 70:956–965

Boesch C, Crockford C, Herbinger I, Wittig R, Moebius Y, Normand E (2008) Intergroup conflicts among chimpanzees in Tai National Park: lethal violence and the female perspective. Am J Primatol 70:519–532

Campbell CJ (2006) Lethal intragroup aggression by adult male spider monkeys (*Ateles geoffroyi*). Am J Primatol 68:1197–1201

Chagnon NA (1997) Yanomamö. Harcourt College, Fort Worth, TX

Cheney DL (1987) Interactions and relationships between groups. In: Smuts BB, Cheney DL, Seyfarth RM, Wrangham RW, Struhsaker TT (eds) Primate societies. University of Chicago Press, Chicago, pp 267–281

Cheney DL, Seyfarth RM (1987) The influence of intergroup competition on the survival and reproduction of female vervet monkeys. Behav Ecol Sociobiol 21:375–386

Collins R (2008) Violence: a micro-sociological theory. Princeton University Press, Princeton

Cooper MA, Aureli F, Singh M (2004) Between-group encounters among bonnet macaques (*Macaca radiata*). Behav Ecol Sociobiol 56:217–227

Cords M (2002) Friendship among adult female blue monkeys (*Cercopithecus mitis*). Behaviour 139:291–314

Cords M (2007) Variable participation in the defense of communal feeding territories by blue monkeys in the Kakamega Forest, Kenya. Behaviour 144:1537–1550

Crofoot MC (2008) Intergroup competition in white-faced capuchin monkeys (*Cebus capucinus*). PhD thesis, Harvard University, Cambridge, MA

Crofoot MC, Gilby IC, Wikelski MC, Kays RW (2008) Interaction location outweighs the competitive advantage of numerical superiority in *Cebus capucinus* intergroup contests. Proc Natl Acad Sci USA 105:577–581

Dittus WPJ (1986) Sex differences in fitness following a group take-over among toque macaques: testing models of social evolution. Behav Ecol Sociobiol 19:257–266

Dunbar RIM (1988) Primate social systems. Croom Helm, London
Fashing PJ (2001) Male and female strategies during intergroup encounters in guerezas (*Colobus guereza*): evidence for resource defense mediated through males and a comparison with other primates. Behav Ecol Sociobiol 50:219–230
Ferguson RB (2000) The causes and origins of 'primitive warfare': on evolved motivations for war. Anthropol Q 73:159–170
Fry DP (2006) The human potential for peace: an anthropological challenge to assumptions about war and violence. Oxford University Press, New York
Gat A (2006) War in human civilization. Oxford University Press, New York
Gilby IC, Eberly LE, Wrangham RW (2008) Economic profitability of social predation among wild chimpanzees: individual variation promotes cooperation. Anim Behav 75:351–360
Goodall J (1986) The chimpanzees of Gombe: patterns of behavior. Harvard University Press, Cambridge, MA
Gould L, Sussman RW, Sauther ML (2003) Demographic and life-history patterns in a population of ring-tailed lemurs (*Lemur catta*) at Beza Mahafaly Reserve, Madagascar: a 15-year perspective. Am J Phys Anthropol 120:182–194
Gros-Louis J, Perry S, Manson JH (2003) Violent coalitionary attacks and intraspecific killing in wild white-faced capuchin monkeys (*Cebus capucinus*). Primates 44:341–346
Harris TR (2005) Roaring, intergroup aggression, and feeding competition in black and white colobus monkeys (*Colobus guereza*) at Kanyawara, Kibale National Park, Uganda. PhD thesis, Yale University, New Haven, NJ
Harris TR (2006a) Between-group contest competition for food in a highly folivorous population of black and white colobus monkeys (*Colobus guereza*). Behav Ecol Sociobiol 61: 317–329
Harris TR (2006b) Within- and among-male variation in roaring by black and white colobus monkeys (*Colobus guereza*): what does it reveal about function? Behaviour 143:197–218
Harris TR (2007) Testing mate, resource and infant defence functions of intergroup aggression in non-human primates: issues and methodology. Behaviour 144:1521–1535
Harrison MJS (1983) Territorial behaviour in the green monkey, *Cercopithecus sabaeus*: seasonal defense of local food supplies. Behav Ecol Sociobiol 12:85–94
Hinde RA (1993) Aggression and war: individuals, groups, and states. In: Tetlock PE, Husbands JL, Jervis R, Stern PC, Tilly C (eds) Behavior, society, and international conflict. Oxford University Press, Oxford, pp 8–70
Hood LC, Jolly A (1995) Troop fission in female *Lemur catta* at Berenty Reserve, Madagascar. Int J Primatol 16:997–1015
Horiuchi S (2008) High population density promotes the evolution of ownership. Ecol Res 23:551–556
Huntingford FA, Turner AK (1987) Animal conflict. Chapman & Hall, London
Janson CH (1985) Aggressive competition and individual food consumption in wild brown capuchin monkeys (*Cebus apella*). Behav Ecol Sociobiol 18:125–138
Jolly A, Pride E (1999) Troop histories and range inertia of *Lemur catta* at Berenty, Madagascar: a 33-year perspective. Int J Primatol 20:359–373
Jolly A, Dobson A, Rasamimanana HM, Walker J, O'Connor S, Solberg M, Perel V (2002) Demography of *Lemur catta* at Berenty Reserve, Madagascar: effects of troop size, habitat and rainfall. Int J Primatol 23:327–353
Kaufmann JH (1983) On the definitions and functions of dominance and territoriality. Biol Rev 58:1–20
Keegan J (1993) A history of war. Alfred A Knopf, New York
Kelly RC (2000) Warless societies and the origin of war. University of Michigan Press, Ann Arbor
Kelly RC (2005) The evolution of lethal intergroup violence. Proc Natl Acad Sci USA 102:15294–15298
Kinnaird MF (1992) Variable resource defense by the Tana River crested mangabey. Behav Ecol Sociobiol 31:115–122

Kitchen DM (2004) Alpha male black howler monkey responses to loud calls: effect of numeric odds, male companion behaviour and reproductive investment. Anim Behav 67:125–139
Kitchen DM (2006) Experimental test of female black howler monkey (*Alouatta pigra*) responses to loud calls from potentially infanticidal males: effects of numeric odds, vulnerable offspring, and companion behavior. Am J Phys Anthropol 131:73–83
Kitchen DM, Beehner JC (2007) Factors affecting individual participation in group-level aggression among non-human primates. Behaviour 144:1551–1581
Kitchen DM, Cheney DL, Seyfarth RM (2004a) Factors mediating inter-group encounters in savannah baboons (*Papio cynocephalus ursinus*). Behaviour 141:197–218
Kitchen DM, Horwich RH, James RA (2004b) Subordinate male black howler monkey (*Alouatta pigra*) responses to loud calls: experimental evidence for the effects of intra-group male relationships and age. Behaviour 141:703–723
Koenig A (2000) Competitive regimes in forest-dwelling Hanuman langur females (*Semnopithecus entellus*). Behav Ecol Sociobiol 48:93–109
Koyama N (1991) Troop division and inter-troop relations of ring-tailed lemurs (*Lemur catta*) at Berenty, Madagascar. In: Ehara A, Kimura K, Takenaka O, Iwamonto M (eds) Primatology today. Elsevier Science, Amsterdam, pp 173–176
Koyama N, Nakamichi M, Ichino S, Takahata Y (2002) Population and social dynamics changes in ring-tailed lemur troops at Berenty, Madagascar between 1989–1999. Primates 43:291–314
Majolo B, de Bortoli Vizioli A, Schino G (2008) Costs and benefits of group living in primates: group size effects on behaviour and demography. Anim Behav 76:1235–1247
Manson JH, Wrangham RW (1991) Intergroup aggression in chimpanzees and humans. Curr Anthropol 32:369–390
Maruhashi T, Saito C, Agetsuma N (1998) Home range structure and inter-group competition for land of Japanese macaques in evergreen and deciduous forests. Primates 39:291–301
Maynard Smith J, Parker GA (1976) The logic of asymmetric contests. Anim Behav 24:159–175
Muller MN, Wrangham RW (2004) Dominance, aggression and testosterone in wild chimpanzees: a test of the 'challenge hypothesis'. Anim Behav 67:113–123
Nishida T, Haraiwa-Hasegawa M, Hasegawa T, Takahata Y (1985) Group extinction and female transfer in wild chimpanzees in the Mahale National Park, Tanzania. Z Tierpsychol 67:284–301
Nunn CL, Deaner RO (2004) Patterns of participation and free riding in territorial conflicts among ringtailed lemurs (*Lemur catta*). Behav Ecol Sociobiol 57:50–61
Nunn CL, Lewis RJ (2001) Cooperation and collective action in animal behaviour. In: Noë R, van Hooff JARAM, Hammerstein P (eds) Economics in nature. Cambridge University Press, Cambridge, pp 42–66
Pride RE (2005a) Optimal group size and seasonal stress in ring-tailed lemurs (*Lemur catta*). Behav Ecol 16:550–560
Pride RE (2005b) High faecal glucocorticoid levels predict mortality in ring-tailed lemurs (*Lemur catta*). Biol Lett 1:60–63
Pride RE, Felantsoa D, Randriamboavonjy R, Randriambelona R (2006) Resource defense in *Lemur catta*: the importance of group size. In: Jolly A, Sussman RW, Koyama N, Rasamimanana H (eds) Ringtailed lemur biology: *Lemur catta* in Madagascar. Springer, New York, pp 208–232
Robinson JG (1988) Group-size in wedge-capped capuchin monkeys, *Cebus olivaceus,* and the reproductive success of males and females. Behav Ecol Sociobiol 23:187–197
Roscoe P (2007) Intelligence, coalitional killing, and the antecedents of war. Am Anthropol 109:485–495
Saito C, Sato S, Suzuki S, Sugiura H, Agetsuma N, Takahata Y, Sasaki C, Takahashi H, Tanaka T, Yamagiwa J (1998) Aggressive intergroup encounters in two populations of Japanese macaques (*Macaca fuscata*). Primates 39:303–312
Sherrow HM, Amsler SJ (2007) New intercommunity infanticides by the chimpanzees of Ngogo, Kibale National Park, Uganda. Int J Primatol 28:9–22

Silk JB (2007) The adaptive value of sociality in mammalian groups. Phil Trans Roy Soc B 362:539–559

Smith ML (2007) Territories, corridors, and networks: a biological model for the premodern state. Complexity 12:28–35

Sponsel LE (1996) The natural history of peace: a positive view of human nature and its potential. In: Gregor T (ed) A natural history of peace. Vanderbilt University Press, Nashville, TN, pp 95–125

Srikosamatara S (1987) Group size in wedge-capped capuchin monkeys (*Cebus olivaceus*): vulnerability to predators, intragroup and intergroup feeding competition. PhD thesis, University of Florida, Gainesville

Starin ED (1991) Socioecology of the red colobus monkey in the Gambia with particular reference to female-male differences and transfer patterns. PhD thesis, City University of New York, New York

Steenbeek R (1999) Tenure related changes in wild Thomas's langurs. I. Between-group interactions. Behaviour 136:595–625

Steenbeek R, van Schaik CP (2001) Competition and group size in Thomas's langurs (*Presbytis thomasi*): the folivore paradox revisited. Behav Ecol Sociobiol 49:100–110

Struhsaker TT (1967) Social structure among vervet monkeys (*Cercopithecus aethiops*). Behaviour 29:83–121

Sugiura H, Saito C, Sato S, Agetsuma N, Takahashi H, Tanaka T, Furuichi T, Takahata Y (2000) Variation in intergroup encounters in two populations of Japanese macaques. Int J Primatol 21:519–535

Suzuki S, Noma N, Izawa K (1998) Inter-annual variation of reproductive parameters and fruit availability in two populations of Japanese macaques. Primates 39:313–324

Takahata Y, Suzuki S, Okayasu N, Sugiura H, Takahashi H, Yamagiwa J, Izawa K, Agetsuma N, Hill D, Saito C, Sato S, Tanaka T, Sprague D (1998) Does troop size of wild Japanese macaques influence birth rate and infant mortality in the absence of predators? Primates 39:245–251

Takahata Y, Koyama N, Ichino S, Miyamoto N, Nakamichi M (2006) Influence of group size on reproductive success of female ring-tailed lemurs: distinguishing between IGFC and PFC hypotheses. Primates 47:383–387

Takahata Y, Koyama N, Ichino S, Miyamoto N, Nakamichi M, Soma T (2008) The relationship between female rank and reproductive parameters of the ringtailed lemur: a preliminary analysis. Primates 49:135–138

Thayer BA (2004) Darwin and international relations: on the evolutionary origins of war and ethnic conflict. University Press of Kentucky, Kentucky

Turney-High HH (1949) Primitive war: its practice and concepts. University of South Carolina Press, Columbia, SC

Valderrama Aramayo MX (2002) Reproductive success and genetic population structure in wedge-capped capuchin monkeys. PhD thesis, Columbia University, New York

Valero A, Schaffner CM, Vick LG, Aureli F, Ramos-Fernandez G (2006) Intragroup lethal aggression in wild spider monkeys. Am J Primatol 68:732–737

van der Dennen JMG (1995) The origin of war: the evolution of a male-coalitional reproductive strategy. Origin Press, Groningen, NL

van Schaik CP (1983) Why are diurnal primates living in groups? Behaviour 87:120–144

Wadley RL (2003) Treachery and deceit: parallels in tribal and terrorist warfare. Stud Conflict Terrorism 26:331–345

Wallace RB (2008) Towing the party line: territoriality, risky boundaries and male group size in spider monkey fission-fusion societies. Am J Primatol 70:271–281

Warner WL (1958) A black civilization: a social study of an Australian tribe. Harper, New York

Watts DP, Muller MN, Amsler SJ, Mbabazi G, Mitani JC (2006) Lethal inter-group aggression by chimpanzees in Kibale National Park, Uganda. Am J Primatol 68:161–180

Wich SA, Assink PR, Becher F, Sterck EHM (2002a) Playbacks of loud calls to wild Thomas langurs (Primates; *Presbytis thomasi*): the effect of location. Behaviour 139:65–78

Wich SA, Assink PR, Becher F, Sterck EHM (2002b) Playbacks of loud calls to wild Thomas langurs (Primates; *Presbytis thomasi*): the effect of familiarity. Behaviour 139:79–87

Williams JM, Oehlert GW, Carlis JV, Pusey AE (2004) Why do male chimpanzees defend a group range? Anim Behav 68:523–532

Wilson ML, Wrangham RW (2003) Intergroup relations in chimpanzees. Annu Rev Anthropol 32:363–392

Wilson ML, Hauser MD, Wrangham RW (2001) Does participation in intergroup conflict depend on numerical assessment, range location, or rank for wild chimpanzees? Anim Behav 61:1203–1216

Wilson ML, Wallauer WR, Pusey AE (2004) New cases of intergroup violence among chimpanzees in Gombe National Park, Tanzania. Int J Primatol 25:523–549

Wrangham RW (1980) An ecological model of female-bonded primate groups. Behaviour 75:262–300

Wrangham RW (1999) Evolution of coalitionary killing. Yearb Phys Anthropol 42:1–30

Wrangham RW, Peterson D (1996) Demonic males: apes and the origins of human violence. Houghton Mifflin, Boston

Wrangham RW, Wilson ML, Muller MN (2006) Comparative rates of violence in chimpanzees and humans. Primates 47:14–26

Wrangham RW, Crofoot MC, Lundy R, Gilby I (2007) Use of overlap zones among group-living primates: a test of the risk hypothesis. Behaviour 144:1599–1619

Wright Q (1942) A study of war. University of Chicago Press, Chicago

Yamagiwa J (2008) History and present scope of field studies on *Macaca fuscata yakui* at Yakushima Island, Japan. Int J Primatol 29:49–64

Zegwaard G (1959) Headhunting practices of Netherlands New Guinea. Am Anthropol 61:1020–1041

Zhao Q-K (1997) Intergroup interactions in Tibetan macaques at Mt. Emei, China. Am J Phys Anthropol 104:459–470

Zhao Q-K (1999) Responses to seasonal changes in nutrient quality and patchiness of food in a multigroup community of Tibetan macaques at Mt. Emei. Int J Primatol 20:511–524

Zinner D, Hindahl J, Kaumanns W (2001) Experimental intergroup encounters in lion-tailed macaques (*Macaca silenus*). Primate Rep 59:77–92

Chapter 9
Why War? Motivations for Fighting in the Human State of Nature

Azar Gat

Abstract The chapter addresses the causes of fighting among hunter-gatherers, whose way of life represents 99.5% of the history of the genus *Homo* and about 90% of that of *Homo sapiens sapiens*. Based on anthropological observations on the behavior of extant and recently extinct hunter-gatherer societies, compared with animal behavior, the chapter begins with somatic and reproductive causes. It proceeds to demonstrate that other motives, such as dominance, revenge, the security dilemma, and "pugnacity," originally arose from the somatic and reproductive competition. Rather than being separate, all motives come together in an integrated motivational complex, shaped by the logic of evolution and natural selection.

9.1 Introduction

In contrast to long-held Rousseauite beliefs that reached their zenith in the 1960s with the writings of Konrad Lorenz (1966) and Niko Tinbergen (1968), widespread deadly violence within species – including humans (Keeley 1996, Gat 1999, 2006; LeBlanc and Register 2003) – has been found to be the norm in nature. What are the evolutionary rewards that can make this highly dangerous activity worthwhile? This question relates to the age-old philosophical and psychological inquiry into the nature of the basic human system of motivation. Numerous lists of basic needs and desires have been put together over the centuries (Hobbes *Leviathan*, Chap. 6, Maslow 1970; Burton 1990), but in the absence of an evolutionary perspective, they have always had something arbitrary and trivial about them. They lacked

A. Gat
Department of Political Science, Tel-Aviv University, Tel-Aviv, Israel
e-mail: Azargat@post.tau.ac.il

P.M. Kappeler and J.B. Silk (eds.), *Mind the Gap*,
DOI 10.1007/978-3-642-02725-3_9, © Springer-Verlag Berlin Heidelberg 2010

a unifying regulatory rationale that would suggest the reason why the various needs and desires came into being, or the way in which they related to one another. Arguing that the human motivational system as a whole should be approached from the evolutionary perspective, I examine what can be meaningfully referred to as the "human state of nature", the 99.5% of the genus *Homo*'s evolutionary history in which humans lived in small kin groups as hunter-gatherers. In this "state of nature," people's behavior patterns are generally to be considered as having been evolutionarily adaptive. They form the biological inheritance that we later carried with us throughout history, when this inheritance constantly interacted with our staggering cultural development.

Although I shall survey the reasons for warfare among hunter-gatherers one by one, it is not my intention to provide yet another "list" of elements. Instead, I seek to show how the various reasons come together in an integrated motivational complex, shaped by the logic of evolution and natural selection for billions of years, including the two million-year history of our genus, *Homo*, and the tens of thousands of years of our species, *H. sapiens sapiens*. The aspects discussed include the pursuit of subsistence resources and reproduction as ultimate causes, and behaviors relating to dominance, revenge, the security dilemma, the supernatural, and playfulness, as proximate and subordinate causes that arise from the first.

9.2 Subsistence Resources

Competition over resources is a prime cause of aggression and deadly violence among humans, as in other animals. The reason for this is that food, water, and, to a lesser degree, shelter against the elements are tremendous selection forces. As Darwin, following Malthus, explained, living organisms, including humans, tend to propagate rapidly. Their numbers are checked only by the limited resources of their particular ecological habitats and by all sorts of competitors, such as conspecifics, animals of other species which have similar consumption patterns, predators, parasites, and pathogens.

When their environments suddenly expand, an unusual event in nature, demographic growth is dramatic. One of the best known examples is the rapid proliferation of Old World wildlife into new territories in the wake of the European age of discovery. Humans propagated equally dramatically in similar circumstances. As recently as several tens of thousands of years ago, the small groups that crossed over from Asia into North America propagated into hundreds of thousands and millions of people, filling up the Americas. Similarly, the small "founder groups" that arrived in the Pacific islands during the last two millennia, in most cases no more than a few dozens of people landing on each island, rapidly filled up their new habitats, increasing in numbers to thousands and tens of thousands.

Such dramatic "breaking of the barriers" was rare, however. Contrary to the Rousseauite imagination, humans, and animals, did not live in a state of primordial plenty. Even in lush environments, plenty is a misleading notion, for it is relative, first, to the number of mouths that have to be fed. The more resource-rich a region is, the more people it attracts from outside, and the greater the internal population growth that takes place. As Malthus pointed out, a new equilibrium between resource volume and population size would eventually be reached, recreating the same tenuous ratio of subsistence that was the fate of most preindustrial societies throughout history. The concept of "territoriality," which became popular in the 1960s (Ardrey 1966; Lorenz 1966; Tinbergen 1968), should be more subtly defined in this light. Among hunter-gatherers, territories varied dramatically in size – territorial behavior itself gained or lost in significance – in direct relation to the resources and resource competition. The same applies to high population density, another popular explanation in the 1960s for violence. Except in the most extreme cases, it is mainly in relation to resource scarcity that population density functions as a trigger for fighting. Otherwise, Tokyo and the Netherlands would have been among the most violent places on earth (Durham 1976; Dyson-Hudson and Smith 1978; Mueller 1983; Huntingford and Turner 1987; de Waal 1996).

Competition over resources existed in most hunter-gatherer cultures and sometimes escalated into conflict, mostly among, but occasionally also within, groups. This competition was largely about nourishment, the basic and most critical somatic activity of all living creatures, which often causes dramatic fluctuations in their numbers. Resource competition, and conflict, is not, however, a given quantity but a highly modulated variable. Resource competition and conflict change over time and place in relation to the varying nature of the resources available and of human population patterns in diverse ecological habitats. The basic question, then, is what the main scarcities, stresses, and hence, objects of human competition, are in any particular circumstances.

In extreme cases, such as the mid-Canadian arctic, where resources were highly diffused and human population density was very low, resource competition and conflict barely existed. In arid and semi-arid environments, like those of Central Australia, where human population density was also very low, water holes were often the main cause of resource competition and conflict. They were critical in times of drought, when whole groups of Aborigines are recorded to have perished. For this reason, however, there was a tendency to control them, also violently, even when stress was less pressing (Meggitt 1965b, p 42). In well-watered environments, where there was no water shortage and hence, no water competition, food often became the chief cause of resource competition and conflict, especially in times of stress, and also in expectation of and preparation for stress (Ember and Ember 1992: 242–262; also Hamilton 1975: 146). As Lournados (1997: 33) writes with respect to Aboriginal Australia: "In southwestern Victoria, competition between groups involved a wide range of natural resources, including territory, and is recorded by many early European observers throughout Victoria." Lournados's next sentence shows that his "competition" also includes "combat."

The nature of the food in question varied with the environment. Still, it was predominantly meat of all sorts that was hotly contested among hunter-gatherers. This fact, which is simply a consequence of nutritional value, is discernible throughout nature. Herbivores rarely fight over food, for the nutritious value of grass is too low for effective monopolization. Fruit, roots, seeds, and some plants that are considerably more nutritious than grass are often the object of competition and fighting, both among animals and humans. Meat, however, represents the most concentrated nutritional value in nature and is the object of the most intense resource competition: hence, the inherent state of competition and conflict found between Stone Age human hunters.

Let us understand more closely the evolutionary calculus that can make the highly dangerous activity of fighting over resources worthwhile. In our affluent societies, it might be difficult to comprehend how precarious people's subsistence in premodern societies was (and still is). The specter of hunger and starvation was ever-present. Affecting both mortality and reproduction, they constantly trimmed down population numbers. Thus, struggle over resources was very often evolutionarily cost-effective. The benefits of fighting also had to be matched against possible alternatives (other than starvation). One of them was to move elsewhere. This, of course, often happened, especially if one's enemy was much stronger, but this strategy had clear limitations. By and large, there were no "empty spaces" for people to move to. In the first place, the quality of space is not uniform, and the best, most productive habitats were normally already taken. One could be forced out to less hospitable environments, which may also be already populated by other less fortunate people. Indeed, finding empty niches required exploration, which again might involve violent encounters with other human groups. Furthermore, a move meant leaving a habitat with whose resources and dangers the group's members were intimately familiar, and traveling into uncharted environments. Such a change could involve heavy costs. Moreover, giving in to pressure from outside might establish a pattern of victimization. Encouraged by its success, the alien group might repeat and even increase its pressure. A strategy of conflict concerns not only the object presently in dispute but also the whole pattern of future relations. Standing for one's own might, in fact, mean lessening the occurrence of conflict in the future. No less, and perhaps more, than actual fighting, conflict is about deterrence.

Having discussed the possible benefits and alternatives of fighting, deterrence brings us to the costs side. Conflict becomes an evolutionarily more attractive strategy for those who resort the it to lower their risk of incurring serious bodily harm and death. Consequently, displays of strength and threats of aggressive behavior are the most widely used weapons in conflict, both among animals and humans. Furthermore, when humans, and animals, do resort to deadly violence, they mostly do so under conditions in which the odds are greatly tilted in their favor (Crofoot and Wrangham, this volume). Among animals, it is mostly the defenseless young, chicks and eggs that fall victim to deadly violence, whereas adult animals are very cautious of fighting to the finish with their peers for fear of self-injury. Among hunter-gatherer and other prestate societies, it is not the open-pitched battle but the raid and ambush that constitute the principal and, by far, the most lethal form of warfare. Asymmetrical fighting is the norm in nature, including the human state of nature (Gat 1999).

9.3 Reproduction

The struggle for reproduction is largely about access to sexual partners. There is a fundamental asymmetry between males and females in this respect, which runs through most of nature. At any point in time, a female can be fertilized only once. Consequently, evolutionarily speaking, she must take care to make the best of it. It is quality rather than quantity that she seeks. She must select the male who looks the best equipped for survival and reproduction, so that he will impart his genes, and his qualities, to the offspring. In those species, like the human, where the male also contributes to the raising of the offspring, his skills as a provider and his loyalty are other crucial considerations. In contrast to the female, there is theoretically almost no limit to the number of offspring a male can produce. He can fertilize an indefinite number of females, thus multiplying his own genes in the next generations. The main brake on male sexual success is competition from other males.

All this, of course, is only an abstract, around which sexual strategies in nature are highly diverse (Symons 1979; Daly and Wilson 1983; Ridley 1994; Buss and Malamuth 1996). Some species are highly polygynous; yet access to females can be more evenly spread, all the way down to pair-bonding. However, although pair-bonding reduces, it by no means terminates, male competition. In pair-bonding systems, the quality of the female partner also gains significance. If the male is restricted to one partner, it becomes highly important for him as well to choose the partner with the best reproductive qualities he can get: young, healthy, and optimally built for bearing offspring; that is, in sexual parlance, the most attractive female.

The need to take care of very slowly maturing offspring, which required sustained investment by both parents, turned humans towards pair-bonding. However, competition over the best female partners remains. Furthermore, humans, and men in particular, are not strictly monogamous. In the first place, males tend to have more than one wife when they can. Only a minority can, however. Although in most known human societies polygyny was legitimate, only a select few well-to-do men were able to support, and thus have, the extra wives and children. Second, in addition to official or unofficial wives, men tend to search for extramarital sexual liaisons with other women, married or unmarried.

How does all this affect human violent conflict and fighting? The evidence across the range of hunter-gatherer peoples tells the same story. Within the tribal groupings, women-related quarrels and violence were rife, often constituting the principal category of violence, and occasionally escalating to blood feuds and homicide. Incidents were caused by competition among suitors, by women's abduction and forced sex, by broken promises of marriage, and by jealous husbands suspicious of their wives' fidelity. Between groups, the picture was not very different and was equally uniform (but see Chapais, this volume). Warfare regularly involved the stealing of women, who were then subjected to multiple rape, or taken for marriage, or both.

So, hunter-gatherer fighting commonly involved the stealing and raping of women, but was this the cause or a side effect of hunter-gatherer fighting? In recent anthropological literature, this question has been posed by R.B. Ferguson in respect to Yanomamo warfare. Ferguson (1995: 355–358), who holds that warfare is caused by material reasons alone, has disputed Napoleon Chagnon's claim that the Yanomamo fought primarily for women. Chagnon (1977: 123, 146), for his part, dismissed the materialist position, enlisting the testimony of Yanomamo men who had told him amused: "Even though we like meat, we like women a whole lot more!" However, even Chagnon wavered on occasions on whether Yanomamo warfare was really *about* women.

The Yanomamo are hunters and horticulturalists rather than pure hunter-gatherers. However, the fundamental question in dispute is relevant to pure hunter-gatherers as well. As argued here, this is a pointless question that has repeatedly led anthropologists to a dead end. It artificially takes out and isolates one element from the wholeness of the human motivational complex that may lead to warfare, losing sight of the overall rationale that underpins these elements. It is as if one were to ask what people are *really* after when they go to the supermarket: meat, bread, or milk (Ferguson 2000; Gat 2000). Both somatic and reproductive elements are present in humans; moreover, both these elements are intimately interconnected, for people must feed, find shelter, and protect themselves in order to reproduce successfully. Conflict over resources was at least partly conflict over the ability to acquire and support women and children, and to demonstrate that ability in advance, in order to rank worthy of the extra wives. Brian Hayden (1994) has advanced an anthropological model whereby simple resources in resource-rich societies are accumulated and converted to luxury items in an intensified competition for status, prestige, and power (see Plourde and Henrich, this volume). He could add women to the list of converted goods. Thus, competition over women can lead to warfare *indirectly* as well as directly. As with mass and energy in Einstein's equations, resources, reproduction, and, as we shall see, status, are interconnected and interchangeable in the evolution-shaped complex that motivates people. Motives are mixed, interacting, and widely refracted. Nonetheless, it is the purpose of this chapter to show that this seemingly immense complexity and inexhaustible diversity can be traced back to a central core, shaped by the evolutionary rationale.

Wealth, status, matrimonial success, and power were interconnected among the "big men" of northern Australia (Hart and Pilling 1964: 18, 50). The same pattern applied to the "big men" (umialik) of the Eskimo hunter-gatherers of the Alaskan coast: "If he [an umialik] had more than one wife, his ties of blood and marriage were greater than those of others, and he could depend on many persons for support. Furthermore, by being an umialik he was a person whose opinions the others respected" (Oswalt 1967, p 178; also Burch 1974, p 6). A positive feedback mechanism was in operation. Chagnon (1979) has shown one way this mechanism worked with the Yanomamo, and Keen (1988: 290) has independently detected the same pattern among the Australian hunter-gatherers. The largest clans in a tribe, those comprising more siblings and cousins, acted on the principle of kin solidarity *vis-à-vis* the rest of the tribe. They moved on to increase their advantage by controlling the leadership positions, resources, and marriage opportunities at the expense of the others. As a result, large clans tended to dominate a tribe, politically and demographically, over time. The notion that there is a self- and mutually reinforcing tendency which works in favor of the rich, mighty, and successful, facilitating their access to the "good things of life," goes back a long way.

Polygyny was a significant factor in many hunter-gatherer societies. Australia constitutes our best laboratory. Its size, near complete isolation, and ecological diversity make it far superior to other, more recently studied and more publicized cases that are mostly confined to arid environments. Polygyny was legitimate among all the Aborigines tribes of Australia and highly desired by the men. However, comparative studies among the tribes show that men with only one wife comprised the largest category among married men, often the majority. Men with two wives comprised the second largest category. The percentage of men with three or more wives fell sharply, to around 10–15% of all married men, with the figures declining with every extra wife (Meggitt 1965a; Long 1970). To how many wives could the most successful men aspire? There was a significant environmental variation here. In the arid Central Desert, four, five, or six wives were the top. Five or six was also the top figure mentioned by Buckley for the Aborigines living in the region of Fort Philip (Melbourne) in the south-east in the early nineteenth century. However, in the more rich and productive parts of Arnhem Land and nearby islands in the north, a few men could have as many as 10–12 wives, and in some places, in the most extreme cases, even double that number. There was a direct correlation between resource density, resource accumulation and monopolization, social ranking, and polygyny (Berndt and Berndt 1964: 172; Hart and Pilling 1964, pp 17–18, 50; Meggitt 1965b, pp 78, 80–81; Morgan 1980, p 58; Keen 1982; Lournados 1988, p 151–152).

Data from other hunter-gatherer societies reveal a similar picture. Resource scarcity reduced social differentiation, including in marriage, but did not eliminate it. The leaders of the Aka Pygmies were found to be more than twice as polygynous as ordinary people, and to father more children (Betzig et al. 1991, p 410). Among the !Kung of the arid Kalahari Desert, polygyny was limited, but 5% of married men still had two wives (Daly and Wilson 1988: 285). Women-related feuds were the main cause of homicide among them. In the extremely harsh conditions of the mid-Canadian arctic, where resources were scarce and diffused, fighting over resources barely existed. Because of the resource scarcity, marriages among the native Eskimo were also predominantly monogamous. One study registered only three polygynies out of 61 marriages. Still, wife-stealing was a widespread, probably the main, cause of homicide and "blood feuds" among the Eskimos (Betzig et al. 1991). "A stranger in the camp, particularly if he was traveling with his wife, could become easy prey to the local people. He might be killed by any camp fellow in need of a woman" (Daly and Wilson 1988, p 222; citing Balikci 1970, p 182). Among the Eskimos of the more densely populated Alaskan Coast, abduction of women was a principal cause of warfare. Polygyny, too, was more common among them, although restricted to the few (Oswalt 1967, pp 178, 180, 182, 185, 187, 204;

Burch and Correll 1972, p 33; Dickemann 1979, p 363; Symons 1979, p 152; Nelson 1983, p 292, 327–329; Irwin 1990, pp 201–202). Strong *Ingalik* ("big men") often had a second wife, and "there was a fellow who had five wives at one time and seven at another. This man was a great fighter and had obtained his women by raiding" (Betzig et al. 1991, p 410).

The resource-rich environment of the Northwest Coast accentuated resource competition and social ranking. Conflict over resources was therefore intense. However, resource competition was not disassociated from reproduction, but constituted, in fact, an integral whole with it. Women are not even mentioned in R. B. Ferguson's elaborate materialist study of Northwest Coast Indian Warfare (1984). Nonetheless, they were there. Most natives of the Northwest Coast were monogamous. However, the rich, strong, and powerful were mostly polygynous. The number of wives varied from tribe to tribe, but "a number" or "several" is normally quoted, and up to 20 wives are mentioned in one case. The household of such successful men is repeatedly described as having been very substantial and impressive indeed. Furthermore, as is universally the case, the mainly female slaves taken in the raids and working for their captors also shared their masters' bed (Drucker 1951, p 301, 1965, p 54; Krause 1970, p 154; Rosman and Rubel 1971, pp 16–17, 32, 110; Donald 1997, p 73).

Naturally, the increase in the number of a man's wives generally correlated with his reproduction rate (number of children). Statistics for hunter-gatherers, beyond those already cited, are scarce, and most of the following derives from simple horticulturalists who may have had more impressive reproductive skews. Among the Xavante horticulturalists of Brazil, for example, 16 of the 37 adult males in one village (74 out of 184 according to a larger survey) had more than one wife. The chief had five, more than any other man. He fathered 23 surviving offspring who constituted 25% of the surviving offspring in that generation. Shinbone, a most successful man among the Yanomamo of the Orinoco basin, had 43 children. His brothers were also highly successful, so Shinbone's father had 14 children, 143 grandchildren, 335 great grandchildren, and 401 great-great grandchildren, at the time of the research (Chagnon 1979; Symons 1979, p 143; Daly and Wilson 1983, pp 88–89, 332–333). Again, women are such a prominent motive for competition and conflict because reproductive opportunities are a very strong selection force indeed.

To be sure, this does not mean that people always want to maximize the number of their children. Although there is some human desire for children per se and a great attachment to them follows once they are born, it is mainly the desire for sex – Malthus's "passion" – which functions in nature as the powerful biological proximate mechanism for maximizing reproduction. As humans, and other living creatures, normally engage in sex throughout their fertile lives, they have a vast reproductive potential, which, before the introduction of effective contraception, mainly depended on resource availability for its realization.

Polygyny (and female infanticide) created a scarcity of women and increased men's competition for, and conflict over, them (Divale and Harris 1976).

In conjunction with the other motives surveyed here, this was a major reason for the high violent mortality rate among hunter-gatherers. Among Aboriginal Australian tribes, about 30% of the Murngin adult males are estimated to have died violently, and similar findings have been recorded for the Tiwi. The Plains Indians showed a deficit of 50% for the adult males in the Blackfoot tribe in 1805 and 33% deficit in 1858 (but by the nineteenth century, they already possessed guns and horses), while during the reservation period the sex ratio rapidly approached 50–50. Among the Eskimos of the central Canadian arctic, where group warfare was practically nonexistent, the rate of violent deaths, in the so-called "blood feuds" and "homicide," was estimated at one per 1,000 persons per year, ten times the 1990 US rate which is the highest in the developed world. Among the !Kung of the Kalahari Desert, dubbed the "harmless people," there were 22 cases of homicide in the period of study, 1963–1969; 19 of the victims were males, as were all of the 25 killers. This amounts to a rate of 0.29 person per 1,000 per year, and had been 0.42 before the coming of state authority, 3–4 times higher than the 1990 US rate (Gat 2006, pp 129–132, for references to this and the following paragraphs).

The data for prestate agriculturalists is basically the same. Among the Yanomamo, about 15% of the adults died as a result of inter and intragroup violence: 24% of the males and 7% of the females. The Waorani (Auca) of the Ecuadorian Amazon hold the registered world record: more than 60% of adult deaths were caused by feuding and warfare. Among the many peoples in Highland New Guinea, violent mortality estimates are very similar: among the Dani, 28.5% of the men and 2.4% of the women; among the Enga, 34.8% of the adult males; among the Goilala, whose total population was barely over 150, there were 29 (predominantly men) killed during a period of 35 years; among the Lowland Gebusi, 35.2% of the adult males and 29.3% of the adult females. Archeology unearths similar findings. In the Neolithic site of Madisonville, Ohio, 22% of the adult male skulls had wounds and 8% were fractured.

Another consequence of sexual deprivation in young adult males is their marked restlessness, risk-taking behavior, and belligerency. Young adult males are genetically inclined to greater risk-taking, for their matrimonial status-quo is highly unsatisfactory. They still have to conquer their place in life. Thus, they have always been the most natural recruits for violent action and war. Male murder rates peak in both London and Detroit – although 40 times higher in the latter – at the age of 25 (Daly and Wilson 1983, pp 92–97, 297–301; Jones 1993, p 92).

The interconnected competition over resources and reproduction is the *root* cause of conflict and fighting in humans, as in all other animal species. Other causes and expressions of fighting in nature, and the motivational and emotional mechanisms associated with them, are derivative of, and subordinate to, these primary causes, and *originally* evolved this way in humans as well. This, of course, does not make them any less "real" but only explains their function in the evolution-shaped motivational complex, and, thus, how they came into being. It is to these "second-level" causes and motivational mechanisms, directly linked to the first, that we now turn.

9.4 Dominance: Rank, Power, Status, Prestige

Among social animals, possessing higher rank in the group promises one a greater share in the communal resources, such as hunting spoils, and better reproductive opportunities. While there is considerable diversity among species, rank is hotly contested for that reason. It is the strong, fierce, and – among our sophisticated cousins, the chimpanzees – also the "politically" astute, that win status by the actual and implied use of force. Rivalry for rank and domination in nature is, then, a proximate means in the competition over resources and reproduction (Watts, this volume).

In determining one's status, image and perception have always been as important as tangible reality. Thus, both overt and subtler displays of worth are a constant human activity. It is limited only by the desire to avoid the provocation of a negative social response, because other people as well jealously guard their honor in the social competition for esteem. In traditional societies in particular, people were predisposed to go to great lengths in defense of their honor. The slightest offense could provoke violence. Where no strong centralized authority existed, one's honor was a social commodity of vital significance, affecting both somatic and reproductive prospects.

Does this mean that what people who strive for leadership or esteem "really" want is sexual opportunity or resources? Not necessarily. Wanting is subjective, and mentally it can be genuinely disassociated from ultimate evolutionary aims. For instance, people widely desire love and sex for their own sake rather than for the resulting offspring, whom they often positively, and even desperately, do not want. In the same way, the pursuit of rank and esteem in humans, as with animals, was closely associated with better somatic and reproductive prospects, and evolved as a proximate means for achieving them, even though the evolutionary aim often lacked conscious expression. Again, to remove all too prevalent misunderstandings regarding the evolutionary rationale, the argument, of course, is not that these behavior patterns are a matter of conscious decision and complex calculation conducted by flies, mice, lions, or even humans. It is simply that those who failed to behave adaptively became decreasingly represented in the next generations, and their maladaptive genes, responsible for their maladaptive behavior, were consequently selected against. The most complex structural engineering and behavior patterns have thus evolved in, and program, even the simplest organisms, including those lacking any consciousness (Dawkins [1976], 1989, pp 96, 291–292).

As with competition over women, competition over rank and esteem could lead to violent conflict indirectly as well as directly. For instance, even in the simplest societies people desired ornamental, ostentatious, and prestige goods. Although these goods are sometimes lumped together with subsistence goods, their social function and significance are entirely different. Body and clothes ornamentation are designed to enhance physically desirable features that function everywhere in nature as cues for health, vigor, youth, and fertility (Darwin [1871], 1962, pp 467–468, Low 1979, pp 462–487, Diamond 1992: Chap. 9). For example, articial coloring is used to enhance eye, lip, hair, and skin color; natural – and by extension, added – symmetrical, orderly, and refined features signal good genes, good nourishment, and high-quality physical design; tall and magnificent headgear enhances one's size. It is precisely on these products of the "illusions industry" – cosmetics, fashion, and jewellery – that people everywhere spend so much money. Furthermore, where some ornaments are scarce and therefore precious, the very fact that one is able to afford them indicates wealth and success: hence, the source of what economist Thorstein Veblen, referring to early twentieth century American society, called "conspicuous consumption." In Stone Age societies as well, luxury goods, as well as the ostentatious consumption of ordinary ones, became in themselves objects of desire as symbols of social status. For this reason, people may fight for them.

Indeed, plenty and scarcity are relative not only to the number of mouths to be fed but also to the potentially ever-expanding and insatiable range of human needs and desires. Human competition increases with abundance – as well as with deficiency – taking more complex forms and expressions, widening social gaps, and enhancing stratification. While the consumption capacity of simple, subsistence, products is inherently limited, that of more refined, lucrative ones is practically open-ended. One can simply move up the market.

9.5 Revenge: Retaliation to Eliminate and Deter

Revenge is one of the major causes of fighting cited in anthropological accounts of prestate societies. Violence was activated to avenge injuries to honor, property, women, and kin. If life was taken, revenge reached its peak, often leading to a vicious circle of death and counter-death.

How is this most prevalent, risky, and often bloody behavior pattern to be explained? From the evolutionary perspective, revenge is retaliation that is intended either to destroy an enemy or to foster deterrence against him, as well as against other potential rivals. This, of course, applies to nonphysical and nonviolent, as well as to physical and violent action. If one does not pay back on an injury, one may signal weakness and expose oneself to further injuries not only from the original offender but also from others. A process of victimization might be created. I suspect that experts would be able tell us that a similar behavioral pattern occurs, if only rudimentarily, within other social species (Aureli et al. 1992). All the same, humans have far longer memories than do animals, and, thus, revenge – the social settling of accounts with those who offended them – assumes a wholly new level with them. Of course, depending on one's overall assessment of the stakes and relative balance of power and if the challenger is much stronger than oneself, it is equally common for one to accept in silence an injury and the consequences of reduced status. This rationale applies wherever there is no higher authority that can be relied upon for protection, that is, in the so-called anarchic systems. In modern societies, it thus applies to the wide spheres of social relations in which the state or other authoritative bodies do not intervene. In prestate societies, however, it applied far more widely to the basic protection of life and property.

But are people not driven to revenge by blind rage rather than by calculation? I raise this typical question only in order to reiterate the point which is all too often misunderstood with respect to the evolutionary rationale. Basic emotions evolved, and are tuned the way they are, in response to very long periods of adaptive selective pressures (Fessler and Gervais, this volume). They are proximate mechanisms in the service of somatic and reproductive purposes. To work, they do not need to be conscious, and the vast majority of them indeed are not – in humans – let alone in animals. Thus, the instinctive desire to strike back is a basic emotional response which evolved precisely because those who struck back – of course, within the limits mentioned above – were generally more successful in protecting their own. Indeed, this rationale is remarkably supported by the famous computerized game that found tit-for-tat the most effective strategy a player can adopt (Axelrod 1984).

Tit-for-tat poses a problem. One's offender cannot always be eliminated. Furthermore, the offender has kin who will avenge him, and it is even more difficult to eliminate them as well. In many cases, tit-for-tat becomes a negative loop of retaliation and counter-retaliation from which it is very hard to exit. One original offense may produce a pattern of prolonged hostility. Thus, retaliation might produce escalation rather than annihilation or deterrence. In such cases, fighting seems to feed on, and perpetuate, itself, bearing a wholly disproportional relation to its "original" cause. People become locked into conflict against their wishes and best interests. It is this factor that has always given warfare an irrational appearance that seems to defy a purely utilitarian explanation.

How can this puzzle be explained? In the first place, it must again be stressed that both the original offense and the act of retaliation arise from a fundamental state of interhuman competition that carries the potential of conflict, and is consequently fraught with suspicion and insecurity. Without this basic state of somatic and reproductive competition and potential conflict, retaliation as a behavior pattern would not have evolved. Indeed, sometimes revenge is merely a pretext for conflict emanating from more fundamental reasons. However, while explaining the root cause of retaliation, this does not in itself account for retaliation's escalation into what often seems to be a self-defeating cycle. A prisoner's dilemma-like situation is responsible for the emergence of such cycles. In the absence of an authority that can enforce mutually beneficial cooperation on people, or at least minimize their damages, the cycle of retaliation is often their only rational option, though, exposing them to very heavy costs, is not their best option.

Like any game, the prisoner's dilemma is predicated on its assumptions. It has proven so fruitful because it has been found that many situations in real life exhibit elements of the dilemma. Indeed, the prisoner's dilemma is of great relevance when explaining the war complex as a whole and not only that of revenge and retribution. Still, it ought to be emphasized that not all violent conflicts or acts of revenge fall under the special terms of the prisoner's dilemma. In the context of a fundamental resource scarcity, if one is able to eliminate, decisively weaken, or subdue the enemy, and consequently reap most of the benefits, then this strategy is better for one's interests than a compromise. It is only when such a decisive result cannot be achieved that conditions similar to those specified by the prisoner's dilemma come into play.

9.6 Power and the Security Dilemma

Revenge or retaliation is an active reaction to an injury, emanating from a competitive and, hence, potentially conflictual basic state of relations. However, as Hobbes saw (*Leviathan*, Chap. 13), the basic condition of competition and potential conflict, which gives rise to endemic suspicion and insecurity, invites not only reactive but also preemptive response, which further magnifies mutual suspicion and insecurity. It must be stressed that the source of the potential conflict here is again of a "second level." It does not necessarily arise directly from an actual conflict over the somatic and reproductive resources themselves, but from the fear, suspicion, and insecurity that the potential of those "first-level" causes for conflict creates. Potential conflict can thus breed conflict. When the "other" must be regarded as a potential enemy, his very existence poses a threat, for he might suddenly attack one day. For this reason, one must take precautions and increase one's strength as much as possible. The other side faces a similar security problem and takes similar precautions.

Things do not stop with precautionary and defensive measures, because such measures often inherently possess some offensive potential, indirectly or directly. Indirectly, a defended home base may have the effect of freeing one for offensive action with less fear of a counter-strike – it reduces mutual deterrence. Directly, a defensive alliance, for example, may be transformed into an offensive one. Thus, the measures that one takes to increase one's security in an insecure world often decrease another's security and vice versa.

What are the consequences of this so-called "security dilemma"? (Herz 1950; Jervis 1978). In the first place, it tends to escalate arms races. Arms races between competitors take place throughout nature. Through natural selection, they produce faster cheetahs and gazelles; deer with longer antlers to fight one another; more devious parasites and viruses and more protected "hosts." Many of these arms races involve very heavy costs to the organisms, which would not have been necessary if it were not for the competition. This, for example, is the reason why trees have trunks. Trees incur the enormous cost involved in growing trunks only because of their intense struggle to outgrow other trees in order to get sunlight. As with humans, competition is most intense in environments of plenty, where more competitors can play and more resources be accumulated. This is why trees grow highest in the dense forests of the water-rich tropical and temperate climates.

Arms races often have paradoxical results. The continuous and escalating effort to surpass one's rival may prove successful, in which case the rival is destroyed or severely weakened, and the victor reaps the benefits. However, in many cases, every step on one side is matched by a counter-step on the other. Consequently, even though each side invests increasing resources in the conflict, neither gains an advantage. This is called, after one of Alice's puzzles in Lewis Carroll's *Through the Looking-Glass*, the "Red Queen effect": both sides run faster and faster only to find themselves remaining in the same place. Arms races may, thus, become a prisoner's dilemma. If the sides gave up the hope of outpacing each other and winning the contest, they could at least save themselves the heavy costs incurred, which anyway cancel each other out. However, they are often unable to stop the race, because of suspicion, faulty communication, and inability to verify what exactly the other side is doing.

Thus arms races are, in general, the natural outcome of competition. The special feature of arm races created by the security dilemma is that their basic motivation on both sides is defensive. Again, one way to stop the spiral is to find a means to reduce mutual suspicion. Marriage ties used to be a standard measure for achieving this aim in all premodern societies (Chapais, this volume). Fostering familiarity and demonstrating good will through mutual friendly visits and ceremonial feasts were other prominent universal measures. For all that, suspicion and insecurity are difficult to overcome for the reasons already mentioned. Furthermore, even ostensibly friendly overtures sometimes turned out to be treacherous. However, there is another way to reduce the insecurity. Although both sides in the security dilemma may be motivated by defensive concerns, they may choose to actively preempt their opponents; that is, take not only defensive precautions but attack in order to eliminate or severely weaken the other side. Indeed, this option in itself makes the other side even more insecure, making the security dilemma more acute. Warfare can thus become a self-fulfilling prophecy. Since full security is difficult to achieve, history demonstrates that constant warfare can be waged, conquest carried afar, and power accumulated, all truly motivated by security concerns, "for defense." Of course, in reality motives are often mixed, with the security motive coexisting with a quest for gain.

The basic condition of interhuman competition and potential conflict thus creates "second-level" causes for warfare, arising from the first, such as the cycle of revenge and the security dilemma. This does not mean that actual competition over somatic and reproductive resources has to exist on every particular occasion for the security dilemma to flare up. Still, it is the prospect of such competition that stands behind the mutual insecurity, and the stronger the competition and potential conflict, the more intense the security dilemma will grow.

9.7 World-View and the Supernatural

But what about the world of culture that after all is our most distinctive mark as humans? Do not people kill and get killed for ideas and ideals? From the Stone Age on, the spiritual life of human communities has been imbued with supernatural beliefs, sacred cults and rituals, and the practice of magic. Here, the difference between humans and other animals is the most marked, even if rudimentary culture forms already manifest themselves in primates and hominids. It should be noted, however, that the capacity for culture itself evolved as a biological adaptation, and its various forms undergo evolutionary selection, both biological and cultural.

The evolutionary status of religion is beyond our scope here. Like warfare, religion is a complex phenomenon which is probably the result of several different interacting factors. Some scholars believe religion to be detrimental for survival and hold that it emerged as a "bug," "parasite," or "virus" on *H. sapiens sapiens*' advanced intellectual "software" (Dawkins [1976], 1989, pp 189–201, 329–331, 2006; Bowker 1995; Boyer 2001). By contrast, functionalist theorists, from Emile Durkheim on, have argued that religion's main role was in fostering social cohesion, inter alia in war (Durkheim 1965; Ridley 1996: 189–193; Wilson 2002; Hayden 2003). In evolutionary terms, this means that in those groups in which common ritual and cult ceremonies were more intensive, social cooperation became more habitual and more strongly legitimized, which probably translated into an advantage in warfare.

But how did the hunter-gatherers' supernatural beliefs and practices affect the reasons for conflict and fighting? I argue that on the whole they added to, sometimes accentuating, the reasons we have already discussed. The all-familiar glory of the gods, let alone missionary quests, never appear as reasons for hunter-gatherers' warfare. These will appear later in human cultural evolution. The supernatural reason for fighting among hunter-gatherers most cited by anthropologists is fear and accusations of sorcery. It should be noted, however, that these did not appear randomly, but were directed against people whom the victim of the alleged sorcery felt had reasons to want to harm him. This, of course, does not necessarily mean that they really did. It certainly does not mean that these people actually did harm the victim by witchcraft. What it does mean is that competition, potential conflict, animosity, and suspicion were conducive to fears and accusations of sorcery. To further clarify the point, it is not that these "imagined" fears and accusations did not add to the occurrence of deadly violence beyond the "real" or potentially "real" causes that underlie them. They certainly did. But, to a greater degree than with the security dilemma, the paranoia here reflects the running amok of real, or potentially real, fears and insecurity, thus further exacerbating and escalating the war complex.

Supernatural elements sometimes came into play in connection with motives for warfare other than fear and insecurity. For instance, trespassing was often regarded in hunter-gatherer societies as an offense against a group's sanctified territory. In other cases, an act of sacrilege against the clan's totem was regarded as an insult to the clan itself. In both these instances, the supernatural element functioned as a sanctified symbol of less imagined goods: resources and honor. The totem was thus like an emblem or flag. Of course, in some cases, supernatural reasons were evoked as mere pretexts for other motives. However, even when they were not, the supernatural elements added an extra dimension to existing motives, taken from the realm of the spiritual and sanctified.

9.8 Playfulness, Adventurism, Ecstasy

For all that we have said about the evolution-shaped aims of warfare, do not people sometimes fight for no particular purpose, just for the fun of it, as a game, an outlet, arising from sheer pugnacity?

Playing and sports have often been regarded – indeed, defined – as purposeless, expressive, pure fun activity. What is its evolutionary logic? After all, it is an activity that consumes a great deal of energy for no apparent gain. In reality, though, its purpose is physical exercise and behavioral training for the tasks of life, such as hunting, escaping predators and natural dangers, fighting, nurturing, and social cooperation in all these. For this reason, in all mammalian species (distinctive for their learning ability and playing activity), it is the young who exhibit the most active and enthusiastic play behavior, compared with the more mature and experienced (Fagan 1981; Smith 1984; Huntingford and Turner 1987: 198–200). Since adaptive behaviors are normally encouraged by emotional gratifications, play and sport are generally enjoyable.

So, games and sports serve, among other things, as preparation for fighting. In this light, fighting may even be perpetrated in rare cases as playful training for more serious fighting. However, is fighting sometimes not perpetrated only for evoking the sort of emotional gratifications associated with play or sport behavior? Do emotional gratifications sometimes not take on an end of their own in perpetrating fighting? I claim that they do, but as an extension rather than a negation of the evolutionary logic.

In the first place, it should be borne in mind that even wholly playful or "expressive" fighting behavior developed within a general evolutionary context in which conflict was normal and fighting a distinct possibility and, therefore, a deeply rooted behavior pattern. In this respect, wholly "purposeless" violence is a "misplaced" or "misactivated" expression of a "normal," evolution-shaped behavior. We shall return to this in a moment. Second, as with respect to accusations of sorcery, it should be noted that even seemingly purposeless violence is not purely random. It is much more often directed against aliens or competitors than against perceived friends. Thus again, it is often an extension of, or over-reaction to, a state of competition and potential conflict.

Still, allowing that some "purposeless," "expressive" violence does exist, at least marginally, what does it mean to describe such behavior as "misplaced," or "misactivated"? Surely, the intention is not to pass any sort of value judgment. Rather, the terms describe behavior which, while having an evolutionary root, is expressed out of its evolutionarily "designed" context, and thus is typically also maladaptive. But if so, how does it survive? In reality, maladaptive traits *are* constantly selected against. For this reason, their prevalence remains marginal. Still, they do exist. It is not only that natural selection is perpetual because of mutations, the unique gene recombination that occurs with every new individual, and changing environmental conditions; the main reason is that no mechanism, whether purposefully designed by humans or blindly by natural selection, is ever perfect, 100% efficient, or fully tuned. Like any other design, the products of natural selection, for all their marvels, vary greatly in their level of sophistication, have limitations, flaws, and "bugs," can only operate in a proximate manner, and are, thus, far from optimal. The only requirement they are bound to meet is that they are good enough to survive in a given environment and facing given competitive challenges. The emotional mechanisms controlling violence have all the above limitations. Thus, they can be triggered or "misactivated" into "purposeless," "expressive," "spontaneous," or "misdirected" violence. However, like overeating or sleeplessness – to give more familiar examples – such behavior should be understood as a range of deviation from an evolutionarily shaped norm.

Ecstatic behavior is another case in point. Ecstasy is a feeling of elation and transcendence produced by an increasing flow of hormones such as adrenaline, serotonin, and dopamine. It reduces body sensitivity to pain and fatigue, raises its energy to a high pitch, and lowers normal inhibitions. In nature, ecstatic behavior can be produced during extreme bodily exertion, often associated with struggle and fighting. However, humans very early on found ways to arouse it artificially for the feel-good effect itself, for instance, through rhythmic dance or by the use of narcotic substances. In some cases, narcotic substances were consumed before fighting and in preparation for it; a few shots of alcohol before an assault was ordinary practice in most armies until not very long ago. However, in other cases, the ecstatic condition itself can breed violence; again, drunkenness greatly contributes to the occurrence of violence in many societies. Furthermore, in some cases, the sequence is reversed, with fighting entered into in order to produce ecstatic sensations. For example, in addition to "ordinary" reasons, such as money, females, social esteem, and so forth, this motivation plays a prominent role – often in conjunction with alcohol consumption – in perpetrating "purposeless" youth gangs' violence. Again, what we have here is a mostly maladaptive outgrowth and deviation from an evolution-shaped behavioral pattern.

9.9 Cooperation in Fighting

Fighting in the human state of nature is carried out at the individual and group levels. Cooperation in fighting takes place among family, clan, and tribe (regional group) members. In principle, there are strong advantages to cooperation. In warfare, for example, there is a strong advantage to group size (Crofoot and Wrangham, this volume). However, the problem with cooperation throughout nature is that one has a clear incentive to "free ride" – reap the benefits of cooperation while avoiding one's share in the costs. Three or four different mechanisms overlap to secure a measure of cooperation in hunter-gatherer groups.

First, as the theory of inclusive fitness predicts, people risk their lives in support of close kin, with whom they share more genes. Family members tend to support one another in disputes and clashes with members of other families. In interclan rivalry, clans which are intermarried are likely to support one another against other clans. Companions for raids, the most common form of fighting among hunter-gatherers, come mainly from one's family and clan. The members of regional groups and confederation of regional groups, numbering in the hundreds and more, are not as closely related as family and clan members, and yet (weaker) cooperation among them takes place, particularly in conflict against alien regional groups. In part, the same logic that, in J.B.S. Haldane's famous formulation, makes it evolutionarily beneficial to sacrifice one's life in order to save more than two siblings or eight cousins, and take risks at even lower ratios, holds true for 32 second cousins, 128 third cousins, or 512 fourth cousins. This, in fact, is pretty much what a regional group is. Moreover, although not every member of the regional group is a close kin of all the others, the regional group is a dense network of close kinship through marriage ties (Chapais, this volume). Marriage links criss-cross the regional group, making families and clans ready to take risks in support of one another. Since most marriages take place within the regional group, there is a wide gap between the "us" of the tribe and outsiders (Hamilton 1975, p 144; Silverman 1987, p 113; Bowles 2006). Hunter-gatherers only felt safe to go where they had kin.

Secondly, social cooperation can be sustained in groups that are intimate enough to allow mutual surveillance and social accounting. If detected, a "free rider" faces the danger of being excluded, "ostracized," from the system of cooperation, which is on the whole beneficial to him. People not only keep a very watchful eye for "cheaters" and "defectors," but in comparison with other animal species, they also have very long memories. They would help other people on the assumption that they would get similar help in return, but are likely to cease cooperating if the expected return fails to arrive. This is the basis for the so-called "reciprocal altruism" in human relations, which explains most of human seeming altruism towards non-kin (Trivers 1971; Alexander 1987; Frank 1988; Ridley 1996). The regional group is small enough to have dense kinship networks, as well as for all its members to know one another, to be in contact with them, and to hold them to account.

Thirdly, apart from biology, humans have culture, and are differentiated by their cultures. This is a human universal that set humans far apart from other animals. As culture, particularly among hunter-gatherers, was local and thus closely correlated with kinship, cultural identity became a strong predictor of kinship (Irwin 1987, p 131–156). Moreover, culture sharing is also crucial for human social cooperation. Cooperation is dramatically more effective when cultural codes, above all language, are shared (Silk and Boyd, this volume). Like genes, culture changes over time, only much faster. In Australia, for example, where the time depth of the Aboriginal population measures in tens of thousands of years, lingual diversity among the hundreds of regional groups or "dialect tribes" was great. There were more than 200 different languages and even more dialects (Lournados 1997, p 38). The tribal groupings, differing from their neighbors in their language and customs, were thus the most effective frameworks of social cooperation for their members. Outside them, people would find themselves in a great disadvantage. Therefore, shared culture in a world of cultural diversity further increases the stake of a regional group's members in their group's survival. This factor may not have been sufficiently recognized in the literature. The regional group is bound together by mutually reinforcing and overlapping ties of kinship, social cooperation, and cultural distinctiveness. Hence, the phenomenon of "ethnocentrism," a human universal that started at the level of the hunter-gatherer regional group and would be expanded onto larger ethnic groupings later in history.

Fourthly, there is the contentious issue of group selection. Modern evolutionary theory centers on individual or gene survival, with cooperation explained by the principles of "kin selection" and "reciprocal altruism." However, according to an older view, first raised as a possibility by Darwin and now affecting a comeback, biological selection takes place not only at the individual or gene level but also among groups. A group which is biologically endowed with greater solidarity and with individual willingness to sacrifice for the group would defeat less cohesive groups. In rejection of this view, it used to be claimed that genes for self-sacrifice on behalf of the group would have the effect of annihilating those who possessed them much faster than aiding them through improved group survival, and that "cheaters" would proliferate. However, a modulated multilevel selection, working through the individual, family, and larger group levels, is supported by mathematical modeling (Hamilton 1975; Levitt 1980; Wilson and Sober 1994, 1998; Hamilton 1996; Wilson and Wilson 2007).

As can be inferred from Bowles (2006), one should guard against a sharp *empirical* distinction between kin selection and group selection. For in reality, throughout the vast majority of human evolutionary history, groups were anyhow small kin-groups. The extended family group of a few dozens, the basic human group, consisted of close kin. Even the regional group of a few hundreds consisted of medium-range kin cris-crossed by marriage ties. Truly large societies of non- (or remote-) kin emerged only very recently, with agriculture and civilization.

9.10 Conclusion: Fighting in the Evolutionary State of Nature

The hunter-gatherer way of life covers 99.5% of the history of the genus *Homo*, and more than 90% of the history of the species *H. sapiens*. Agriculture and the state are recent cultural inventions, starting in the most pioneering groups of our species only some 10,000 and 5,000 years ago, respectively, and having little effect on the human genome. Thus, to speak in a meaningful manner about human nature is to address human adaptations to the human natural habitats, which are responsible for the human biological inheritance.

Conflict and fighting in the human state of nature, as in the state of nature in general, were fundamentally caused by competition. While violence is evoked and suppressed by powerful emotional stimuli, it is not a primary, "irresistible" drive; it is a highly tuned, both innate and optional, evolution-shaped tactic, turned on and off in response to changes in the calculus of survival and reproduction. It can be activated by competition over scarce resources, as scarcity and competition are the norm in nature because of the tendency of organisms to propagate rapidly when resources are abundant. Deadly violence is also regularly activated in competition over women, directly as well as indirectly, when men compete over resources in order to be able to afford more women and children.

From these primary somatic and reproductive aims, other, proximate and derivative, "second-level," aims arise. The social arbiters within the group can use their position to reap somatic and reproductive advantages and hence the competition for – and conflict over – esteem, prestige, power, and leadership, as proximate goods. There are highly complex interactions at work here, which are, however, underpinned by a simple evolutionary rationale. An offense or injury will often prompt retaliation, lest it persists and turns into a pattern of victimization. Tit-for-tat may end in victory or a compromise, but it may also escalate, developing into a self-perpetuating cycle of strikes and counter-strikes, with the antagonists locked in conflict in a sort of prisoner's dilemma situation.

Similarly, in a state of potential conflict, security precautions are called for, which may take on defensive but also offensive or preemptive character. The security dilemma variant of the prisoner's dilemma breeds arms races that may produce an advantage to one side but often merely produces a "Red Queen" effect, by which both sides escalate their resource investment only to find themselves in the same position *vis-á-vis* one another. Organisms can cooperate, compete, or fight to maximize their survival and reproduction. Sometimes, fighting is the most promising choice for at least one of the sides. At other times, however, fighting, while being their rational choice, is not their best one.

Competition and conflict are, thus, "real" in the sense that they arise from genuine scarcities among evolution-shaped self-propagating organisms and can end in vital gains for one and losses for the other. At the same time, they are often also "inflated," self-perpetuated, and mutually damaging, because of the logic imposed on the antagonists by the conflict itself in an anarchic, unregulated environment. In a way, this justifies both of the prevalent polarized attitudes to war: the one that sees it as a serious business for serious aims and the other that is shocked by its absurdity.

Finally, a few comments on the evolutionary perspective that underpins this study. As our grand scientific theory for understanding nature, evolutionary theory does not compete with scholarly constructs such as psychoanalytic theories in explaining motivation; rather, evolutionary theory may encompass some of their main insights within a comprehensive interpretative framework. For instance, Freud, Jung, and Adler were divided over the elementary drive which each posited as the underlying regulating principle for understanding human behavior. These were respectively: sex; creativeness and the quest for meaning; and the craving for superiority. All these drives, in fact, come together and interact within the framework of evolutionary theory, which also explains their otherwise mysterious origin. Evolutionary theory explains how long-cited motives for fighting – like William Graham Sumner's (1968: 212) hunger, love, vanity, and fear of superior powers – came into being and how they hang together and interconnect.

Some readers may wonder why evolutionary theory should be presented here as different from and superior to other scholarly approaches. Indeed, it is because evolutionary theory is nature's *immanent* principle rather than an artificial analytical construct. It is the only nontranscendent mechanism for explaining life's complex design. This mechanism is blind natural selection in which at every stage those who are endowed with the most suitable qualities for surviving and reproducing survive. There is no reason for their survival other than that they proved successful in the struggle for survival. "Success" is not defined by any transcendent measurement but by the immanent logic of the evolutionary process.

This brings us to another widespread cause of resistance to "sociobiology," the belief that it upholds biological determinism in a subject which is distinctively determined by human culture. For once humans developed agriculture, they set in motion a continuous chain of developments that have taken us far away from our evolutionary natural way of life. Original, evolution-shaped, innate human wants, desires, and proximate behavioral and emotional mechanisms now express themselves in radically altered, "artificial" conditions. In the process, they have been greatly modified, assuming novel and diverse appearances. At the same time, however, cultural evolution has not operated on a "clean slate," nor has it been capable of producing simply "anything." Its multifarious and diverse forms have been built on a clearly recognizable deep core of innate human propensities. It has been working on a human physiological and psychological "landscape" deeply grooved by long-evolved inborn predispositions. Cultural takeoff took place much too recently to affect the human genome in any significant way (except for some well known aspects such as genes for lactose absorption, disease resistance, and a few other cases of strong selection) (Cavalli-Sforza and Feldman 1981; Lumsden and Wilson 1981; Boyd and Richerson 1985; Durham 1991; Richerson and Boyd 2005). Genetically, we are virtually the same people as our Stone Age forefathers and are endowed with the same predispositions. With cultural evolution, all bets are not off – they are merely hedged.

Unfortunately, space is too limited for a discussion of how the motives for human fighting and fighting itself have endured, and how they have been affected, by cultural evolution, through history. Interested readers are referred to my book (Gat 2006, Chaps. 12 and 17).

References

Alexander R (1987) The biology of moral systems. Aldine de Gruyter, New York

Ardrey R (1966) The territorial imperative. Atheneum, New York

Aureli F, Cozzolino R, Cordischi C, Scucchi S (1992) Kin-oriented redirection among Japanese macaques: an expression of a revenge system? Anim Behav 44:283–291

Axelrod R (1984) The evolution of cooperation. Basic Books, New York

Balikci A (1970) The Netsilik eskimo. Natural History Press, Garden City, NY

Berndt RM, Berndt CH (1964) The world of the first Australians. Angus & Robertson, London

Betzig LL, Denton RK, Rodseth L (1991) Comments on Bruce Knauft, violence and sociality in human evolution. Curr Anthropol 32:391–428

Bowker J (1995) Is god a virus: genes, culture and religion. SPCK, London

Bowles S (2006) Group competition, reproductive leveling, and the evolution of human altruism. Science 314:1569–1572

Boyd R, Richerson PJ (1985) Culture and the evolutionary process. University of Chicago Press, Chicago

Boyer P (2001) Religion explained: the evolutionary origins of religious thought. Basic Books, New York

Burch ES (1974) Eskimo warfare in Northwest Alaska. Anthropol Pap Univ Alaska 16:1–14

Burch ES, Correll TC (1972) Alliance and conflict: inter-regional relations in North Alaska. In: Guemple DL (ed) Alliance in eskimo society. University of Washington, Seattle, pp 17–39

Burton J (1990) Conflict: human needs theory. Macmillan, London

Buss D, Malamuth N (1996) Sex, power, conflict: evolutionary and feminist perspectives. Oxford University Press, New York

Cavalli-Sforza LL, Feldman MW (1981) Cultural transmission and evolution: a quantitative approach. Princeton University Press, Princeton

Chagnon NA (1977) Yanomamo: the fierce people, 2nd edn. Holt, New York

Chagnon NA (1979) Is reproductive success equal in egalitarian societies? In: Chagnon N, Irons W (eds) Evolutionary biology and human social behavior: an anthropological perspective. Duxbury, North Scituate, MA, pp 385–401

Daly M, Wilson M (1983) Sex, evolution, and behavior. Willard Grant, Boston, MA

Daly M, Wilson M (1988) Homicide. Aldine de Gruyter, New York

Darwin C [1871] (1962) The origin of the species and the descent of man. The Modern Library, New York

Dawkins R (2006) The god delusion. Houghton Mifflin, Boston, MA

Dawkins R [1976] (1989) The selfish gene. Oxford University Press, Oxford

de Waal FBM (1996) Good natured: the origins of right and wrong in humans and other animals. Harvard University Press, Cambridge, MA

Diamond J (1992) The rise and fall of the third chimpanzee. Vintage, London

Dickemann M (1979) Female infanticide, reproductive strategies, and social stratification: a preliminary model. In: Chagnon N, Irons W (eds) Evolutionary biology and human social behavior. Duxbury, North Scituate, MA, pp 321–367

Divale WT, Harris M (1976) Population, warfare and the male supremacist complex. Am Anthropol 78:521–538

Donald L (1997) Aboriginal slavery on the Northwest Coast of North America. University of California Press, Berkeley

Drucker P (1951) The Northern and Central Nootkan tribes. Smithsonian, Washington, DC

Drucker P (1965) Cultures of the North Pacific Coast. Chandler, San Francisco

Durham WH (1976) Resource competition and human aggression. Part I: a review of primitive war. Q Rev Biol 51:385–415

Durham WH (1991) Coevolution: genes, culture, and human diversity. Stanford University Press, Stanford

Durkheim E (1965) The elementary forms of religious life. Free Press, New York

Dyson-Hudson R, Smith EA (1978) Human territoriality: an ecological reassessment. Am Anthropol 80:21–41

Ember CR, Ember M (1992) Resource unpredictability, mistrust, and war: a cross-cultural study. J Conflict Resolut 36:242–262

Fagan R (1981) Animal play behavior. Oxford University Press, New York

Ferguson RB (1995) Yanomami warfare. School of American Research, Santa Fe, NM

Ferguson RB (2000) The causes and origins of 'primitive warfare': on evolved motivations for war. Anthropol Q 73:159–164

Frank R (1988) Passions within reason: the strategic role of the emotions. Norton, New York

Gat A (1999) The pattern of fighting in simple, small scale, pre-state societies. J Anthropol Res 55:563–583
Gat A (2000) Reply to Ferguson. Anthropol Q 73:165–168
Gat A (2006) War in human civilization. Oxford University Press, Oxford
Hamilton WD (1975) Innate social aptitude of man: an approach from evolutionary genetics. In: Fox R (ed) Biosocial anthropology. Wiley, New York, pp 133–153
Hamilton WD (1996) Narrow roads of gene land. WH Freeman, Oxford
Hart CWM, Pilling AR (1964) The Tiwi of North Australia. Holt, Rinehart & Winston, New York
Hayden B (1994) Competition, labor, and complex hunter-gatherers. In: Burch ES, Ellanna LJ (eds) Key issues in hunter-gatherer research. Berg, Oxford, pp 223–239
Hayden B (2003) Shamanism, sorcerers and saints: a prehistory of religion. Smithsonian, Washington, DC
Herz JH (1950) Idealist internationalism and the security dilemma. World Polit 2:157–180
Huntingford F, Turner A (1987) Animal conflict. Chapman, London
Irwin CJ (1987) A study in the evolution of ethnocentrism. In: Reynolds V, Falger VSE, Vine I (eds) The sociobiology of ethnocentrism. Croom Helm, London, pp 131–156
Irwin CJ (1990) The inuit and the evolution of limited group conflict. In: van der Dennen JMG, Falger VSE (eds) Sociobiology and conflict. Chapman, London, pp 189–226
Jervis R (1978) Cooperation under the security dilemma. World Polit 30:167–214
Jones S (1993) The language of the genes. Anchor Books, New York
Keeley LH (1996) War before civilization. Oxford University Press, New York
Keen I (1982) How some murngin men marry ten wives: the marital implications of matrilateral cross-cousin structures. Man 17:620–642
Keen I (1988) Yolngu religious property. In: Ingold T, Riches D, Woodburn J (eds) Hunter and gatherers. Berg, New York and Oxford, pp 272–307
Krause A (1970) The Tlingit Indians. University of Washington, Seattle
LeBlanc SA, Register KE (2003) Constant battles: the myth of the peaceful noble savage. St Martin, New York
Levitt PR (1980) The genetics of altruism. Academic, New York
Long J (1970) Polygyny, acculturation and contact: aspects of aboriginal marriage in Central Australia. In: Berndt RM (ed) Australian aboriginal anthropology. University of Western Australia Press, Nedland, pp 292–304
Lorenz K (1966) On aggression. Methuen, London
Lournados H (1988) Palaeopolitics: resource intensification in aboriginal Australia and Papua New Guinea. In: Ingold T, Riches D, Woodburn J (eds) Hunter and gatherers, vol I, History, evolution and social change. Berg, New York, pp 148–160
Lournados H (1997) Continent of hunter-gatherers: new perspectives in Australian prehistory. Cambridge University Press, Cambridge
Low DS (1979) Sexual selection and human ornamentation. In: Chagnon NA, Irons W (eds) Evolutionary biology and human social behavior: an anthropological perspective. Duxbury, North Scituate, MA, pp 462–487
Lumsden CJ, Wilson EO (1981) Genes, mind, and culture. Harvard University Press, Cambridge, MA
Maslow AH (1970) Motivation and personality. Harper, New York
Meggitt MJ (1965a) Marriage among the Walbiri of Central Australia: a statistical examination. In: Berndt RM, Berndt CH (eds) Aboriginal man in Australia. Angus & Robertson, Sidney, pp 146–159
Meggitt MJ (1965b) Desert people: a study of the Walbiri Aborigines of Central Australia. University of Chicago, Chicago
Mueller CW (1983) Environmental stressors and aggressive behavior. In: Geen RG, Donnerstein EI (eds) Aggression: theoretical and empirical reviews, vol 2. Academic, New York, pp 51–76
Oswalt WH (1967) Alaskan eskimos. Chandler, Washington, DC

Richerson PJ, Boyd R (2005) Not by genes alone: how culture transformed human evolution. University of Chicago Press, Chicago

Ridley M (1994) The red queen: sex and the evolution of human nature. Macmillan, New York

Ridley M (1996) The origins of virtue: human instincts and the evolution of cooperation. Viking, New York

Rosman A, Rubel P (1971) Feasting with mine enemy: rank and exchange among northwest coast societies. Columbia University Press, New York

Silverman I (1987) Inclusive fitness and ethnocentrism. In: Reynolds V, Falger V, Vine I (eds) The sociobiology of ethnocentrism. Croom Helm, London, pp 112–117

Smith PK (1984) Play in animals and humans. Blackwell, Oxford

Sumner WG (1968) War, reprinted from his war and other essays (1911). In: Bramson L, Goethals G (eds) War: studies from psychology, sociology, anthropology. Basic Books, New York, pp 205–227

Symons R (1979) The evolution of human sexuality. Oxford University Press, New York

Tinbergen N (1968) On war and peace in animals and man. Science 160:1411–1418

Trivers RL (1971) The evolution of reciprocal altruism. Q Rev Biol 46:35–57

Wilson DS (2002) Darwin's cathedral: evolution, religion, and the nature of society. University of Chicago, Chicago

Wilson DS, Sober E (1994) Reintroducing group selection to the human behavioral sciences. Behav Brain Sci 17:585–654

Wilson DS, Sober E (1998) Unto others: the evolution and psychology of unselfish behavior. Harvard University Press, Cambridge, MA

Wilson DS, Wilson EO (2007) Rethinking the theoretical foundations of sociobiology. Q Rev Biol 82:327–348

Part V
Foundations of Cooperation

Chapter 10
From Grooming to Giving Blood: The Origins of Human Altruism

Joan B. Silk and Robert Boyd

Abstract Cooperation plays an important role in the lives of most primates, including ourselves. However, the magnitude and scope of cooperation varies considerably across taxa: callitrichids pool efforts to rear a pairs' offspring, male langurs jointly challenge resident males for access to groups of females, female baboons groom one another equitably, and male chimpanzees exchange support for mating opportunities. All of these forms of cooperation have analogs in human societies, but humans cooperate in more diverse contexts, with a wider range of partners, and at larger scales than other primates. The evolutionary foundations of cooperation in nonhuman primates – kinship, reciprocity, and mutualism – also generate cooperation in human societies, but cooperation in human societies may also be supported by group-level processes that do not exist in other primate species. The human capacities for culture may have created novel evolutionary forces that altered the selective benefits derived from cooperation.

10.1 Introduction

Humans are exceptionally altruistic creatures. We honor promises, make donations to charity, vote in elections, recycle bottles, give blood, participate in political protests, punish cheaters, and go to war. We are moved by prosocial sentiments, such as empathy and compassion, that influence our responses to others in need, and moral emotions, such as a concern for fairness, that shape our judgments about what we should do in particular situations. Although other animals can be altruistic, our species is unusual because our altruistic impulses extend to people who lie outside the circle of close kin and beyond networks of reciprocating partners. This suggests

J.B. Silk (✉) and R. Boyd
Department of Anthropology, University of California, Los Angeles, CA, USA
e-mail: jsilk@anthro.ucla.edu, rboyd@anthro.ucla.edu

P.M. Kappeler and J.B. Silk (eds.), *Mind the Gap*,
DOI 10.1007/978-3-642-02725-3_10, © Springer-Verlag Berlin Heidelberg 2010

that the evolutionary processes that are thought to underlie altruistic cooperation in other animals, kin selection, contingent reciprocity, and biological market processes, may not be sufficient to account for the patterning of altruism in humans.

How were humans transformed from smart, sociable, and cooperative apes into other-regarding altruists? In this chapter, we attempt to provide some answers to this question. We begin with a discussion of the evolutionary processes that underlie altruism in nature, and then review the distribution, scope, and limits of altruism in nonhuman primates. We also review a growing body of data that provides insight about the nature of preferences that underlie cooperation in nonhuman primates. This information provides the context for understanding the similarities and differences among humans and other primates, and for considering the kinds of selective processes that may have played a role in the evolution of altruism in human societies. Then, we discuss the evolutionary processes that may have favored the evolution of altruistic, other-regarding creatures like ourselves.

10.2 The Evolution of Altruism

Evolutionary biologists define altruism as any behavior that reduces the genetic fitness of the actor and increases the genetic fitness of the recipient. There is considerably less consensus among evolutionary biologists about the definition of cooperation. Sometimes, cooperation is used as a synonym for altruism, and sometimes it is used to encompass any type of coordinated mutually beneficial behavior. Here, we will adopt the narrower definition of cooperation as a synonym for altruism, and we will use the terms cooperation and altruism interchangeably.

Natural selection is not expected to favor indiscriminate altruism because altruists always bear the costs of the altruistic behaviors that they perform on behalf of others, so the average fitness of an allele that increases the likelihood of performing altruistic behaviors will be lower than the average fitness of the non-altruistic allele. In order for altruism to evolve, there must be some cue that causes altruists to direct benefits selectively to other altruists. In nature, three types of cues seem important: recent common descent, proximity in viscous populations, and previous behavior.

Selection can favor altruism toward close relatives because recent common descent provides a reliable cue of genetic similarity. Individuals that are descended from the same ancestors have some probability of inheriting copies of the same genes. Thus, individuals who carry genes that are associated with altruistic behavior are more likely to have relatives who carry copies of the same genes than individuals drawn at random from the population. If individuals can identify their relatives and preferentially behave altruistically toward them, they will tend to confer benefits on individuals who also carry copies of the the genes that lead to altruistic behavior. Selection can also favor indiscriminate altruism toward other individuals if limited dispersal in viscous populations causes neighbors to be more closely related than chance would dictate even though they do not share a recent common ancestor.

Both these processes are often lumped together under the heading of kin selection (Hamilton 1964). What has come to be known as Hamilton's rule predicts that altruism will be favored when $br > c$. The quantities b and c represent the benefits and costs associated with the altruistic act. The quantity r measures how much the possession of a particular gene in one individual predicts the presence of the same gene in a second individual. Kin selection relies on the existence of a cue that allows individuals to direct altruism toward kin, and natural selection has produced a variety of mechanisms for kin recognition (rodents: Holmes and Mateo 2007; primates: Widdig 2007; social insects: Gamboa 2004; amphibians: Blaustein and Waldman 1992; birds: Komdeur and Hatchwell 1999).

The same basic logic underlies the theory of reciprocal altruism that was first introduced by Trivers (1971) and later formalized by Axelrod and Hamilton (1981). Reciprocal altruism is a form of contingent reciprocity in which the past behavior of other group members provides a cue about whether they carry alleles that lead to altruistic behavior. When individuals interact repeatedly, contingent altruistic strategies can arise. In the first interaction, an individual who carries the gene that leads to altruistic behavior provides help, and continues to help only if its partner reciprocates. If individuals follow this tit-for-tat rule, then contingent altruists will channel help toward other altruists after the first interaction. These kinds of contingent strategies can be sustained as long as $(1-1/t)\ b>c$, where b is the benefit derived from the other's helpful act, c is the cost of the helpful act, and t is the expected number of interactions between the two. It is not possible to satisfy this inequality when $t = 1$, so multiple interactions are required for contingent reciprocity to be favored. It is also easier to satisfy the inequality when the ratio of benefits to costs is high.

Although the logic underlying kin selection is fundamentally similar to the logic underlying contingent reciprocity, the outcome can be quite different because there are multiple equilibria. When the conditions $(1-1/t)\ b>c$ are fulfilled, contingent reciprocity can persist as an evolutionarily stable strategy (ESS), but unconditional defection is also an ESS, as are a variety of other strategies. In a world of unconditional defectors, contingent reciprocators will not prosper because they will invariably encounter partners who do not cooperate. In order for mutually beneficial strategies like tit-for-tat to evolve, there must be some factor that shifts the balance in their favor and makes it more likely that the population will move toward a cooperative equilibrium.

The solution to this problem was provided by Axelrod and Hamilton (1981). When pairs of related individuals interact, the odds of encountering another individual with the same rare strategy are substantially increased. If the benefits obtained over time by a cooperating pair are sufficiently greater than the benefits obtained by two unconditional defectors, then these rare cooperators can compensate for the fact that they do poorly when they are paired with unconditional defectors. Quite small amounts of relatedness may allow reciprocating strategies to invade a population composed of unconditional defectors. Axelrod and Hamilton termed this the "ratchet effect," because small amounts of relatedness can ramp up cooperation. Although the ratchet effect is powerful, it is also quite restrictive: it only works when groups are small (Boyd and Richerson 1988).

Below, we briefly review empirical evidence which suggests that kin selection and contingent reciprocity have shaped the distribution of altruistic behavior in nonhuman primate groups.

10.3 Kin Biases in Behavior

Primatologists have produced a rich body of information about maternal kin biases in behavior over the last 30 years (reviewed in Silk 2002, 2005; Chapais and Berman 2007). Perhaps the best way to summarize this extensive literature is to say that female primates behave as though they understand the logic of Hamilton's rule. In nearly every species in which females live in groups with their relatives, there are pronounced nepotistic biases among females in the distribution of altruistic behaviors, such as grooming, coalitionary support, and food sharing. Moreover, the most costly forms of altruism, including reproductive suppression and defense against higher ranking opponents, are limited to very close kin.

The most extreme form of nepotism occurs in cooperatively breeding groups of marmosets and tamarins. Marmoset and tamarins, members of the subfamily Callitrichinae, live in small territorial groups (French 1997; Tardif 1997). Cooperatively breeding callitrichid groups typically contain only one breeding pair, who are assisted by several nonbreeding adults, who may be same-sexed siblings of the breeding pair, and mature offspring from previous litters (Dietz 2004; French 1997; Tardif 1997). Breeding females typically give birth to fraternal twins and can produce two litters per year (in contrast, most other primates give birth to singletons at considerably longer intervals). After females give birth, nonbreeding group members provide extensive help carrying and provisioning infants. In golden lion tamarins, the species for which we have the most complete data in the wild, coresident adult males are generally close kin, but only one male sires offspring (Dietz 2004).

Nepotism is also a pronounced feature of behavior in the well-studied Cercopithecine societies, which include baboons, macaques, and vervets. In these species, mothers support their immature daughters when they are involved in conflicts with members of lower-ranking families, and daughters acquire rank positions just below their mothers (Silk 2002, 2005). Females form matrilineal dominance hierarchies in which all members of one matriline rank above or below all members of other matrilines. This process has long-lasting impacts on females because matrilineal dominance hierarchies are remarkably stable over time. High-ranking females have priority of access to resources, including food and water, and generally reproduce more successfully than lower-ranking females.

Male philopatry characterizes a much smaller set of primate species, including chimpanzees, bonobos, spider monkeys, muriquis, and wooly spider monkeys (Pusey and Packer 1987). Unlike males in most other primate species, males in these species form close ties with one another. For example, chimpanzee males spend a considerable amount of time in parties with other males, and males groom, hunt, share meat, aid, and patrol the borders of their territories with one another (Mitani 2005; Muller and Mitani 2005). In chimpanzee communities, males tend to form close relationships with their maternal brothers when they are available (Nishida 1979; Goodall 1986; Langergraber et al. 2007), but many males do not have brothers in their groups and kinship does not seem to be a necessary ingredient of close relationships among male chimpanzees (Langergraber et al. 2007).

Until recently, analyzes of nepotistic biases in favor of paternal kin were complicated by uncertainties about paternity (at least on the part of observers), which made it impossible to identify the paternal kin. The development of molecular genetic techniques for assessing paternity and noninvasive methods for obtaining genetic material from free-ranging animals now allow primatologists to identify the sires and to study the effects of paternal kinship on the distribution of altruistic behavior in primate groups (reviewed by Widdig 2007).

In baboon and rhesus macaque groups, females are more likely to associate with and groom paternal half sisters than unrelated females (Widdig et al. 2001, 2002; Smith et al. 2003; Silk et al. 2006a). In general, females show considerably stronger preferences for the maternal half sisters than for their paternal half sisters. This may reflect some degree of uncertainty about paternal relatedness or differences in the value of relationships with maternal and paternal sisters. If social bonds reinforce alliances, then maternal sisters may be more valuable allies than paternal sisters. Widdig et al. (2006) found that female rhesus macaques do not selectively support their paternal half sisters in agonistic encounters, but they do avoid intervening against them. Paternal kinship does not seem to play an important role in the distribution of altruistic behavior among adult male chimpanzees (Langergraber et al. 2007), young chimpanzees (Lehmann et al. 2006), juvenile mandrills (Charpentier et al. 2007), or white-faced capuchins (Perry et al. 2008).

10.4 Cooperation Among Reciprocating Partners

Until recently, most analyzes of the patterning of cooperation among unrelated individuals were based on the assumption that these interactions were the product of some form of contingent reciprocity. However, some researchers have argued that primates lack the cognitive ability to keep track of interactions with multiple partners across time (de Waal 2000; Barrett and Henzi 2002, 2005) and have cognitive biases, such as a preference for immediate rewards, that constrain the evolution of contingent reciprocity (Stevens and Hauser 2004; Stevens et al. 2005). Skepticism about the plausibility of contingent reciprocity as a strategic option for primates has led to interest in alternative processes, particularly the biological market model (Noë 2005, 2006). In biological markets, transactions are influenced by economic forces, such as supply and demand; trading partners act as buyers and sellers, negotiating prices for commodities based on their value to each party and the availability of alternative trading partners. Because buyers and sellers both obtain immediate benefits from their exchanges, there is no need to develop long-term relationships with partners.

The theoretical foundations of these approaches are quite different, but it is difficult to disentangle these processes empirically. For example, positive correlations between the amount of grooming given and received might be the product of contingent reciprocity or the outcome of trade in a biological market. Similarly, the absence of correlations in the amount of support given and received might mean that contingent reciprocity is not operating, that support is being exchanged for some other commodity (and there is some mechanism for enforcing trades), or that the frequency of interactions provides a poor index of the benefits accrued or costs incurred. Below, we summarize the pattern of interactions among unrelated individuals, although it is not always clear whether these patterns of exchange are the product of contingent reciprocity or market forces.

10.4.1 Patterns of Exchange and Interchange

Monkeys and apes generally spend the most time grooming those from whom they receive the most grooming, although the magnitude of the correlations in grooming given and received varies considerably (Schino and Aureli 2008). In addition, in some species, grooming within dyads is associated with the distribution of other commodities, such as access to infants (Henzi and Barrett 2002) and food (de Waal 1997), and with the distribution of some types of services, including support (Schino 2001; Watts 2002; Schino and Aureli 2008), participation in border patrols (Watts and Mitani 2001), and mating opportunities (Duffy et al. 2007).

Several recent studies indicate that grooming is more evenly balanced across multiple bouts than within single bouts (Frank and Silk 2009a; Gomes et al. 2009; Schino et al. 2009). For example, grooming among pairs of adult female baboons was significantly more evenly balanced over an 18-month study period than within single grooming bouts (Frank and Silk 2009a). These results strongly suggest that monkeys and apes are able to keep track of the distribution of grooming given and received over substantial periods of time.

Several lines of evidence suggest that cooperation may be limited to cooperating partners. First, female baboons and male chimpanzees form the strongest and most enduring social relationships with those that groom them most equitably, and this holds for both related and unrelated females (Silk et al. 2006b; Mitani 2009). We do not know whether equitable grooming relationships are more likely to be maintained across time, or whether relationships among close associates become more equitable over time. Either way, female baboons and male chimpanzees selectively maintain relationships with those that groom them most equitably.

Several factors seem to influence the distribution of cooperation within dyads. Grooming is often directed up the dominance hierarchy in macaque groups, as low-ranking females groom their partners more than they are groomed in return

(Schino 2001). This imbalance may exist because low-ranking females trade grooming for support as Seyfarth (1977) originally suggested. It may also exist because low-ranking females use grooming to appease high-ranking partners and reduce the risk of harassment from them. Barrett and her colleagues have shown that such trades become more imbalanced when the risk of aggression from higher-ranking females increases (Barrett et al. 1999). In some cases, females may use grooming to obtain access to other kinds of resources. For reasons that are not entirely clear, females are highly motivated to handle other females' infants, and often use grooming as a means to this end (Maestripieri 1994; Henzi and Barrett 2002). In some groups, females spend more time grooming mothers of newborns when few other infants are present than when many infants are available (Henzi and Barrett 2002; Manson 1999), but in others, the supply of infants does not seem to influence the grooming behavior (Frank and Silk 2009b). Similarly, female baboons groom higher-ranking mothers more than lower-ranking mothers in some groups (Henzi and Barrett 2002), but not in others (Frank and Silk 2009a).

Several naturalistic experiments have been designed to detect contingencies in cooperative behavior. Wild vervet monkeys were more attentive to the tape-recorded distress calls of unrelated group members if they had been groomed recently by the caller than if they had not been groomed recently by the same monkey (Seyfarth and Cheney 1984). In contrast, grooming among closely related monkeys did not influence the likelihood of responding to distress calls. Similarly, when disputes over food were instigated by researchers, long-tailed macaques were more likely to intervene on behalf of monkeys who had recently groomed them than monkeys who had not groomed them (Hemelrijk 1994). Grooming also enhances feeding tolerance among chimpanzees (de Waal 1997). In this experiment, chimpanzees were provisioned with leafy branches and all food transfers were recorded. In addition, grooming behavior *before* the provisioning event was monitored. The chimpanzees were more tolerant to individuals that had previously groomed them than they were to other individuals, and the effects of previous grooming were most pronounced for pairs that did not frequently groom at other times.

More formal laboratory experiments that were designed to assess how individuals respond to the helpful or unhelpful behavior of their partners have generated mixed results (see Silk 2007a for a review). In some cases, researchers have detected biases that favor partners who have provided help in previous trials (e.g., Cronin and Snowdon 2008; de Waal and Berger 2000; Hauser et al. 2003), while in others little or no evidence of contingency has been detected (Brosnan et al. 2009; Melis et al. 2008).

10.5 Limits of Altruism in Primate Groups

In nonhuman primate groups, cooperative interactions involve relatively small numbers of familiar individuals, often close relatives or reciprocating partners. Grooming involves pairs of individuals, coalitionary aggression may involve several individuals, and sizable numbers of individuals may participate in intergroup encounters.

Responses to strangers and members of neighboring groups generally range from passive avoidance to active hostility (see Crofoot and Wrangham this volume). Members of different groups rarely groom, and there are no reports of groups forming coalitions against other groups. In contrast, even human hunter-gatherer societies can orchestrate cooperative activities involving hundreds, sometimes thousands, of individuals. In market economies, goods and services are traded among strangers.

In human societies, people who violate social norms, break rules, or commit crimes are often punished by other group members. Punishment of this sort is altruistic because the individuals who impose sanctions on transgressors incur costs, while the benefits flow to the entire community. Hauser and Marler (1993a, b) provided the first evidence for altruistic punishment in primate groups. Rhesus macaques give distinctive calls when they find desirable foods. Hauser and Marler hid piles of coconut, a rhesus treat, in the monkeys' home range, and monitored what happened when these caches were discovered. In some cases, the monkeys who discovered the piles of coconut called, and in other cases, they remained silent. Female macaques were more likely to be attacked by other monkeys, particularly young males, if they remained silent after finding food than if they gave food calls. The authors hypothesized that females were being punished for attempting to conceal the location of these prized foods. This could constitute a form of altruistic punishment if the screams of the victim alert other group members to the site of the food, giving many animals an opportunity to profit from the aggressor's actions, or if harassment reduces the likelihood that the victims will remain silent after finding food in the future.

Subsequent work on food calling in white-faced capuchins (Gros-Louis 2004) provides an alternate interpretation for aggression in this context. Capuchins who called after finding food were less likely to be approached by others than monkeys that remained silent. In addition, individuals who gave food calls when they were approached by higher-ranking animals were less likely to receive aggression than monkeys who did not call. Gros-Louis (2004) suggests that food calls may function to establish the ownership of resources and signal the owners' willingness to defend them. This would explain why monkeys are especially likely to call when they are approached by high-ranking monkeys, who might challenge them for possession of food items.

10.6 Motives Underlying Altruism in NHPs

It is easy to perceive similarities between the altruistic behaviors we observe in primate groups and some forms of altruistic interactions in humans. A chimpanzee who is being groomed looks very much like someone getting a good massage – deeply relaxed and contented. When a juvenile baboon nestles in the lap of a male who is defending him from harassment, he seems to be relieved and reassured. Despite these parallels, it is not clear whether the motives and sentiments that underlie altruism are the same in humans and other species. In humans, altruism seems to be motivated at least in part by an understanding of others' thoughts and desires, concern for the welfare of others, and a preference for outcomes that benefit others (Fehr and Fischbacher 2003; Henrich et al. 2006). We may also be motivated by a concern for reputation that makes us want others to think that we are generous, fair, or charitable (Haley and Fessler 2005, Fessler and Gervais this volume). Below, we discuss the evidence for empathy and other regarding preferences in nonhuman primates. For a discussion of the evidence for fairness, see Lakshminarayan and Santos, this volume.

10.6.1 Empathy

Until recently, discussions of empathy in other primates were based on anecdotal reports of one individual helping another or reacting to another animal's distress (Silk 2007b). Although these events are intriguing, anecdotal data present several problems. First, interpretations of singular events are based on subjective impressions of observers, and are very difficult to verify. Second, observers may be more likely to notice and remember incidents that suggest that primates are empathetic than they are to take note when they seem oblivious or indifferent. Third, observers may be more likely to offer more anthropomorphic interpretations of the behavior of some animals, such as apes or dogs, than others.

This has led researchers to try to devise more systematic ways to assess the empathic responses of other primates. One study capitalized on the fact that macaques and baboons display elevated rates of self-directed behaviors, such as scratching, when they are under stress. Rates of self-directed behaviors rise sharply after monkeys are threatened or harassed by other group members (Aureli and van Schaik 1991; Castles and Whiten 1998). If monkeys experience empathy, then mothers would be expected to experience distress when their infants are distressed. However, Japanese macaque females showed no obvious signs of stress when their infants were harassed (Schino et al. 2004). Moreover, mothers did not approach their infants or offer comfort after their infants were victimized.

These results are consistent with the results from a study of mothers' reactions when their infants were exposed to danger (Cheney and Seyfarth 1990). In these experiments, mothers had learned that a dangerous or frightening object, such as a model of a snake, was concealed in a box in their enclosure. Mothers made no effort to stop their infants or warn them of danger when their infants, who were ignorant of danger, approached the box. Maternal indifference in these situations strongly suggests that monkeys may not have the capacity for empathy.

Female macaques may not respond to their infants' distress or protect them from potential dangers because they do not have a well-developed understanding of others' knowledge, feelings, and desires (Tomasello and Call 1997). Apes seem to have a more complete understanding of others' minds, and stronger claims are made about their capacity for empathy (Preston and de Waal 2002; de Waal 2003).

The capacity for empathy might underlie chimpanzees' responses to the victims of aggression. Third-party affiliation after conflicts has been documented in a number of chimpanzee populations (reviewed in Silk 2007a). De Waal and van Roosmalen (1979) hypothesized that these interactions are designed to alleviate the distress of the victims of aggression, and applied the label "consolation." This functional interpretation implies that actors are motivated by empathy for the victim (Palagi et al. 2006; Fraser et al. 2008), but studies designed to evaluate the function and effectiveness of third-party affiliation after conflicts have produced conflicting results. Palagi et al. (2006) found that chimpanzees did not selectively console kin or close associates, as might be expected if consolation is based on empathy. More recently, Fraser et al. (2008) have found that chimpanzees are significantly more likely to offer consolation to valued social partners than to other group members. Consolation effectively reduced self-directed behavior in one group of chimpanzees (Fraser et al. 2008), but not in another (Koski and Sterck 2007). The inconsistency among these results makes it difficult to draw strong conclusions about the function of consolation.

10.6.2 Other Regarding Preferences

The difficulties associated with identifying the sentiments that underlie altruistic behavior have led researchers to borrow techniques developed by behavioral economists to investigate the nature of social preferences in primates. In these experimental studies, subjects are faced with decisions about how to allocate resources to themselves and others. The choices that they make in these situations provide insight about their social preferences.

For example, in one set of experiments, chimpanzees were presented with the opportunity to deliver food rewards to themselves and/or other individuals. To implement their choices, the animals manipulated simple mechanical apparatuses that were baited with food. One of the options provided identical food rewards to the actor and to the occupant of the other enclosure (the 1/1 option); and the other option provided a food reward only to the actor (the 1/0 option). Individuals might prefer the 1/1 option because they have prepotent biases toward larger numbers of rewards (regardless of the distribution), so a control condition was included in which there was no potential recipient present. If individuals are concerned about the welfare of others, they are expected to prefer the 1/1 option over the 1/0 option, and this preference is expected to be stronger when another individual is present than when the actor is alone. Chimpanzees at three different sites, using four different apparatus, were as just as likely to choose the 1/1 option when another chimpanzee was present as when they were alone (Silk et al. 2005; Jensen et al. 2006; Vonk et al. 2008). It is possible that actors did not choose the 1/1 option more often because they were unaware that their partners preferred this outcome (Warneken et al. 2007). However, Vonk et al. (2008) found that potential recipients made begging gestures before the actor had made a choice in some trials, and clearly directed their begging gestures toward the side of the apparatus from which they would obtain rewards. Begging did not increase the likelihood that the actor would deliver rewards.

These experiments with chimpanzees have been followed with similar experiments in other species. In each case, actors were offered choices with different payoff structures. Capuchins demonstrate prosocial preferences in two different experimental paradigms (de Waal et al. 2008; Lakshminarayanan and Santos 2008), while results for two cooperatively breeding species are mixed. Cooperatively breeding common marmosets, *Callithrix jacchus*, were presented with an opportunity to deliver food to another individual, but received no reward themselves (0/0 vs. 0/1). The marmosets were significantly more likely to choose the 0/1 option when another marmoset was present than when they were alone (Burkart et al. 2007), satisfying the criterion for prosocial preferences. However, different results were obtained for another cooperatively breeding callitrichid, the cotton-top tamarin, *Saguinus oedipus* (Cronin et al. 2009). In this case, cotton-top tamarins did not distinguish between partner-present and partner-absent trials for two different payoff distributions (1/1 vs. 1/0; 0/1 vs. 0/0).

In another experimental study, Cronin and Snowdon (2008) evaluated tamarins' willingness to deliver rewards to their mates when (a) both parties got rewards, (b) both parties were rewarded on alternate days, (c) both parties shared access to a monopolizable reward, and (d) one partner could deliver rewards to its partner in repeated trials. The tamarins were most likely to solve the task, and deliver rewards to their partners, when both got rewards (100% trials), and they were least likely to solve the task when they did not receive rewards themselves (46%). Although helping was not extinguished completely when actors did not obtain rewards, latency to solve the task increased and there was more variability across pairs.

Chimpanzees seem to be more inclined to provide instrumental help to humans and other group members than to deliver food rewards to them. In two sets of experiments, chimpanzees retrieved objects that human experimenters could not reach (Warneken and Tomasello 2006; Warneken et al. 2007), and their performance was not influenced by the availability of food rewards (Warneken et al. 2007). In another experiment, chimpanzees were given an opportunity to provide help to other group members. In this experiment, the door to an adjacent room was fastened by a chain. The actor could remove a peg and release the door, but could not enter the room. The potential beneficiary could not reach the peg, but could enter the room. During experimental trials, a bowl of food (visible only to the potential beneficiary) was placed in this room. In control trials, food was placed in another room that was inaccessible to both chimpanzees. During all trials in which food was placed in the accessible room, potential beneficiaries oriented toward the door of the room, while they oriented toward the door to the other room in only half of the control trials. Moreover, actors were significantly more likely to release the door to the accessible room in experimental trials than in control trials. Thus, actors helped potential beneficiaries get into the locked room, and were sensitive to cues about their desires.

These experiments indicate that chimpanzees understand others' needs and desires, but do not provide unambiguous information about the motivations that underlie their behavior. In these situations, chimpanzees could be motivated by other regarding preferences or expectations of delayed reciprocity.

It is not easy to explain why these experiments lead to prosocial behavior in marmosets and capuchins, but not in tamarins or chimpanzees, and why chimpanzees display prosocial preferences when they are given an opportunity to provide instrumental help to others at some cost to themselves, but not when they are given an opportunity to provide food to others at no cost to themselves. Having a large brain and sophisticated knowledge of others' thoughts and desires is apparently neither necessary nor sufficient for the development of other regarding preferences. Burkart et al. (2007) suggested that cooperative breeding in callitrichids and humans might generate prosocial preferences, but this explanation does not fit the prosocial results for capuchins or the absence of prosocial preferences in tamarins. Warneken and his colleagues (Warneken and Tomasello 2006; Warneken et al. 2007) have speculated that chimpanzees may not display prosocial preferences in these experiments because they perceive food as a limited resource, and are not predisposed to provide food to others. This might explain why capuchins, which are remarkably tolerant of scrounging, display prosocial preferences. But it does not explain why prosocial responses are seen in marmosets, but not in tamarins, as food transfers play an important role in both taxa (Brown et al. 2004). It is possible that chimpanzees view all interactions involving food as zero-sum games because food supplies are limited in nature (Warneken et al. 2007), and have selfish preferences about food. If such preferences biased their behavior in these experiments, they would be expected to show consistent preferences for the 1/0 option; instead they choose the 1/1 and 1/0 option with equal frequencies.

10.7 The Origins of Other Regarding Preferences in Humans

When the human lineage diverged from the great ape lineage 5–7 million years ago, our ancestors were probably something like modern chimpanzees – smart, sociable, and cooperative. They would have been helpful to group members in some situations, and hostile to strangers. They might have exchanged goods, services, and favors with reciprocating partners, but were not unconditionally altruistic and may have had limited capacities for empathy and sense of fairness. To understand the origins and evolution of group-level cooperation and generalized other regarding preferences in humans, we need to consider two related questions. First, why did selection favor the evolution of the group-level cooperation and other regarding preferences in ancestral human populations, but not in other closely related species of primates? What evolutionary processes sustain these other regarding preferences in human groups?

To answer this question, it is useful to consider how the social organization and subsistence strategies of human foraging societies differ from those of other primates. Like the cooperatively breeding callitrichids, humans often get considerable help from others in rearing their offspring (Hrdy 2005; Burkart et al. 2007). Humans make more use of resources that are difficult to obtain and complicate to process than other primates do (Kaplan et al. 2000, 2003). We also rely more heavily on social learning to acquire the knowledge, skills, and information we need to make a living (see Whiten this volume). Finally, warfare has played an important role in the history of human societies (Gat this volume), but lethal intergroup conflict is absent in nonhuman primate groups, with the exception of chimpanzees (Crofoot and Wrangham this volume). All of these factors have been implicated in the evolution of other regarding preferences of humans. Here, we focus on cooperative breeding, complex foraging, and cultural evolution. For a more complete discussion of the role of warfare in the evolution of human societies see Gat (this volume) and Crofoot and Wrangham (this volume).

10.7.1 Cooperative Breeding

Hrdy (2005) hypothesizes that the high costs of producing and supporting slow-growing human children favored the development of extensive allomaternal care networks, which included fathers, grandmothers, and older siblings. In societies with high levels of infant mortality, alloparental care was an integral element of females' reproductive strategies. Hrdy considers humans to be cooperative breeders because multiple individuals contribute to children's care. This definition conflates taxa in which there is only one breeding pair who are assisted by nonbreeding helpers (e.g., wild dogs, meerkats) with taxa in which there are multiple breeding females who share some maternal tasks and may be assisted by other group members (e.g., lions, banded mongoose). We reserve the term cooperative breeders for the former, and use the term communal breeders for the latter. By this definition, callitrichids are cooperative breeders and humans are communal breeders.

Hrdy suggests that the ability to engage caretakers and elicit investment would be advantageous for infants, and this would favor the evolution of cognitive capacities that allow young children to assess the intentions and predict the responses of others. Over the course of our evolutionary history, selection elaborated these capacities to produce empathy and a well-developed theory of mind. According to this argument, delayed maturation and cooperative/communal breeding coevolved, and both of these developments preceded the marked expansion of brain size in humans and the origin of other regarding preferences.

Burkart et al. (2007) have also emphasized the link between cooperative breeding and other regarding preferences in humans and marmosets. They suggest that "...unsolicited prosociality, which arose in the context of provisioning, carrying, and sharing was then generalized toward the sharing of information and psychological states." When unsolicited prosociality was added to the the ape-like brain of our ancestors, it precipitated a "cascade of further developments" including language, teaching, and the development of other regarding preferences and group-level cooperation.

Generalized prosocial preferences in small groups of closely related individuals could evolve through kin selection. Callitrichid groups are typically small and group members are closely related to the infants that they care for. Although there is some evidence that altruistic responses are not limited to kin and not all group members are equally altruistic (Burkart et al. 2007), it is possible that there are few opportunities for prosocial behavior toward nonrelatives in the wild. Thus, group-level cooperation evolved through kin selection.

It is more difficult to invoke the same argument for the evolution of other regarding preferences in humans. Contemporary human foragers live in larger and more complex groups than cooperatively breeding monkeys do. Allomaternal care in such groups is typically nepotistic, and grandmothers are the most common caregivers for children (reviewed by Hrdy 2005). Male provisioning and direct care of infants may be a form of parenting effort or mating effort (Anderson et al. 1999a, b; Marlowe 1999a,b); in both cases, mens' contributions to childcare linked to their own fitness benefits. Selective pressures favoring allomaternal care and communal breeding in human groups have not produced indiscriminant altruism toward children, and it seems difficult to link communal breeding directly to the emergence of other regarding preferences in human groups.

The similarity in the responses of marmosets and humans in the prosocial task may arise because group-level cooperation is favored in both taxa, not because both species are cooperative/communal breeders. Kin selection may favor group-level cooperation in marmosets while other forces may generate group-level cooperation in humans. Below we consider two possible mechanisms underlying group-level cooperation in human societies.

10.7.2 Complex Foraging

The cooperative breeding hypothesis does not explain why humans mature more slowly in relation to their body size than other primates and why human infants require so much more care than the infants of other primates. Kaplan and his colleagues suggest that humans mature slowly because it takes a long time to acquire the knowledge and skills that human food foragers need to make a living (Kaplan et al. 2000, 2003). Human foragers rely heavily on complex foraging skills, including hunting and extractive foraging, and exploit a much wider range of resources using a larger repertoire of tools and techniques than other primates do. Comparative data indicate that extracted and hunted foods account for about 5% of chimpanzee diets, while these types of food account for about 90% of the diet of human hunter-gatherers. Kaplan and his colleagues emphasize the fact that humans specialize on resources, such as meat, roots, and nuts, that are rare and patchy, but provide rich sources of nutrients.

The reliance on complex foraging techniques may have favored economic interdependence within families and groups. If foraging skills are difficult to master, it makes sense to develop foraging specializations. Such diversification, by sex, age, or ability, will pay off if specialists share the products of their foraging efforts. Sexual division of labor is a universal feature of human foraging societies, as men mainly hunt and women mainly gather. Sharing may also buffer the economic risks associated with hunting (Winterhalder 1986). On some days, hunters return with carcasses large enough to feed many people, but on other days, they come back empty-handed. Sharing provides one way to insure against such risks. Sharing networks that extend beyond family or household, buffer risk even further (Gurven 2004).

According to this argument, the importance of complex foraging techniques in human subsistence strategies may have favored delayed development, extended periods of parental provisioning, division of labor within families, and the formation of extended sharing networks. In hunter-gatherer groups, resource exchanges are influenced by the dynamics of contingent reciprocity, as well as by the norms of fairness (Gurven 2004, 2006). To explain the origins of prosocial preferences, however, we need to take the argument one step further, and consider the role of cultural evolution.

10.7.3 Cultural Evolution

Complex foraging strategies may be linked to the evolution of social learning and the capacity for culture. Early Pleistocene hominins occupied a considerably wider range of habitats than any contemporary apes do today. The knowledge and subsistence technology required for complex foraging varies greatly from one habitat to another, and it would have become more and more difficult for individuals to acquire this information on their own. Social learning allows human populations to gradually accumulate useful knowledge as individuals learn from others, make modest improvements, and pass this accumulated knowledge on. This kind of cumulative cultural change can give rise to complex habitat-specific adaptations much more rapidly than genetic evolution can (Boyd and Richerson 1985, 1996, McElreath this volume). Although we may have underestimated the social learning capacities of chimpanzees and other primates in the past (Whiten this volume), there is no doubt that humans rely on social learning to a much greater extent than any other primates do.

The cultural transmission of information may have been especially important for our ancestors during the Middle and Upper Pleistocene. During this period, world temperatures fluctuated widely. At some points, average world temperatures changed as much as 10°C in 1,000 years (Richerson et al. 2001). In this kind of world, the ability to make rapid adjustments to changing conditions would have provided strong selection for the evolution of cultural capacities.

Social learning may have enabled humans to adapt to changing conditions, but it also had the potential to generate considerable cultural variation among groups. Social interactions commonly give rise to multiple adaptive equilibria. (a nonbiological analog of this process would be conventions about which side of the road to drive on: it is equally efficient to drive on the right or the left, but essential that everyone follows the same rule.) Systems of reciprocity, reputation management, and punishment can stabilize a vast range of behaviors including ones that lead to large-scale cooperation (Axelrod 1986; Boyd and Richerson 1992; Nowak and Sigmund 1998; Henrich and Boyd 1998; Panchanathan and Boyd 2004). Adaptive processes, including both individual learning and the tendency to imitate successful individuals which generates conformist biases, will cause local populations to evolve toward different equilibria. This tendency will be counteracted by the flow of ideas between groups, just as genetic variation among groups is counteracted by migration. However, if individuals adopt the ideas and traditions of their new groups, then cultural variation among groups will be maintained. Cultural adaptation can proceed much more quickly than genetic adaptation, so it is likely that as cultural adaptation became more and more important, the amount of variation in behavior and social organization among human groups also increased (Richerson and Boyd 2005).

Increased variation between groups could have had important effects on the cultural evolution of group-beneficial traits. To understand the why, it is helpful to adopt the formulation derived by Price (1970) which partitions genetic evolutionary change into two components:

$$\Delta p \propto \underbrace{V_{\mathrm{G}}\,\beta_{\mathrm{G}}}_{\text{between groups}} + \underbrace{\overline{V_{\mathrm{W}}\,\beta_{\mathrm{W}}}}_{\text{within groups}}$$

The effect of selection on the change in frequency of a gene, Δp, is proportional to the amount of genetic variation between groups (V_{G}) and the amount of genetic variation within groups (V_{W}). Behaviors are beneficial to the group when the behavior increases group fitness, or $\beta_{\mathrm{G}} > 0$. If it is costly to the individual, $\beta_{\mathrm{W}} < 0$. When behaviors are beneficial to the group and costly to individuals (as is the case for altruistic traits), then the outcome will depend on the relative magnitude of the variance within and between groups. When groups are large, selection is weak, and rates of migration are not too low, V_{W} will greatly exceed V_{G} (Rogers 1990), and group selection cannot overcome opposing individual selection. These conditions often hold in nature, and group selection is generally not thought to be an important force.

Group selection can play a more important role when there are multiple stable equilibria. To see how this works, consider a situation in which there are two alleles, A and B, and homozygotes have higher fitness than heterozygotes ($W_{\mathrm{AA}} > W_{\mathrm{BB}} > W_{\mathrm{AB}}$). This means that when A is common, B will not be able to invade because $W_{AB} < W_{AA}$; when B is common, A will not be able to invade because $W_{AB} < W_{BB}$. In this situation, there is a paradoxical result: when B is common, individual selection will not favor the A allele, even though it confers a fitness advantage. However, group selection can lead to the spread of the B allele when there are multiple stable equilibria. Imagine that a large population is divided into a number of separate groups: A is common in some groups, and B is common in other groups. There is little variance within groups, so V_W will be low. Selection favors A in some populations and B in other populations, so the average value of β_W across groups will also be small. This means that the within-group component of the Price equation will approach zero, and selection within groups will have little impact on the frequency of A and B. If the A allele has higher average fitness, then group selection can favor the spread of the A allele. The A allele may spread if carriers of the A allele are more successful in forming new groups, or if groups with lower frequencies of the A allele are more likely to become extinct and are replaced by individuals carrying the A allele (Boyd and Richerson 2002). Thus, when there are multiple stable equilibria and selection is strong compared to migration, selection will preserve the variation among groups and favor the evolution of traits that increase group fitness.

Although the Price equation was formulated to describe the effects of selection on genetic traits, the same basic logic applies to cultural variants. When there are multiple stable equilibria, processes that reduce the amount of cultural variation within groups will reduce the within-group component of variation and strengthen the forces of selection among groups. This will favor the evolution of group-beneficial cultural traits that increase the competitive ability of groups. Competition between groups will favor the spread and elaboration of cooperative cultural norms that makes groups larger, more productive, and more successful in conflicts with neighboring groups. Cooperative cultural norms may be maintained by concerns about reputation, desire to maintain reciprocity, or fear of costly sanctions. This, in turn, may have favored the development of new prosocial emotions, such as compassion, guilt, and shame. Individuals who lack these new social emotions would have violated the prevailing norms more often, and as a result they may have been punished, denigrated, denied access to community resources, or ostracized. Cooperation and group identification in intergroup conflict may have set up an arms race that drove social evolution to progressively greater extremes of in-group cooperation. Eventually, human populations came to resemble the hunter-gathering societies of the ethnographic record. These societies are egalitarian and political power is diffuse. People readily punish others for transgressions of social norms, even when their own personal interests are not directly at stake.

These new motivations did not replace the motivational biases that we inherited from our primate ancestors. We still care strongly about our own welfare, are biased in favor of our relatives, and place special trust in reciprocating partners. However, we are also moved by broader loyalties to clan, tribe, class, caste, and nation. In some cases, these loyalties conflict: we wince when we pay our taxes, although we realize that our taxes support schools, hospitals, and other worthy institutions.

10.8 Conclusions

In the last 5 million years or so, a smart, sociable, and cooperative forest ape was transformed into a slow growing, highly adaptable, technologically sophisticated, other-regarding altruist. Although much of the ancient ape remains within us, we differ from other apes in our exceptionally slow life history, our reliance on social learning and cultural adaptations, and in our development of group beneficial social norms and social preferences. Our understanding of this transformation is based on work from many academic disciplines. Primatologists have accumulated a wealth of information about the distribution of altruistic behavior in primate groups, which provides a broad comparative framework for understanding the roots of cooperation. Evolutionary theorists have developed a rich body of theory about the evolution of altruism and cultural evolution, which has enabled us to understand the dynamics of these processes in a more rigorous way. Finally, developmental psychologists, behavioral economists, human behavioral ecologists, and evolutionary psychologists have contributed a diverse set of methods and extensive empirical evidence about motivations, preferences, and behavior of contemporary people. This body of work has helped us to define the continuities in the behavior of humans and other primates, and to illuminate the gaps. It is clear that our understanding of the origins of human altruism is incomplete, but we have begun to get some traction on the problem.

References

Anderson KG, Kaplan H, Lancaster JB (1999a) Paternal care by genetic fathers and stepfathers. I. Reports from Albuquerque men. Evol Hum Behav 20:405–431

Anderson KG, Kaplan H, Lam D, Lancaster JB (1999b) Paternal care by genetic fathers and stepfathers. II. Reports by Xhosa high school students. Evol Hum Behav 20:433–451

Aureli F, van Schaik CP (1991) Post conflict behaviour in long-tailed macaques (*Macaca fasicularis*). II. Coping with the uncertainty. Ethology 89:101–114

Axelrod R (1986) An evolutionary approach to norms. Am Polit Sci Rev 80:1095–1111

Axelrod R, Hamilton WD (1981) The evolution of cooperation. Science 211:1390–1396

Barrett L, Henzi SP (2002) Constraints on relationship formation among female primates. Behaviour 139:263–289

Barrett L, Henzi SP (2005) Monkeys, markets and minds: biological markets and primate sociality. In: Kappeler PM, van Schaik CP (eds) Cooperation in primates and humans: mechanisms and evolution. Springer, Berlin, pp 209–232

Barrett L, Henzi SP, Weingrill T, Lycett J, Hill RA (1999) Market forces predict grooming reciprocity in female baboons. Proc Roy Soc Lond B 266:665–670

Blaustein AR, Waldman B (1992) Kin recognition in anuran amphibians. Anim Behav 44:207–221

Boyd R, Richerson PJ (1985) Culture and the evolutionary process. University of Chicago Press, Chicago

Boyd R, Richerson PJ (1988) The evolution of reciprocity in sizable groups. J Theor Biol 132:337–356

Boyd R, Richerson PJ (1992) Punishment allows the evolution of cooperation (or anything else) in sizable groups. Ethol Sociobiol 13:171–195

Boyd R, Richerson PJ (1996) Why culture is common, but cultural evolution is rare. Proc Br Acad 88:77–93

Boyd R, Richerson PJ (2002) Group beneficial norms spread rapidly in a structured population. J Theor Biol 215:287–296

Brosnan SF, Silk JB, Henrich J, Mareno MC, Lambeth SP, Schapiro SJ (in press) Chimpanzees (*Pan troglodytes*) do not develop contingent reciprocity in an experimental task. Anim Cogn

Brown GR, Almond REA, van Bergen Y (2004) Begging, stealing, and offering: food transfer in nonhuman primates. Adv Study Behav 34:265–295

Burkart JM, Fehr E, Efferson C, van Schaik CP (2007) Other-regarding preferences in a nonhuman primate: common marmosets provision food altruistically. Proc Natl Acad Sci USA 104:19762–19766

Castles DL, Whiten A (1998) Post-conflict behaviour of wild olive baboons. II. Stress and self-directed behaviour. Ethology 104:148–160

Chapais B, Berman CM (2007) Kinship and behavior in primates. Oxford University Press, Oxford

Charpentier MJE, Peignot P, Hossaert-Mckey M, Wickings EJ (2007) Kin discrimination in juvenile mandrills, *Mandrillus sphinx*. Anim Behav 73:37–45

Cheney DL, Seyfarth RM (1990) Attending to behaviour versus attending to knowledge: examining monkeys' attribution of mental states. Anim Behav 40:742–753

Cronin KA, Snowdon CT (2008) The effects of unequal reward distributions on cooperative problem solving by cotton-top tamarins (*Saguinus oedipus*). Anim Behav 75:245–257

Cronin KA, Schroeder KKE, Rothwell ES, Silk JB, Snowdon CT (in press) Cooperatively breeding cottontop tamarins (*Saguinus oedipus*) do not donate rewards to their long-term mates. J Comp Psychol

de Waal FBM (1997) The chimpanzee's service economy: food for grooming. Evol Hum Behav 18:375–386

de Waal FBM (2000) Attitudinal reciprocity in food sharing among brown capuchin monkeys. Anim Behav 60:253–261

de Waal FBM (2003) On the possibility of animal empathy. In: Manstead T, Frijda N, Fischer A (eds) Feelings and emotions: the Amsterdam symposium. Cambridge University Press, Cambridge, pp 379–399

de Waal FBM, Berger ML (2000) Payment for labour in monkeys. Nature 404:563

de Waal FBM, van Roosmalen A (1979) Reconciliation and consolation among chimpanzees. Behav Ecol Sociobiol 5:55–66

de Waal FBM, Leimgruber K, Greenberg AR (2008) Giving is self-rewarding for monkeys. Proc Natl Acad Sci USA 105:13685–13689

Dietz JM (2004) Kinship structure and reproductive skew in cooperatively breeding primates. In: Chapais B, Berman CM (eds) Kinship and behavior in primates. Oxford University Press, Oxford, pp 223–241

Duffy KG, Wrangham RW, Silk JB (2007) Male chimpanzees exchange political support for mating opportunities. Curr Biol 17:R586–R587

Fehr E, Fischbacher U (2003) The nature of human altruism. Nature 425:785–791

Frank RE, Silk JB (2009a) Impatient traders or contingent reciprocators? Evidence for the extended time course of grooming exchanges in baboons. Behaviour 136:1123–1135

Frank RE, Silk JB (2009b) Grooming exchange between mothers and non-mothers: the price of natal attraction in wild baboons (*Papio anubis*). Behaviour 136:889–906

Fraser ON, Stahl D, Aureli F (2008) Stress reduction through consolation in chimpanzees. Proc Natl Acad Sci USA 105:8557–8662

French JA (1997) Proximate regulation of singular breeding in callitrichid primates. In: Solomon NG, French JA (eds) Cooperative breeding in mammals. Cambridge University Press, Cambridge, pp 34–75

Gamboa GJ (2004) Kin recognition in eusocial wasps. Ann Zool Fennici 41:789–808

Gomes CM, Mundry R, Boesch C (2009) Long-term reciprocation of grooming in wild West African chimpanzees. Proc Roy Soc Lond B 276:699–706

Goodall J (1986) The chimpanzees of Gombe: patterns of behavior. Belknap Press, Cambridge, MA
Gros-Louis J (2004) The function of food-associated calls in white-faced capuchin monkeys, *Cebus capucinus*, from the perspective of the signaller. Anim Behav 67:431–440
Gurven M (2004) To give and to give not: the behavioral ecology of human food transfers. Behav Brain Sci 27:543–583
Gurven M (2006) The evolution of contingent cooperation. Curr Anthropol 47:185–192
Haley KJ, Fessler DMT (2005) Nobody's watching? Subtle cues affect generosity in an anonymous economic game. Evol Hum Behav 26:245–256
Hamilton WD (1964) The genetical evolution of social behavior. I and II. J Theor Biol 7:1–52
Hauser MD, Marler P (1993a) Food-associated calls in rhesus macaques (*Macaca mulatta*). I. Socioecological factors. Behav Ecol 4:194–205
Hauser MD, Marler P (1993b) Food-associated calls in rhesus macaques (*Macaca mulatta*). II. Costs and benefits of call production and suppression. Behav Ecol 4:206–212
Hauser MD, Chen MK, Chen F, Chuang E (2003) Give unto others: genetically unrelated cotton-top tamarins monkeys preferentially give food to those who altruistically give food back. Proc Roy Soc Lond B 270:2363–2370
Hemelrijk CK (1994) Support for being groomed in long-tailed macaques, *Macaca fasicularis*. Anim Behav 48:479–481
Henrich J, Boyd R (1998) The evolution of conformist transmission and the emergence of between-group differences. Evol Hum Behav 19:215–241
Henrich J, McElreath R, Barr A, Ensminger J, Barrett C, Bolyanatz A, Cardenas JC, Gurven M, Gwako E, Henrich N, Lesorogol C, Marlowe F, Tracer D, Ziker J (2006) Costly punishment across human societies. Science 312:1767–1770
Henzi SP, Barrett L (2002) Infants as a commodity in a baboon market. Anim Behav 63: 915–921
Holmes WG, Mateo JM (2007) Kin recognition in rodents: issues and evidence. In: Wolff JO, Sherman PW (eds) Rodent societies: an ecological and evolutionary perspective. University of Chicago Press, Chicago, pp 216–228
Hrdy SB (2005) Evolutionary context of human development: the cooperative breeding model. In: Carter CS, Ahnert L, Grossman KE, Hrdy SB, Lamb ME, Porgest SW, Sacher N (eds) Attachment and bonding: a new synthesis. MIT, Boston, MA, pp 9–32
Jensen K, Hare B, Call J, Tomasello M (2006) What's in it for me? Self-regard precludes altruism and spite in chimpanzees. Proc Roy Soc Lond B 273:1013–1021
Kaplan HS, Hill KR, Lancaster JB, Hurtado AM (2000) A theory of human life history evolution: diet, intelligence, and longevity. Evol Anthropol 9:156–185
Kaplan H, Lancaster JB, Robson A (2003) Embodied capital and the evolutionary economics of the human lifespan. In: Carey JR, Tuljapurkar S (eds) Life span: evolutionary, ecological and demographic perspectives. The Population Council, New York, Population and Development Review 29(Suppl):152–182
Komdeur J, Hatchwell BJ (1999) Kin recognition: function and mechanism in avian societies. Trends Ecol Evol 14:237–241
Koski SE, Sterck EHM (2007) Triadic postconflict affiliation in captive chimpanzees: does consolation console? Anim Behav 73:133–142
Lakshminarayanan VR, Santos LR (2008) Capuchin monkeys are sensitive to others' welfare. Curr Biol 18:R999–R1000
Langergraber KE, Mitani JC, Vigilant L (2007) The limited impact of kinship on cooperation in wild chimpanzees. Proc Natl Acad Sci USA 104:7786–7790
Lehmann J, Fickenscher G, Boesch C (2006) Kin biased investment in wild chimpanzees. Behaviour 143:931–955
Maestripieri D (1994) Social structure, infant handling, and mother styles in group-living Old World monkeys. Int J Primatol 15:531–553
Manson JH (1999) Infant handling in wild *Cebus capucinus*: testing bonds between females? Anim Behav 57:911–921

Marlowe F (1999a) Male care and mating effort among Hadza foragers. Behav Ecol Sociobiol 46:57–64

Marlowe F (1999b) Showoffs or providers? The parenting effort of Hadza men. Evol Hum Behav 20:391–404

Melis AP, Hare B, Tomasello M (2008) Do chimpanzees reciprocate received favours? Anim Behav 76:951–962

Mitani JC (2005) Reciprocal exchange in chimpanzees and other primates. In: Kappeler PM, van Schaik CP (eds) Cooperation in primates and humans: mechanisms and evolution. Springer, Berlin, pp 107–119

Mitani JC (2009) Male chimpanzees form enduring and equitable social bonds. Anim Behav 77:633–640

Muller MN, Mitani JC (2005) Conflict and cooperation in wild chimpanzees. Adv Study Behav 35:275–331

Nishida T (1979) The social structure of chimpanzees of the Mahale mountains. In: Hamburg DA, McCown ER (eds) The great apes. Benjamin Cummings, Menlo Park, CA, pp 73–121

Noë R (2005) Digging for the roots of trading. In: Kappeler PM, van Schaik CP (eds) Cooperation in primates and humans: evolution and mechanisms. Springer, Berlin, pp 233–261

Noë R (2006) Cooperation experiments: coordination through communication versus acting apart together. Anim Behav 71:1–18

Nowak MA, Sigmund K (1998) Evolution of indirect reciprocity by image scoring. Nature 393:573–577

Palagi E, Cordoni G, Borgognini Tarli S (2006) Possible roles of consolation in captive chimpanzees (*Pan troglodytes*). Am J Phys Anthropol 129:105–111

Panchanathan K, Boyd R (2004) Indirect reciprocity can stabilize cooperation without the second-order free rider problem. Nature 432:499–502

Perry S, Manson JH, Muniz L, Gros-Louis J, Vigilant L (2008) Kin-biased social behaviour in wild adult female white-faced capuchins, *Cebus capucinus*. Anim Behav 76:187–199

Preston SD, de Waal FBM (2002) Empathy: its ultimate and proximate bases. Behav Brain Sci 25:1–72

Price GR (1970) Selection and covariance. Nature 227:520–521

Pusey AE, Packer C (1987) Dispersal and philopatry. In: Smuts BB, Cheney DL, Seyfarth RM, Wrangham RW, Struhsaker TT (eds) Primate societies. University of Chicago Press, Chicago, pp 250–266

Richerson PJ, Boyd R (2005) Not by genes alone: how culture transformed human evolution. University of Chicago Press, Chicago

Richerson PJ, Boyd R, Bettinger RL (2001) Was agriculture impossible during the Pleistocene but mandatory during the Holocene? A climate change hypothesis. Am Antiq 66:387–411

Rogers AR (1990) Group selection by selective emigration: the effects of migration and kin structure. Am Nat 135:398–413

Schino G (2001) Grooming, competition and social rank among female primates: a meta-analysis. Anim Behav 62:265–271

Schino G, Aureli F (2008) Grooming reciprocation among female primates: a meta-analysis. Biol Lett 4:9–11

Schino G, Gemiani S, Rosati L, Aureli F (2004) Behavioral and emotional response of Japanese macaque (*Macaca fuscata*) mothers after their offspring receive an aggression. J Comp Psychol 118:340–346

Schino G, Di Giuseppe F, Visalberghi E (2009) The time frame of partner choice in the grooming reciprocation of *Cebus apella*. Ethology 115:70–76

Seyfarth RM (1977) A model of social grooming among adult female monkeys. J Theor Biol 65:671–698

Seyfarth RM, Cheney DL (1984) Grooming, alliances, and reciprocal altruism in vervet monkeys. Nature 308:541–543

Silk JB (2002) Kin selection in primate groups. Int J Primatol 23:849–875

Silk JB (2005) Practicing Hamilton's rule: kin selection in primate groups. In: Kappeler PM, van Schaik CP (eds) Cooperation in primates and humans: mechanisms and evolution. Springer, Berlin, pp 25–46

Silk JB (2007a) The strategic dynamics of cooperation in primate groups. Adv Study Behav 37: 1–42

Silk JB (2007b) Empathy, sympathy, and prosocial preferences in primates. In: Dunbar RIM, Barrett L (eds) Oxford handbook of evolutionary psychology. Oxford University Press, Oxford, pp 115–126

Silk JB, Brosnan SF, Vonk J, Henrich J, Povinelli DJ, Richardson AS, Lambeth SP, Mascaro J, Schapiro SJ (2005) Chimpanzees are indifferent to the welfare of unrelated group members. Nature 437:1357–1359

Silk JB, Altmann J, Alberts SC (2006a) Social relationships among adult female baboons (*Papio cynocephalus*). I. Variation in the strength of social bonds. Behav Ecol Sociobiol 61:183–195

Silk JB, Alberts SC, Altmann J (2006b) Social relationships among adult female baboons (*Papio cynocephalus*). II. Variation in the quality and stability of social bonds. Behav Ecol Sociobiol 61:197–204

Smith KL, Alberts SC, Altmann J (2003) Wild female baboons bias their social behaviour towards paternal half-sisters. Proc Roy Soc Lond B 270:503–510

Stevens JR, Hauser MD (2004) Why be nice? Psychological constraints on the evolution of cooperation. Trends Cogn Sci 8:60–65

Stevens JR, Cushman FA, Hauser MD (2005) Evolving the psychological mechanisms for cooperation. Annu Rev Ecol Syst 36:499–518

Tardif SD (1997) The bioenergetics of parental behavior and the evolution of alloparental care in marmosets and tamarins. In: Solomon NG, French JA (eds) Cooperative breeding in mammals. Cambridge University Press, Cambridge, pp 11–33

Tomasello M, Call J (1997) Primate cognition. Oxford University Press, Oxford

Trivers RL (1971) The evolution of reciprocal altruism. Q Rev Biol 46:35–57

Vonk J, Brosnan SF, Silk JB, Henrich J, Richardson AS, Lambeth SP, Schapiro SJ, Povinelli DJ (2008) Chimpanzees do not take advantage of very low cost opportunities to deliver food to unrelated group members. Anim Behav 75:1757–1770

Warneken F, Tomasello M (2006) Altruistic helping in human infants and young chimpanzees. Science 311:1301–1303

Warneken F, Hare B, Melis AP, Hanus D, Tomasello M (2007) Spontaneous altruism by chimpanzees and young children. PLoS Biol 5:1414–1420

Watts DP (2002) Reciprocity and interchange in the social relationships of wild male chimpanzees. Behaviour 139:343–370

Watts DP, Mitani JC (2001) Boundary patrols and intergroup encounters in wild chimpanzees. Behaviour 138:299–327

Widdig A (2007) Paternal kin discrimination: the evidence and likely mechanisms. Biol Rev 82:319–334

Widdig A, Nürnberg P, Krawczak M, Streich WJ, Bercovitch FB (2001) Paternal relatedness and age proximity regulate social relationships among adult female rhesus macaques. Proc Natl Acad Sci USA 98:13769–13773

Widdig A, Nürnberg P, Krawczak M, Streich WJ, Bercovitch FB (2002) Affiliation and aggression among adult female rhesus macaques: a genetic analysis of paternal cohorts. Behaviour 139:371–391

Widdig A, Streich WJ, Nürnberg P, Croucher PJP, Bercovitch FB, Krawczak M (2006) Paternal kin bias in the agonistic interventions of adult female rhesus macaques (*Macaca mulatta*). Behav Ecol Sociobiol 61:205–214

Winterhalder B (1986) Diet choice, risk, and food sharing in a stochastic environment. J Anthropol Archaeol 5:369–392

Chapter 11
Evolved Irrationality? Equity and the Origins of Human Economic Behavior

Venkat Lakshminarayanan and Laurie R. Santos

Abstract While the economic approach to human decision-making characterizes our choices in terms of how we maximize utility, in many cases, such an approach fails to predict the decisions we actually make. Specifically, preference-biases such as loss-aversion and reference-dependence demonstrate that decision-makers make relative (rather than absolute) comparisons in judging the quality of their rewards. In the case of inequity aversion, decision-makers take into account not just their own earnings, but also the quality of another individual's reward. In this chapter, we discuss an evolutionary approach to uncovering the origins of these irrational economic strategies. To do this, we review recent experiments showing that nonhuman primates possess a variety of economic tendencies previously thought to be unique to humans. The existence of economic decision-making biases outside of our species implies that ancestral organisms may have possessed these same tendencies – suggesting that humans' biased economic decisions may actually have been adaptive in evolutionarily ancient environments, even though they might be characterized as irrational in contemporary economic settings.

11.1 *Homo economicus*: Model Subject or Tall Tale?

One of the major challenges facing humans from all cultural backgrounds is how to make good decisions. In the human species, decision-making seems remarkably complicated – judging the potential benefit of even simple choices, like walking to the theater to see a movie, can be incredibly tough. Is there a better movie playing

---

V. Lakshminarayanan (✉) and L.R. Santos
Department of Psychology, Yale University, New Haven, CT, USA
e-mail: venkat.lakshminarayanan@yale.edu, laurie.santos@yale.edu

P.M. Kappeler and J.B. Silk (eds.), *Mind the Gap*,
DOI 10.1007/978-3-642-02725-3_11, © Springer-Verlag Berlin Heidelberg 2010

somewhere else? Am I being charged a fair price for the ticket? Is the person in the next theater enjoying himself more than I am? And how does seeing the movie compare to something else I could be doing with my time, like re-reading my favorite novel? The answers to each of these questions will, undoubtedly, affect the value of my choice to go to the movies. In addition, even posing these kinds of questions enlists a vast array of sophisticated mental operations: taking others' perspectives, making comparisons to other similar situations, imagining the consequences of other hypothetical actions, and so on. Making even very simple decisions is therefore a tough cognitive task, not just for our own species but for other nonhuman creatures as well.

The question of how decision-makers should go about making their decisions is one that has plagued a number of academic fields for some time. One of the most heralded answers to this tough question comes from the field of classical economics. The economic approach thinks of the rational decision-maker as someone who maximizes his utility in any decision-making scenario. Economists begin by assuming that a decision-maker has certain stable preferences, and that he makes decisions that optimally satisfy these preferences. If a decision-maker did indeed make decisions based on factors that were consistent across many contexts and possessed preferences that were not heavily influenced by the way he processed information, then one would conveniently be able to interpret his behavior according to this utility-maximizing approach. A perfectly rational subject such as this – a *Homo economicus* – would therefore be expected to choose to go to the movies whenever this action would deliver more utility than the other options that *Homo economicus* would have to sacrifice in order to attend the film.

Notably, several important economic phenomena – such as pricing behavior in markets – are, in fact, neatly consistent with the *Homo economicus* utility-maximizing model.[1] Despite the potential advantages of a traditional economic approach in such cases, a similarly notable body of empirical research suggests that humans do not always behave like aspiring *Homo economici*. Many phenomena that are common in human decision-making – for example, giving anonymously to charity, helping a stranger cross the street, or switching your preferences based on context – do not necessarily make sense from the perspective of perfect utility maximization (see reviews in Kahneman et al. 1982; Camerer 1998). Indeed, when considering real human behavior, using *Homo economicus* as a normative standard leads to strong and often unrealistic assumptions about how

[1] It's not just humans that sometimes obey the maxims of utility-maximization. Indeed, researchers have previously identified economically-rational feeding strategies in a variety of organisms (for reviews, see Glimcher 2003; Krebs and Davies 1993). To take one famous example, great tits behave as though they are optimizing their foraging utility when they make decisions about remaining in (or leaving) a feeding patch (Cowie 1977).

people ought to act: for example, the assumption that decision-makers always selfishly maximize their own wealth level (which would presumably conflict with anonymously giving money away), or the assumption that preferences are stable regardless of how information is presented. Studies conducted by social psychologists, anthropologists, and economists suggest that, instead of behaving like *Homo economicus*, decision-makers' choices are constrained in predictable ways: we sacrifice self-interest to invest in fairness norms (see Kahneman et al. 1986) and punish wrongdoers (Henrich et al. 2004), and will change our decisions systematically based on the wording of different problems (Kahneman and Tversky, 2000). In short, humans do not actually behave like classical economists would predict, even when their own self-interest is at stake.

These seemingly "irrational" human tendencies can be considered puzzling not just from a classical economic perspective, but also from an evolutionary one. Presumably, the decision-making behavior of modern humans has been shaped over time by natural selection: the economic strategies observed today are thus likely to have been shaped over generations of competition for scarce resources. Because rational decision-making behavior plausibly increases an organism's chance of survival, one would assume that our human ancestors might indeed have figured out ways to optimally maximize their expected payoffs, which would thus lead to maximizing the ultimate evolutionary currency – survival and reproductive success. In this way, our deviation from the optimal normative model of expected utility maximization poses not just an economic dilemma, but also an evolutionary one.

How then did humans develop such irrational strategies for making decisions? Some researchers have attempted to explain the origins of human irrationality through an adaptive (or, following Tinbergen's terminology, functional) approach that examines whether purportedly irrational strategies actually can work to increase survival and reproductive success (e.g., Gigerenzer et al. 1999; Gigerenzer and Selten 2001). Here, we approach the problem of human irrationality from a slightly different level of analysis. Specifically, we take a phylogenetic approach to these seemingly irrational behaviors, one that focuses on whether these tendencies exist not just in human decision-making but also in the decision-making of our close primate relatives. We argue that people violate the standard assumptions of rational choice because their economic decision-making reflects a set of mental processes that are actually evolutionarily quite ancient, even though they conflict with certain norms of economic rationality. In this sense, economic irrationalities may have originated in ancestrally related species, and may therefore emerge independently of human-specific cultural experiences.

In this chapter, we examine the evolutionary origins of several key features of our "irrational" economic behavior. We first review each of these phenomena in human economic behavior and then investigate the evolutionary origins of these phenomena by exploring whether similar features exist in the decision-making of closely related primates. We then examine what each of these phenomena means for the evolutionary origins of human economic behavior more generally.

11.2 The Irrationality of Human Preferences

One of the standard assumptions made by rational accounts of human decision-making is the idea that decision-makers should have preferences that are stable over time and consistent across contexts. Indeed, much of the field of economics rests on assumption that preferences are stable enough to be modeled and used to make formal predictions. Unfortunately, despite the usefulness of assumptions concerning the stability of our preferences, there is a growing body of empirical work demonstrating that humans do not obey these classic assumptions. Much of the early empirical evidence for these preference violations came from the ground-breaking work of Daniel Kahneman and Amos Tversky (e.g., Kahneman and Tversky 1979; Tversky and Kahneman 1981, 1986). Beginning in the 1970s, these researchers developed elegant empirical demonstrations of situations in which decision-makers reliably failed to make rational choices.

One of Kahneman and Tversky's most famous early observations was that people's choices seem to vary based on the context in which they are made. Such context-specificity in choice seems to violate a central assumption of rational choice, the invariance assumption, namely, that people should treat any presentation of the same information identically. Rather than always choosing the option with the highest expected payoff, people seem to make different decisions when problems are described in different ways. Take, for example, the two scenarios first presented by Kahneman and Tversky (1979):

Scenario 1: Gains
You have been given $1,000. You are now asked to choose between:
a. 50% chance to receive another $1,000 and 50% chance to receive nothing [16% of subjects choose this option]
b. Receiving $500 with certainty [84% of subjects choose this option]

Scenario 2: Losses
You have been given $2,000. You are now asked to choose between:
a. 50% chance to lose $1,000 and 50% chance to lose nothing [69% of subjects choose this option]
b. Losing $500 with certainty [31% of subjects choose this option]

Note first that the two scenarios are simply alternate formulations of the same final outcomes: each scenario involves a choice between either a risky final payoff of either $2,000 or $1,000 [A] or a sure final payoff of $1,500. People may vary in how much they'd like to gamble by taking choice A over choice B, but they should not have a different preference for risk-taking in Scenario 1 and Scenario 2, since the two scenarios actually present participants with the same two choices. But as shown above, people did show different preferences across the two scenarios. When the final payoffs were represented or "framed" in terms of a gain, as in Scenario 1, the majority of people chose to go for the safe option. In contrast, when the payoffs were framed as losses, most people chose the riskier option. Kahneman and

Tversky labeled this phenomena the "reflection effect": when given a safe and risky choice with the same average payoff, people paradoxically choose the risky gamble when the options are framed to emphasize losses, but not when the same choices are presented as gains These examples demonstrate the puzzling result that when subjects are given alternate formulations of the same set of choices, their preferences flip from being risk-averse to being risk-seeking.

The reflection effect nicely illustrates two central features of human economic choice that violate the assumptions of a *Homo economicus* model. The first feature is that human economic choice is based on *relative* rather than absolute payoffs. Kahneman and Tversky described this phenomenon in terms of what they called a "reference point" bias; people seem to evaluate different options in regards to a particular (usually arbitrary) *reference point* (e.g., one's current asset position in a particular experimental gamble, etc.). The second feature concerns the fact that people treat options differently depending on whether their choices lead to positive (gains) or negative (losses) outcomes relative to their reference point. As the scenarios above demonstrate, people tended to be risk-averse when dealing with outcomes that are gains relative to their reference point – they chose sure smaller gains over larger riskier gains – but became risk-seeking when dealing with losses – they preferred a risky chance not to have any loss over a sure small loss. Kahneman and Tversky famously described this phenomenon in terms loss aversion – people work harder to avoid losses than they do to seek out equally sized gains.[2] As Kahneman and Tversky (1979) observed in the above scenario, the disutility that decision-makers experience from losses tends to be greater than the utility they experience with identically sized gains. Such loss aversion makes relative changes in the negative direction far more salient than negative changes in the positive direction (see Kahneman and Tversky 2000 for review).

These two features of human economic choice – reference dependence and loss aversion – explain many real-world irrationalities in human economic decision-making. For example, investors are reluctant to sell real estate (Genesove and Mayer 2001) or stocks (Odean 1998) for less than their buying price, even when doing so would be profitable. Making choices based on reaching predetermined reference-levels also leads to poor business decisions in the real world. The behavioral economist Colin Camerer famously observed that reference dependence and loss aversion lead New York cab drivers to work less hard than they should when presented with more lucrative conditions (such as on rainy days when the supply of potential clients is higher) in part because they tend to work until they hit

[2] Kahneman and Tversky (1979, 1981) developed a formal descriptive account of choice behavior known as Prospect Theory to explain reference dependence and loss aversion. Prospect Theory posits that people represent choices in terms of their *value*. Values are described in relative terms and are measured as losses or gains relative to a specified (yet often arbitrary) reference point. Because of loss aversion, there is a kink in the value function at the reference point, such that receiving a gain of a certain size (e.g., gaining \$100) is associated with an increase in value that is far smaller (i.e., about half as large) than the decrease in value associated with an identically sized loss (e.g., losing \$100).

a certain reference payoff level; this leads cab drivers to mistakenly work longer days when the supply of clients is low and shorter days when the supply of clients is higher. Finally, reference dependence and loss aversion are thought to be responsible for an irrational phenomenon known as the endowment effect, a bias in which people overvalue objects they own. The endowment effect can lead to real-world problems in bartering and bilateral trade (see Thaler 1980). In one famous example, Kahneman et al. (1990) gave a group of participants one of two identically priced items – a coffee mug or a box of pens. They then allowed subjects to barter with each other for the alternative item. Few objects seem to trade hands in part because owners tended to demand a larger price to sell or trade their object than non-owners were willing to pay to buy it. Kahneman and colleagues interpreted this phenomenon in terms of loss aversion: owners think of bartering in terms of losing their owned object. Losses affect well-being more severely than equally sized gains, which can lead an owner to systematically over-estimate how much money he should demand when selling his possession. In fact, Kahneman and colleagues observed that owners sometimes demand nearly twice as much as buyers are willing to pay in order to give up an owned item.

In summary, humans decision-making appears to exhibit two phenomena that violate the tenets of rational utility-maximization and lead to irrationalities in economic markets. First, and most critically, human think about choices in *relative terms*; rather than focusing on our absolute payoffs, we instead make decisions with reference to arbitrary information that is typically irrelevant to the problem at hand. Second, we are *more sensitive to losses* relative to reference points than we are to equally sized gains. Such loss aversion can lead us to think about problems differently when they are framed in negative versus positive terms.

11.3 Irrational Equity-Seeking and the Emergence of Human Fairness Norms

The framing effects discussed above are not the only situations in which people exhibit loss aversion and relative preference switches across contexts. Another situation in which people exhibit strikingly irrational tendencies to forgo earnings is when thinking about their own payoffs relative to those of others. Behavioral economists have long observed that one of the most salient reference points for our own decisions is "what others are getting." Indeed, researchers have observed that people are often willing to give up much in terms of their absolute happiness and wealth in order to be better off than others. In one striking example, Solnick and Hemenway (1998) gave people the choice between earning \$50,000 while others are earning \$25,000 or earning \$100,000 while others are earning \$250,000. Surprisingly, almost half of the respondents were willing to take the first choice, despite the fact that this choice would cut one's actual wealth level and purchasing-power in half.

Human decision-makers' sensitivity to the relative payoffs of others is central to another way that human economies violate self-interested utility-maximization:

we care a bit more than a *Homo economicus* should about equity. For example, Kahneman et al. (1986) identified a number of economic scenarios in which people's interpretation of what should be done depends not on a decision's profitability, but instead on a decision's perceived equitability. Consider the following scenario:

> *A small company employs several workers and has been paying them average wages. There is severe unemployment in the area and the company could easily replace its current employees with good workers at a lower wage. The company has been making money. The owners reduce the current workers' wages by 5%.*

Kahneman and colleagues observed that 77% of people found the company's action in this case to be unacceptable. Interestingly, people's perceptions seem to change depending on the payoff received by the company; when the researchers changed the above scenario to read that the company "has been losing money," only 32% of people thought the company's wage cuts were unacceptable. In this way, people seem to use fairness constraints over economic constraints when judging the acceptability of economic decisions; as in the above, people judge a wage-cut for the employees as more acceptable when the company is also experiencing a loss than in cases when it is not.

Our human concern for equitable outcomes seems to lead to a number of real-world violations of selfish utility-maximization. First, people often give more than is minimally required in experimental giving or trust games. One such scenario involves a well-known economic game known as the dictator game (e.g., Camerer 2003; Henrich et al. 2004). In this game, two anonymous individuals play a one-shot game in which one individual, the dictator, must decide how to split a cash offer (e.g., $10). Because the game is anonymous and played only once, there is little incentive from a utility maximization perspective for the dictator to give any of the cash allotment to the second player. Nevertheless, participants do not selfishly keep all of the money. Indeed, most people give nonzero offers with many people splitting the allotment 50–50.

A second set of cases in which people exhibit a preference for fair behaviors concerns altruistic punishment – instances in which people are willing to sacrifice individual gain in order to punish wrongdoers (e.g., Kahneman et al. 1986; Zizzo and Oswald 2001; Camerer 2003; Fehr and Rockenbach 2003). Using experimental markets, researchers have found time after time that participants are willing to exact costly punishment even in cases where these behaviors yield no material gain for the punisher or gift-giver (Fehr and Gächter 2002). For example, consider the ultimatum game, a modified form of the dictator game. In this game, the first individual (the proposer) can again split the money any way he sees fit, but the second individual (the receiver) can then choose either to accept or reject the proposer's offer. If the receiver rejects the offer, then neither the proposer nor the receiver receives any money. From the perspective of self-interest, one might expect receivers to accept any nonzero offer. After all, any offer – even one of $0.01 – is better than nothing. It, therefore, follows that we would expect extremely low offers from proposers in anonymous one-shot games. In contrast, researchers robustly observe that offers lower than 25% are nearly always rejected, and proposers frequently offer much higher than this threshold (Güth et al. 1982).

Clearly, neither the proposer nor the receiver is maximizing his own personal payoffs in this anonymous one-shot game. If they were, then proposers would give away as little as possible (e.g., $0.01 out of a possible $10), and receivers would be willing to accept these extremely small offers. Nevertheless, the fact that receivers deviate from this profit-maximizing strategy indicates that they may actually be reacting to the relative fairness of proposers' offers. Indeed, it seems that participants exhibit a cognitive capacity to detect the unfairness of certain rewards – namely, they possess the ability to compare their reward to that of another individual, to react negatively to unfair cases, and to forego small offers in order to punish those who created the inequity.[3]

11.4 The Evolution of Primate Economic Strategies: Monkey Markets

As reviewed above, a wealth of empirical work has established that loss-aversion – which requires sensitivity to others' rewards, and at least a minimal capacity to understand the relative quality of a reward – is robust in adult humans. A similarly large body of work has demonstrated that human decision-makers attend to and obey fairness constraints when making choices. Are these capacities specific to our own species? Or do other species possess similar tendencies? Unfortunately, less work to date has addressed whether similar biases are shared across the animal kingdom and, more specifically, whether they are present in the decision-making strategies of other species within the primate order. The remainder of this chapter will review the work that has been done to date exploring whether similar irrationalities exist in the decision-making of closely related primates.

To begin, do primates share our human economic sensitivity to relative rewards and payoffs? Researchers have known for some time that primates seem to evaluate their rewards not merely in absolute terms, but also on the basis of expectations. In a famous early example, Tinklepaugh (1928) presented macaques monkeys with a memory game in which they had to wait for a hidden reward over a delay period. During some trials, he switched the kind of hidden reward while the monkey was waiting. When monkeys found a reward that was smaller than the one they had originally seen hidden, they reacted negatively (see Watanabe 1996; Santos et al. 2002 for a more recent version of this hiding expectancy task). Monkeys were happy to accept a small reward (e.g., a slice of lettuce) if they were expecting a small reward, but refused to eat a small reward if they were expecting a tastier one

[3] Interestingly, recent work suggests that the cognitive precursors of inequity aversion emerge quite early in human development (Fehr et al. 2008; Lobue et al. in press). For example, using a simplified ultimatum game, Fehr et al. (2008) showed that children as young as eight years exhibit an aversion to inequity in anonymous one-shot games.

(e.g., a banana). These results suggest that monkeys might also evaluate payoffs in terms of reference points.

To get at this issue more carefully, we (Chen et al. 2006) decided to explore the possibility that primates might be reference-dependent in a context more similar to that of human economic choices. Specifically, we examined whether primates would evaluate relative payoffs in a real economic market – one that used "monetary" gambles similar to the scenarios presented to humans. To this end, we trained a group of capuchin monkeys (*Cebus apella*) to trade small metal tokens with human experimenters in order to obtain food (see Westergaard et al. 1998; Brosnan and de Waal 2003, 2004; Addessi et al. 2007 for similar token trading methodologies). In each study, we gave monkeys a "wallet" of small metal tokens, and reinforced them with food rewards for handing these tokens back to a human experimenter. Once subjects understood that they could "buy" food with tokens, we looked at whether they exhibited the same kinds of strategies as humans do in their experimental markets. We gave monkeys a limited budget of tokens and measured how they allocated these tokens across two experimenters, each of whom offered different kinds of foods at different prices. In this way, we could establish a measure of the monkeys' "preference" for each of the experimenters, and could thus explore how those preferences changed based on prices and framing.

We first explored whether monkeys paid attention to the "price" of different kinds of food. In our experiment, monkeys could use each token to purchase either an apple piece from one experimenter or a grape from a second experimenter. In this way, the experimenters initially sold different goods at equal prices. Because monkeys like apples and grapes equally, they spent approximately half of their token budget on each good. Once monkeys had gotten used to this price, we introduced what economists call a "price shock," and observed whether the monkeys adjusted their spending. Essentially, one of the two goods – say, apples – went on sale such that a single token now bought two units of the same item. We predicted that if monkeys reacted rationally to the price shock – that is, if their purchasing habits were related to the price of the goods, then they should change their pattern of purchasing in order to buy more of the cheaper good. Our subjects did just this, showing that their monkey token economy shared at least some features of human economies.

In a second study, we established that capuchins rationally spend more resources to obtain a greater reward. To do so, we adapted our experimental market so that the two experimenters provided the same type of reward for a token – apples – but differed in the number of apple pieces they delivered and the frequency that they provided these rewards. We again gave subjects a choice between two experimenters with whom they could trade their limited budget of tokens. The first experimenter initially displayed two pieces of apple before receiving a token from the subject, but then when paid sometimes removed one of these pieces and gave the monkey only one piece of apple. In contrast, the second experimenter always displayed one piece of apple and delivered exactly this quantity in exchange for a token (neither adding nor subtracting from it). This resulted in a situation in which the risky experimenter provided an average payoff that was one and half times the payoff of the safe experimenter. Given this choice between a gamble with an average payoff of one and a half apple pieces and a sure choice of only one apple, capuchins preferred the gamble with the larger expected value (Chen et al. 2006). In this way, capuchins behaved rationally in at least some aspects of our trading market: they spent more resources on a cheaper food than an equally valuable but more expensive alternative, and chose options which maximize expected value even when this required choosing a risky choice over a safe one.

11.5 Primate Economic Irrationalities: Do Monkeys Exhibit Human Economic Biases?

Having established these basic similarities with our own markets, we then investigated whether capuchins shared some of the irrational preferences that commonly influence how people allocate resources – in particular, reference dependence and loss aversion. Specifically, we wanted to see whether capuchins base their choices not only on the absolute outcome of their choice but also on their initial expectation. To begin investigating whether capuchins share our human irrational purchasing tendencies, we again used our experimental two-trader market but this time did not vary either the kind or amount of food that was given to the subjects; both experimenters delivered food of *the same expected value*. Instead, we varied whether the experimenters added to or subtracted from the amount of food that they initially showed to the subject. Sometimes the experimenters gave more apples than they originally showed to the subject, and sometimes they gave less. This allowed us to explore whether subjects base their choices on the expectation (or *reference point*) created before each trade. In one study, we gave the monkeys a choice between two experimenters who both delivered the same average expected payoff of one and a half pieces of apples. For one experimenter, this average payoff was less than the monkeys' expectation. This first experimenter began every trade by showing the monkey two pieces of apple, but when paid either delivered two pieces of apple as shown or removed one to deliver only a single apple piece. The second experimenter, in contrast, gave more on average than the monkeys expected. This second experimenter began by showing the monkey only one piece of apple, but then when paid either delivered that one piece as promised or added a second piece for a payoff of two apple pieces. Despite the fact that the average payoff was identical across the two experimenters, our monkeys traded more tokens with the second experimenter, the one who offered more than the monkeys' initial expectation. This pattern of performance suggests that capuchins take into account reference points, just as humans do.

But do capuchins also exhibit loss aversion? To investigate this, we again had both experimenters deliver an identical final payoff (in this case, a single apple piece). The first experimenter, however, initially began by displaying two pieces of apple at the start of each trial, but then subtracted one piece before giving the monkey his payoff. The second experimenter began each trial with a single piece of apple, and provided exactly this quantity. Again, despite the fact that the final outcome of trading was the same across the two experimenters, the monkeys traded more of their tokens with the second experimenter and avoided the first experimenter who initially offered a larger quantity and then subtracted during the trade (Chen et al. 2006). Like humans, monkeys appear to avoid choices that involve perceived losses.

In a final demonstration of how capuchin monkeys share our biased choices, we gave the monkeys an opportunity to be the "owner" of one good that they could barter for an equally priced good. Our aim was to see whether monkeys, like humans, exhibited an endowment effect, overvaluing the goods that they owned. To do so, we (Lakshminarayanan et al. 2008) set up a new market in which capuchins received foods that they could either keep and eat, or trade away for an equally valuable alternative. In the first phase of our experiment, we found two foods that the capuchins preferred equally. That is, given a small budget of tokens, they spent approximately half of their tokens on one these goods (e.g., cereal pieces) and half of their tokens on the other good (e.g., fruit pieces). We then made the monkeys owners of one of the two goods, say cereal, which they could then trade for the equally valuable fruit pieces. Like humans, capuchins preferred to retain the kind of food they owned, and rarely traded away food from their budget even for an equally preferred kind of food. Even across control conditions in which monkeys were compensated for any perceived transaction costs, our capuchin participants failed to trade goods they owned for equally preferred or slightly more valuable goods. Brosnan and colleagues (Brosnan et al. 2007, 2008) reported a similar endowment effect in chimpanzees (*Pan troglodytes*). Chimpanzees also refused to trade owned objects for equally valued alternatives. Like humans, capuchins and chimpanzees seem to avoid losing owned objects even if it means foregoing the gain of an equally valued alternative object.

The capuchin trading studies highlight the fact that at least one species of primate seems to share our human economic biases. Like humans, capuchin monkeys tested in an economic market exhibited reference dependence, loss aversion, and endowment effect. They too appear to evaluate their payoffs in relative terms and weigh losses more heavily than equally sized gains. The demonstration that other primates react similarly to humans in these market studies raises the question of whether nonhuman primates also share our relative evaluations in the social domain. Do primates also compare their payoffs to those of others? Do they react negatively when they get less than their equitable share? In short, do primates share our human concerns for equity and fairness considerations?

11.6 Are Primates Irrationally Equity-Seeking?

Excitingly, recent work in primate cognition has devoted considerable attention to the question of whether primates share a sense of equity and fairness (see reviews in de Waal 1996, 2008; Brosnan 2006; Silk 2007, Silk and Boyd this volume). Much of this work has examined whether primates are also averse to inequitable outcomes. In a landmark series of studies, Brosnan and colleagues (Brosnan and de Waal 2003; Brosnan et al. 2005; Brosnan 2006; van Wolkenten et al. 2007) investigated whether two primate species – capuchins and chimpanzees – would forego an otherwise desired food reward that was delivered unfairly in a trading task. In the original capuchin study (Brosnan and de Waal 2003), monkeys watched an experimenter trade with a conspecific and pay that conspecific either a low-valued (e.g., cucumber) or high-valued food reward (e.g., a grape). After seeing the conspecific's payoff, the subject monkey got its own chance to trade and was paid only the low-valued reward. Although capuchins were happy with a low-valued cucumber payoff when the other monkey also got a cucumber, capuchins rejected the cucumber when the other monkey got a grape. Surprisingly, in this and other studies (e.g., Brosnan et al. 2005; Brosnan 2006; van Wolkenten et al. 2007), capuchins and chimpanzees were willing to reject an otherwise desired food reward if they previously observed another conspecific obtaining a better reward for the same amount of work (but see Dubreuil et al. 2006; Roma et al. 2006 for failures to replicate this effect with a different paradigm). In a similar vein, capuchins have been shown to spontaneously share food with the individuals that helped to work for that food (de Waal and Berger 2000).

Taken together, capuchins and chimpanzees tested in token trading tasks seem to act in ways that are consistent with notions of inequity aversion, even when doing so results in losing an otherwise valued piece of food (e.g., Brosnan and de Waal 2003). Like humans, these primate species seem to evaluate their own payoff relative to that of others and react negatively when others "earn" more than they do for equal amounts of work. To summarize, then, basic components of equity – for example, judging one's own reward against the payoffs of others – are present in several primate species.

That said, there does appear to be some potentially important differences between the equity seeking strategies observed in human and nonhuman primates. As reviewed above, humans often act in ways that promote equity not just for their own outcomes, but also for those of others. In the parlance of experimental economics, humans are averse not just to *disadvantageous inequity* – situations in which an actor gets relatively less than other individuals – but also to *advantageous inequity* – situations in which the actor gets relatively more than other individuals. Human decision-makers often act to reduce inequity even in cases in which they themselves are not negatively effective (e.g., providing equitable offers in one-shot dictator games, giving to charity, working for social justice, etc.). To date, there is relatively little evidence that any nonhuman primate does the same. Brosnan (2006) reported, for example, that capuchins who received the grape in her task never rejected it or offered to share it with the monkey who received the unfair cucumber payoff. In fact, she noted that higher-paid monkeys occasionally stole rejected cucumbers from their lesser-paid cagemates. Failures to observe advantageous inequity aversion in these and other tasks (see Silk and Boyd this volume for an elegant review of striking demonstrations of primates' failures to provide equitable rewards to others even at low cost to themselves) suggest that humans may be unique in their advantageous inequity concerns.

11.7 Conclusions About the Evolution of Human Economic Strategies

Like humans, nonhuman primates also violate at least some of the standard tenets of expected utility maximization. Several primate species show economic preferences that are consistent with loss aversion, reference dependence, and disadvantageous inequity aversion. To summarize, nonhuman primates seem to share many human-like irrational decision-making biases, despite the fact that making choices in this way would not lead to economically rational choices in the long run.

The fact that researchers have observed decision-making biases, such as inequity-aversion, loss-aversion, and prosocial-giving outside of our species, is striking for at least two reasons. First, these findings suggest that despite the fact that human irrational decision-making violates the norms of rational economic behavior, these behaviors seem to represent cognitive strategies that have been in place for millions of years of evolutionary history. Even though loss-aversion and inequity norms may not contribute straightforwardly to better economic choices, these strategies have still been around for quite a long time. Indeed, results demonstrating economic irrationalities and fairness-like norms in capuchin monkeys suggest that similar biases may have existed in our common ancestor with New World primates over 35 million years ago.

Second, the existence of several of these biases in nonhuman primates may provide a valuable hint for understanding *why* these biases may be so persistent and robustly demonstrated in our own species. Although the biases we have discussed in this chapter are typically considered non-normative from a classical economic perspective, several researchers have argued that such irrational strategies may in fact be rather smart in certain contexts (e.g., Gigerenzer and Goldstein 1996; Gigerenzer et al. 1999; Gigerenzer and Selten 2001). Gigerenzer and colleagues, for example, have long speculated that seemingly "irrational" decision-making biases may allow decision-makers to quickly navigate what would normally be cognitively-taxing complicated computations. In this way, the irrational strategies we have observed may be better from an evolutionary perspective than originally thought in part because they are faster and cheaper than a more rational approach. Our observation that such strategies are evolutionarily ancient provides credence to this view, and suggests that researchers might need to rethink what should be considered the best normative model of decision-making.

Finally, we end with an exciting implication of the recent work on primate irrationalities. Although the nature of nonhuman primate economic strategies is still far from understood, researchers now have a variety of methods in place that can be tweaked to investigate many aspects of primates' social and economic preferences.

These methods create exciting possibilities for future research across human and nonhuman species. The hope, then, is that the next decade of work in this area can determine not just whether primates share different human economic behaviors, but also whether the mechanisms that underlie these behaviors are also shared broadly across the primate order.

References

Addessi E, Crescimbene L, Visalberghi E (2007) Do capuchin monkeys (*Cebus apella*) use tokens as symbols? Proc Roy Soc Lond B 274:2579–2585

Brosnan SF (2006) Nonhuman species' reactions to inequity and their implications for fairness. Soc Justice Res 19:153–185

Brosnan SF, de Waal FBM (2003) Monkeys reject unequal pay. Nature 425:297–299

Brosnan SF, de Waal FBM (2004) Socially learned preferences for differentially rewarded tokens in the brown capuchin monkey (*Cebus apella*). J Comp Psychol 118:133–139

Brosnan SF, Schiff HC, de Waal FBM (2005) Tolerance for inequity may increase with social closeness in chimpanzees. Proc Roy Soc Lond B 272:253–258

Brosnan SF, Jones OD, Lambeth SP, Mareno MC, Richardson AS, Schapiro SJ (2007) Endowment effects in chimpanzees. Curr Biol 17:1704–1707

Brosnan SF, Grady MF, Lambeth SP, Schapiro SJ, Beran MJ (2008) Chimpanzee autarky. PLoS One 3:e1518. doi:doi:10.1371/journal.pone.0001518

Camerer CF (1998) Bounded rationality in individual decision making. Exp Econ 1:163–183

Camerer CF (2003) Behavioral game theory: experiments in strategic interaction. Princeton University Press, Princeton, NJ

Chen MK, Lakshminarayanan V, Santos LR (2006) The evolution of our preferences: evidence from capuchin monkey trading behavior. J Polit Econ 114:517–537

Cowie RJ (1977) Optimal foraging in great tits (*Parus major*). Nature 268:137–139

de Waal FBM (1996) Good natured: the origins of right and wrong in humans and other animals. Havard University Press, Cambridge, MA

de Waal FBM (2008) Putting the altruism back into altruism: the evolution of empathy. Annu Rev Psychol 59:279–300

de Waal FBM, Berger ML (2000) Payment for labour in monkeys. Nature 404:563

Dubreuil D, Gentile MS, Visalberghi E (2006) Are capuchin monkeys (*Cebus apella*) inequity averse? Proc Roy Soc Lond B 273:1223–1228

Fehr E, Gächter S (2002) Altruistic punishment in humans. Nature 415:137–140

Fehr E, Rockenbach B (2003) Detrimental effects of sanctions on human altruism. Nature 422:137–140

Fehr E, Bernhard H, Rockenbach B (2008) Egalitarianism in young children. Nature 454:1079–1083

Genesove D, Mayer C (2001) Loss aversion and seller behavior: evidence from the housing market. Q J Econ 116:1233–1260

Gigerenzer G, Goldstein DG (1996) Reasoning the fast and frugal way: models of bounded rationality. Psychol Rev 103:650–669

Gigerenzer G, Selten R (2001) Bounded rationality: the adaptive toolbox. MIT Press, Cambridge, MA

Gigerenzer G, Todd PM, ABC Research Group (1999) Simple heuristics that make us smart. Oxford University Press, New York

Glimcher PW (2003) Decisions, uncertainty, and the brain: the science of neuroeconomics. MIT Press, Cambridge, MA

Güth W, Schmittberger R, Schwarze B (1982) An experimental analysis of ultimatum bargaining. J Econ Behav Organ 3:367–388

Henrich J, Boyd R, Bowles S, Camerer C, Fehr E, Gintis H (2004) Foundations of human sociality: economic experiments and ethnographic evidence from fifteen small-scale societies. Oxford University Press, Oxford

Kahneman D, Tversky A (1979) Prospect theory: an analysis of decision under risk. Econometrica 47:263–292

Kahneman D, Tversky A (2000) Choices, values, and frames. Cambridge University Press, Cambridge

Kahneman D, Slovic P, Tversky A (1982) Judgment under uncertainty: heuristics and biases. Cambridge University Press, Cambridge

Kahneman D, Knetsch JL, Thaler RH (1986) Fairness as a constraint on profit seeking: entitlements in the market. Am Econ Rev 76:728–741

Kahneman D, Knetsch JL, Thaler RH (1990) Experimental tests of the endowment effect and the coase theorem. J Polit Econ 98:1325–1348

Krebs JR, Davies NB (1993) An introduction to behavioural ecology. Oxford University Press, Oxford

Lakshminarayanan V, Chen MK, Santos LR (2008) Endowment effect in capuchin monkeys. Phil Trans Roy Soc B 363:3837–3844

Lobue V, Nishida T, Chiong C, Deloache J, Haidt J (in press) When getting something good is bad: Even 3-year-olds react to inequality. Social Development

Odean T (1998) Are investors reluctant to realize their losses? J Finance 5:1775–1798

Roma PG, Silberberg A, Ruggiero AM, Suomi SJ (2006) Capuchin monkeys, inequity aversion, and the frustration effect. J Comp Psychol 120:67–73

Santos LR, Sulkowski GM, Spaepen GM, Hauser MD (2002) Object individuation using property/kind information in rhesus macaques (*Macaca mulatta*). Cognition 83:241–264

Silk JB (2007) Empathy, sympathy, and prosocial preferences in primates. In: Dunbar RIM, Barrett L (eds) Oxford handbook of evolutionary psychology. Oxford University Press, Oxford, pp 115–126

Solnick SJ, Hemenway D (1998) Is more always better? A survey on positional concerns. J Econ Behav Organ 37:373–383

Thaler RH (1980) Toward a positive theory of consumer choice. J Econ Behav Organ 1:39–60

Tinklepaugh OL (1928) An experimental study of representative factors in monkeys. J Comp Psychol 8:197–236

Tversky A, Kahneman D (1981) The framing of decisions and the psychology of choice. Science 211:453–458

Tversky A, Kahneman D (1986) Rational choice and the framing of decisions. J Bus 59:251–278

van Wolkenten M, Brosnan SF, de Waal FBM (2007) Inequity responses of monkeys modified by effort. Proc Natl Acad Sci USA 104:18854–18859

Watanabe M (1996) Reward expectancy in primate prefrontal neurons. Nature 382:629–632

Westergaard GC, Liv C, Chavanne TJ, Suomi SJ (1998) Token mediated tool-use by a tufted capuchin monkey (*Cebus apella*). Anim Cogn 1:101–106

Zizzo DJ, Oswald A (2001) Are people willing to pay to reduce others' incomes? Ann Econ Stat 63–64:39–62

Chapter 12
From Whence the Captains of Our Lives: Ultimate and Phylogenetic Perspectives on Emotions in Humans and Other Primates

Daniel M.T. Fessler and Matthew Gervais

Let's not forget that the little emotions are the great captains of our lives and we obey them without realizing it.

Vincent van Gogh

Abstract We outline an evolutionary approach to emotions intended to spur further research on the subject in humans and nonhumans alike. Combining adaptationist, comparative, and phylogenetic analyses, we seek to illuminate the functions that emotions fulfill, the reasons why they take the forms that they do, and the extent to which they are shared across species. Using similar logic, we distinguish between emotions and attitudes, cognitive representations of other actors that are both informed by, and potentiate, emotions. Employing select emotions as illustrations, we discuss a taxonomy of emotions. We begin with emotions that address adaptive challenges common across animals, and which require minimal cognitive capacities, features that make it likely that they are widely shared across species. Next, we consider emotions involved in elementary sociality, a category further elaborated in emotions playing a role in parenting and pair-bonding. In light of the importance of dyadic cooperative relationships in primate societies, we describe a set of emotions undergirding such relationships that we expect to be shared by human and nonhuman primates. To a more limited degree, we expect pan-primate similarities with regard to vicarious emotions, those wherein the individual experiencing the emotion is affected only indirectly by the eliciting event. The greater range and complexity of human social relationships, including the human

D.M.T. Fessler (✉)
Center for Behavior, Evolution & Culture, Department of Anthropology, University of Anthropology, Los Angeles, CA, USA
e-mail: dfessler@anthro.ucla.edu

M. Gervais
Center for Behavior, Evolution & Culture, Department of Anthropology, University of California Los Angeles, Los Angeles, CA, USA
e-mail: mgervais@ucla.edu

P.M. Kappeler and J.B. Silk (eds.), *Mind the Gap*,
DOI 10.1007/978-3-642-02725-3_12, © Springer-Verlag Berlin Heidelberg 2010

propensity to essentialize cultural groups, extend the class of vicarious emotions beyond anything evident in nonhumans. Finally, underscoring the importance of culture in human evolution, we examine moral emotions elicited by norm violations, a pattern unique to humans.

12.1 Introduction

It has long been recognized that an evolutionary perspective is useful in investigating emotions. In this chapter, we employ two complementary applications of evolutionary theory. One is adaptationism, viewing emotions as discrete adaptations for behavior regulation that evolved in response to challenges repeatedly confronting organisms over evolutionary time (Frijda 1986; Nesse 1990; Lazarus 1991; Ekman 1992; Tooby and Cosmides 2008). Each emotion is elicited by cues to the presence of a recurring fitness-relevant challenge or opportunity, and each coordinates information-processing, motivational, and physiological systems to respond adaptively. The second application is phylogenetic and comparative, examining (1) the actual or expected taxonomic distribution of emotion systems based upon patterns of descent, the distribution of the relevant adaptive problems, and cognitive capacities, and (2) the logic of each system's transformation through descent as a result of the interaction of system evolvability, system affordances, and the structure of adaptive problems. Neither of these applications of evolutionary theory to the study of emotions has yet been fully realized, and the synergy resulting from their integration is often overlooked. Our goal is to further develop these approaches, to demonstrate their complementarity, and to employ them in examining some aspects of emotions that contribute to the "gap" between humans and other primates.

Darwin's ([1872] 1955) pioneering work on emotional expressions employed a comparative perspective in order to: (1) substantiate the utility of general explanatory principles by application to all species; (2) demonstrate human descent from nonhuman ancestors; (3) evaluate the innateness of human expressions by showing similarity to other species; and (4) explain some human characters as vestigial traces of ancestral forms. The first two goals have been achieved. Ekman and other modern students of expressions advanced the third goal, and many human facial expressions are now considered pan-cultural. This leaves Darwin's fourth goal. While extreme vestigialism has rightly fallen from favor, many emotions do exhibit evidence of deep histories of successive modification. We will present several examples of emotional systems apparently evolved for one purpose that were subsequently modified to serve a different purpose. In some cases, the properties of an ancestral emotional system appear to have "preapted" it (Gould and Vrba 1982) to the task demands of a newly arisen adaptive problem (e.g., Rozin et al. 1997; Fessler 2004; Gervais and Wilson 2005). In a number of cases, an extant emotional action tendency has been put to novel use by modifying the eliciting conditions (cf. Tooby and Cosmides 1989; Rozin 1996; Keltner and Anderson 2000). Yet, even with subsequent secondary selection, this process does not necessarily erase the legacy of past selection; substantial residues of former function sometimes remain, and it is only with reference to ancestral traits that such features become intelligible. Applying this consideration when reverse engineering a trait ("reverse tinkering"; Andrews et al. 2002; Gangestad and Simpson 2007) illuminates the kludge-like structure of emotions that have served multiple functions over evolutionary time.

As the above discussion suggests, phylogenetic and comparative approaches both constrain and inform adaptationist hypotheses (see Maestripieri 2003b; Gangestad and Simpson 2007; Gosling and Graybeal 2007). In addition to controlling for nonindependence of trait correlations in comparative research (Nunn and Barton 2001), "tree thinking" helps to specify the time frame within which the target properties of a trait were selected, while suggesting an ancestral state from which the trait was derived. This refines the task of reverse engineering by rephrasing the ultimate question as: Why did this form emerge given the form from which it emerged? A phylogenetically informed comparative approach can help to parse a trait into aspects that were impacted by selection pressures operating during particular periods in the lineage. This approach highlights the process of secondary selection through which (*contra* Gould and Vrba 1982; Gould 1991) an extant system is not merely exapted (put to new use), but is further subjected to specializing selection for that new function. This process, which we term cooptation, has rarely been foregrounded in the study of the mind.

Just as comparative and phylogenetic approaches offer benefits to students of human emotions, so too can adaptationist perspectives enhance the study of nonhuman primate emotions. Because emotion is difficult to quantify, most primatologists have sidestepped the study of emotion, although this is beginning to change (e.g., de Waal 1996; Aureli and Schaffner 2002; Maestripieri 2003a). The conjunction of adaptationist and phylogenetic analyses indicates where one can expect to find homologues of human emotions. It should be possible to predict the taxonomic distribution of emotion traits through a consideration of (1) the phylogenetic relationships among species, (2) the selection pressures thought to produce and maintain a given trait, and (3) the cognitive capacities that constitute prerequisites for the trait. This generates a taxonomy of emotions, ranging from ancient emotions expected to be present in all vertebrates, to conserved pan-primate emotions, to derived emotions expected to be unique to our species. Because explorations of primate emotions are in their infancy, we cannot conduct rigorous tests of our phylogenetic hypotheses. However, there is reason for optimism, as investigators are beginning to employ new methods to probe primate motives (e.g., Silk et al. 2005; Warneken et al. 2007; de Waal 2008; Lakshminarayanan and Santos 2008, Silk and Boyd, this volume), and obstacles to the study of primate emotions are not insurmountable. Below, we sketch a set of phylogenetically informed adaptationist proposals intended to provide starting points for future studies of emotions in both humans and nonhuman primates. This is not an exhaustive description of these species' emotional repertoires. Rather, our goal is to demonstrate the form that we think such descriptions should take, and to spark further discussion.

A note on terminology: because cultures differentially emphasize or ignore facets of the pan-human spectrum of emotions, any language's emotion lexicon is but a crude gloss for the underlying phenomena. Accordingly, the emotion terms used here should be viewed as heuristic labels for adaptations that are not necessarily isomorphic with Anglo-American folk psychology.

12.2 Ancient, Relatively Conserved Emotions

All vertebrates confront a common set of elementary adaptive challenges, including avoiding injury and disease, and finding mates. The core motivational systems that evolved in ancestral vertebrates in response to these selective pressures have likely been largely conserved in descendant taxa. Because of their ancient origins, the cognitive prerequisites for these adaptations are relatively limited. In general, attention to simple exogenous or endogenous cues suffices to identify the eliciting conditions. Many of these eliciting cues are sufficiently uniform across circumstances to allow canalization, obviating learning. Below, we sketch some of the relevant adaptive challenges and corresponding emotions. Then, we consider how natural selection refined these basic building blocks

Avoiding imminent threats to life and limb is a fundamental determinant of fitness; *fear* marks the presence of such threats, motivating flight as a principal behavioral outcome, with fighting as a secondary outcome when the threat is animate and escape is appraised as impossible. The neural underpinnings of fear appear to be highly conserved, certainly among mammals, and arguably among vertebrates (Panksepp 1998; Braithwaite and Boulcott 2007; Öhman et al. 2007). Fear expressions closely paralleling those present in humans are recognizable in many primates (Parr et al. 2007). This uniformity potentially reflects selection having conserved an adaptive form. Human experimental results support Darwin's ([1872] 1955) supposition that many facial expressions are functional beyond their signal value, as the fear expression enhances perceptual acuity and reaction times to threat (Susskind et al. 2008).

The avoidance of toxins and pathogens is another elementary adaptive challenge. In humans, *disgust* is involved in the rejection of ingestible contaminants and avoidance of cues reliably associated with pathogen risk (e.g., rotting flesh or feces). Oral contact is a powerful disgust elicitor (see Fessler and Haley 2006), and disgust reactions involve gastrointestinal rejection, suggesting that preventing intoxication was the original function of this emotion. Disgust was likely subsequently coopted for prophylaxis by extending elicitors to include contact with cues of disease (see Kelly in prep). Human disgust can also be elicited by representational contamination (transfer of disgust-eliciting status through contact with an elicitor absent perceptible changes) and a variety of symbolically mediated events. The extension of disgust to such cues leads some to argue that disgust is a culturally constructed defense against existential anxiety (Rozin et al. 2000). However, this position overlooks design features evident in the avoidance of contact with stimuli ecologically associated with disease transmission. Because microbes cannot be detected by the human eye, yet spread through physical proximity, it is beneficial that the set of cues that elicits disgust includes representational contamination (Curtis and Biran 2001; Fessler and Haley 2006, Kelly in prep).

An ingestion rejection system is found in all mammals. Across primates, we predict that the degree of elaboration of this system (i.e., number of eliciting cues; flexibility of response) will vary as a function of carnivory, as meat is a potent source of disease. Similarly, we expect some form of pathogen-avoidance mechanism to be found in all mammals; because sociality is a determinant of disease risk, these mechanisms should vary as a function of sociality. In contrast, because it requires abstract conceptual abilities, we expect representational contamination to be quite rare. In a number of species, investigators have documented behaviors possibly explicable as prophylaxis, including avoidance of parasitized conspecifics, and grooming and feces avoidance (see Nunn and Altizer 2006: 159–170 for review on primates). However, it is not known whether such behavior is motivated by a disgust-like emotion, as it is in humans.

Ultimately, reproduction, not survival, determines fitness, so we expect all sexually reproducing species to be equipped with an emotion akin to *lust*, the principal motivator of sexual behavior. Although the frequency, form, and elicitors for sexual behavior vary across mammals, the ubiquity of goal-oriented behavior leading to copulation is consistent with such an emotion being widespread. There are several parallels in the task demands associated with mate-seeking and foraging (appetitive cycles; resource-seeking behavior; discrimination among resource options), and, in mammals, there is overlap between the neurophysiological systems regulating proceptive sexual behavior and those regulating ingestive behavior (reviewed in Fessler 2003). This suggests that systems regulating ingestion may have constituted building blocks from which systems regulating sexual reproduction were subsequently constructed. Disgust, antithetical to both hunger and lust, plays a central role in human inbreeding avoidance (Lieberman et al. 2003, 2007; Fessler and Navarrete 2004); likewise, women's sexual disgust sensitivity increases around ovulation, possibly functioning to reduce contact with suboptimal reproductive partners at peak fertility (Fessler and Navarrete 2003). Most primates exhibit marked inbreeding avoidance (Paul and Kuester 2004; Muniz et al. 2006), and it is possible that a similar emotional system is involved. Indeed, we suggest that both the ingestion regulation and sexual regulation facets of disgust are pan-mammalian.

12.3 Emotions Associated with Elementary Sociality

As illustrated by disgust at the prospect of sex with individuals identified as close kin, emotion elicitation is contingent on appraisal, the process of construing the nature and meaning of a situation (Scherer 1999). Sociality introduces a complex set of adaptive challenges because there are many ways in which conspecifics can affect fitness, hence many distinct appraisals and consequent emotional responses should attend social interaction. At the most elementary level, because competition for resources is a key determinant of fitness, we expect distinct emotions, with corresponding appraisals, to address competition. From the perspective of a resource holder, an attempt to displace one from a resource should be appraised as a transgression, the infliction of an unwelcome (i.e., fitness-reducing) cost, or threat thereof. In humans, transgressions elicit *anger*, motivating aggressive retaliation when feasible. While varying in degree across species, many social animals exhibit aggressive responses to actions that investigators plausibly interpret as transgressions. Hence, an elementary anger-like emotion and its eliciting appraisal are probably both ancient and widely shared. This conclusion is supported by the extensive conservation across mammals of the neurophysiological systems underlying aggression (e.g., Dierick and Greenspan 2007).

In humans (particularly young men), anger often motivates disproportionate responses to transgression. Disproportionate responses may be advantageous because they serve as a signal that transgressing is dangerous, generating reputational benefits that deter further transgressions (reviewed in Fessler in press). Reputation-based strategies such as this require repeated interactions, the ability to identify and recall individual actors, and the capacity to acquire information as a third-party observer. All of these features are likely present among many social primates (Cheney and Seyfarth 2005), and age-related patterns of impulsive aggression may parallel those found in humans (cf. Fairbanks et al. 2004). Thus, it is likely that the central features of the anger adaptation are a shared ancestral trait among primates.

While the question of applying the appropriate appraisal to a given event is never trivial, this task is particularly complex in the social domain because the range of fitness consequences of social interactions is so large. This complexity is multiplied by the fact that, due to repeat interactions, appraisals of current social actions hinge on past events. We suggest that the substantial informational demands of applying the appropriate appraisal to social events are managed, in part, through *attitudes*, durable, hierarchically organized representations of previously appraised traits and relational outcomes that potentiate differential emotional readiness toward others. Attitudes and emotions are reciprocally related, as attitudes are updated by emotions, and subsequently help to regulate emotions by shaping appraisals. Responsive to the fitness-relevant traits and behaviors of other individuals, attitudes are continuously adjusted over time, thus constituting summary representations that proxy the future fitness implications of all past interactions. Attitudes allow individuals to represent their current relations to others without the need for explicit bookkeeping or recall of all encounters. Via their role in appraisals, these representations adaptively regulate current behavior. Attitudes can thus be conceptualized as "internal regulatory variables," in the sense that they functionally translate past appraisals into current behavior regulation (see Tooby and Cosmides 2008). This proposal obviates the need to posit complex cognitive operations in some relational domains (the potential complexities of which are discussed in Cords 1997; Silk 2003; see also Aureli and Schaffner 2002; Aureli and Whiten 2003). Likewise, congruent with extensive social psychological findings on implicit attitudes (Greenwald and Banaji 1995), our position does not require positing problematic conscious processes.

The interaction of appraisals, attitudes, and emotions is evident in the case of *envy*, the emotion associated with the goal of displacing a resource holder. In humans, envy is elicited by the appraisal that, in a zero sum situation, the target possesses an advantage to which the actor is entitled. The target is therefore represented as a rival, and there is an enduring attitude of hostility toward the target (see Smith and Kim 2007) such that the actor is willing to inflict costs on the target. Likewise, potentiated by this attitude, attempts by the target to retain or increase control of the resource elicit anger from the actor. Envy is thus usefully contrasted with *admiration*, in which pursuit of non-zero-sum advantages enjoyed by another does not involve an appraisal of entitlement, an attitude of hostility, or anger in response to the target's attempts to control the resource (more on admiration later). Conflict over zero sum resources is central to much social behavior, hence, we suggest that the core components of the envy system are both ancient and widely shared. Observations in multiple species of distress at inequitable distributions (reviewed in Brosnan 2006) are congruent with this suggestion.

Competition and hostile intent are also central to *jealousy*. Deriving from a proprietary attitude toward a relationship partner and a corresponding appraisal of potential interlopers as transgressors, this emotion motivates attempts to maintain exclusivity by warding off rivals and restricting the partner's options (reviewed in Smith and Kim 2007). While sociality does not necessitate the formation of discrete relationships, well-differentiated relationships exist in many primate species (see Silk 2007). If such relationships are widespread, jealousy may be widely shared. Elicitors for jealousy will depend on the nature of the threats posed to a relationship. The utility of biparental care in humans leads to sex-specific adaptive challenges, namely the possibility of female extra-pair copulation (leading to misallocation of male parental investment) and male abandonment (leading to reduced female access to resources). A growing literature investigates the corresponding sex-specific relative importance of sexual and emotional infidelity as elicitors of mating jealousy (reviewed in Haselton and Ketelaar 2006). While extensive biparental care is rare among primates, we expect the same logic of an actor-centered appraisal of different threats to a relationship to apply across species.

In most social animals, conflicts establish and maintain dominance rank, which then determines priority of access to resources. As Darwin ([1872] 1955) suggested, ethology provides clues to both the phylogeny and ultimate function of emotions, and this is particularly evident in the case of dominance interactions. Displays often precede, and sometimes obviate, conflict, and threat displays that reduce the costs of conflict are a pervasive aspect of social behavior in primate groups. In most species, body size is a key determinant of success in combat, and, correspondingly, threat displays generally involve an exaggeration of body size. Direct attention (staring) is often a feature of such displays (reviewed in Fessler 2004). In some cases, it is beneficial to acknowledge subordinate status, and appeasement displays are counterpoints to threat displays. Appeasement displays generally involve an attempt to minimize apparent body size and direct attention away from the aggressor. In humans, threat displays are associated with an emotion we term *proto-pride*, and appeasement displays are associated with *proto-shame* (see Fessler 2007;

see also Tracy and Matsumoto 2008). Proto-pride is elicited when an actor appraises herself as occupying a superior position in a social hierarchy and interacting with a subordinate in a context in which the disparity in status is relevant; conversely, when the complementary appraisal occurs, the subordinate feels proto-shame. In each case, the emotion enhances a corresponding attitude that represents in enduring fashion the disparities in status. For the dominant, this attitude includes a sense of entitlement relative to the other, while the converse is true for the subordinate. In keeping with their opposite hedonic valences, this pair of emotions motivates striving for power and social position (Fessler 2007). Given the clear homologies with display behaviors in other primates, and the ubiquity of the relevant adaptive problem across social animals, it is likely that proto-pride and proto-shame are pan-primate; indeed, the core features of these emotions and related appraisals and attitudes may be shared by all mammals, and possibly by most vertebrates.

12.4 Emotions Associated with Parenting and Pair-Bonding

Parental behavior is a defining feature of mammals, although the patterns and extent of parental care vary greatly across species. Parental behavior must be underlain by a discrete motivational system on the part of the parent, complemented by a corresponding system in the offspring (Maestripieri 2003a). We expect these motivations to be modulated by complementary attitudes, what we term *parental attachment* and *offspring attachment,* that represent the fitness value of offspring and parent to one another. These attitudes shape appraisals of actions, determining the circumstances that elicit the corresponding emotions of *parental love* and *offspring love*. For example, positive parental attachment leads a mother to appraise her infant's midnight rooting as affiliation, eliciting parental love, rather than as a transgression that would elicit anger. Offspring develop positive attachment as a result of the experience of receiving succor. In contrast, because offspring impose costs on parents without immediate compensation, the building of positive parental attachment is achieved in part through the pump-priming effects of an emotion (*natal attraction*) that makes interaction with infants intrinsically rewarding, a phenomenon well-documented in primates (e.g., Silk 1999). It appears that, in humans, natal attraction transforms into parental love in part as a result of positive infant responses to parental overtures. Given the general primate trend of reductions in the importance of olfactory cues for parent-offspring bonding, and increases in the importance of behavioral cues (Broad et al. 2006), similar patterns may obtain across primates.

Two different systems appear to underlie human mate selection, courtship, and long-term pair- bonding. *Limerance* (Tennov 1979) is an emotion characterized by intrusive ideation concerning a prospective mate, attention to indications of reciprocation, and a motivation to be near and make a positive impression on the target individual. Both E. Pillsworth and R. Kurzban (pers. comm.) propose that the intrusive and obsessive nature of limerance are explicable in terms of the need to signal commitment to the target party, given the risk of defection. The most persuasive signals are costly, taking the form of resource provisioning, time allocation, and the public spurning of alternate potential mates. Once a stable mateship has formed, the benefits of signaling are reduced: once both parties have a concrete interest (e.g., offspring) in extended cooperation, it is adaptive to value the other party's welfare, and provide benefits noncontingently. Although investigators (e.g., Hatfield 1988) claim that an emotion, *companionate love,* replaces limerance in pair-bonded couples, much of the phenomenon thus labeled is an attitude rather than an emotion – the actual emotion is only present during punctuated events in which displays of affection reaffirm mutual valuation, reinforcing this attitude. Pair-bonding and biparental investment occur in a variety of mammals, and there is some commonality in the neurophysiological systems that underlie these behaviors (Curley and Keverne 2005; Broad et al. 2006). Thus, it is possible that human limerance and companionate love are complex manifestations of a basic mammalian potential that has been further developed in pair-bonded species.

12.5 Emotions Regulating Dyadic Cooperative Relationships

Companionate love and the attitude with which it is intertwined motivate altruistic behavior toward committed partners. Although pair bonds exist in a limited number of primate species, long-term affiliative relationships are more common (Silk 2007). In humans, a number of emotions play key roles in motivating behavior in affiliative and cooperative relationships. *Gratitude* follows the receipt of a welcome benefit provided by another party, motivating reciprocation (Trivers 1971), and enhancing the attitude toward the other. Gratitude thus facilitates the initiation and maintenance of cooperative relationships (McCullough et al. 2008). Gratitude is subjectively and behaviorally differentiated from a sense of indebtedness (McCullough et al. 2008). This is understandable in functional terms, as gratitude marks an increased estimation (summarized in the attendant positive attitude) of the potential long-term utility of the relationship, while indebtedness stresses the short-term burden of repayment, indexing a different type of relationship. Many primates differentiate among individuals and act in light of past interactions, exhibiting durable alliances and affiliative behaviors (van Schaik and Aureli 2000). This suggests that a gratitude-like mechanism may be both widely shared and of considerable antiquity (Bonnie and de Waal 2004).

Whether by mistake or due to the temptations of short-term rewards for defection, individuals can also inflict costs on their valued partners. If individuals who commit such acts perceive that they have damaged their partners' attitude toward themselves and this is disadvantageous given the utility of the relationship, ameliorative action is called for. *Guilt* is the prototypical emotion elicited when harm is done to an ally (Baumeister et al. 1994; Tangney 1998). Guilt motivates apologies and, importantly, reparations, compensating the partner for damages and signaling the individual's commitment to the relationship (Trivers 1971). Humans employ theory-of-mind reasoning in contemplating harm done to another; although other primates may not do likewise, this aspect of human guilt may be an extension of the basic phenomenon rather than an intrinsic feature, as many animals appear to calibrate costs inflicted on others. We therefore expect a guilt-like mechanism to be present in many primates, consistent with the observation that conflict among allies is sometimes followed by reconciliatory behavior (reviewed in Flack and de Waal 2000). Lastly, in humans, if reparations are not possible, guilt motivates penance, infliction of self-imposed costs that signal that the individual does not pursue self-interest at the expense of partners. Consistent with the symbolic framing upon which such behavior rests, we expect penance to be absent among nonhumans.

Selection of prospective partners often precedes the exchange of benefits. Several emotions mark the positive evaluation of an individual as a prospective cooperative partner; paralleling pair bonding, an attitude summarizing the value of the relationship is informed by these emotions. *Affiliative liking* is a response to individuals who simultaneously possess valuable attributes and share with the evaluator sufficient commonalities as to constitute useful partners (commonalities are important because they facilitate coordination and enhance the likelihood of shared objectives; Tooby and Cosmides 1996; McElreath et al. 2003). This emotion builds *amicability*, an attitude summarizing the potential utility of the target as a cooperative partner. *Admiration* resembles affiliative liking, and similarly enhances amicability, but is elicited by individuals who possess a greater preponderance of valuable attributes relative to commonalities, an asymmetry that forces the evaluator to invest relatively more in the relationship. *Pity* may be elicited by potential allies who are currently incapacitated and cannot contribute to a cooperative relationship. Pity motivates the actor to provide aid, eliciting gratitude from the incapacitated individual (Cottrell and Neuberg 2005; cf. Trivers 1971 on sympathy).

Alliances that yield extensive benefits over a long period must begin with a positive appraisal of the other's value as a cooperative partner, and a motivational stance that entails willingness to provide benefits. It is therefore likely that affiliative liking and amicability exist in many social animals, constituting fairly ancient traits. Because dominance hierarchies create asymmetries in power between potential cooperators, we also expect some version of admiration to be present in many animals capable of calibrating the provision of benefits in light of relative status. Primates seem to exhibit this capacity, and observers have reported obsequious affiliation directed at dominants by young subordinates (A. Pusey pers. comm., Walters and Seyfarth 1987). Evidence for pity is considerably weaker, as individuals seem more likely to avoid an injured or sick group member than to provide aid (e.g., Goodall 1986; but see also Preston and de Waal 2002). It is unclear whether this is because other primates lack the requisite theory of mind capacity, cannot judge the probability of recovery and future usefulness of potential allies, or other factors.

Grief is the emotion felt at the death or loss of partners. The magnitude of grief appears proportional to the impact of the loss on one's fitness, suggesting that grief is adaptive to the extent that it motivates individuals to seek out replacement partners. Some primates do show marked physiological responses to losses of preferred companions (Engh et al. 2006), and make efforts to expand their social networks in the weeks that follow loss.

Not all incapable individuals evoke pity, as not all constitute potential allies worth rehabilitating (Kurzban and Leary 2001). Additionally, previously beneficial relationships can later prove unprofitable. Such individuals are excluded from future alliances and the benefits thus produced, and are often exploited. Although a considerable literature explores *contempt* as an emotion, findings are inconsistent. We suggest this is because contempt is an attitude, not an emotion (Gervais 2009) – it is a representation of the evaluation of an individual as having no value as a current or future ally. As such, it plays a central role in social event appraisal and mediates emotion systems. Without any interest in the welfare of another, any cost imposed by that party is appraised as a transgression, evoking anger. Likewise, any risk of actual or symbolic contagion stemming from association elicits disgust. A lack of interest in the welfare of another also undermines guilt, as damage done does not warrant demonstration of positive valuation; in turn, this mutes anticipatory inhibitions that prevent doing harm. The lack of motivation not to hurt another is compounded by a lack of empathy (where empathy can be viewed as a trans-emotional mechanism for assessing the needs of others), as the needs of the contemned are of no interest to the contemnor. Finally, the contemnor experiences no grief at the death of the condemned, as this event does not reduce the contemnor's fitness.

The absence of a prosocial attitude toward some individuals will occur in any species capable of discriminative affiliation. However, while contempt exists whenever valuation of another's welfare has not been raised above zero (what we term *minimal contempt*), contempt can also arise through active diminution of valuation. This occurs when established relationships break down (a possibility in even minimally cooperative species), or upon unfavorable social appraisal in species in which baseline conspecific valuation is greater than zero. A positive default valuation should scale with the possibility of cooperation. For example, we expect an elevated baseline in male chimpanzees owing to the importance of alliances and intergroup defense (Watts 2006). Note that evidence of prosocial behavior in non-human primates (e.g., de Waal 2008; Lakshminarayanan and Santos 2008; but see also Vonk et al. 2008) does not speak to the question of baseline valuations, as subjects in such experiments have histories of prior interaction during which valuations may have been raised. In contrast, evidence that captive chimpanzees spontaneously help unfamiliar humans (Warneken et al. 2007) provides indirect evidence of a positive baseline valuation, at least in this particular interspecific context. Positive default valuation likely reaches its extreme in humans, owing to a history of intergroup competition and dependence on transmitted culture and cooperation (Brewer and Campbell 1976; Richerson and Boyd 2005). In humans, welfare valuation can be readily downregulated in response to derived cues of low value such as cultural difference (see McElreath et al. 2003) or nonconformity (see Rozin et al. 1999). The combination of the readiness with which this occurs, the subjectively negative affect that marks such attitude change, and the role of contempt in potentiating anger and disgust likely explains why contempt is often miscategorized as an emotion.

12.6 Linked Fate and Vicarious Emotions

Lickel et al. (2005) use the term *vicarious* to describe emotions elicited when events that befall another are treated as if they befell the self (see also Rydell et al. 2008). While possibly phylogenetically linked to emotional contagion, elicitation of vicarious emotions hinges on knowledge of the involvement of others in events, rather than exposure to others' emotion displays. Lickel et al. identify two axes, interdependence and shared identity, that determine vicarious elicitation. We suggest that Lickel et al.'s criteria for interdependence – interaction, joint goals, and shared norms – are constituents of human cooperation. Cooperation links the fates of the participants: the greater the investment in, and payoffs from, cooperative ventures, the more the events that affect one member of the venture also impact other members. Hence, calibrated for degree of cooperative interdependence, it is adaptive to respond to such events vicariously. This process is likely undergirded by attitudes that capture the degree of alignment of interests entailed in cooperation; in turn, these attitudes generate appraisals of events befalling others that parallel appraisals of events befalling the self.

Reduced by the coefficient of relatedness, it is also adaptive to react to events affecting kin as if they affected oneself, since one's own fitness is at least partially aligned with that of one's kin. Kinship is separate from cooperation, and hence from action interdependence in Lickel et al.'s (2005) sense – while kin-recognition mechanisms may rely on propinquity and interaction as cues of relatedness, kin should experience vicarious emotions even when the level of social interaction and degree of shared goals and norms is low. While some form of cooperation is found in many mammals, nepotistic behavior predates this, suggesting that kin-based vicarious reactions were co-opted long ago for use in the cooperative domain. We expect that any primate emotion experienced in an individual fashion will also be experienced in a vicarious fashion given the proper elicitors, just as we expect to find attitudinal proxies of fitness alignment resulting from either kinship or cooperative interdependence.

Shared identity, Lickel et al.'s (2005) second axis whereby vicarious emotions are elicited, can also be understood as a manifestation of an underlying process whereby the fates of individuals become linked. Humans attribute a shared essence to members of groups, such as ethnies, that exhibit distinctive cultural markers. This essence constitutes grounds for inductive reasoning regarding the actions and attributes of group members, a process thought to occur due to the utility of such markers in predicting behavior (Gil-White 2001). Because essentializing supports inductive generalization, the fates of members of an essentialized group can become intertwined, as actions by one group member can be taken as indicative of the character of all group members, influencing outsiders' attitudes. Essentializing and inductive generalization are likely matters of degree – some categories of individuals are seen as sharing a great deal of essence, while others less so. Like kinship and degree of cooperation, social identity is thus a variable property that influences vicarious emotion elicitation.

Noting that essentialism is adaptive in interacting with animals, Gil-White (2001) suggests that essentialist social cognition derives from this ancestral trait. We concur, adding that the coexistence of multiple hominid species during human evolution may have facilitated a process of cooptation, wherein the phenotypic markers used in essentialist reasoning were expanded to include culturally transmitted traits, as interspecific cultural diffusion might have blurred the line between biological and cultural phenotypes. Given both the utility of inductive generalizations and evidence of primate antipredator strategies that involve a degree of learning and are directed at specific species or genera (e.g., Seyfarth et al. 1980), it is plausible that the ancestral interspecific form of essentialist reasoning is widely shared. In contrast, the use of symbols to mark cultural affiliation and reinforce cooperation, the related importance of essentialist reasoning, and the resulting role of shared identity as an elicitor of vicarious emotions are all uniquely human.

12.7 Norm-Based Emotions

Humans differ from other primates in the extent and importance of cooperation, a feature tightly linked to the degree to which norms regulate behavior, as socially transmitted standards define goals, actions, and social relations that promote prosociality and enhance coordination (McElreath et al. 2003). Both our species' exploitation of cooperation and our reliance on norms are likely undergirded by a set of uniquely human emotion systems.

Paralleling the extension of eliciting conditions through appraisal modification that occurs in vicarious emotion systems, in *moral outrage, moral disgust,* and *moral approbation*, events that do not involve the actor elicit emotions as if they did: moral outrage and moral disgust are, respectively, anger and disgust elicited by others' norm-violating actions; moral approbation is a gratitude-like emotion elicited by others' exemplary performance of normative ideals (Fessler and Haley 2003). Moral outrage motivates inflicting costs on the norm violator as if in retribution; moral disgust motivates avoiding the norm violator as if in contamination avoidance (Gutierrez and Giner-Sorolla 2007); and moral approbation motivates providing a benefit to the norm-embodier as if in response to a benefit received. As in vicarious emotions, in each case, the emotion modifies an attendant attitude toward the target despite no direct interaction. However, whereas in vicarious emotions a connection to the self is made via another person, in these emotions, the connection is solely via the norm at issue.

Socially transmitted standards for behavior play a much more restricted role in the lives of nonhuman primates, and evidence for norm enforcement by disinterested parties is very limited (see Flack et al. 2005). In contrast, although the proportions in which they occur likely differ across cultures, moral outrage, moral disgust, and moral approbation are all readily observable in any human society (e.g., see Henrich et al. 2006).

There are three basic kinds of explanations for the coevolution of cultural norms and norm based emotions. Boyd and Richerson (2002) contend that punishment can stabilize any behavior, and cultural group selection favors the predominance of group-beneficial norms (see Silk and Boyd, this volume). Norm violators are punished, as are those who fail to punish norm-violators, and this generates selective pressure favoring the propensity to treat norms as extensions of the self, as this reduces the frequency with which the actor is punished. A second view holds that norm-based emotions are a product of natural selection acting directly on individuals (e.g., Haley and Fessler 2005; Kurzban et al. 2007). Here, actors compete in a marketplace of prospective allies; because conformists are predictable, adherence to cultural norms increases an actor's attractiveness in this regard. Actors can advertise their adherence to cultural norms by punishing norm violators and rewarding those who exemplify norms, with the reputational benefits thus gained outweighing the costs of these actions. The fitness advantages of inclusion in cooperative ventures thus favor emotion systems in which norms are treated as extensions of the self. Finally, a third view holds that unique features of human sociality, such as intergroup conflict and reproductive leveling, provided the necessary conditions for biological group selection in humans. Group selection favored the evolution of prosocial motivations, including norm-based emotions that motivate third-party punishing and rewarding behaviors (e.g., Gintis et al. 2003; Bowles 2006).

Although at present it is difficult to determine which of the above accounts accurately describes hominid evolution, or whether some additional account is needed, it is, nonetheless, clear that humans possess the motivational architecture upon which these perspectives converge. The aversive emotion *shame* is elicited by the appraisal that others are aware of one's failure to conform to important norms, while the rewarding emotion *pride* is elicited by the appraisal that others are aware of one's success in exemplifying important norms (reviewed in Fessler 2007). Even holding aside the limited role of norms among non-human primates, we expect shame and pride to be uniquely human, as they are contingent on sophisticated theory-of-mind reasoning.

Shame and pride exemplify the evolutionary process of co-optation and modification: despite employing the same display behaviors, qualia, and action tendencies as proto-shame and proto-pride (see above), the key eliciting conditions differ, as proto-shame and proto-pride focus exclusively on relative position in a social hierarchy, do not rely on norms as evaluative criteria for behavior, and do not involve theory-of-mind reasoning (Fessler 2007). The action tendencies associated with shame similarly reveal its kludge-like structure. Although parties offended by norm violations are best mollified by apologies and public commitments to future conformity, shame paradoxically interferes with these behaviors by motivating flight and hiding, tactics more appropriate to dealing with dominants than to placating judgmental community members. Proto-shame and proto-pride, ancestral pan-primate rank-related emotions, were modified in the hominid line. With the rise in importance of norms, coercive force (dominance) was largely replaced by freely-granted deference (prestige) as a determinant of social position (Henrich and Gil-White 2001). Apparently, because prestige is contingent on the opinions of others, existing emotions that motivated rank-striving behavior were modified by selection so as to include an assessment of others' evaluations of the actor. Whether to avoid punishment, enhance inclusion in cooperative ventures, or generate non-contingent group-beneficial behavior, these emotions were then further refashioned, shifting the focus away from rank, and toward norm adherence. Contemporary humans exhibit all three forms of these emotions – the ancestral forms that focus on dominance-based rank, the intermediate derived forms that focus on prestige, and the final derived forms that focus on norm adherence.

All three forms of shame and pride can be experienced vicariously. As we would expect to also be true among primates that engage in coalitional aggression, vicarious proto-pride or proto-shame should be experienced whenever an ally enjoys victory or suffers defeat relative to a rival. Vicarious prestige-based pride and shame, though limited to humans, should exhibit a similar pattern. Finally, whenever a person linked to the actor via a shared social identity succeeds or fails with regard to normative standards, to the extent that others will engage in inductive generalizing, those successes or failures will influence how others treat the actor, hence the actor should experience vicarious pride or shame (see Lickel et al. 2005). The primary exception to the latter pattern occurs when identity sharing is incomplete, in which case an actor can distance herself from a norm violator in order to manage third parties' assessments; in this case, moral outrage, rather than vicarious shame, may be elicited (cf. Haley 2002).

Shame and pride illustrate the extensive re-working that can be achieved through processes of co-optation and modification. However, as the vicarious versions of these emotions demonstrate, substantial functional changes can also occur merely by extending an emotion's eliciting conditions. We propose that an emotion that we term *normative guilt* is elicited by norm violations absent a harmed relationship partner – within the worldviews in which they are defined, many sins do not harm other people, yet the sinner experiences guilt nonetheless. This extension is made possible by our elaborately developed ability to manipulate representations of social others. In some manifestations of normative guilt, the represented partner is a culturally-constructed nonexistent agent (a deity, ancestor, etc.; cf. Darwin [1871] 1909: 115–116 on remorse); in others, the imagined partner is a representation of the actor's kin network or cooperative group; and, in the most abstract manifestation, the (only dimly imagined) partner is a representation of society as a whole. In each case, violating a norm leads actors to undertake reparations or penance as if, by doing so, they mitigated the harm done, or signaled their future reliability. Normative guilt thus illustrates one of the most profound disjunctions between ourselves and our non-human primate relatives, namely the nature of the internal representations that constitute the informational environment in which emotions operate. Despite our common origins, the complexity of humans' internal representations, and the fundamentally cultural nature of those representations, create a gap between our emotions and those of other primates, a gap that bears minding indeed.

References

Andrews PW, Gangestad SW, Matthews D (2002) Adaptationism – how to carry out an exaptationist program. Behav Brain Sci 25:489–504

Aureli F, Schaffner CM (2002) Relationship assessment through emotional mediation. Behaviour 139:393–420

Aureli F, Whiten A (2003) Emotions and behavioral flexibility. In: Maestripieri D (ed) Primate psychology: the mind and behaviour of human and nonhuman primates. Harvard University Press, Cambridge, MA, pp 289–323

Baumeister RF, Stillwell AM, Heatherton TF (1994) Guilt – an interpersonal approach. Psychol Bull 115:243–267

Bonnie KE, de Waal FBM (2004) Primate social reciprocity and the origin of gratitude. In: Emmons RA, McCullough ME (eds) The psychology of gratitude. Oxford University Press, New York, pp 213–229

Bowles S (2006) Group competition, reproductive leveling, and the evolution of human altruism. Science 314:1569–1572

Boyd R, Richerson P (2002) Solving the puzzle of human cooperation. In: Levinson S (ed) Evolution and culture. MIT, Cambridge, MA, pp 105–132

Braithwaite VA, Boulcott P (2007) Pain perception, aversion and fear in fish. Dis Aquat Organ 75:131–138

Brewer MB, Campbell DT (1976) Ethnocentrism and intergroup attitudes: East African evidence. Sage, Beverly Hills, CA

Broad KD, Curley JP, Keverne EB (2006) Mother-infant bonding and the evolution of mammalian social relationships. Phil Trans Roy Soc Lond B 361:2199–2214

Brosnan SF (2006) Nonhuman species' reactions to inequity and their implications for fairness. Soc Justice Res 19:153–185

Cheney DL, Seyfarth RM (2005) Social complexity and the information acquired during eavesdropping by primates and other animals. In: McGregor PK (ed) Animal communication networks. Cambridge University Press, New York, pp 583–603

Cords M (1997) Friendships, alliances, reciprocity, and repair. In: Whiten A, Byrne RW (eds) Machiavellian intelligence. II. Extensions and evaluations. Cambridge University Press, New York, pp 24–49

Cottrell CA, Neuberg SL (2005) Different emotional reactions to different groups: a sociofunctional threat-based approach to 'prejudice'. J Pers Soc Psychol 88:770–789

Curley JP, Keverne EB (2005) Genes, brains and mammalian social bonds. Trends Ecol Evol 20:561–567

Curtis VA, Biran A (2001) Dirt, disgust, and disease – is hygiene in our genes? Perspect Biol Med 44:17–31

Darwin C [1871] (1909) The descent of man and selection in relation to sex. D. Appleton, New York

Darwin C [1872] (1955) The expression of the emotions in man and animals. Philosophical Library, New York

de Waal FBM (1996) Good natured: the origins of right and wrong in humans and other animals. Harvard University Press, Cambridge, MA de Waal FBM (2008) Putting the altruism back into altruism: the evolution of empathy. Annu Rev Psychol 59:279–300

Dierick HA, Greenspan RJ (2007) Serotonin and neuropeptide F have opposite modulatory effects on fly aggression. Nat Genet 39:678–682

Ekman P (1992) An argument for basic emotions. Cogn Emot 6:169–200

Engh AL, Beehner JC, Bergman TJ, Whitten PL, Hoffmeier RR, Seyfarth RM, Cheney DL (2006) Behavioural and hormonal responses to predation in female chacma baboons (*Papio hamadryas ursinus*). Proc Roy Soc Lond B 273:707–712

Fairbanks LA, Jorgensen MJ, Huff A, Blau K, Hung YY, Mann JJ (2004) Adolescent impulsivity predicts adult dominance attainment in male vervet monkeys. Am J Primatol 64:1–17

Fessler DMT (2003) No time to eat: an adaptationist account of periovulatory behavioral changes. Q Rev Biol 78:3–21

Fessler DMT (2004) Shame in two cultures: implications for evolutionary approaches. J Cogn Cult 4:207–262

Fessler DMT (2007) From appeasement to conformity: evolutionary and cultural perspectives on shame, competition, and cooperation. In: Tracy JL, Robins RW, Tangney JP (eds) The self-conscious emotions: theory and research. Guilford Press, New York, pp 174–193

Fessler DMT. Madmen: an evolutionary perspective on anger and men's violent responses to transgression. In: Potegal M, Stemmler G, Spielberger CD (eds) Handbook of anger: constituent and concomitant biological, psychological, and social processes. Springer, New York (in press)

Fessler DMT, Haley KJ (2003) The strategy of affect: emotions in human cooperation. In: Hammerstein P (ed) Genetic and cultural evolution of cooperation. MIT, Cambridge, MA, pp 7–36

Fessler DMT, Haley KJ (2006) Guarding the perimeter: The outside-inside dichotomy in disgust and bodily experience. Cogn Emot 20:3–19

Fessler DMT, Navarrete CD (2003) Domain-specific variation in disgust sensitivity across the menstrual cycle. Evol Hum Behav 24:406–417

Fessler DMT, Navarrete CD (2004) Third-party attitudes toward sibling incest – evidence for westermarck's hypotheses. Evol Hum Behav 25:277–294

Flack JC, de Waal FBM (2000) 'Any animal whatever' – Darwinian building blocks of morality in monkeys and apes. J Conscious Stud 7:1–29

Flack JC, de Waal FBM, Krakauer DC (2005) Social structure, robustness, and policing cost in a cognitively sophisticated species. Am Nat 165:E126–E139

Frijda NH (1986) The emotions. Cambridge University Press, New York

Gangestad SW, Simpson JA (2007) Whither science of the evolution of mind? In: Gangestad SW, Simpson JA (eds) The evolution of mind: fundamental questions and controversies. Guilford Press, New York, pp 397–437

Gervais M (2009) Distinguishing emotions and attitudes in the study of contempt: conceptual, functional, and phylogenetic considerations. Unpubl ms

Gervais M, Wilson DS (2005) The evolution and functions of laughter and humor: a synthetic approach. Q Rev Biol 80:395–430

Gil-White FJ (2001) Are ethnic groups biological 'species' to the human brain? Essentialism in our cognition of some social categories. Curr Anthropol 42:515–554

Gintis H, Bowles S, Boyd R, Fehr E (2003) Explaining altruistic behavior in humans. Evol Hum Behav 24:153–172

Goodall J (1986) Social rejection, exclusion, and shunning among the Gombe chimpanzees. Ethol Sociobiol 7:227–236

Gosling SD, Graybeal A (2007) Tree thinking: a new paradigm for integrating comparative data in psychology. J Gen Psychol 134:259–277

Gould SJ (1991) Exaptation – a crucial tool for an evolutionary psychology. J Soc Issues 47:43–65

Gould SJ, Vrba ES (1982) Exaptation – a missing term in the science of form. Paleobiology 8:4–15

Greenwald AG, Banaji MR (1995) Implicit social cognition – attitudes, self-esteem, and stereotypes. Psychol Rev 102:4–27

Gutierrez R, Giner-Sorolla R (2007) Anger, disgust, and presumption of harm as reactions to taboo-breaking behaviors. Emotion 7:853–868

Haley KJ (2002) With us or against us? Reputational psychology and the punishment of norm violations. Master of Arts, University of California, Los Angeles

Haley KJ, Fessler DMT (2005) Nobody's watching? Subtle cues affect generosity in an anonymous economic game. Evol Hum Behav 26:245–256

Haselton MG, Ketelaar T (2006) Irrational emotions or emotional wisdom? The evolutionary psychology of emotions and behavior. In: Forgas JP (ed) Hearts and minds: affective influences on social cognition and behavior. Psychology Press, New York, pp 21–40

Hatfield E (1988) Passionate love and companionate love. In: Sternberg RJ, Barners ML (eds) The psychology of love. Yale University Press, New Haven, pp 191–217

Henrich J, Gil-White FJ (2001) The evolution of prestige: freely conferred deference as a mechanism for enhancing the benefits of cultural transmission. Evol Hum Behav 22:165–196

Henrich J, McElreath R, Barr A, Ensminger J, Barrett C, Bolyanatz A, Cardenas JC, Gurven M, Gwako E, Henrich N, Lesorogol C, Marlowe F, Tracer D, Ziker J (2006) Costly punishment across human societies. Science 312:1767–1770

Kelly DR. The philosophy and psychology of disgust. MIT, Cambridge, MA (in prep)

Keltner D, Anderson C (2000) Saving face for Darwin: the functions and uses of embarrassment. Curr Dir Psychol Sci 9:187–192

Kurzban R, Leary MR (2001) Evolutionary origins of stigmatization: the functions of social exclusion. Psychol Bull 127:187–208

Kurzban R, DeScioli P, O'Brien E (2007) Audience effects on moralistic punishment. Evol Hum Behav 28:75–84

Lakshminarayanan VR, Santos LR (2008) Capuchin monkeys are sensitive to others' welfare. Curr Biol 18:R999–R1000

Lazarus RS (1991) Emotion and adaptation. Oxford University Press, New York

Lickel B, Schmader T, Curtis M, Scarnier M, Ames DR (2005) Vicarious shame and guilt. Group Proc Intergroup Relat 8:145–157

Lieberman D, Tooby J, Cosmides L (2003) Does morality have a biological basis? An empirical test of the factors governing moral sentiments relating to incest. Proc Roy Soc Lond B 270:819–826

Lieberman D, Tooby J, Cosmides L (2007) The architecture of human kin detection. Nature 445:727–731

Maestripieri D (2003a) Primate psychology. Harvard University Press, Cambridge, MA

Maestripieri D (2003b) The past, present, and future of primate psychology. In: Maestripieri D (ed) Primate psychology. Harvard University Press, Cambridge, MA, pp 1–16

McCullough ME, Kimeldorf MB, Cohen AD (2008) An adaptation for altruism? The social causes, social effects, and social evolution of gratitude. Curr Dir Psychol Sci 17:281–285

McElreath R, Boyd R, Richerson PJ (2003) Shared norms and the evolution of ethnic markers. Curr Anthropol 44:122–129

Muniz L, Perry S, Manson JH, Gilkenson H, Gros-Louis J, Vigilant L (2006) Father-daughter inbreeding avoidance in a wild primate population. Curr Biol 16:R156–R157

Nesse RM (1990) Evolutionary explanations of emotions. Hum Nat 1:261–289

Nunn CL, Altizer S (2006) Infectious diseases in primates: behavior, ecology, and evolution. Oxford University Press, New York

Nunn CL, Barton RA (2001) Comparative methods for studying primate adaptation and allometry. Evol Anthropol 10:81–98

Öhman A, Carlsson K, Lundqvist D, Ingvar M (2007) On the unconscious subcortical origin of human fear. Physiol Behav 92:180–185

Panksepp J (1998) Affective neuroscience: the foundations of human and animal emotions. Oxford University Press, New York

Parr LA, Waller BM, Vick SJ (2007) New developments in understanding emotional facial signals in chimpanzees. Curr Dir Psychol Sci 16:117–122

Paul A, Kuester J (2004) The impact of kinship on mating and reproduction. In: Chapais B, Berman CM (eds) Kinship and behavior in primates. Oxford University Press, New York, pp 271–291

Preston SD, de Waal FBM (2002) Empathy: its ultimate and proximate bases. Behav Brain Sci 25:1–72

Richerson PJ, Boyd R (2005) Not by genes alone: how culture transformed human evolution. University of Chicago Press, Chicago

Rozin P (1996) Towards a psychology of food and eating: from motivation to module to model to marker, morality, meaning, and metaphor. Curr Dir Psychol Sci 5:18–24

Rozin P, Haidt J, McCauley CR, Imada S (1997) Disgust: preadaptation and the cultural evolution of a food-based emotion. In: Macbeth HM (ed) Food preferences and taste: continuity and change. Berghahn, Oxford, pp 65–82

Rozin P, Lowery L, Imada S, Haidt J (1999) The CAD triad hypothesis: a mapping between three moral emotions (contempt, anger, disgust) and three moral codes (community, autonomy, divinity). J Pers Soc Psychol 76:574–586

Rozin P, Haidt J, McCauley CR (2000) Disgust. In: Lewis M, Haviland-Jones JM (eds) Handbook of emotions, 2nd edn. Guilford Press, New York, pp 637–653

Rydell RJ, Mackie DM, Maitner AT, Claypool HM, Ryan MJ, Smith ER (2008) Arousal, processing, and risk taking: consequences of intergroup anger. Pers Soc Psychol Bull 34:1141–1152

Scherer KR (1999) Appraisal theory. In: Dalgleish T, Power MJ (eds) Handbook of cognition and emotion. Wiley, New York, pp 637–663

Seyfarth RM, Cheney DL, Marler P (1980) Monkey responses to three different alarm calls: evidence of predator classification and semantic communication. Science 210:801–803

Silk JB (1999) Why are infants so attractive to others? The form and function of infant handling in bonnet macaques. Anim Behav 57:1021–1032

Silk JB (2003) Cooperation without counting: the puzzle of friendship. In: Hammerstein P (ed) Genetic and cultural evolution of cooperation. MIT, Cambridge, MA, pp 37–54

Silk JB (2007) Social components of fitness in primate groups. Science 317:1347–1351

Silk JB, Brosnan SF, Vonk J, Henrich J, Povinelli DJ, Richardson AS, Lambeth SP, Mascaro J, Schapiro SJ (2005) Chimpanzees are indifferent to the welfare of unrelated group members. Nature 437:1357–1359

Smith RH, Kim SH (2007) Comprehending envy. Psychol Bull 133:46–64

Susskind JM, Lee DH, Cusi A, Feiman R, Grabski W, Anderson AK (2008) Expressing fear enhances sensory acquisition. Nat Neurosci 11:843–850

Tangney JP (1998) How does guilt differ from shame? In: Bybee J (ed) Guilt and children. Academic, San Diego, pp 1–17

Tennov D (1979) Love and limerence: the experience of being in love. Stein and Day, New York

Tooby J, Cosmides L (1989) Adaptation versus phylogeny: the role of animal psychology in the study of human behavior. Int J Comp Psychol 2:105–118

Tooby J, Cosmides L (1996) Friendship and the banker's paradox: other pathways to the evolution of adaptations for altruism. In: Runciman WG, Maynard Smith J, Dunbar RIM (eds) Evolution of social behaviour patterns in primates and man: proceedings of the British Academy, vol 88. Oxford University Press, New York, pp 119–143

Tooby J, Cosmides L (2008) The evolutionary psychology of the emotions and their relationship to internal regulatory variables. In: Lewis M, Haviland-Jones JM, Barrett LF (eds) Handbook of emotions, 3rd edn. Guilford Press, New York, pp 114–137

Tracy JL, Matsumoto D (2008) The spontaneous expression of pride and shame: evidence for biologically innate nonverbal displays. Proc Natl Acad Sci USA 105:11655–11660

Trivers RL (1971) The evolution of reciprocal altruism. Q Rev Biol 46:35–57 van Schaik CP, Aureli F (2000) The natural history of valuable relationships in primates. In: Aureli F, de Waal FBM (eds) Natural conflict resolution. University of California Press, Berkeley, pp 307–333

Vonk J, Brosnan SF, Silk JB, Henrich J, Richardson AS, Lambeth SP, Schapiro SJ, Povinelli DJ (2008) Chimpanzees do not take advantage of very low cost opportunities to deliver food to unrelated group members. Anim Behav 75:1757–1770

Walters JR, Seyfarth RM (1987) Conflict and cooperation. In: Smuts BB, Cheney DL, Seyfarth RM, Wrangham RW, Struhsaker TT (eds) Primate societies. University of Chicago Press, Chicago, pp 306–317

Warneken F, Hare B, Melis AP, Hanus D, Tomasello M (2007) Spontaneous altruism by chimpanzees and young children. PLoS Biol 5:1414–1420

Watts DP (2006) Conflict resolution in chimpanzees and the valuable-relationships hypothesis. Int J Primatol 27:1337–1364

Part VI
Language, Thought & Communication

Chapter 13
Primate Communication and Human Language: Continuities and Discontinuities

Dorothy L. Cheney and Robert M. Seyfarth

Abstract Here, we review some questions in the production and perception of nonhuman primate vocalizations, focusing on three related issues. First, flexible vocal production separates humans not only from nonhuman primates but also from most other mammals and birds. The rarity of learned, modifiable call production in most mammals suggests that important changes in the mechanisms governing human phonation occurred relatively recently in human evolution, after the divergence of our hominid ancestors from the common ancestors of humans and chimpanzees. Second, while exclusive focus on vocal production reveals clear differences between humans and most other species, a broader examination of call usage and perception paints a more complicated picture, with both similarities and differences between the two groups. Third, an analysis of vocal production and perception draws attention to the very different mechanisms that underlie the behavior of signalers and recipients, even when they are involved in the same communicative event. Nonhuman primates have only a small repertoire of acoustically fixed vocalizations. However, because calls are individually distinctive and each call type is predictably linked to a particular social context, this limited call repertoire can, nonetheless, provide listeners with an open-ended, highly modifiable, and cognitively rich set of meanings, allowing them to construct "narratives" of unseen events. However, although nonhuman primates and other animals seem capable of thinking, as it were, in simple sentences, this ability does not motivate them to speak in sentences. Their knowledge remains largely private. We suggest that long before our ancestors spoke in sentences, they had a language of thought in which they represented the world – and the meaning of call sequences – in terms of actors, actions, and those who are acted upon. The linguistic revolution occurred

D.L. Cheney (✉)
Department of Biology, University of Pennsylvania, Philadelphia, PA, USA
e-mail: cheney@sas.upenn.edu

R.M. Seyfarth
Department of Psychology, University of Pennsylvania, Philadelphia, PA, USA
e-mail: seyfarth@psych.upenn.edu

P.M. Kappeler and J.B. Silk (eds.), *Mind the Gap*,
DOI 10.1007/978-3-642-02725-3_13, © Springer-Verlag Berlin Heidelberg 2010

when our ancestors began to express this tacit knowledge, and to use their cognitive skills in speaking as well as listening.

13.1 Introduction

Darwin (1871) drew attention to a dichotomy in the vocal communication of animals that had perplexed philosophers and naturalists for at least 2,000 years. In marked contrast to human language, he wrote, animal vocalizations appeared to be involuntary expressions of emotion and movement: "When the sensorium is strongly excited, the muscles of the body are generally thrown into violent action; and as a consequence, loud sounds are uttered, . . . although the sounds may be of no use" ([1871] 1981: 83). Two pages later, however, Darwin also acknowledged: "That which distinguishes man from the lower animals is not the understanding of articulate sounds, for, as every one knows, dogs understand many words and sentences. . . . Nor is it the mere capacity of connecting definite sounds with definite ideas; for it is certain that some parrots, which have been taught to speak, connect unerringly words with things, and persons with events" ([1871] 1981: 85).

As a contemporary example of the dichotomy first identified by Darwin, consider Rico, a border collie dog who learned the names of more than 200 different toys (Kaminski et al. 2004). Rico was able to learn and remember the names of new toys by process of exclusion, or "fast mapping," and – like small children – used gaze and attention to guide word learning. But of course, Rico never learned to *say* any of the words he learned. In this respect, his limited vocal production and extensive comprehension are similar to those of human-trained sea lions (Schusterman et al. 2002) and dolphins (Herman et al. 1993).

The dichotomy between call production and perception is puzzling because listeners are also signalers. The same animal that has no problem understanding a word like "ball" is completely unable to articulate it. Indeed, while there are fundamental differences between animal communication and human language in the context of call production, in the context of call perception, these differences seem much less obvious. So why do not animals articulate their mental representations of the world and other individuals more often, and in more contexts? Why does their knowledge remain largely private?

Here, we review some questions in the production and perception of nonhuman primate vocalizations, focusing in particular on the communication of free-ranging baboons (*Papio hamadryas* spp.) (for a more detailed discussion, see Cheney and Seyfarth 2007). For those interested in comparing animal vocal communication with human language, three conclusions emerge. First, flexible vocal production separates humans not only from nonhuman primates but also from most other mammals and birds. The rarity of learned, modifiable call production throughout most of the Class *Mammalia* suggests that important changes in the mechanisms governing human phonation occurred relatively recently in human evolution, after the divergence of our hominid ancestors from the common ancestors of humans and chimpanzees. Second, while an exclusive focus on vocal production reveals clear differences between humans and most other species, a broader examination of call usage and perception paints a more complicated picture, with both similarities and differences between the two groups. Third, an analysis of vocal production and perception draws attention to the very different mechanisms that underlie the behavior of signalers and recipients, even when they are involved in the same communicative event. Nonhuman primates have only a small repertoire of acoustically fixed vocalizations. However, because calls are individually distinctive and each call type is predictably linked to a particular social context, this limited call repertoire can nonetheless provide listeners with an open-ended, highly modifiable, and cognitively rich set of meanings. Listeners can acquire a huge number of messages from a finite number of signals.

13.2 Call Production and Development

Monkeys and apes have a small repertoire of calls that show relatively little modification in their acoustic properties during development. The development of vocal production is largely unaffected by variations in auditory experience or rearing (see Hammerschmidt and Fischer 2008 for a recent review). For example, male baboons produce loud two-syllable *alarm* wahoos when they encounter lions or leopards. *Alarm* wahoos are acoustically similar to the *contest* wahoos that males give during competitive contests with other males, but the two types of wahoo differ according to a number of acoustic measures (Fischer et al. 2002). Similarly, the *alarm* barks given by females and juveniles are acoustically similar to the *contact* barks that baboons give when they become lost or separated from their companions. Again, however, there are subtle acoustic differences between the two bark types that allow them to be distinguished by ear (Fischer et al. 2001a).

Unlike predator alarm calls, which depend in a fairly simple way on the type of predator or the degree of danger, the vocalizations given by animals during social interactions are elicited by a more complex array of factors that may include both the immediate social context and the history of interactions between the particular individuals involved. Baboon grunts offer an example.

Grunts are the baboons' most common vocalization. They are individually distinctive, given in a variety of non-aggressive circumstances, and seem to comprise two acoustically graded types (Owren et al. 1997; Rendall 2003). The *move* grunt is typically given in the context of group movement. Like similar calls given by other primate species, *move* grunts function to alert other individuals to the signaler's intentions to travel in a particular direction. By contrast, the *infant* grunt is given during many sorts of friendly interactions and functions as a signal of benign intent. *Infant* grunts are most commonly given in the context of infant handling, but they also occur during grooming, other friendly behavior, and reconciliation. The two grunt subtypes thus differ in the specificity of the stimuli that elicit them. *Move* grunts are linked to a specific context. By comparison, *infant* grunts are given in a wide variety of friendly situations. Whereas *infant* grunts function to signal benign intent toward one specific individual, *move* grunts broadcast the signaler's intentions to many individuals. *Move* and *infant* grunts exemplify the variation found in the baboon vocal repertoire, and indeed in the vocal repertoires of many other primates. Some calls are tightly linked to a relatively narrow context, whereas others are used in a wider variety of circumstances. Some calls are directed at a specific individual, whereas others are more widely broadcast.

The acoustic intergradation in many nonhuman primate vocalizations may be caused by gradation in the caller's arousal or emotional state. Analyses of baboons' *contact* and *alarm* barks, *contest* and *alarm* wahoos, and *move* and *infant* grunts all suggest that acoustic variation is consistent with variation in the caller's emotions (Fischer et al. 2001a, 2002, 2004; Rendall 2003, see also Jürgens 1995). In each of these cases, however, graded calls whose production may be determined largely by the signaler's emotions are nonetheless perceived as discretely different vocalizations (Rendall et al. 1999). And once listeners have recognized that each of these discretely different signals is predictably linked to a particular event, calls have the potential to convey to listeners a meaning that goes far beyond information about the signaler's emotional state.

In marked contrast to children, who learn to produce and comprehend thousands of new words during their first three years of life, monkeys and apes rarely modify their vocal repertoires by adding new sounds. Although some primates make subtle modifications in their vocalizations as a result of experience (Hauser 1992; Elowson and Snowdon 1994; Mitani and Brandt 1994; Crockford et al. 2004) and can modify the loudness of their calls through auditory feedback (Hage et al. 2006), a baboon in Kenya produces more or less the same sounds in the same contexts as a baboon in Botswana. This conclusion follows not only from research on many primate species (Seyfarth and Cheney 1997a) but also from a cross-fostering experiment involving two closely related species of macaques. In this experiment, juveniles that had been fostered into a group of macaques from another species continued to produce their own species' calls, despite being physically capable of producing their adoptive species' calls (Owren et al. 1993). Monkeys seem genetically predisposed to give particular calls in particular contexts.

This is not to say that nonhuman primate call production is involuntary. In both the field and the laboratory, nonhuman primates can control their vocalization and choose to either vocalize or remain silent.. After behaving aggressively toward a subordinate, for example, a baboon may give a "reconciliatory" grunt to her opponent or she may not (Cheney et al. 1995b; Cheney and Seyfarth 1997). Similarly, when capuchin monkeys (*Cebus capucinus*) find food, they may call or remain silent (Gros-Louis 2004). Even in highly emotional circumstances like encounters with predators, some individuals give alarm calls at high rates, others call less often, and still others remain silent (Cheney and Seyfarth 1990). In more controlled laboratory settings, the timing, duration, and rate of calling by monkeys can be brought under operant control (Pierce 1985; Egnor et al. 2007). Clearly, primates can control whether they vocalize or not depending upon variations in both the ecological, social, and acoustic environments.

Within a given context, nonhuman primates can also make subtle modifications in the acoustic structure of their calls (reviewed by Hammerschmidt and Fischer 2008, Seyfarth and Cheney in press). To cite just one example, wild chimpanzees in Uganda give long, elaborate pant-hoots either alone or in "choruses" with others. When two individuals have called together several times, the acoustic features of their pant-hoots begin to converge (Mitani and Brandt 1994; Mitani and Gros-Louis 1998). Apparently, individuals modify the acoustic structure of their calls depending upon auditory experience (see also Crockford et al. 2004).

Many other studies suggest that, whereas the basic acoustic structure of nonhuman primate calls is innately determined, the fine spectrotemporal features of vocalizations can be modified (for reviews see Seyfarth and Cheney 1997a, in press, Egnor and Hauser 2004; Hammerschmidt and Fischer 2008). Some features are more easily modified than others. As Janik and Slater (1997) first pointed out, temporal features like call duration and amplitude are more easily modified than frequency parameters. The distinction between relatively innate and more modifiable components of phonation is important, because it has significant implications for future research on the neurobiology of primate vocal production (see Egnor and Hauser 2004; Hammerschmidt and Fischer 2008 for further discussion).

In their relatively fixed vocal production, nonhuman primates are typical of most mammals and even the great majority of birds. In their 1997 review, Janik and Slater found evidence for learned, modifiable vocal production in only three orders of birds, cetaceans, harbor seals, and humans. Although we may yet be surprised by novel evidence of vocal imitation (e.g., Poole et al. 2005) or creative call combinations (Zuberbuhler 2002; Crockford and Boesch 2003; Arnold and Zuberbuhler 2006), for the moment it appears that the ability to modify vocal production depending upon experience is comparatively modest.

13.3 Call Perception

Whenever there is a predictable relation between a particular call type and a specific social context, a vocalization has the potential to inform nearby listeners about objects or events. The underlying mechanisms are irrelevant. A tone that informs a rat about the imminence of a shock, an alarm call that informs a vervet about the presence of an eagle, or a scream that informs a baboon that her offspring is involved in a fight all have the potential to provide a listener with precise information if they are predictably associated (Rescorla 1988) with a narrow range of events. The widely different mechanisms that lead to this association have no effect on the signal's potential to inform (Seyfarth and Cheney 2003).

But while they have this potential, do animal vocalizations really provide listeners with information about who is doing what to whom? The study of vocal perception in animal communication is fraught with both practical and conceptual difficulties because we cannot interview our subjects. As a result, the only way to determine the "meaning" of a signal is to examine the recipient's response – a very crude measure of whether the recipient has or has not acquired information.

In some cases, animals respond in qualitatively different ways to different vocalizations, and the characteristics of their responses suggest that each call type conveys specific information. Consider, for example, the responses of Diana monkeys (*Cercopithecus diana*) to a playback of different alarm calls. When Diana monkeys hear the growl of a leopard, they respond with their own "leopard alarm"; when they hear the shriek of an eagle, they give their own "eagle alarm" (Zuberbuhler et al. 1999).

In many other cases, the lack of qualitatively different responses complicates the interpretation of call meaning. Two methods help to circumvent this problem, at least partially. First, if observations suggest that some generic response, like looking toward the speaker, will be the only reaction elicited by a playback experiment, scientists can design a matched pair of trials, alike in all but one respect, and then compare subjects' responses under two different conditions. A consistent difference in the duration of response may permit inferences about the different sorts of information conveyed by different calls, or by the same call under different conditions. Alternatively, subjects may show no immediate response to playback of a call but their subsequent behavior may nonetheless be affected. Having heard a particular call from individual X, for example, a subject may be more likely to approach X in the next 30 min than if no call, or the call of a different individual, had been played. Such longer-term changes in behavior also allow one to make inferences, albeit indirect, about the meaning of a specific vocalization (Seyfarth and Cheney 1997b, for examples see Cheney and Seyfarth 2007).

Playback experiments on a wide variety of species have now demonstrated that many nonhuman primate calls function referentially, providing listeners with information about what is happening and who is involved. This is true both of calls that are acoustically very different and calls that are acoustically graded. Baboons, for example, respond differently to *alarm* and *contest* wahoos, *alarm* and *contact* barks, and *move* and *infant* grunts (Rendall et al. 1999; Fischer et al. 2001b; Kitchen et al. 2003). Signaler identity and context also play crucial roles in informing listeners about what is occurring. Whatever the mechanisms underlying call production, therefore, listeners extract meaning from the calls they hear.

The call-meaning relationship in the listener's mind is interesting in several respects. First, it constitutes an arbitrary association between a sound and the thing for which it stands. There is nothing about the sound of a Diana monkey's *leopard* alarm call, for example, that sounds like a leopard, and nothing about the sound of a monkey's *eagle* alarm that would obviously link it to an eagle. In much the same way, there is nothing in the acoustic details of baboons' *alarm* and *contest* wahoos that would help a listener learn that one is given in response to predators while another is given during male–male aggression.

Second, the meaning of each call is defined not just by its relation to an object in the world but also by its relation to other calls in the monkey's repertoire. A male Diana monkey's *leopard* alarm is similar in meaning to a leopard's growl and a female's *leopard* alarm, but different in meaning from all of three eagle-associated calls. In the monkey's mind, there exists a kind of semantic space in which the three leopard-associated sounds are closely linked in one cluster, whereas the three eagle-associated sounds are closely linked in another.

This leads to a third conclusion, that primate calls are acoustic units linked to particular concepts. When a Diana monkey hears a male's *leopard* alarm, she appears to form a mental representation of the call's meaning. Then, when she hears a leopard's growl coming from the same location shortly thereafter, she forms a second representation and compares the two calls on the basis of their meaning. Her lack of response to the now redundant growl is based on this assessment. The female, in other words, has a concept – a kind of mental image – of a leopard. The concept can be activated by any one of three quite different sounds that are linked together based on their shared meaning (e.g., Cohen et al. 2006). The concept is also amodal or multimodal, involving a combination of visual and auditory information (Gil-da-Costa et al. 2004; Ghazanfar et al. 2005).

As another example, consider the phenomenon of individual recognition by voice, which has been amply demonstrated in many species and underlies many of our playback experiments (e.g., Snowdon 1986; Rendall et al. 1996; Owren et al. 1997). Baboons clearly recognize other group members from their voices alone, regardless of whether they are giving a grunt, a *contact* bark, or a threat-grunt, and regardless of whether they are vocalizing in a calm or in an agitated manner. Despite wide variation in the acoustic cues that mark a call as a particular individual's, and the fact that the calls of one individual may grade acoustically into the calls of another, listeners still link each call with a unique individual in a discrete, categorical fashion. Individual recognition occurs in so many contexts, with so many vocalizations, that it is hard to escape the impression that listeners have a mental representation, or concept, of each group member as an individual. If monkeys were human, we would call this a concept of *person*.

In sum, whereas call production in primates is relatively fixed, the cognitive mechanisms that underlie call perception are considerably more complex. Underlying primates' assessment of call meaning is a rich conceptual structure, in which calls are linked both to objects and relations in the world and to other calls in the species' repertoire. When responding to calls, monkeys act as if they recognize individuals and have concepts like *leopard*, *eagle*, *close associate*, and so on. The contrast between impoverished production and rich, conceptually based perception argues strongly against the view that a concept cannot be acquired unless it is instantiated in one's language (reviewed by Gleitman and Papafragou 2005). Monkeys and apes have many concepts for which they have no words.

13.4 Syntax

There is little evidence for rule-governed syntax in the calls of nonhuman primates. Recent work by Zuberbuhler and colleagues on the alarm calls of forest monkeys provide intriguing evidence that the presence of one call type can modify the meaning of another (Zuberbuhler 2002; Arnold and Zuberbuhler 2006; Clarke et al. 2006), and a study by Crockford and Boesch (2003) suggests that a call combination in chimpanzees may carry new meaning that goes beyond the meaning of the individual calls themselves, but these rare exceptions meet few of the definitions of human syntax.

Nonetheless, monkeys and apes *hear* different calls in combination all the time. These calls are usually given by different individuals, allowing listeners to construct narratives about what is happening and who is involved. In assessing call sequences, listeners attend simultaneously to the signalers' identities, call type, the rank and kinship of the signalers, and the temporal and spatial juxtaposition of different individuals' calls. Baboons, for example, respond much more strongly to a call sequence that suggests a reversal in the female dominance hierarchy than one that is consistent with it (Cheney et al. 1995a; Bergman et al. 2003). When played a threat-grunt, scream sequence in which a high-ranking female – say, Sylvia – threat-grunts and a lower-ranking female – say, Hannah – screams, they show little if any response. Their responses are much stronger, however, if they hear a sequence that appears to violate their knowledge of the female dominance hierarchy; for example, Hannah threat-grunts and Sylvia screams. Although the callers' identities and the call types are the same, subtle changes in the elements of the sequence cause its meaning to change fundamentally. Listeners also seem to assume a causal relationship between calls that are closely juxtaposed in time: Hannah's threat-grunts *caused* Sylvia to scream. These assessments are done instantaneously and probably large unconsciously. They are based on discrete properties, such as the signalers' identities, kinship, and rank, that are combined in a combinatorial system and that encode propositional information: for example, A is threatening B; A is mating with B; and so on. All of these features are also present in language.

13.5 Attributing Intentions to Signalers

During conversation, humans routinely make inferences about the motives and beliefs of their intended recipients. Baboons, too, seem to recognize the intended recipient of someone else's calls.

Baboon groups are noisy, tumultuous societies, and a baboon could not manage her social interactions if she interpreted every vocalization she heard as directed at her. Inferences about the "directedness" of vocalizations are probably often mediated by gaze direction and relatively simple contingencies. Even in the absence of visual signals, however, baboons are able to make inferences about the intended recipient of a call based on their knowledge of a signaler's identity and the nature of recent interactions. For example, when females were played the "reconciliatory grunt" of a recent aggressor within minutes after being threatened, they were more likely to approach their former opponent and to tolerate their opponent's approaches than after hearing either no grunt or the grunt of another dominant female unrelated to their opponent (Cheney and Seyfarth 1997). Call type was also important, because females avoided their recent opponent if they heard her threat-grunt rather than her reconciliatory grunt (Engh et al. 2006). Thus, baboons use their memory of recent interactions to make inferences about whether a call is being directed at themselves or at some other individual.

In some cases, these inferences are complex and indirect, and call upon baboons' knowledge of the kinship relationships of other group members. For example, when female baboons were played the threat-grunts of their aggressor's relative soon after being threatened, they avoided members of their aggressor's matriline. In contrast, when they heard the same threat-grunts in the absence of aggression, they ignored the call and acted as if they assumed that the call was directed at someone else (Wittig et al. 2007a). Similarly, when subjects heard the "reconciliatory" grunt of their aggressor's relative after a fight, they were more likely to approach both their aggressor and the relative whose grunt they had heard (Wittig et al. 2007b). They did not do so, however, if they had heard the "reconciliatory" grunt of another, unrelated female, nor did they approach their aggressor's other relatives. Here again, subjects behaved as if they believed that a grunt from their aggressor's relative must be directed at *them*, as a consequence of the fight. What is especially interesting in these experiments is that subjects inferred that they were the target of the vocalization even though they had not recently interacted with the signaler, but with her relative. They could only have done so if they recognized the close bond that existed between the two females.

In primates, faces and voices are the primary means of transmitting social signals, and monkeys recognize the correspondence between facial and vocal expressions (Ghazanfar and Logothetis 2003). Presumably, visual and auditory signals are somehow combined to form a unified, multimodal percept in the mind of a monkey. In a study using positron emission tomography (PET), Gil-da-Costa et al. (2004) showed that when rhesus macaques hear one of their own species' vocalizations, they exhibit neural activity not only in areas associated with auditory processing but also in higher-order visual areas, including STS. Auditory and visual areas also exhibit significant anatomical connections (Poremba et al. 2003).

Ghazanfar et al. (2005) explored the neural basis of sensory integration using the *coos* and *grunts* of rhesus macaques as stimuli. They found clear evidence that cells in certain areas of the auditory cortex are more responsive to bi-modal (visual and auditory) presentation of species-specific calls than to unimodal presentation. Although significant integration of visual and auditory information occurred in trials with both vocalizations, the effect of cross-modal presentation was greater with *grunts* than with *coos*. The authors speculate that this may occur because *grunts* are usually directed toward a specific individual in dyadic interactions, whereas *coos* tend to be broadcasted generally to the group at large. The greater cross-modal integration in the processing of grunts may therefore have arisen because, in contrast to listeners who hear a *coo*, listeners who hear a *grunt* must immediately determine whether or not the call is directed at them.

In sum, when deciding "Who, me?" upon hearing a vocalization, baboons must take into account the identity of the signaler (who is it?), the type of call given (friendly or aggressive?), the nature of their prior interactions with the signaler (were they aggressive, friendly, or neutral?), and the correlation between past interactions and future ones (does a recent grooming interaction lower or increase the likelihood of aggression?). Learned contingencies, doubtless, play a role in these assessments. But because listeners' responses depend on simultaneous consideration of all of these factors, this learning is likely to be both complex and subtle.

Although baboons and other monkeys may be able to recognize other individual's intentions when inferring whether or not they are the target of another individual's call, there is no evidence that they take into account their audience's knowledge or beliefs when producing or assessing calls. Both monkeys and apes give alarm calls, for example, without any apparent recognition of whether listeners are ignorant or already informed about the presence of a predator (reviewed by Cheney and Seyfarth 2007). Similarly, although the "food calls" of capuchin monkeys (Gros-Louis 2004) and the pant hoots of chimpanzees (Clark and Wrangham 1994) attract others to food, signalers show no evidence of recognizing whether their audience is already aware of the presence of food. Baboons often give *contact* barks when separated from others. When several individuals are calling simultaneously, it often appears that they are answering each other's calls in order to inform others of the group's location. Playback experiments suggest, however, that baboons call primarily with respect to their own separation from the group, not their audience's. They "answer" others when they themselves are separated, and they often fail to respond to the calls of even their offspring when they themselves are in close proximity to other group members (Cheney et al. 1996; Rendall et al. 2000). In this respect, the vocalizations of monkeys and apes are very different from human speech, where we routinely take into account our audience's beliefs and knowledge during conversation (Grice 1957).

13.6 Primate Communication and the Evolution of Language

The vocal communication of nonhuman primates is very different from human language, especially in the domain of call production. At the same time, however, comparisons between primate communication and human language have tended to focus on the differences, ignoring some of the intriguing continuities in perception and cognition.

As already noted, the striking difference between production and comprehension in animal communication is puzzling because producers are also perceivers: why should an individual who can deduce an almost limitless number of meanings from the calls of others be able to produce only a limited number of calls of her own? The difference may arise because call production depends on mechanisms of phonation, which are largely innate, whereas comprehension depends on mechanisms of learning (classical and operant conditioning), which are considerably more malleable. But this explanation offers no answer to a crucial question: Why has natural selection so rarely acted to favor flexible vocal production? Here, we offer some speculations as they apply to human and nonhuman primates.

At some point in our evolutionary history – probably after the divergence of the evolutionary lines leading to chimpanzees and bonobos on the one hand and humans on the other hand (Enard et al. 2002) – our ancestors developed much greater control over the physiology of vocal production. As a result, vocal output became both more flexible and considerably more dependent on auditory experience and imitation (Lieberman 1991; Fitch 2007). What selective pressures might have given rise to these physiological changes?

Vocal communication in nonhuman primates lacks three features that are abundantly present in human language: the ability to generate new words, lexical syntax, and a theory of mind. By the latter, we mean the ability of both speakers and listeners to make attributions about each others' beliefs, knowledge, and other mental states (Grice 1957). These are the simplest, most basic features that distinguish human and nonhuman primate vocal production, and it is with these traits that speculations about the evolution of language must start. At the earliest stages of language evolution, we need not worry about the more complex properties of language that probably came later – properties like case, tense, subject–verb agreement, open- and closed-class items, recursion, long-distance dependency, subordinate clauses, and so on.

How might the ability to generate new words, lexical syntax, and a theory of mind have evolved: simultaneously, in response to the same selective pressures, or more serially, in some particular order? We propose that the evolution of a theory of mind preceded language, creating the selective pressures that gave rise to the ability to generate new words and lexical syntax, and to the flexibility in vocal production that these two traits would have required (Cheney and Seyfarth 2005, 2007).

There is no evidence in nonhuman primates for anything close to the large vocal repertoire we find even in very young children. Similarly, nonhuman primates provide few examples of lexical syntax. By contrast, there is growing evidence that both Old World monkeys (Flombaum and Santos 2005; Engh et al. 2006; Cheney and Seyfarth 2007) and apes (Hare et al. 2001; Tomasello et al. 2005; Buttelmann et al. 2007) may possess rudimentary abilities to attribute motives or knowledge to others, and engage in simple forms of shared attention and social referencing. Taken together, these data suggest that a rudimentary theory of mind appeared among primates long before flexible vocal production, the ability to generate new words, and lexical syntax.

A rudimentary theory of mind seems to be crucially important for word learning in young children. Beginning as early as 9–12 months, children exhibit a nascent understanding of other individuals' motives, beliefs, and desires, and this skill forms the basis of a shared attention system that is essential for early word learning (Bloom and Markson 1998; Tomasello 2003). One-year old children understand implicitly that words can be mapped onto objects and actions. Crucially, this understanding is accompanied by a kind of "social referencing" in which the child uses other people's direction of gaze, gestures, and emotions to assign labels to objects (Baldwin 1991, reviewed in Pinker 1994; Fisher and Gleitman 2002).

Gaze and attention also facilitate word learning in dogs and other animals. Children, however, rapidly surpass the simpler forms of shared attention and word learning demonstrated by animals. Long before they begin to speak in sentences, young children develop implicit notions of objects and events, actors, actions, and those that are acted upon. Moreover, in contrast to monkeys, apes, and other animals, 1-year-old children are highly motivated to share what they know with others (Tomasello and Carpenter 2007). While animals are concerned with their own goals and knowledge, young children are concerned with making their thoughts and knowledge publicly available. The acquisition of a theory of mind thus creates a cognitive environment that drives the acquisition of new words and new grammatical skills. Indeed, results suggest that children could not increase their vocabularies or learn grammar as rapidly as they do if they did not have some prior notion of other individuals' mental states (Pinker 1994; Fisher and Gleitman 2002; Tomasello 2003). In sum, data on children provide an excellent illustration of how a theory of mind can *drive* language development.

By contrast, it is much more difficult to imagine how our ancestors could have learned new words or grammatical rules if they were unable to attribute mental states to others. The lack of syntax in nonhuman primate vocalizations cannot be traced to an inability to recognize argument structure – to understand that an event can be described as a sequence in which an agent performs some action on an object. Baboons, for example, clearly distinguish between a sequence of calls indicating that Sylvia is threatening Hannah, as opposed to Hannah is threatening Sylvia. Nor does the lack of syntax arise because of an inability to mentally represent descriptive verbs, modifiers, or prepositions. In captivity, a variety of animals, including dolphins (Herman et al. 1993), sea lions (Schusterman and Krieger 1986), and African gray parrots (Pepperberg 1993), can be taught to understand and in some cases even to produce verbs, modifiers, and prepositions. Even in their natural behavior, nonhuman primates and other animals certainly seem capable of *thinking* in simple sentences. However, this ability does not motivate them to *speak* in sentences. Their knowledge remains largely private.

This limitation may arise because nonhuman primates and other animals cannot distinguish between what they know and others know and cannot recognize, for example, that an ignorant individual might need to have an event explained to them. As a result, although they may mentally tag events as argument structures, they fail to map these tags into a communicative system in any stable or predictable way. Because they cannot attribute mental states like ignorance to others, and are unaware of the causal relation between behavior and beliefs, monkeys and apes do not actively seek to explain or elaborate upon their thoughts. As a result, they are largely incapable of inventing new words or of recognizing when thoughts should be made explicit.

We suggest, then, that long before our ancestors spoke in sentences, they had a language of thought in which they represented the world – and the meaning of call sequences – in terms of actors, actions, and those who are acted upon. The linguistic revolution occurred when our ancestors began to express this tacit knowledge, and to use their cognitive skills in speaking as well as listening. The prime mover behind this revolution was a theory of mind that had evolved to the point where its possessors did not just recognize other individuals' goals, intentions, and even knowledge – as monkeys and apes already do – but were also motivated to share their own intentions, beliefs, and knowledge with others. It led to a mind that was motivated to make public thoughts and knowledge that had previously remained private. The evolution of a theory of mind spurred the evolution of words and grammar. It also provided the selective pressure for the evolution of the physiological adaptations that enabled vocal modifiability. Whatever the selective pressures that prompted this change, the complex suite of skills that we call human speech built upon mental computations that had their origins and foundations in social interactions.

References

Arnold K, Zuberbuhler K (2006) Language evolution: compositional semantics in primate calls. Nature 441:303

Baldwin DA (1991) Infants' contribution to the achievement of joint reference. Child Dev 62:875–890

Bergman TJ, Beehner JC, Cheney DL, Seyfarth RM (2003) Hierarchical classification by rank and kinship in baboons. Science 302:1234–1236

Bloom P, Markson L (1998) Capacities underlying word learning. Trends Cogn Sci 2:67–73

Buttelmann D, Carpenter M, Call J, Tomasello M (2007) Enculturated chimpanzees imitate rationally. Dev Sci 10:F31–F38

Cheney DL, Seyfarth RM (1990) How monkeys see the world: inside the mind of another species. University of Chicago Press, Chicago

Cheney DL, Seyfarth RM (1997) Reconciliatory grunts by dominant female baboons influence victims' behaviour. Anim Behav 54:409–418

Cheney DL, Seyfarth RM (2005) Constraints and preadaptations in the earliest stages of language evolution. Ling Rev 22:135–159

Cheney DL, Seyfarth RM (2007) Baboon metaphysics: the evolution of a social mind. University of Chicago Press, Chicago

Cheney DL, Seyfarth RM, Silk JB (1995a) The responses of female baboons (*Papio cynocephalus ursinus*) to anomalous social interactions: evidence for causal reasoning? J Comp Psychol 109:134–141

Cheney DL, Seyfarth RM, Silk JB (1995b) The role of grunts in reconciling opponents and facilitating interactions among adult female baboons. Anim Behav 50:249–257

Cheney DL, Seyfarth RM, Palombit R (1996) The function and mechanisms underlying baboon 'contact' barks. Anim Behav 52:507–518

Clark AP, Wrangham RW (1994) Chimpanzee arrival pant-hoots: do they signify food or status? Int J Primatol 15:185–205

Clarke E, Reichard UH, Zuberbühler K (2006) The syntax and meaning of wild gibbon songs. PLoS One 1:e73. doi:10.1371/journal.pone.0000073

Cohen YE, Hauser MD, Russ BE (2006) Spontaneous processing of abstract categorical information in the ventrolateral prefrontal cortex. Biol Lett 2:261–265

Crockford C, Boesch C (2003) Context-specific calls in wild chimpanzees (*Pan troglodytes verus*): analysis of barks. Anim Behav 66:115–125

Crockford C, Herbinger L, Vigilant L, Boesch C (2004) Wild chimpanzees produce group-specific calls: a case for vocal learning? Ethology 110:221–243

Darwin C [1871] (1981) The descent of man and selection in relation to sex. Princeton University Press, Princeton

Egnor SER, Hauser MD (2004) A paradox in the evolution of primate vocal learning. Trends Neurosci 27:649–654

Egnor SER, Wickelgren JG, Hauser MD (2007) Tracking silence: adjusting vocal production to avoid acoustic interference. J Comp Physiol A 193:477–483

Elowson AM, Snowdon CT (1994) Pygmy marmosets, *Cebuella pygmaea*, modify vocal structure in response to changed social environment. Anim Behav 47:1267–1277

Enard W, Przeworski M, Fisher SE, Lai CSL, Wiebe V, Kitano T, Monaco AP, Pääbo S (2002) Molecular evolution of FOXP2: a gene involved in speech and language. Nature 418:869–872

Engh AL, Hoffmeier RR, Cheney DL, Seyfarth RM (2006) Who me? Can baboons infer the target of vocalisations? Anim Behav 71:381–387

Fischer J, Hammerschmidt K, Cheney DL, Seyfarth RM (2001a) Acoustic features of female chacma baboon barks. Ethology 107:33–54

Fischer J, Metz M, Cheney DL, Seyfarth RM (2001b) Baboon responses to graded bark variants. Anim Behav 61:925–931

Fischer J, Hammerschmidt K, Cheney DL, Seyfarth RM (2002) Acoustic features of male baboon loud calls: influences of context age and individuality. J Acoust Soc Am 111:1465–1474

Fischer J, Kitchen DM, Seyfarth RM, Cheney DL (2004) Baboon loud calls advertise male quality: acoustic features and their relation to rank, age, and exhaustion. Behav Ecol Sociobiol 56:140–148

Fisher C, Gleitman LR (2002) Language acquisition. In: Pashler HF, Gallistel CR (eds) Stevens' handbook of experimental psychology, vol 3, Learning and motivation. Wiley, New York, pp 445–496

Fitch WT (2007) The evolution of language: a comparative perspective. In: Gaskell G (ed) Oxford handbook of psycholinguistics. Oxford University Press, Oxford, pp 787–804

Flombaum JI, Santos LR (2005) Rhesus monkeys attribute perceptions to others. Curr Biol 15:447–452

Ghazanfar AA, Logothetis NK (2003) Facial expressions linked to monkey calls. Nature 423:937–938

Ghazanfar AA, Maier JX, Hoffman KL, Logothetis NK (2005) Multisensory integration of dynamic faces and voices in rhesus monkey auditory cortex. J Neurosci 25:5004–5012

Gil-da-Costa R, Braun A, Lopes M, Hauser MD, Carson RE, Herskovitch P, Martin A (2004) Toward an evolutionary perspective on conceptual representation: species-specific calls activate visual and affective processing systems in the macaque. Proc Natl Acad Sci USA 101:17516–17521

Gleitman LR, Papafragou A (2005) Language and thought. In: Holyoak K, Morrison R (eds) Cambridge handbook of thinking and reasoning. Cambridge University Press, Cambridge, pp 633–661

Grice HP (1957) Meaning. Philos Rev 66:377–388

Gros-Louis J (2004) The function of food-associated calls in white-faced capuchin monkeys, *Cebus capucinus*, from the perspective of the signaller. Anim Behav 67:431–440

Hage SR, Jürgens U, Ehret G (2006) Audio-vocal interaction in the pontine brainstem during self-initiated vocalization in the squirrel monkey. Eur J Neurosci 23:3297–3308

Hammerschmidt K, Fischer J (2008) Constraints in primate vocal production. In: Oller DK, Griebel U (eds) The evolution of communicative creativity: complexity, creativity, and adaptability in human and animal communication. MIT Press, Cambridge, MA, pp 93–119

Hare B, Call J, Tomasello M (2001) Do chimpanzees know what conspecifics know? Anim Behav 61:139–151

Hauser MD (1992) Articulatory and social factors influence the acoustic structure of rhesus monkey vocalizations: a learned mode of production? J Acoust Soc Am 91:2175–2179

Herman LM, Pack AA, Morrel-Samuels P (1993) Representational and conceptual skills of dolphins. In: Roitblat HL, Herman LM, Nachtigall PE (eds) Language and communication:

comparative perspectives. Comparative cognition and neuroscience. Lawrence Erlbaum Associates, Hillsdale, NJ, pp 403–442

Janik VM, Slater PJB (1997) Vocal learning in mammals. Adv Study Behav 26:59–99

Jürgens U (1995) Neuronal control of vocal production in human and nonhuman primates. In: Zimmermann E, Newman JD, Jürgens U (eds) Current topics in primate vocal communication. Plenum Press, New York, pp 199–206

Kaminski J, Call J, Fischer J (2004) Word learning in a domestic dog: evidence for 'fast mapping'. Science 304:1682–1683

Kitchen DM, Cheney DL, Seyfarth RM (2003) Female baboons' responses to male loud calls. Ethology 109:401–412

Lieberman P (1991) Uniquely human: the evolution of speech, thought, and selfless behavior. Harvard University Press, Cambridge, MA

Mitani JC, Brandt KL (1994) Social factors influence the acoustic variability in the long-distance calls of male chimpanzees. Ethology 96:233–252

Mitani JC, Gros-Louis J (1998) Chorusing and call convergence in chimpanzees: tests of three hypotheses. Behaviour 135:1041–1064

Owren MJ, Dieter JA, Seyfarth RM, Cheney DL (1993) Vocalizations of rhesus (*Macaca mulatta*) and Japanese (*M. fuscata*) macaques cross-fostered between species show evidence of only limited modification. Dev Psychobiol 26:389–406

Owren MJ, Seyfarth RM, Cheney DL (1997) The acoustic features of vowel-like grunt calls in chacma baboons (*Papio cynocephalus ursinus*): implications for production processes and functions. J Acoust Soc Am 101:2951–2963

Pepperberg IM (1993) A review of the effects of social interaction on vocal learning in African grey parrots (*Psittacus erithacus*). Neth J Zool 43:104–124

Pierce JD Jr (1985) A review of attempts to condition operantly alloprimate vocalizations. Primates 26:202–213

Pinker S (1994) The language instinct. William Morrow and Sons, New York

Poole JH, Tyack PL, Stoeger-Horwarth AS, Watwood S (2005) Animal behaviour:elephants are capable of vocal learning. Nature 434:455–456

Poremba A, Saunders RC, Crane AM, Cook M, Sokoloff L, Mishkin M (2003) Functional mapping of the primate auditory system. Science 299:568–572

Rendall D (2003) Acoustic correlates of caller identity and affect intensity in the vowel-like grunt vocalizations of baboons. J Acoust Soc Am 113:3390–3402

Rendall D, Rodman PS, Emond RE (1996) Vocal recognition of individuals and kin in free-ranging rhesus monkeys. Anim Behav 51:1007–1015

Rendall D, Seyfarth RM, Cheney DL, Owren MJ (1999) The meaning and function of grunt variants in baboons. Anim Behav 57:583–592

Rendall D, Cheney DL, Seyfarth RM (2000) Proximate factors mediating 'contact' calls in adult female baboons (*Papio cynocephalus ursinus*) and their infants. J Comp Psychol 114:36–46

Rescorla RA (1988) Pavlovian conditioning: it's not what you think it is. Am Psychol 43:151–160

Schusterman RJ, Krieger K (1986) Artificial language comprehension and size transposition by a California sea lion (*Zalophus californianus*). J Comp Psychol 100:348–355

Schusterman RJ, Reichmuth Kastak C, Kastak D (2002) The cognitive sea lions: meaning and memory in the lab and in nature. In: Bekoff M, Allen C, Burghardt G (eds) The cognitive animal: empirical and theoretical perspectives on animal cognition. MIT Press, Cambridge, MA, pp 217–228

Seyfarth RM, Cheney DL (1997a) Some general features of vocal development in nonhuman primates. In: Husberger M, Snowdon CT (eds) Social influences on vocal development. Cambridge University Press, Cambridge, pp 249–273

Seyfarth RM, Cheney DL (1997b) Behavioral mechanisms underlying vocal communication in nonhuman primates. Anim Learn Behav 25:249–267

Seyfarth RM, Cheney DL (2003) Signalers and receivers in animal communication. Annu Rev Psychol 54:145–173

Seyfarth RM, Cheney DL. Primate vocal communication. In: Platt M, Ghazanfar AA (eds) Primate neuroethology. Oxford University Press, Oxford (in press)

Snowdon CT (1986) Vocal communication. In: Mitchell G, Erwin J (eds) Comparative primate biology, vol 2A, Behavior, conservation, and ecology. Alan R Liss, New York, pp 495–530

Tomasello M (2003) Constructing a language: a usage-based theory of language acquisition. Harvard University Press, Cambridge, MA

Tomasello M, Carpenter M (2007) Shared intentionality. Dev Sci 10:121–125

Tomasello M, Carpenter M, Call J, Behne T, Moll H (2005) Understanding and sharing intentions: the origins of cultural cognition. Behav Brain Sci 28:675–691

Wittig RM, Crockford C, Wikberg E, Seyfarth RM, Cheney DL (2007a) Kin-mediated reconciliation substitutes for direct reconciliation in female baboons. Proc Roy Soc Lond B 274:1109–1115

Wittig RM, Crockford C, Seyfarth RM, Cheney DL (2007b) Vocal alliances in chacma baboons, *Papio hamadryas ursinus*. Behav Ecol Sociobiol 61:899–909

Zuberbuhler K (2002) A syntactic rule in forest monkey communication. Anim Behav 63:293–299

Zuberbuhler K, Cheney DL, Seyfarth RM (1999) Conceptual semantics in a nonhuman primate. J Comp Psychol 113:33–42

Chapter 14
Language, Lies and Lipstick: A Speculative Reconstruction of the African Middle Stone Age "Human Revolution"

Chris Knight

> *We have need of lies in order to conquer this reality, this "truth," that is, in order to live – that lies are necessary in order to live is itself a part of the terrifying and questionable character of existence... if there is anything to be worshipped it is appearance that must be worshipped... the lie – and not the truth – is divine!*
>
> F. Nietzsche (1968: 451, 523)

Abstract Ovulation in humans is well concealed, leaving menstruation salient as an external sign of fertility. Extant hunter-gatherers package this information in ways designed to prevent philanderer males from exploiting it to their advantage. This culminates in human symbolic culture – a digital world of institutional facts designed to conceal and reconstruct selected brute facts of biology. Coalitions of females, supported by male kin, first began managing the public representation of menstruation, using cosmetics to scramble the fertility information made available to outgroup males. Selected for their brilliance and redness, ochre pigments such as those found at Blombos Cave in South Africa match the expectations of this model of symbolic cultural origins. As costly signals, high-quality cosmetics reliably indicate the strength of a woman's commitments and corresponding coalitionary support. Investor males can benefit by colluding with women's cosmetic "lies."

14.1 Digital Minds in an Analog World

Language has been described as a "mirror of mind." Chomsky attributes this exciting idea to "the first cognitive revolution" inspired by Descartes among others in the seventeenth century. "The second cognitive revolution" – triggered in large

C. Knight
Professor of Anthropology, Comenius University, Bratislava, Slovakia
e-mail: chris.knight@live.com

P.M. Kappeler and J.B. Silk (eds.), *Mind the Gap*,
DOI 10.1007/978-3-642-02725-3_14, © Springer-Verlag Berlin Heidelberg 2010

measure by Chomsky's own work – is taken to have been a twentieth century rediscovery of these earlier insights into the nature of language and mind. In 1660, the renowned Port Royal grammarians (Arnauld and Lancelot 1972 [1660], p 27) celebrated "*this marvelous invention of composing out of twenty-five or thirty sounds that infinite variety of expressions which, whilst having in themselves no likeness to what is in our mind, allow us to disclose to others its whole secret, and to make known to those who cannot penetrate it all that we imagine, and all the various stirrings of our soul.*"

For Descartes himself, however, this "marvelous" capacity was no mere human invention: "the seat of the soul" was the pineal gland (Descartes 1991 [1640], p 143). In Chomsky's twentieth century reformulation, the relevant organ becomes "*that little part of the left hemisphere that is responsible for the very specific structures of human language*" (Chomsky in Piattelli-Palmarini 1980, p 182). As Pinker (1999: 287) puts it: "*We have digital minds in an analog world. More accurately, a part of our minds is digital.*"

But if "a part of the mind is digital," how did it ever get to be that way? Under what Darwinian selection pressures and by what conceivable mechanisms might a digital computational module become installed in an otherwise analog primate brain? Can natural selection acting on an analog precursor transform it incrementally into a digital one? Is such an idea even logically coherent?

If these were easy questions, the origins of language – recently dubbed the "hardest problem in science" (Christiansen and Kirby 2003) – might long ago have been solved. Chomsky accepts Darwinism in principle, but doubts its direct relevance to this particular problem. In his view (Chomsky 2005, p 12), the "leap" to language "*was effectively instantaneous, in a single individual, who was instantly endowed with intellectual capacities far superior to those of others, transmitted to offspring and coming to predominate . . .*" He considers the language faculty to be "surprisingly perfect" – just as we might expect had it been designed by "a divine architect" (Chomsky 1996, p 30). Of course, Chomsky is no creationist. But otherwise supportive Darwinians have criticized him for suggesting an apparent miracle – forgetting, perhaps, that Chomsky's guiding principle is internal consistency, not conformity with the rest of science. "*In fact,*" writes Chomsky (2005, p 12) in justifying his "Great Leap Forward" narrative, "*it is hard to see what account of human evolution would not assume at least this much, in one or another form.*" Chomsky is informing us that language, as he defines it, cannot gradually have evolved.

Chomsky (2005, pp 11–12) explains: An elementary fact about the language faculty is that it is a system of discrete infinity. Any such system is based on a primitive operation that takes n objects already constructed, and constructs from them a new object: in the simplest case, the set of these n objects. Call that operation Merge. Either Merge or some equivalent is a minimal requirement. With Merge available, we instantly have an unbounded system of hierarchically structured expressions. The simplest account of the "Great Leap Forward" in the evolution of humans would be that the brain was rewired, perhaps by some slight mutation, to provide the operation Merge, at once laying a core part of the basis for what is found at that dramatic "moment" of human evolution...

Merge, then, is more than an empirical necessity: it is a logical one. It is the procedure central to any conceivable system of "discrete infinity." Merge is recursive: it means combining things, combining the combinations and combining these in turn – in principle to infinity. Chomsky suggests that a "slight mutation" may have allowed a single human ancestor to accomplish this cognitive operation for the first time. No matter how we imagine the physical brain, according to Chomsky, the transition to Merge is instantaneous, not gradual. This is because discrete infinity – "the infinite use of finite means" – either is or is not. What sense is there in trying to envisage "nearly discrete" objects being combined in "nearly infinite" ways? A moment's thought should remind us that when the objects to be arranged are subject to even limited blending, the range of combinatorial possibilities crashes to a limited set. In short, for Merge to work, the elements available for combination must be abstract digits, not concrete sounds or gestures. Combining a sob with a cry would not be an example of Merge. Neither would we call it Merge if a chimpanzee happened to combine, say, a bark with a scream (Crockford and Boesch 2005).

14.2 Analog Minds in a Digital World

One way to escape the conundrums inseparable from Chomsky's position – conundrums central to the recent explosion of debates on language origins and very well documented by Botha (2003) – might be to keep the essential idea, but to reverse the underlying philosophy. Humans have analog minds in a digital world. More accurately, just *a certain part* of our world is digital. We are at one with our primate cousins in being immersed in ordinary material and biological reality – Pinker's "analog world." But unlike them, we have woven for ourselves an additional environment that is digital through and through. This second environment that we all inhabit is sometimes referred to as the "cognitive niche" in nature, but the evolutionary psychologists who invented this expression (Tooby and DeVore 1987) did so in pursuit of their own particular agenda. Adherents of the "cognitive revolution" but attempting to marry a reluctant Chomsky to their own mentalist version of Darwin, they are committed to minimizing the intrinsically social, cultural, and institutional nature of the digital semantic representations made available to our brains. The expression "cognitive niche" may have explanatory value, but not if the purpose is to prioritize "nature" at the expense of what social anthropologists and archeologists term "symbolic culture." Contrary to Tooby and DeVore (1987), the "cognitive niche" does not exist anywhere "in nature." No one has ever found such a niche "in nature." As Tomasello (1999) points out, distinctively human cognition is inseparably bound up with the evolution of *culture*. The "cognitive niche," to be precise, exists only as an internal feature of *human symbolic culture*.

So what exactly is this thing called "symbolic culture?" Following the philosopher Searle (1996), let us begin by drawing a distinction between "brute facts" and "institutional facts." Birth, sex, and death are facts anyway, irrespective of what people think or believe. These, then, are brute facts. Phenomena such as legitimacy, marriage, and inheritance, however, *are facts only if people believe in them.* Suspend the belief and the facts correspondingly dissolve. Although institutional facts rest on human belief, that does not make them mere distortions or hallucinations. Take two five-pound banknotes. Their monetary equivalence to one ten pound note is not merely a subjective belief: it is an objective, indisputable fact. But now imagine a collapse of confidence in the currency. Suddenly, the various bits of paper are worthless – the former facts have dissolved.

Institutional facts are not necessarily dependent on verbal language: one can play chess, use an abacus, or change money without using language. The relevant digits are then the chess pieces, beads, or coins that function as markers in place of linguistic signs. Facts of this kind – the intricacies of the global currency system, for example – are patently nonphysical and nonbiological. We may think of them as internal features of an all-encompassing game of "let us pretend." Needless to say, institutional facts presuppose a brain with certain innate capacities, syntactical language being one possible manifestation of those capacities. But as Tomasello (2006) points out, explaining distinctively human cognition by invoking "language" is circular and unhelpful: it is precisely the language that we need to explain.

When people coin a new word – "spam" to mean "bulk e-mail" is a recent example – it becomes established as an institutional fact. Whether linguistic or nonlinguistic, facts of this kind develop ontogenetically out of the distinctively human capacity for mindreading, joint attention, and "let us pretend." The underlying formula is "Let X count for us as Y" (Searle 1996). Using a broomstick to signify "horse" is, in principle, no different from using "spam" to signify "bulk e-mail." When children learn the meanings of words, they succeed not thanks to a word-learning module dedicated exclusively to this task but by drawing on more fundamental and empirically verifiable features of social and nonsocial cognition (Bloom 2000). In particular, learning the meaning of a word presupposes the ability to correlate perspectives, grasping others' referential intentions. It is this imaginative ability – the ability to infer and share intentions and goals – that distinguishes human cognition so radically from that of apes (Tomasello et al. 2005).

Of course, it is always possible to term this critical ability as "language." This might seem helpful if you consider language to be an innate mechanism operating independently of the rest of cognition or of any institutional setting. Chomsky does hold this view, treating language as a faculty no different, in principle, from walking or stereoscopic vision. Pinker sets out from essentially the same position: language, he says, should be studied on the model of echolocation in bats or stereoscopic vision in primates. Distancing himself from Chomsky, however, Pinker insists that language is specifically designed for a social function – namely, communicating thoughts. Pinker explores how "words and rules" are continuously invented and reinvented for this purpose. In Searle's terms, the entities thereby produced are "institutional facts." Pinker calls them "inventions." But if they are indeed inventions, Chomsky's foundational assumption must be wrong. Language cannot be understood simply as a biological object. It operates on an entirely different level of organizational complexity from walking or stereoscopic vision – mechanisms, which, after all, do not require institutional arrangements in order to work.

What would language consist of in the absence of institutional facts? What meaning would language have to a child deprived of "words and rules?" According to Chomsky, the first human to be endowed with language used it to "articulate to itself its thoughts." As he explains (Chomsky 2002: 148): "*Actually you can use language even if you are the only person in the universe with language, and in fact it would even have adaptive advantage. If one person suddenly got the language faculty, that person would have great advantages; the person could think, could articulate to itself its thoughts, could plan, could sharpen, and develop thinking as we do in inner speech, which has a big effect on our lives. Inner speech is most of speech. Almost all the use of language is to oneself, and it can be useful for all kinds of purposes (it can also be harmful, as we all know): figure out what you are going to do, plan, clarify your thoughts, whatever. So if one organism just happens to gain a language capacity, it might have reproductive advantages, enormous ones. And if it happened to proliferate in a further generation, they all would have it.*"

But if communication were inessential, what need was there for any kind of external transmission via phonology? And if there was no such transmission, how could syntax have interfaced between Phonetic Form and Logical Form? After all, there would have been no Phonetic Form. Finally, if we accept that language can exist when stripped of this interface – when stripped of syntax as Chomsky (2005) himself defines it – in what sense does the residue deserve to be called "language"? Why not just call it "mentalese" or "thought"?

Pinker (1999: 287) concludes his book on "the ingredients of language": "*It is surely no coincidence that the species that invented numbers, ranks, kinship terms, life stages, legal and illegal acts, and scientific theories also invented grammatical sentences and regular past tense forms.*" Confusing correlation with causation, Pinker here treats the supposedly digital concepts intrinsic to human cognitive nature as responsible for the legalistic distinctions so characteristic of symbolic culture. Note, however, that the digital concepts he actually mentions here – whether linguistic or nonlinguistic – belong without exception not to individual cognition but to the realm of *agreements* and *institutions*. This is surely no coincidence – after all, we possess no evidence that language would be possible at all outside such institutional settings. Reversing Chomsky – and correspondingly reversing the whole idea of "digital minds in an analog world" – we can conclude that "doing things with words" (cf. Austin 1978 [1955]) is more than just activating a biological organ. To produce "speech acts" (Searle 1969) is to make moves in a nonbiological realm – a realm of facts whose existence depends entirely on collective belief.

14.3 The Evolution of Deep Social Mind

Evolutionary psychologists often refer to the evolution of "deep social mind" (Whiten 1999). By this, they mean the kind of mind that cannot be restricted to one individual. Deep social mind is recursive – mind as represented in other minds and as it represents to itself such representations. There is a subtle difference between this idea and the theory that thought is dependent on language. "*No support can be found for the view that words are necessary for thought*," writes Bloom (2000) in his exhaustive study of how children learn the meanings of words. But if words are not necessary for thought, in what sense can "language" be said to be necessary?

To appreciate why it is so unhelpful to privilege language as the source of uniquely human cognition, let us take the case of pointing (Tomasello 2006). Intentional pointing begins in children at about 14 months; chimpanzees never reach this stage. Pointing would seem to be a relatively simple activity, not requiring much in terms of computational hardware. Since it appears so simple, why do not chimps do it?

One answer might be that Universal Grammar is required – and chimps do not have Universal Grammar. But that would surely be absurd: pointing does not depend on any kind of grammar. It is true that whatever cognitive abilities enable pointing are necessary to enable talking as well, but that is no excuse for attributing evolutionary priority to language. Tomasello (2006: 520) concludes: "*To explain human cognitive uniqueness, many theorists invoke language. This contains an element of truth, because only humans use language and it is clearly important to, indeed constitutive of, uniquely human cognition in many ways. However, ... asking why only humans use language is like asking why only humans build skyscrapers, when the fact is that only humans, among primates, build freestanding shelters at all. And so for my money, at our current level of understanding, asking why apes do not have language may not be our most productive question. A much more productive question, and one that can currently lead us to much more interesting lines of empirical research, is asking the question why apes do not even point.*"

So why do not apes point? Tomasello offers a social explanation. Regardless of whatever mindreading abilities apes possess, in their natural environment, they lack any *motive* to correlate perspectives or share goals. They are, by nature, incorrigibly competitive Machiavellian social strategists. Only quite peculiarly cooperative creatures motivated to share *goals* and *intentions* could have any reason to point – or any reason to go yet further and invent "words and rules."

When fictional representations are given public and observable form – as in a game of "let us pretend" – language has started to evolve. Scaled up from the level of children's games and extended across society as a whole, "let us pretend" may generate complex systems of ritual and religion (Durkheim 1947 [1915], Knight 1998, 1999, 2000a,b, 2009b; Power 1999, 2000). The morally authoritative intangibles internal to a symbolic community – that is, to a domain of "institutional facts" – are always on some level digital. This has nothing to do with the supposedly digital genetic architecture of the human brain. The explanation is less mystical. It is simply that institutional facts depend entirely on social agreement – and you cannot reach agreement on a slippery slope. What would it mean if the Queen in her official capacity were to "open Parliament," but only slightly? Or if a couple who had just made their wedding vows were pronounced man and wife – but only "more or less"?

What applies to royal and religious edicts also equally applies to speech acts, in general. Chomsky notwithstanding, semantic distinctions are social and institutional, not individual or innate. Take the classic case of basic color terms. All humans, in all cultures, discriminate perceptually among an immense variety of different hues. But while actual colors can be directly perceived – and while innate biases play a key role in determining which regions of the spectrum are picked out – it need hardly be stressed that digital color *categories* operate on a quite different level. Knowing that the spectrum is segmented into two, three, or some other limited set of "colors" – "the seven colors of the rainbow," for example – requires access to the relevant institutional conventions. Basic color terms – English "red" and "green," for example – map directly to these simplified abstractions; they do not and could not possibly map to the vastly more complex features of the human visual system as such (Davidoff et al. 1999; Davidoff 2001; Steels and Belpaeme 2005). To summarize: by definition, anything perceptible can be evaluated and identified through direct sensory input – in other words, on the basis of innate perceptual mechanisms. But institutional intangibles are inaccessible to the senses. Being invisible, intangible and in a fundamental sense unreal, they can be narrowed down and agreed upon only through a process in which abstract possibilities are successively eliminated. "Discrete infinity" captures the recursive principle involved.

The sound system of a language – its phonology – is prototypically digital. It is no more possible to compromise between the *t* and the *d* of *tin* versus *din* than to compromise between 11:59 and 12.00 on the face of a digital clock. Of course, categorical perception is common enough in nature. But the meaningless contrastive phonemes of human language comprise only one digital level out of the two that are essential if meanings are to be conveyed at all. Combining and recombining phonemes – "phonological syntax," as it is called by ornithologists (e.g., Marler 1998) who study the digital phenomenon in songbirds – would be informationally irrelevant if it did not interface with a second digital level, which is the one necessary if *semantic* meanings are to be specified. No animal species has access to this second level of digital structure. It would, therefore, be inconceivable and, in principle, useless anyway for an animal to make use of syntactical operations – whether Merge or anything else – in order to interface between the two digital levels. The explanation is that animals inhabit just their own biological world, and therefore do not have access to the extra digital level. It is the nature and evolution of the entire second level – the level of symbolic culture – that has proved so difficult a puzzle. Explaining "the Great Leap Forward" as an outcome of "Merge" is a parsimonious solution (Chomsky 2005), but only in the sense that explaining it as an outcome of divine intervention might seem persuasive in terms of parsimony although less so in terms of relevance or testability.

14.4 A Darwinian Solution

The alternative is to conceptualize the language capacity as one remarkable manifestation of a "play capacity" continuous with its primate counterparts, but let loose among humans in a manner not open to other animals (cf. Jespersen 1922; Huizinga

1970 [1949], Knight 2000b). The development of play and the development of language in children are widely recognized as isomorphic. Pretend-play and language have the same critical period, the same features of intersubjectivity and joint attention, the same triadic (do you see what I see?) referential structure, and the same cognitive expressivity and independence of external stimuli. It is unlikely that these parallels are a pure coincidence (Bruner et al. 1976; Trevarthen 1979; Tomasello 2003).

"Digital infinity" corresponds to what developmental psychologists might recognize as a children's game – in this case, "let us play infinite trust." Take any patent fiction and let us run with it and see where it leads. Metaphorical usage is an example of this. A metaphor "is, literally, a false statement" (Davidson 1979). React on a literal level – as an autistic person might do – and the signaler is rebuffed, denied the freedom to "lie." But most of us do not react in this unimaginative and unsympathetic way: by accepting the patent fiction and sharing in it, we can construct it as truth on a higher level – truth for "our own" joint purposes of conceptualization and communication. A red shirt is not necessarily "blood" – but identifying it that way might pick out one garment from a limited set. As literal falsehoods become gradually conventionalized, one possible trajectory is that they crystallize out as "dead metaphors" – familiar lexical items whose origins in vivid metaphor have become forgotten. Grammatical markers and associated constructions are historical outcomes of essentially similar processes that are now well understood (Meillet 1903; Lakoff and Johnson 1980; Heine et al. 1991; Gentner et al. 2001; for the best recent overview, see Deutscher 2005).

If all this is accepted, it follows that for words and rules to evolve, humans must trust one another sufficiently to find value in patent falsehoods. It is for social reasons that nonhuman primates are unable to do this. Chimpanzees, like other primates, have solid grounds for attempting to read one another's minds. But like devious spies, they have no reason to allow rivals to reciprocally read back into *their own* minds. Sometimes, a human may direct a pointing gesture or other cooperative signal at a captive chimp. Typically, the recipient gets confused, as if unable to grasp the helpful intent (Hare and Tomasello 2004). In the wild, resistance to deception would prevent such signals from being accepted on trust. Among other consequences, such insistence on reliability blocks metaphor – and in so doing blocks the elaboration of abstract analogical thought of the kind so characteristic of humans (Lakoff and Johnson 1980). If animals do not talk, therefore, it is not because they lack the requisite digital module installed inside their brains. The explanation is more simple: they live in a Darwinian world. By "a Darwinian world," I mean a competitive world that is *not* subject to group-level moral regulation (Knight 2008). Animals value signals to the extent that they are dependable, hence hard to fake. Body language alone has this property. Reflecting cognitive states, intentionally produced symbols are "head language." If they are not found in nature, it is because such things are evidently false.

The social factors that allow metaphorical usage in humans are equally the ones permitting digital concepts to emerge. Hard-to-fake indices such as laughs, sobs, cries, and so forth must be evaluated for intrinsic quality on an analog scale. It is not theoretically possible to evaluate abstract digits in this way. Probe a digit for its quality and it ceases to function as a digit. Analog evaluation, by the same token, rules out the possibility of Merge. But now let us examine the reverse situation. Regardless of innate cognitive architecture, the contrastive possible intentions behind a communicative fiction are *in principle* immune to analog evaluation. "Discrete infinity" becomes unavoidable in this context because linguistic signs are "honest fakes" – literal irrelevancies and falsehoods, significant only as cues to the intentions underlying them. Since communicative intentions are intangibles, deciding between them is digital by reason of logical necessity, not because the brain or any part of it is innately digital.

Durkheim (1947, p 421) long ago observed, "*Animals know only one world, the one which they perceive by experience, internal as well as external. Men alone have the faculty of conceiving the ideal, of adding something to the real. Now where does this singular privilege come from?*"

Maynard Smith and Szathmáry (1995) offered a bold Darwinian answer to Durkheim's question, citing Rousseau and viewing the puzzle of language origins as inseparable from the wider problem of explaining the emergence of life governed by morally binding contracts. Their "major transitions" paradigm is ambitious and conceptually unifying, assuming no unbridgeable chasm between natural and social science. The same applies to the paradigm being developed by Steels and his colleagues (Steels et al. 2002; Steels 2006, 2009), who use robots to show how shared lexicons and grammars – patterns far too complex to be installed in advance in each individual brain – spontaneously self-organize through the processes of learning, recruitment, social coordination, and cumulative grammaticalization. Steels (2009) emphasizes that language evolves spontaneously but only under highly unusual conditions – conditions of mutual trust and cooperation that are far removed from Darwinism as usually understood. By maintaining continuity with primate analog minds while introducing novel social factors – factors such as collective enforcement of "the rule of law" (Knight 2008) – we can continue to apply basic principles of Darwinian behavioral ecology to account for the emergence of distinctively human cognition and communication.

"Analog minds in a digital world" is fully compatible with Darwinian evolutionary theory. "Digital minds in an analog world" is not compatible at all. Installation of an innate digital mind – whether instantaneous or gradual – is a *deus ex machina* with nothing Darwinian about it. A model of language evolution, to qualify as scientific, cannot invent fundamental axioms as it goes along. It cannot invoke currently unknown physical or other natural laws. It should be framed within a coherent, well-tried body of theory; it should generate predictions that are testable in the light of appropriate empirical data; and it should enable us to relate hitherto unrelated disciplinary fields. While the *deus ex machina* approach rejects the accumulated achievements of social science, the play/mindreading/joint attention paradigm (Tomasello 1996, 1999, 2003, 2006; Tomasello et al. 2005) has the potential to link the natural and social sciences in "a theory of everything" – a testable theory to explain the origin of language and symbolic culture as a whole.

14.5 The Female Cosmetic Coalitions Model

The specific hypothesis I favor involves menstruation. I am well aware that this topic is usually avoided, being hedged around with our own culture's unspoken avoidances and taboos.

Primate sexual signals represent compromise outcomes of varying degrees of cooperation and conflict between the sexes. Chimpanzee oestrous swellings, for example, are graded signals representing a compromise between (a) the male demand for reliable evidence of ovulation and (b) the requirement of females to withhold accurate information in defense of their own sexual choice (Nunn 1998). In this battle of the sexes, human females apparently achieved victory in at least one respect: ovulation has become effectively concealed. However, concealment of ovulation leaves menstruation salient as the one remaining external evidence of a female's (imminent) fertility (Power and Aiello 1997). In extant hunter-gatherer societies, neither sex is oblivious to the potential philandering opportunities such as the information represents. The Mbuti "*consider that any couple that really wants children should 'sleep with the moon*'" (Turnbull 1993 [1961], p 169). Hadza informants generally view conception as occurring immediately after menstruation (Marlowe 2004). "*A woman is considered most fertile during menstruation*" writes Lewis (2002: 109) of the Mbendjele. "*Sexual intercourse at this time is widely held to be a sure way of ensuring pregnancy begins. Menstrual blood is the symbol par excellence of human fertility.*" Adequate statistics are lacking, but on theoretical grounds, I would predict that hunter-gatherers everywhere will be found to hold similar views.

Imagine a world in which selfish gene Darwinism was the only law. Menstruation would then act as a starting gun for sexual conflict. Dominant males in pursuance of individual fitness would be tempted to abandon their current partners – already pregnant or nursing – to compete for access to new partners marked out as imminently fertile by their blood. In real life, hunter-gatherers surround the blood with elaborate taboos, as if the aim were precisely to keep philanderers at bay (Knight 1991, 1996). Across Africa, the young menstruant's senior female relatives respond to her condition as if to an immediate threat. Mobilizing the entire community, they subject her to strict supervision and control, celebrating her fertility but at the same time bonding tightly with her and, above all, controlling male access to her. Although menstruation is a biological signal, its salience and significance varying on an analog scale, the logic at work here excludes intermediate states. To acknowledge *any* blood is to publicize the danger represented triggering the full collective response.

Contrary to many western misconceptions, menstrual taboos are not necessarily the evidence of sexist oppression under patriarchal rule. African hunter-gatherers are not noticeably male-dominated. In fact, women typically have much solidarity and power (Turnbull 1993 [1961], Lewis 2002, 2009). Although cultural variability is great, the ubiquity and evident antiquity of menstrual taboos can be explained on a Darwinian basis – as the outcome of an evolutionarily stable strategy pursued

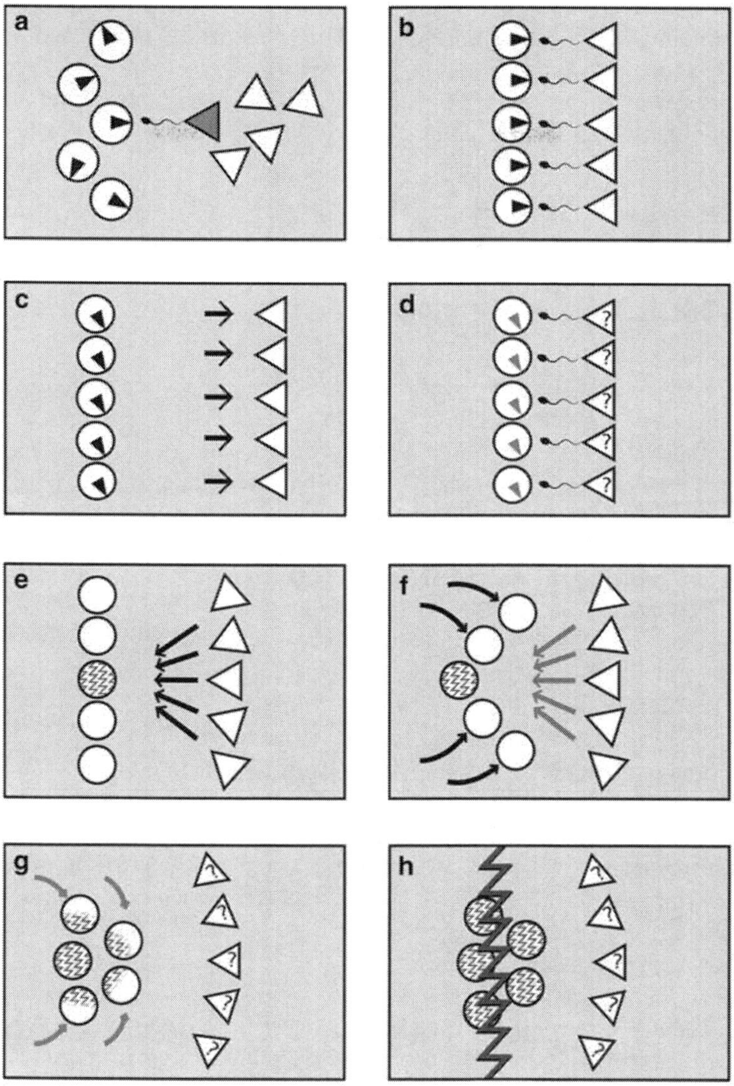

Fig. 14.1 (**a**) Females compete for good genes, staggering signs of ovulation. (**b**) Females compete for mating effort, synchronizing signs of ovulation. (**c**) Males abandon females once ovulation has passed. (**d**) Females counter this problem by concealing ovulation and extending receptivity. (**e**) Menstruation now attracts disproportionate male attention. (**f**) Coalition members respond to this threat by controlling male access to the (imminently) fertile female. (**g**) To prevent males from picking and choosing between them, members of the coalition join forces and "paint up." (**h**) "*The rule of law*" is established as an institutional fact.

for millennia by women in pursuit of their own reproductive interests. Child-burdened mothers require investment from males, not abandonment, harassment, or violence. Socially approved posture, clothing, scarification, and cosmetics are characteristically human ways of "covering up" – concealing, modifying, and scrambling those signals which, if left biologically naked and exposed, might preclude the possibility of rule-governed behavior of any kind. Figure 14.1 presents a schematic outline of the Female Cosmetic Coalitions Model, depicting how coalitionary resistance to philandering in this scenario culminates in the rule of law.

14.6 The Human Revolution

Until recently, most archeologists attributed the emergence of language to a "human revolution" dated to some 45,000 years ago. When *Homo sapiens* began emerging in Africa during the Middle Stone Age, according to this model, cognition and behavior was still "premodern" or "archaic." Only when our species began migrating into Europe, displacing the resident Neanderthals and triggering the "Upper Paleolithic Revolution," did "modern" language and symbolic culture emerge (Mellars and Stringer 1989; Diamond 1992; Klein 1995, 2000; Tattersall 1995).

Over the past decade, it has become apparent that this scenario was an artifact resulting from a Eurocentric sampling of the fossil and archeological records (Mellars et al. 2007). Recent studies by archeologists working in Africa have shown that almost all the cultural innovations dated to around 45,000 years ago in Europe can be found at much earlier dates at one or another site in Africa. Blade and microlithic technology, bone tools, logistic hunting of large game animals, long-distance exchange networks – these and other signs of modern cognition and behavior do not appear suddenly in one package as predicted by the Upper Paleolithic "human revolution" theory. They are found at African sites widely separated in space and time, indicating not a single leap but a much more complex, uneven but broadly cumulative process of biological, cultural, and historical change (McBrearty and Brooks 2000; McBrearty 2007).

Such evidence casts doubt on the notion of a single mutation installing the language faculty in one step and suddenly inaugurating society and history (Bickerton 1990; Chomsky 2005; Klein 1995, 2000). But to reject that particular model of the "human revolution" is not necessarily to reject the basic idea. I prefer to update it in the light of recent archeological evidence, extending the timescale and relocating the whole process within the African Middle Stone Age (Knight et al. 1995; Watts 1999; Mellars et al. 2007; Knight 2009a,b). The human revolution would then take its place in the history of life on earth as a major transition on the model of the emergence of multicellular complexity, the origin of chromosomes, or the first appearance of life itself. In their "major transitions" paradigm, Maynard Smith and Szathmáry (1995, pp 255–309) present the coevolution of symbolic ritual, moral regulation, and language as an example of how Darwinan natural selection may culminate in a revolutionary transition. My aim in this chapter has been to flesh out a story along these lines.

References

Arnauld A, Lancelot C (1972) [1660] Grammaire générale et raisonnée, ou la grammaire de Port-Royal. Réimpression des editions de Paris, 1660 et 1662. Slatkinem, Genève. Quotation translated by Deutscher (2005: 1)
Austin JL (1978) [1955] How to do things with words. Oxford University Press, Oxford
Bickerton D (1990) Language and species. University of Chicago Press, Chicago
Bloom P (2000) How children learn the meanings of words. MIT Press, Cambridge, MA
Botha RP (2003) Unravelling the evolution of language. Elsevier, Oxford
Bruner JS, Jolly A, Sylva K (1976) Play: its role in development and evolution. Basic Books, New York
Chomsky N (1996) Powers and prospects: reflections on human nature and the social order. Pluto Press, London
Chomsky N (2002) On nature and language. Cambridge University Press, Cambridge
Chomsky N (2005) Three factors in language design. Linguist Inq 36:1–22
Christiansen MH, Kirby S (2003) Language evolution: the hardest problem in science? In: Christiansen MH, Kirby S (eds) Language evolution. Oxford University Press, Oxford, pp 1–15
Crockford C, Boesch C (2005) Call combinations in wild chimpanzees. Behaviour 142:397–421
Davidoff J (2001) Language and perceptual categorisation. Trends Cogn Sci 5:382–387
Davidoff J, Davies I, Roberson D (1999) Colour categories in a stone-age tribe. Nature 398: 203–204
Davidson D (1979) What metaphors mean. In: Sacks S (ed) On metaphor. University of Chicago Press, Chicago, pp 29–45
Descartes R (1991) [1640] The philosophical writings of descartes. Vol. III: The Correspondence. Translated by Cottingham JG, Stoothoff R, Murdoch D, Kenny A. Cambridge University Press, Cambridge
Deutscher G (2005) The unfolding of language: an evolutionary tour of mankind's greatest invention. Random House, London
Diamond J (1992) The rise and fall of the third chimpanzee. Vintage, London
Durkheim É (1947) [1915] The elementary forms of the religious life: a study in religious sociology. Translated by Swain JW. Free Press, Glencoe, IL
Gentner D, Holyoak KJ, Kokinov BN (2001) The analogical mind: perspectives from cognitive science. MIT Press, Cambridge, MA
Hare B, Tomasello M (2004) Chimpanzees are more skilful in competitive than in cooperative cognitive tasks. Anim Behav 68:571–581
Heine B, Claudi U, Hünnemeyer F (1991) Grammaticalization: a conceptual framework. University of Chicago Press, Chicago
Huizinga J (1970) [1949] Homo ludens. a study of the play element in culture. Granada, London
Jespersen O (1922) Language: its nature, development and origin. Allen & Unwin, London
Klein RG (1995) Anatomy, behaviour, and modern human origins. J World Prehist 9: 167–198
Klein RG (2000) Archaeology and the evolution of human behavior. Evol Anthropol 9:17–36
Knight C (1991) Blood relations: menstruation and the origins of culture. Yale University Press, New Haven
Knight C (1996) Menstruation. In: Barnard A, Spencer J (eds) Encyclopedia of social and cultural anthropology. Routledge, London, pp 363–364
Knight C (1998) Ritual/speech coevolution: a solution to the problem of deception. In: Hurford JR, Studdert-Kennedy M, Knight C (eds) Approaches to the evolution of language: social and cognitive bases. Cambridge University Press, Cambridge, pp 68–91
Knight C (1999) Sex and language as pretend-play. In: Dunbar RIM, Knight C, Power C (eds) The evolution of culture. Edinburgh University Press, Edinburgh, pp 228–247

Knight C (2000a) Introduction: the evolution of cooperative communication. In: Knight C, Studdert-Kennedy M, Hurford JR (eds) The evolutionary emergence of language: social function and the origins of linguistic form. Cambridge University Press, Cambridge, pp 19–26

Knight C (2000b) Play as precursor of phonology and syntax. In: Knight C, Studdert-Kennedy M, Hurford JR (eds) The evolutionary emergence of language: social function and the origins of linguistic form. Cambridge University Press, Cambridge, pp 99–119

Knight C (2008) Language co-evolved with the rule of law. Mind Soc 7:109–128

Knight C (2009a) Introduction: perspectives on the evolution of language in Africa. In: Botha R, Knight C (eds) The cradle of language. Oxford University Press, Oxford, pp 1–5

Knight C (2009b) Language, ochre and the rule of law. In: Botha R, Knight C (eds) The cradle of language. Oxford University Press, Oxford, pp 280–303

Knight C, Power C, Watts I (1995) The human symbolic revolution: a Darwinian account. Camb Archaeol J 5:75–114

Lakoff G, Johnson M (1980) Metaphors we live by. University of Chicago Press, Chicago

Lewis J (2002) Forest hunter-gatherers and their world: a study of the Mbendjele Yaka Pygmies of Congo-Brazzaville and their secular and religious activities and representations. PhD thesis, University of London

Lewis J (2009) As well as words: Congo Pygmy hunting, mimicry, and play. In: Botha R, Knight C (eds) The cradle of language. Oxford University Press, Oxford, pp 236–256

Marler P (1998) Animal communication and human language. In: Jablonski NG, Aiello LC (eds) The origin and diversification of language. Wattis Symposium Series in Anthropology. Memoirs of the California Academy of Sciences, No. 24. California Academy of Sciences, San Francisco, pp 1–19

Marlowe FW (2004) Is human ovulation concealed? Evidence from conception beliefs in a hunter-gatherer society. Arch Sex Behav 33:427–432

Maynard Smith J, Szathmáry E (1995) The major transitions in evolution. WH Freeman, Oxford

McBrearty S (2007) Down with the revolution. In: Mellars P, Boyle K, Bar-Yosef O, Stringer C (eds) Rethinking the human revolution: new behavioural and biological perspectives on the origin and dipersal of modern humans. McDonald Institute for Archaeological Research, Cambridge, pp 133–151

McBrearty S, Brooks AS (2000) The revolution that wasn't: a new interpretation of the origin of modern human behaviour. J Hum Evol 39:453–563

Meillet A (1903) Introduction à l'étude comparative des langues indo-européennes. Hachette, Paris

Mellars PA, Stringer C (1989) The human revolution: behavioural and biological perspectives on the origins of modern humans. Edinburgh University Press, Edinburgh

Mellars PA, Boyle K, Bar-Yosef O, Stringer C (2007) Rethinking the human revolution: new behavioural and biological perspectives on the origin and dispersal of modern humans. McDonald Institute for Archaeological Research, Cambridge

Nietzsche F (1968) The will to power. Translated by Kaufmann W, Hollingdale RJ, Vintage Books, New York

Nunn CL (1998) The evolution of exaggerated sexual swellings in primates and the graded signal hypothesis. Anim Behav 58:229–246

Piattelli-Palmarini M (1980) Language and learning: the debate between Jean Piaget and Noam Chomsky. MIT Press, Cambridge, MA

Pinker S (1999) Words and rules: the ingredients of language. Weidenfeld & Nicolson, London

Power C (1999) Beauty magic: the origins of art. In: Dunbar RIM, Knight C, Power C (eds) The evolution of culture. Edinburgh University Press, Edinburgh, pp 92–112

Power C (2000) Secret language use at female initiation: bounding gossiping communities. In: Knight C, Studdert-Kennedy M, Hurford JR (eds) The evolutionary emergence of language: social function and the origins of linguistic form. Cambridge University Press, Cambridge, pp 81–98

Power C, Aiello LC (1997) Female proto-symbolic strategies. In: Hager LD (ed) Women in human evolution. Routledge, New York, pp 153–171
Searle JR (1969) Speech acts. An essay in the philosophy of language. Cambridge University Press, Cambridge
Searle JR (1996) The construction of social reality. Penguin, London
Steels L (2006) Experiments on the emergence of human communication. Trends Cogn Sci 10:347–349
Steels L (2009) Is sociality a crucial prerequisite for the emergence of language? In: Botha R, Knight C (eds) The prehistory of language. Oxford University Press, Oxford, pp 36–57
Steels L, Belpaeme T (2005) Coordinating perceptually grounded categories through language: a case study for colour. Behav Brain Sci 28:469–489
Steels L, Kaplan F, McIntyre A, van Looveren J (2002) Crucial factors in the origins of word-meaning. In: Wray A (ed) The transition to language. Oxford University Press, Oxford, pp 252–271
Tattersall I (1995) The fossil trail: how we know what we think we know about human evolution. Oxford University Press, New York
Tomasello M (1996) The cultural roots of language. In: Velichkovsky BM, Rumbaugh DM (eds) Communicating meaning: the evolution and development of language. Erlbaum, Mahwah, NJ, pp 275–307
Tomasello M (1999) The cultural origins of human cognition. Harvard University Press, Cambridge, MA
Tomasello M (2003) Constructing a language: a usage-based theory of language acquisition. Harvard University Press, Cambridge, MA
Tomasello M (2006) Why don't apes point? In: Enfield NJ, Levinson SC (eds) Roots of human sociality: culture, cognition and interaction. Berg, Oxford, pp 506–524
Tomasello M, Carpenter M, Call J, Behne T, Moll H (2005) Understanding and sharing intentions: the origins of cultural cognition. Behav Brain Sci 28:675–735
Tooby J, DeVore I (1987) The reconstruction of hominid behavioral evolution through strategic modeling. In: Kinzey WG (ed) The evolution of human behavior: primate models. SUNY Press, Albany, pp 183–237
Trevarthen C (1979) Communication and cooperation in early infancy: a description of primary intersubjectivity. In: Bullowa MM (ed) Before speech: the beginning of interpersonal communication. Cambridge University Press, New York, pp 321–347
Turnbull C (1993) [1961] The forest people. Random House, London
Watts I (1999) The origin of symbolic culture. In: Dunbar RIM, Knight C, Power C (eds) The evolution of culture. Edinburgh University Press, Edinburgh, pp 113–146
Whiten A (1999) The evolution of deep social mind in humans. In: Corballis MC, Lea SEG (eds) The descent of mind: psychological perspectives on hominid evolution. Oxford University Press, Oxford, pp 173–193

Chapter 15
Brain and Behaviour in Primate Evolution

R.I.M. Dunbar

Abstract Primates have unusually large brains for body size compared to all other vertebrates. Over the years, a number of explanations have been offered for this, leading to some confusion. I use a systems approach to make sense of these suggestions, and this suggests that some are constraints (energetic or neural development explanations), others consequences (generated by windows of opportunity), but only the social hypotheses constitute the real selective pressure for the evolution of brains. The social hypotheses come in two current forms (bonding social groups vs. social learning of foraging skills) that differ in whether predation or food-finding are assumed to be the rate-limiting factor in primate survival. While the standard form of the social brain hypothesis in primates is a quantitative relationship between social group size and brain size, comparative analyses for other mammal and bird taxa reveal that it takes a purely qualitative (i.e., categorical) form in all nonprimates examined so far: species with pairbonded (i.e., monogamous) mating systems have larger brains than all others. I suggest that this difference is due to the fact that anthropoid primates developed bonded social systems early in their evolutionary history. Finally, I consider briefly the implications of these findings for human evolutionary history.

15.1 Introduction

Some three decades ago, Jerison (1973) observed that primates have significantly larger brains for body size than all other vertebrate orders. While considerable progress has been made in understanding brain evolution since then, three classes of explanations still have currency as explanations for the evolution of large brains

R.I.M. Dunbar
Institute of Cognitive and Evolutionary Anthropology, University of Oxford, Oxford, England
e-mail: robin.dunbar@anthro.ox.ac.uk

P.M. Kappeler and J.B. Silk (eds.), *Mind the Gap*,
DOI 10.1007/978-3-642-02725-3_15, © Springer-Verlag Berlin Heidelberg 2010

Table 15.1 Hypotheses that have been proposed to explain the evolution of brain size in primates

Hypothesis	Explanatory focus	Key source reference
Ecology	Frugivory (ephemeral food sources)	Clutton-Brock and Harvey (1980)
	Range size (mental maps)	Clutton-Brock and Harvey (1980)
	Extractive foraging (embedded foods)	Gibson (1986)
	Innovations/tool use (trial-and-error learning)	Reader and Laland (2002)
Social	Innovations (social transmission)	Reader and Laland (2002)
	Machiavellian behaviour	Byrne and Whiten (1988)
	Social bonding	Dunbar (1992, 1998), Barton (1996)
Developmental	Body size (energetic constraints)	Martin (1981)
	Structural constraints	Finlay and Darlington (1995)
	Longevity (energetic constraints)	Barrickman et al. (2008)

among the primates: ecological hypotheses, social hypotheses, and life history hypotheses (Table 15.1). Each of these comes in several alternative versions that can be differentiated by the proximate function that large brains are assumed to mediate. In other words, the differences lie in the mechanisms whereby fitness is maximised, and hence in which components of survival and successful reproduction are rate-limiting.

The ecological hypotheses presume that survival is the principal problem that animals face, and that food acquisition is the rate-limiting process. Traditionally, this has always been interpreted in terms of food-finding (either in the form of a mental mapping issue or in the form of the cognitive demands of frugivory) or of extractive foraging (Parker and Gibson 1979; Clutton-Brock and Harvey 1980; Gibson 1986). More recently, the role of innovations and tool use has been given more prominence, and has received considerable empirical support (Reader and Laland 2002). In addition, the implications of this for extinction risk have not gone unnoticed, especially in the bird literature (Sol et al. 2002, 2005; Shultz et al. 2005).

In contrast, although the social hypotheses have also emphasised the significance of survival, their main contention is that survival is maximised by social means. The key difference between the social hypotheses and the ecological hypotheses is, thus, how these ecological goals are achieved. The former argue that they are achieved socially, the latter that it is purely a consequence of individual trial-and-error learning. Ecologically speaking, survival can be maximised either by solving the problem of food-finding or by minimising predation, and this generates two alternative versions of the social hypothesis. One (formally known as the social brain hypothesis) is that, maintaining coherent groups is the key mechanism for minimising predation risk and that it is the cognitive demands imposed by this that lies at the heart of primate brain evolution (Dunbar 1992, 1998). The other argues that the constraint lies in the capacity to engage in social forms of learning when acquiring new foraging skills (Reader and Laland 2002). Thus, these alternative versions of the social hypothesis differ in whether the rate-limiting process is mortality from predators or mortality from starvation.

Finally, the life history hypotheses emphasise the fact that growing and maintaining large brains is energetically expensive (Aiello and Wheeler 1995), and hence that evolving large brains is, in some way, a consequence (or a cause) of an extended life history. Again, at least two versions can be discerned. One is that large brains correlate with body size in such a way as to offer savings of scale that allow the mother to switch spare energy from herself to fetal brain growth (Martin 1981; Hofman 1983; Armstrong and Bergeron 1985; Isler and van Schaik 2006). In the limiting case, large brains are simply a (nonadaptive) by-product of large body size and require no evolutionary explanation. The other notes that large brains are associated with an extended life history (longer life span, longer gestation, etc.), and argues that large brains are a constraint on the need to evolve a slowed-down life history (Barrickman et al. 2008). A related series of analyses (Finlay and Darlington 1995; Finlay et al. 2001) have emphasised ontogenetic aspects of brain growth: in this case, final brain size depends on the number of cell cycles that can be achieved during the period of brain growth. Since the rate of neurogenesis is constant, the only way to evolve a larger brain is to slow down the rate of development to allow time for more neurogenic cycles. Life history explanations tend not to identify (or even consider) functional explanations as to why large brains might be selectively advantageous. If anything, large brains are often viewed as an accidental by-product of changes in life history.

Evidence has been adduced for all of these hypotheses, but in most cases individual hypotheses have been tested without considering the implications of other hypotheses. General principles of parsimony make this rather confusing situation particularly puzzling: we would not normally expect so many radically different selection pressures to influence the same phenomenon equally. While some previous analyses (e.g. Dunbar 1992; Deaner et al. 2000) have attempted to test between alternative hypotheses for large brain size in primates, they have invariably done so using a series of bivariate analyses (one for each possible independent variable) without forcing the hypotheses into direct competition. Since many of these variables are confounded – not least because some may be emergent properties of others – it was, not too surprisingly, difficult to distinguish between alternative hypotheses. In fact, without an attempt to integrate them into a single explanatory framework, little real progress is likely to be made. One solution to this is the "critical tests approach" advocated by van Schaik (1983): this seeks to put alternative hypotheses into direct contest with each other in such a way that the data are forced to support one and only one hypothesis (e.g. see van Schaik and Dunbar 1990; Dunbar et al. 2002; Calhim et al. 2006). However, a "critical tests approach" assumes that all competing explanations are at the same explanatory level. This need not always be so: some may be functional explanations, while others are concerned more with mechanisms (e.g., constraints). A systems approach will usually be a better approach in such cases. In particular, path analysis offers an alternative way of testing more complex relationships between a set of variables, especially when the variables themselves form part of a nested set of explanations, not all of which are in competition with each other.

15.2 A Systems Approach to Brain Evolution

Given the number of factors that have been advocated as possible (or, indeed, sole) causes of brain evolution, it is perhaps surprising that no serious attempts have been made to integrate all these into a single analytical model in such a way as to test between them and establish why it is that so many factors seem to correlate with brain size. The fact that many variables correlate with brain size ought to prompt us to think systemically in terms of causes, constraints, and consequences. In the interests of parsimony, one ought to assume that there is only one core selection pressure that is principally responsible for the evolution of a given trait (unless and until demonstrated otherwise). An important distinction, therefore, needs to be drawn between (1) the selection pressure that gives rise to a trait (and without whose presence the trait would simply not evolve) and (2) constraints (that have to be resolved in order to allow the trait in question to evolve), (3) secondary selection factors (that push the trait in the same direction, but which of themselves are insufficiently powerful to give rise to the trait in the face of the costs that militate against it), (4) tertiary selection pressures (that impose selection on individual subunits, but not the brain as a whole), and (5) emergent properties ("windows of opportunity" that arise once the trait is in place). All of these are well-known aspects of evolutionary systems, but do not seem to have been kept as well separated as they should have been in discussions of brain evolution.

In considering the factors acting on brain size, it is important to bear in mind both the costs and the benefits. Brain tissue is exceptionally costly (Mink et al. 1981; Aiello and Wheeler 1995; Kaufman 2003; Isler and van Schaik 2006; Karbowski 2007), and any benefits must exceed the costs of evolving and maintaining additional brain tissue. Since the costs are high in this case, the benefits must be as well. Thus, not only do we need to understand which variables are costs (and, thus, act as constraints on brain evolution) and which benefits, but we also need to determine which of the many potential benefits is the core selection factor that has driven brain evolution and which are by-product benefits (emergent properties that provide a supplementary benefit to having a large brain once you have evolved one). This is especially important in the context of primate brain evolution, since it seems often to be assumed that the forces selecting for large brains are constant across taxonomic groups (i.e., one explanation fits all). However, this overlooks Jerison's (1973) original point that primates have significantly larger brains than all other taxa. In other words, we have a size issue that itself needs explaining.

One way of approaching this problem would be to cost out the relative costs and benefits that derive from different sources and calculate the rate of change in payoff against changing brain volume. The rate of change in costs can be calculated fairly easily, since it is, in principle, simply the added energetic costs of a unit of neural tissue. However, calculating the fitness gains from different possible benefits is much less straightforward. Indeed, this may be especially problematic if some of the benefits arise through multilevel selection rather than through direct impact on individual fitness: calculating these indirect fitness components is much more difficult and almost never done in behavioural ecology, except in the special case of kin selection.

Multilevel selection is particularly important in the case of the social brain hypothesis because this explicitly claims that the benefits accrue through reduced predation as a result of more cohesively bonded groups. The proximate goal driving brain size evolution (and hence cognitive ability) in this case is not reduced predation as such (though this may play some role even so: Shultz and Dunbar 2006), but how well bonded the social group itself is. The argument is explicitly a two-step process: large brains allow larger groups to be better bonded, and better bonded groups are more effective at reducing predation (or solving any other ecological survival problem). As with cooperative hunting, group members have to be able to work effectively together *if* they are to gain the benefit that living in a large group offers. It is unlikely that this takes the form of deliberate cooperation (indeed, it is even questionable whether this occurs among cooperative hunters); rather, it probably takes the form of mechanisms that allow group members to ameliorate the stresses that close physical proximity invariably imposes. These include both direct and indirect costs. Direct costs include the impact of dominance on the foraging efficiency of low-ranking animals as well as the temporary infertility that even mild harassment can create by disrupting low-ranking females' menstrual cycles (Dunbar 1980; Abbott et al. 1986). Indirect costs include the familiar ones of longer-day journeys and higher foraging costs (e.g., Lehmann et al. 2008a,b).

Emergent properties (secondary or tertiary benefits) often create an analytical problem in evolutionary explanations: in many cases, these are secondary benefits that are easily confused with the primary benefits that have been responsible for the trait's evolution. Thus, it might be that social coherence was the selection pressure that led to the evolution of large brains; but, once large brains were in place, their generic computational capacities could be used for any number of other purposes (including the kind of smart foraging identified by the innovations hypothesis). In this case, however, we at least have an alternative hypothesis to create a critical test, since the converse causal sequence is, in principle, equally plausible (i.e., that the social benefits of bonding are an emergent property of having a large brain that originally evolved for direct individual-level ecological problem-solving). One way to approach this question is to undertake a phylogenetic analysis in which the two traits and brain size are mapped onto a phylogenetic tree so that the order in which changes in all the traits occurred can be mapped. The aim is to search for correlated evolution between brain size and one trait (the principal selection factor) and lagged evolution between these two and the second trait (the emergent property). It is now possible to do this using statistical methods such as Pagel's (1997) DISCRETE.

A full such analysis has yet to be undertaken, but Pérez-Barbería et al. (2007) have demonstrated that among primates, in contrast to carnivores or ungulates, changes in brain size and sociality are so tightly correlated that there are almost no lag or indirect effects. Although cognitive skills as such were not included in this analysis, it seems safe to conclude on this evidence that it is very unlikely that sociality is a mere by-product of evolving a large brain; rather, one variable must be acting as a constraint on the other and the two are locked in a tight coevolutionary cycle – you cannot change one without forcing a change in the other at the same time. Given this, the most likely interpretation is that brain size evolution is driven by the demands of sociality, and that any other benefits in terms of foraging are a by-product of the added computational power offered by having a large brain.

One reason for assuming the causal arrow must be this way round is that living in large groups incurs costs (as noted above), and a species is unlikely to live in large groups simply because its brain allows it to do so when larger brains have evolved to permit some other behavioural adaptation. In other words, large groups do not come for free if the brain evolves to facilitate smarter foraging, but smarter foraging probably does come for free if the brain evolved to facilitate large groups because smart foraging capitalises on the same inferential cognitive processes as are used in the social domain, but does not in itself incur additional costs.

An alternative is to use a systems approach that explicitly integrates all the components into a single model that specifies directly which factors are causes, which are constraints, and which are emergent properties. Path analysis is particularly helpful in this respect because it allows a number of alternative models to be evaluated against each other using an information index criterion. With group size as the dependent variable, Dunbar and Shultz (2007a) used path analysis to show that the best model tested consisted of just three key predictors: neocortex size, activity pattern (nocturnal vs. diurnal), and range size (Fig. 15.1). It seems unlikely that range size determines group size, so we may assume (as the primate literature

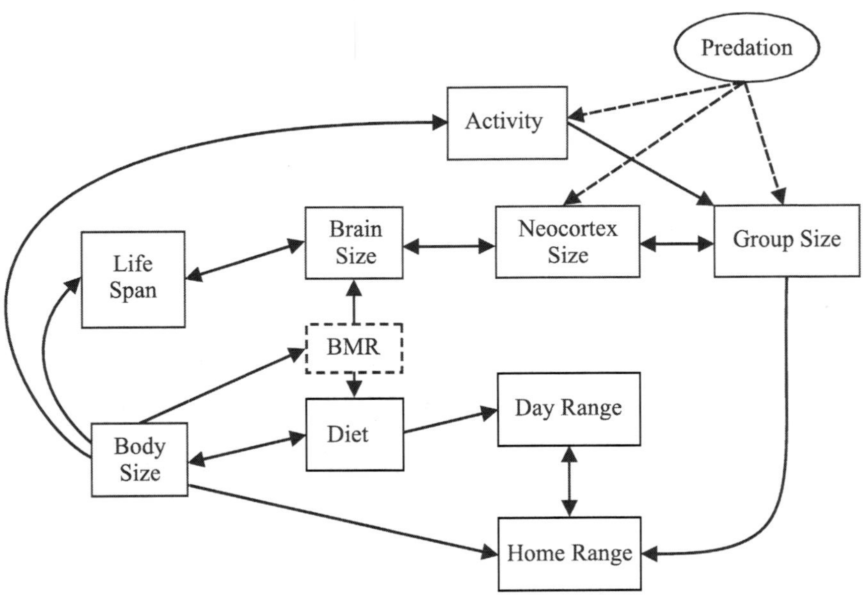

Fig. 15.1 Path analysis of the functional relationships influencing social group size in primates. Path analysis allows us to separate out primary selection factors from those that act indirectly (i.e., constraints). Reprinted with permission from Dunbar and Shultz (2007a)

has always done: Dunbar 1988) that the causal direction in this case is from group size to range size. Thus, only activity period and neocortex volume explicitly influence group size. Since neocortex size and group size are locked in a tight coevolutionary spiral (Pérez-Barbería et al. 2007), the causal arrow, thus, runs both ways in this case. It is important to be clear about what this means: in an evolutionary timescale, it is the demands of increasing group size that select for a larger neocortex, but from the contemporary perspective of primates actually on the ground, it is neocortex size that limits group size.

In principle, path analysis allows us to test between alternative models that rearrange the variables into all possible permutations and combinations of relationships between the variables. However, with the number of variables under consideration in Fig. 15.1, it would be prohibitively time-consuming (and largely unproductive) to consider every conceivable permutation and combination of the structural relationships. Dunbar and Shultz (2007b) looked at the most plausible set and structured models on the basis of findings generated by a minimum adequate model (MAM) analysis, which determined the set of variables that were most predictive of neocortex size. These established the set of primary or first-order relationships for the path analysis. Doing so then allowed them to identify a second layer of factors that function as constraints on brain size evolution: these included several life history variables (total brain size, lifespan, body size) and diet (which might act as a proxy for a number of the ecological hypotheses). Including these in the model, or using them as the only independent variables predicting group size, yields a significantly poorer fit, suggesting that they are constraints that have to be resolved rather than factors that influence group size directly.

In effect, as Finlay and Darlington (1995) pointed out, if you want to have a large neocortex (for whatever reason), you need to find a way to evolve a larger brain to support it. This may involve a shift in diet (to more digestible or higher energy diets) and larger body size, which, in turn, may necessitate an extended life history (mainly because it seems that brain tissue can only be laid down at a slow, constant rate). In other words, it is important to be clear about this: neither the fact that brain tissue is costly (Isler and van Schaik 2006) nor the fact that brain size correlates with a suite of life history variables are explanations for brain evolution; they are constraints that a lineage has to overcome *if* it is to be able evolve a larger brain.

In sum, then, these analyses suggest that the functional demands of large neocortices relate principally to maintaining large social groups. In primates at least, large social groups are demanded by occupancy of predator risky environments (primates: van Schaik 1983; Dunbar 1988; Shultz et al. 2004; ungulates: Adamczak and Dunbar 2008), one contributor to which was the adoption of a diurnal lifestyle. Thus, the causal sequence seems to be as suggested in Fig. 15.2. It is important to understand that the changes indicated by the double arrows represent obligatory coevolutionary changes: the downstream effect cannot be produced unless the upstream effect also occurs. Solving the predation problem, thus, requires a whole suite of variables to be changed more or less simultaneously (Pérez-Barbería et al. 2007).

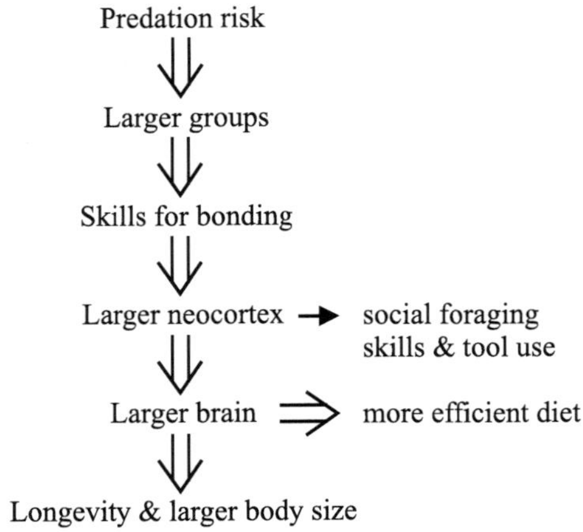

Fig. 15.2 Causal relationships in primate brain evolution, as revealed by the path analysis of Fig. 15.1. *Double arrows*: causal relationships; *Single arrows*: windows of opportunity

This functional benefit of large brain size is given added emphasis by Barrickman et al. (2008), who show that, in primates, reproductive success (in effect, fitness) is mainly a function of longevity rather than instantaneous reproductive rate (see also Clutton-Brock 1988). The capacity to survive is, in part at least, a function of avoiding predation. The important finding here is that instantaneous fecundity and the factors that influence this (such as dominance rank and body condition) play only a secondary role. However, it might still be possible to construe the causal relationship in two alternative ways: maximising fitness requires an extended lifespan and either (1) a large brain is an unavoidable correlate of an extended lifespan or (2) a large brain is necessary to ensure an extended lifespan. The first flies in the face of the costs of brain tissue (~20% of total energy consumption: Aiello and Wheeler 1995), since it offers no reason why brain size should have to correlate with longevity. The second makes sense in the light of Charnov's (Charnov 1993; Charnov and Berrigan 1993) suggestion that minimising predation risk is the main constraint on longevity in primates. But an extended lifespan is a risky gambit in the absence of mechanisms that allow animals to avoid the costs of premature mortality. Although the latter has often been seen in terms of foraging costs, in reality, there is little evidence to suggest that foraging skills influence longevity (at least, above a certain threshold of competence: Altmann 1998) and a lot of evidence to support the claim that predation has a major impact on life expectancy, and that, in primates at least, the principal mechanism for avoiding predators is the formation of cohesive social groups.

15.3 The Structure of Primate Social Groups

Although the social brain hypothesis has often been construed in terms of a quantitative relationship between social group size and some measure of brain volume (Dunbar 1992, 1998; Barton 1996; Barton and Dunbar 1997), in fact it is properly constituted in terms of social complexity, with the group size effect being an emergent property of how well animals handle complex relationships. Indeed, a number of analyses have reported correlations between various indices of behavioural complexity and brain size (Pawłowski et al. 1998; Lewis 2000; Kudo and Dunbar 2001; Byrne and Corp 2004). An important issue arising out of the social brain hypothesis for primate brain evolution thus centres around its implications for how social groups are structured.

Individuals do not have intimate social relationships (i.e., groom) with everyone else in a primate group. Rather, as Kudo and Dunbar (2001) and Lehmann et al. (in press) have shown, each individual has a small number (seldom more than four) of core social partners, and that the size of these grooming cliques correlates very closely with the total group size (Fig. 15.3: $r = 0.81$, $N = 15$, $p < 0.001$). More importantly, although total time devoted to grooming increases more or less linearly with group size in primates (both between and within species: Dunbar 1991; Lehmann et al. 2007a), the additional grooming time in large groups is not distributed proportionately around more group members; rather, individuals often have *fewer* grooming partners in very large groups than they do in small groups, even though they devote more *time* to social grooming (Dunbar 1984, 2003). Evidence in support of this has come from a seminal field study of the endocrinological consequences of network size in baboons: Wittig et al. (2008) showed that females with smaller grooming networks (or those who reduced the size of their networks) showed significantly lower stress responses (indexed as serum corticosteroid titres) to stressful events, such as invasion by males, than those that with larger grooming networks (or those who did not reduce their networks).

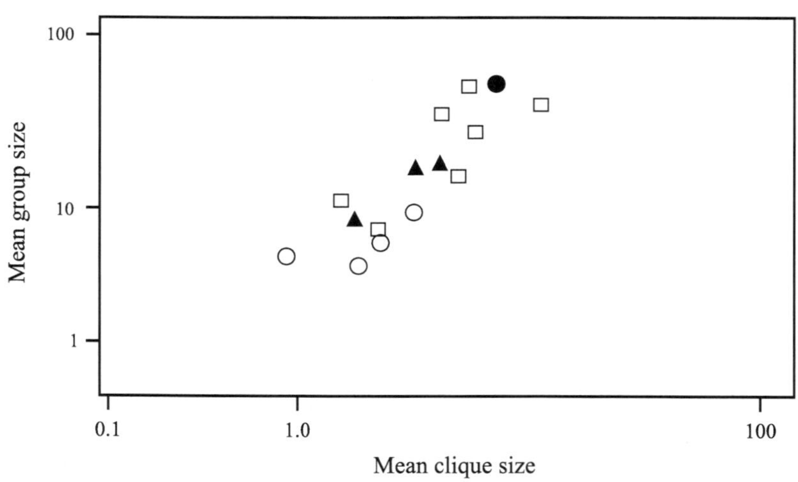

Fig. 15.3 Mean group size plotted against mean grooming clique size for individual primate genera. *Open circles*: Prosimians; *solid triangles*: New World monkeys; *open squares*: Old World monkeys; *solid circle*: Apes. Source: Kudo and Dunbar (2001)

I interpret as reflecting the fact that alliances need to work more effectively in large groups because of the escalating pressure that living with many individuals imposes. As we noted above, female fertility, in particular, declines rapidly with group size in primates (gelada: Dunbar and Dunbar 1977; colobus: Dunbar 1987) as a consequences of stressors that accumulate on lower-ranking females. In order to mitigate these, animals have to invest more heavily in their core relationships as group size increases so as to ensure that, when needed, these relationships do their job of protecting the individuals involved. This might involve active protection (coalitionary support when an ally is attacked) as clearly happens in gelada (Dunbar 1980, 1989), but in many cases may have as much to do with passive support: an observed relationship might imply retaliation by allies later (Datta 1983; Cheney and Seyfarth 1999, 2007; Bergman et al. 2003), and this may be enough to keep other individuals at a sufficient distance to minimise the likelihood that they will impinge on one's wellbeing. A poorly bonded grooming clique may mean that an individual is not buffered against the costs of living in large groups, and these then become unstable and liable to fragmentation.

The fact that the size of these small grooming cliques turns out to be linearly (and rather tightly) related to social group size across the primates may reflect the fact that, as group size increases, it is necessary to have a grooming coalition that is proportionally large enough to balance the stresses that large groups inevitably impose. In effect, primates are engaged in a difficult balancing act that uses grooming cliques as a mechanism to keep other group members just far enough away to avoid these kinds of stresses without driving them away altogether. Without such a mechanism, social groups would fission rapidly and find an equilibrium at a lower size as these stresses drove individuals away. Ultimately, it may be that the real cognitive constraint lies at the level of this inner layer of the social network, and that the cohesion and size of the conventional group is an emergent property of how well animals solve the problem of bonding at this very intimate level.

The intensity of bonding in primate social groups seems to be of a different order to that in most other mammals, and certainly that found in carnivores and artiodactyl ungulates (Shultz and Dunbar 2007). This is clear from Pérez-Barbería et al.'s (2007) analysis of the coevolution of brain size and sociality in these three orders, which showed that the evolutionary coupling of these traits was significantly tighter in primates than the other two mammalian orders. This, combined with Shultz and Dunbar's (2007) finding that anthropoid primates (in particular) have significantly more female-bonded groups than carnivores or ungulates, suggests that primatologists might have been right in their insistence that primate sociality is of a qualitatively different kind to that found in other taxa, the longstanding scepticism of those who study other mammals and birds notwithstanding.

There are obviously two questions here. One is why this transition from a qualitative to a quantitative effect should have happened: what was it about early primate ecology that made bonded groups of this kind especially valuable? The second is why only (anthropoid) primates seem to have made this transition. We have no real idea how to answer either question at present, but it would have to be something that sets anthropoid primates apart from all other taxa, including most (but perhaps not all) prosimians. (In general, prosimians resemble carnivores rather than anthropoid primates in their distribution of social types [Shultz and Dunbar 2007], but there may, nonetheless, be some convergence with anthropoid primates among the more social lemurs).

One plausible explanation for the timing of the evolution of bonded social groups is the switch to a more frugivorous diet in the transition between ancestral prosimians (the Eocene primates) and the lineage leading to modern anthropoid primates, not least because this transition occurs at exactly the right time. In contrast to the characteristic diets of most prosimians (predominantly insects), ungulates (grass) and carnivores (mobile prey), fruits offer the opportunity for spatial concentration and, as a result, a significantly increased risk of competition when individuals are forced to cluster around small but, at least in the short term, relatively stable rich resource packets. The increased within-group competition that would be inevitable under these conditions increases the risk of group fragmentation (thereby neutralising the antipredator benefits of grouping demanded by the simultaneous shift to a more diurnal lifestyle). Frugivorous birds may be able to cope with this by being able to disperse in a form of individual- or pair-based fission–fusion sociality, something that may be possible only because their aerial habit allows them to reduce predation risk. In addition, their small body mass means that relatively large numbers of birds can gather at a fruiting tree before exceeding the local carrying capacity. Anthropoid primates face two disadvantages in this respect: their large body size and the fact that, being plantigrade, they cannot so easily escape predators. Bonded relationships that create functional alliances may have solved this problem by reducing the levels of competition (see also van Schaik 1989), so allowing primate groups to remain together despite these disruptive effects. This might then also explain Barton's (1998) the finding that the grade shift in brain size between the prosimian and anthropoid lineages is associated with the greater importance of colour vision (associated mainly with foraging for fruits).

15.4 Implications for Human Social Evolution

The human lineage has its roots in the ancestral primate lifestyle. It, thus, inherits the same kinds of intense bondedness that, generally speaking, characterises other monkeys and apes. However, human social groups differ from those of most other primates in two key respects: their size and the fact that they have a dispersed (or fission–fusion) form. Although the latter trait is shared with a small number of other primates (spider monkeys and chimpanzees, and the hamadryas and gelada baboons), it is rare enough among the anthropoid primates to appear to be a derived feature and so to require explanation.

There seems little reason to doubt the suggestion that group size has increased steadily over time within the hominid lineage. Modern humans demonstrably live in much larger groupings than any other primate. However, although the size of groups predicted for modern humans from the primate brain size regression equation is close to those actually observed (Dunbar 1993), the changes in brain size within the human fossil record suggest that the kinds of very large groups found in modern humans are a relatively late evolutionary feature (post-0.5 MYA) (Aiello and Dunbar 1993; Dunbar 2004; Dunbar et al. in press). They may well coincide with the major exoduses out of Africa.

On the other hand, the human lineage shares with chimpanzees (and a small number of other mammalian taxa) the distinctive feature of having a series of social groupings that are nested within a large dispersed community. In both cases, these layers have a scaling ratio of approximately 3 (Zhou et al. 2005; Hill et al. 2008). The fact that the largest grouping level is dispersed (often over a wide geographical area) raises questions about the function of these groupings, since they lack the cohesiveness that seems to form such a crucial component of monkey social groups. The provision of protection against predators seems an implausible function for these kinds of dispersed grouping. Indeed, since female body size is a major factor in reducing predation risk, chimpanzees and humans should be able to manage with group sizes that are approximately half those found in baboons (~50 on average: Dunbar 1992). This suggests that the community is not an antipredator defence, but has evolved (or been coopted) for some other function.

The functionality of chimpanzee communities remains unclear, though some form of resource defence (of food trees or of females) remains a likely explanation. One finding of importance for our understanding of human evolution, however, is the fact that, owing to the time costs of travel, chimpanzees can only survive in most of their current geographical range by opting for fission–fusion sociality (Lehmann et al. 2007b, 2008b). Since it is difficult to imagine hominids radically solving this particular problem even with bipedalism, some form of fission–fusion sociality is likely to have continued to be unavoidable throughout human evolution, just as it still is among contemporary hunter-gatherers. While resource defence remains a plausible explanation for large dispersed communities among modern humans, an alternative possibility might be the opportunities created for trading access to keystone resources over a very wide area (Dunbar 1996), especially in the ecologically less predictable habitats at high latitudes. Some evidence to support this comes from analyses of human language distributions. Nettle (1999) showed that both language community size (the number of people who spoke the same language) and language area (the geographical range within which a language was spoken) correlate linearly with latitude in both the New and Old Worlds. He argued that this reflected the need to be able to engage in reciprocal exchange (mediated by a common language) over much wider area in the more unpredictable habitats found at higher latitudes.

Since Zhou et al. (2005; see also Hamilton et al. 2007) showed that human societies form a nested series of grouping levels, it might be possible to deploy the same argument suggested for nonhuman primates as to how these large community-level social groupings are maintained as stable units through time. In other words, the cohesiveness and coherence through time of the large outer community layer (~150 in humans) may depend on the effectiveness with which some key inner grouping layer is maintained through face-to-face interaction. In humans, the tiers of the grouping layers take values that approximate 5, 15, 50, and 150 individuals (Zhou et al. 2005). On the basis of relative size compared to chimpanzees, Kudo and Dunbar (2001) identified the second tier (the 15-person layer) as the equivalent to the nonhuman primate grooming clique in this respect. What is special about this grouping level in humans is not clear, although it has long been recognised in social psychology that groups of size 12–15 have a particular significance where close bondedness (e.g., in team sports) or emotional closeness is required (juries, cabinets, etc.) (Buys and Larsen 1979). As yet, we understand almost nothing of either the dynamics or the function of these grouping levels in humans, and this remains a rich area for future development.

References

Abbott DH, Keverne EB, Moore GF, Yodyinguad U (1986) Social suppression of reproduction in subordinate talapoin monkeys, *Miopithecus talapoin*. In: Else JG, Lee PC (eds) Primate ontogeny, cognition, and social behaviour. Cambridge University Press, Cambridge, pp 329–341

Adamczak V, Dunbar RIM (2008) Variation in the mating system of oribis and its ecological determinants. Afr J Ecol 45:197–206

Aiello LC, Dunbar RIM (1993) Neocortex size, group size and the evolution of language. Curr Anthropol 34:184–193

Aiello LC, Wheeler P (1995) The expensive-tissue hypothesis: the brain and the digestive system in human and primate evolution. Curr Anthropol 36:199–221

Altmann SA (1998) Foraging for survival. University of Chicago Press, Chicago

Armstrong E, Bergeron R (1985) Relative brain size and metabolism in birds. Brain Behav Evol 26:141–153

Barrickman NL, Bastian ML, Isler K, van Schaik CP (2008) Life history costs and benefits of encephalization: a comparative test using data from long-term studies of primates in the wild. J Hum Evol 54:568–590

Barton RA (1996) Neocortex size and behavioural ecology in primates. Proc Roy Soc Lond B 263:173–177

Barton RA (1998) Visual specialization and brain evolution in primates. Proc Roy Soc Lond B 265:1933–1937

Barton RA, Dunbar RIM (1997) Evolution of the social brain. In: Whiten A, Byrne RW (eds) Machiavellian intelligence II: extensions and evaluations. Cambridge University Press, Cambridge, pp 240–263

Bergman TJ, Beehner JC, Cheney DL, Seyfarth RM (2003) Hierarchical classification by rank and kinship in baboons. Science 302:1234–1236

Buys CJ, Larsen KL (1979) Human sympathy groups. Psychol Rep 45:547–553

Byrne RW, Corp N (2004) Neocortex size predicts deception rate in primates. Proc Roy Soc Lond B 271:1693–1699

Byrne RW, Whiten A (1988) Machiavellian intelligence. Oxford University Press, Oxford

Calhim S, Shi J, Dunbar RIM (2006) Sexual segregation among feral goats: testing between alternative hypotheses. Anim Behav 72:31–41

Charnov EL (1993) Life history invariants. Oxford University Press, Oxford

Charnov EL, Berrigan D (1993) Why do female primates have such long lifespans and so few babies? Or life in the slow lane. Evol Anthropol 2:191–194

Cheney DL, Seyfarth RM (1999) Recognition of other individuals' social relationships by female baboons. Anim Behav 58:67–75

Cheney DL, Seyfarth RM (2007) Baboon metaphysics: the evolution of a social mind. University of Chicago Press, Chicago

Clutton-Brock TH (1988) Reproductive success: studies of individual variation in contrasting breeding systems. University of Chicago Press, Chicago

Clutton-Brock TH, Harvey PH (1980) Primates, brains and ecology. J Zool Lond 190:309–323

Datta SB (1983) Relative power and the maintenance of dominance. In: Hinde RA (ed) Primate social relationships: an integrated approach. Blackwell, Oxford, pp 103–112

Deaner RO, Nunn CL, van Schaik CP (2000) Comparative tests of primate cognition: different scaling methods produce different results. Brain Behav Evol 55:44–52

Dunbar RIM (1980) Determinants and evolutionary consequences of dominance among female gelada baboons. Behav Ecol Sociobiol 7:253–265

Dunbar RIM (1984) Reproductive decisions: an economic analysis of gelada baboon social strategies. Princeton University Press, Princeton, NJ

Dunbar RIM (1987) Habitat quality, population dynamics and group composition in colobus monkeys (*Colobus guereza*). Int J Primatol 8:299–329

Dunbar RIM (1988) Primate social systems. Chapman & Hall, London

Dunbar RIM (1989) Reproductive strategies of female gelada baboons. In: Rasa AE, Vogel C, Voland E (eds) The sociobiology of sexual and reproductive strategies. Chapman & Hall, London, pp 74–92

Dunbar RIM (1991) Functional significance of social grooming in primates. Folia Primatol 57:121–131

Dunbar RIM (1992) Neocortex size as a constraint on group size in primates. J Hum Evol 22:469–493

Dunbar RIM (1993) Coevolution of neocortex size, group size and language in humans. Behav Brain Sci 16:681–735

Dunbar RIM (1996) Grooming, gossip and the evolution of language. Faber & Faber, London

Dunbar RIM (1998) The social brain hypothesis. Evol Anthropol 6:178–190

Dunbar RIM (2003) Evolution of the social brain. Science 302:1160–1161

Dunbar RIM (2004) The human story: a new history of mankind's evolution. Faber & Faber, London

Dunbar RIM, Dunbar EP (1977) Dominance and reproductive success among female gelada baboons. Nature 266:351–352

Dunbar RIM, Shultz S (2007a) Understanding primate brain evolution. Phil Trans Roy Soc B 362:649–658

Dunbar RIM, Shultz S (2007b) Evolution in the social brain. Science 317:1344–1347

Dunbar RIM, Cornah L, Daly F, Bowyer K (2002) Vigilance in humans: a test of alternative hypotheses. Behaviour 139:695–711

Dunbar RIM, Gamble C, Gowlett JAG. Human evolution and the archaeology of the social brain. Curr Anthropol (in press)

Finlay BL, Darlington RB (1995) Linked regularities in the development and evolution of mammalian brains. Science 268:1578–1584

Finlay BL, Darlington RB, Nicastro N (2001) Developmental structure in brain evolution. Behav Brain Sci 24:263–308

Gibson KR (1986) Cognition, brain size and the extraction of embedded food resources. In: Else JG, Lee PC (eds) Primate ontogeny, cognition and social behaviour. Cambridge University Press, Cambridge, pp 93–104

Hamilton MJ, Milne BT, Walker RS, Burger O, Brown JH (2007) The complex structure of hunter-gatherer social networks. Proc Roy Soc Lond B 274:2195–2202

Hill RA, Bentley RA, Dunbar RIM (2008) Network scaling reveals consistent fractal pattern in hierarchical mammalian societies. Biol Lett 4:748–751

Hofman MA (1983) Energy metabolism, brain size and longevity in mammals. Q Rev Biol 58:495–512

Isler K, van Schaik CP (2006) Metabolic costs of brain size evolution. Biol Lett 2:557–560

Jerison HJ (1973) Evolution of the brain and intelligence. Academic Press, London

Karbowski J (2007) Global and regional brain metabolic scaling and its functional consequences. BMC Biol 5:18. doi:10.1186/1741-7007-5-18

Kaufman JA (2003) On the expensive tissue hypothesis: independent support from highly encephalized fish. Curr Anthropol 44:705–706

Kudo H, Dunbar RIM (2001) Neocortex size and social network size in primates. Anim Behav 62:711–722

Lehmann J, Korstjens AH, Dunbar RIM (2007a) Group size, grooming and social cohesion in primates. Anim Behav 74:1617–1629

Lehmann J, Korstjens AH, Dunbar RIM (2007b) Fission–fusion social systems as a strategy for coping with ecological constraints: a primate case. Evol Ecol 21:613–634

Lehmann J, Korstjens AH, Dunbar RIM (2008a) Time management in great apes: implications for gorilla biogeography. Evol Ecol Res 10:517–536

Lehmann J, Korstjens AH, Dunbar RIM (2008b) Time and distribution: a model of ape biogeography. Ethol Ecol Evol 20:337–359

Lehmann J, Andrews K, Dunbar RIM (in press). Social networks and social complexity in female-bonded primates. In: Dunbar RIM, Gamble C, Gowlett JAG (eds) Social Brain, Distributed Mind. Oxford University Press, Oxford

Lewis KP (2000) A comparative study of primate play behaviour: implications for the study of cognition. Folia Primatol 71:417–421

Martin RD (1981) Relative brain size and basal metabolic rate in terrestrial vertebrates. Nature 293:57–60

Mink JW, Blumenschine RJ, Adams DB (1981) Ratio of central nervous system to body metabolism in vertebrates – its constancy and functional basis. Am J Physiol 241:R203–R212

Nettle D (1999) Linguistic diversity. Oxford University Press, Oxford

Pagel MD (1997) Inferring evolutionary processes from phylogenies. Zool Scr 26:331–348

Parker ST, Gibson KR (1979) A developmental model for the evolution of language and intelligence in early hominids. Behav Brain Sci 2:367–408

Pawłowski BP, Lowen CB, Dunbar RIM (1998) Neocortex size, social skills and mating success in primates. Behaviour 135:357–368

Pérez-Barbería FJ, Shultz S, Dunbar RIM (2007) Evidence for coevolution of sociality and relative brain size in three orders of mammals. Evolution 61:2811–2821

Reader SM, Laland KN (2002) Social intelligence, innovation and enhanced brain size in primates. Proc Natl Acad Sci USA 99:4436–4441

Shultz S, Dunbar RIM (2006) Chimpanzee and felid diet composition is influenced by prey brain size. Biol Lett 2:505–508

Shultz S, Dunbar RIM (2007) The evolution of the social brain: anthropoid primates contrast with other vertebrates. Proc Roy Soc Lond B 274:2429–2436

Shultz S, Noë R, McGraw WS, Dunbar RIM (2004) A community-level evaluation of the impact of prey behavioural and ecological characteristics on predator diet composition. Proc Roy Soc Lond B 271:725–732

Shultz S, Bradbury RB, Evans KL, Gregory RD, Blackburn TM (2005) Brain size and resource specialisation predict long-term population trends in British birds. Proc Roy Soc Lond B 272:2305–2311

Sol D, Timmermans S, Lefebvre L (2002) Behavioural flexibility and invasion success in birds. Anim Behav 63:495–502

Sol D, Duncan RP, Blackburn TM, Cassey P, Lefebrve L (2005) Big brains, enhanced cognition and response of birds to novel environments. Proc Natl Acad Sci USA 102:5460–5465

van Schaik CP (1983) Why are diurnal primates living in groups. Behaviour 87:120–144 van Schaik CP (1989) The ecology of social relationships amongst female primates. In: Standen V, Foley RA (eds) Comparative socioecology: the behavioral ecology of humans and other mammals. Blackwell, Oxford, pp 195–218

van Schaik CP, Dunbar RIM (1990) The evolution of monogamy in large primates: a new hypothesis and some crucial tests. Behaviour 115:30–62

Wittig RM, Crockford C, Lehmann J, Whitten PL, Seyfarth RM, Cheney DL (2008) Focused grooming networks and stress alleviation in wild female baboons. Horm Behav 54:170–177

Zhou W-X, Sornette D, Hill RA, Dunbar RIM (2005) Discrete hierarchical organization of social group sizes. Proc Roy Soc Lond B 272:439–444

Chapter 16
The Gap is Social: Human Shared Intentionality and Culture

Michael Tomasello and Henrike Moll

Abstract Human beings share many cognitive skills with their nearest primate relatives, especially those for dealing with the physical world of objects (and categories and quantities of objects) in space and their causal interrelations. But humans are, in addition, biologically adapted for cultural life in ways that other primates are not. Specifically, humans have evolved unique motivations and cognitive skills for understanding other persons as cooperative agents with whom one can share emotions, experience, and collaborative actions (shared intentionality). These motivations and skills first emerge in human ontogeny at around one year of age, as infants begin to participate with other persons in various kinds of collaborative and joint attentional activities. Participation in such activities leads humans to construct during ontogeny, perspectival and dialogical cognitive representations.

16.1 Introduction

The gap we are trying to explain is obvious: we humans live in complex societies dependent on complex technologies, symbol systems, and social institutions – whereas other primate species do not. We, humans, are scientifically investigating and writing about them, not they about us.

Following Vygotsky (1978) and Tomasello (1999), the general proposal here is that the human gap is best explained in terms of, ultimately, social (or cultural) factors. That is, human beings are especially sophisticated cognitively not because of their greater individual brainpower, but rather because of their unique ability to put their individual brainpowers together to create cultural practices, artifacts, and institutions – underlain by skills and motivations for shared intentionality – which

M. Tomasello (✉) and H. Moll
Max Planck Institute for Evolutionary Anthropology, Leipzig, Germany
e-mail: tomasello@eva.mpg.de, moll@eva.mpg.de

P.M. Kappeler and J.B. Silk (eds.), *Mind the Gap*,
DOI 10.1007/978-3-642-02725-3_16, © Springer-Verlag Berlin Heidelberg 2010

are then passed along to youngsters as a second line of inheritance in the species, resulting in a ratcheting up of cultural and cognitive complexity over historical time. A child raised alone on a desert island, or even by chimpanzees, would cognitively not be very different from the apes, as its unique adaptation for absorbing culture would be intact but there would be nothing there to absorb.

But it turns out that identifying the cognitive and social-cognitive factors that enable developing human beings to take advantage of the cultural practices, artifacts, and institutions around them is not so easy. The approach in our research group for some years has been to focus on situations and phenomena that are mostly simpler than those in modern adult life, but still have all of the key characteristics for exploring the gap between human and ape cognition. Toward this end, we have focused on young children as representative of the human species, with the idea that they help us to abstract away from the complexities of adult human life and get down to the essentials. For comparison, we have focused on great apes, especially chimpanzees, as humans' closest primate relatives.

What we will do here is three things. First, we will present evidence that the gap is indeed social, drawn especially from a recent large-scale study of the full range of cognitive skills in great apes and human children. Second, we will review some recent studies specifying in more detail the nature of the difference between apes and human children in situations involving shared intentionality in (1) collaboration, (2) communication, and (3) social learning. Finally, we will offer some speculations as to how these small-scale cooperative abilities of shared intentionality scale up into uniquely human cognition and culture.

16.2 Human and Great Ape Cognitive Skills Compared

An obvious hypothesis about the human difference is that human beings simply have bigger brains and so more "general intelligence" than other animals: more memory capacity, greater inferential skills, faster learning, further foresight and planning, finer skills of perceptual discrimination – more and better of everything. And this quantitative difference somehow translates into a qualitative difference in cognitive abilities.

In a recent study, we tested this hypothesis – in comparison with what we called the cultural intelligence hypothesis – by giving a very large test battery (a kind of nonverbal IQ test) to two of humans' closest primate relatives, chimpanzees and orangutans, along with 2-year-old human children (Herrmann et al. 2007). If humans simply have more cerebral computing power and general intelligence, then the children should have differed from the apes uniformly across the different types of tasks. But that was not the result. The result was that the three species were very similar when it came to cognitive skills for dealing with the physical world – problems having to do with space, quantities, and causality – but the human children were much better than the two ape species when it came to cognitive skills for dealing with the social world – problems of imitative learning, gestural communication, and reading intentions.[1] When correlational analyses were done, no general intelligence, or *g*-factor, was found (Herrmann et al. in press). These findings are not compatible with the hypothesis that the cognitive difference between humans and apes is a simple function of more cerebral horsepower.

There have also been a few hypotheses about more specific computational capacities that might make human cognition unique – though they cannot explain these cross-species test battery results either. In one hypothesis, for example, humans are better able to perform multiple tasks simultaneously. In another, humans are better with relational categories and making analogies across different materials. These hypotheses may have some validity, but it is difficult to see how multitasking alone or analogy-making could account for the results of our cross-species study of intelligence, much less such things as linguistic symbols, social institutions, and cultural norms. These are all collective cultural products that are not easily accounted for by simply adding up the computational power of individuals either generally or in specific skills.

Uniquely, human cognitive skills are not simply the result of greater computational power overall, or of some increase in a specialized cognitive ability. Rather, they result from an ability enabling humans to put their heads together, so to speak, in cooperating and communicating with one another in ways that led to the creation of complex cultural products, including both material and symbolic artifacts, such as linguistic symbols. Human children grow up in the midst of these material and symbolic artifacts, and by learning to use them in interaction with others (as well as internalizing these interactions cognitively), they actually create, during ontogeny, evolutionarily new ways of thinking. This hypothesis sets the problem that must be addressed if we want to provide a plausible evolutionary account of how human cognition became, in effect, a collective enterprise.

16.3 Cultural Activities in Humans and Great Apes

Humans' adaptation for living and exchanging information in cultural groups manifests itself in many ways. What we would like to focus on here is the three main classes: (1) small-scale collaborative activities (translating ultimately into large-scale social and cultural institutions); (2) cooperative communication (transforming ultimately into language); and (3) cultural learning (resulting ultimately in cumulative cultural evolution).

[1]One might object that the human subjects had an advantage over the non-human ones in the social tasks, because they interacted with a conspecific vs. a member of a different species. However: (1) many of the tasks were chosen based on previous research showing no difference in performance as a function of conspecific or human interactants (e.g., ape vs. human demonstrators in social learning); and (2) the comfort level of all subjects in the testing situation was assessed and, if anything, the children were most shy (and this measure did not correlate with overall cognitive performance in the main tasks).

16.3.1 Collaboration

Individuals of virtually all primate species engage in group activities on a daily basis. These activities may be considered cooperative in the sense that they coordinate their behavior temporally and spatially with the other animals in the group. However, as in previous theoretical work (Tomasello et al. 2005), here, we want to single out for special attention "shared cooperative activities" – a subtype in which humans routinely engage. In our modified version of Bratman's (1992) characterization, joint or shared cooperative activities are mainly characterized by the three features: (1) the participants in the cooperative activity share a joint goal, to which they are jointly committed; (2) the participants take reciprocal or complementary roles in order to achieve this joint goal; and (3) the participants are generally motivated and willing to help one another to accomplish their role, if needed.

Joint commitment to a goal. One group activity that has been posited as being especially complex is chimpanzee group hunting. Boesch and colleagues (Boesch and Boesch 1989; Boesch and Boesch-Achermann 2000; Boesch 2005) have observed chimpanzees in the Taï forest hunting in groups for arboreal prey, mainly monkeys. In the account of these researchers, the animals take complementary roles in their hunting. One individual, called the driver, chases the prey in a certain direction, while others, the so-called blockers, climb the trees and prevent the prey from changing directions. An ambusher then silently moves in front of the prey, making an escape impossible. Of course, when the hunting event is described with this vocabulary of complementary roles, then it appears to be a joint cooperative activity: complementary roles already imply that there is a joint goal, shared by the role-takers. But the question really is whether this vocabulary is appropriate at all. A more plausible characterization of the hunting event, from our perspective, is as follows: each animal fills whatever spatial position is still available at any given time so that the encircling is accomplished in a stepwise fashion, without any kind of prior plan or agreement to a shared goal or assignment of roles. Then, without pursuing a joint goal or accomplishing a certain role within a higher-order framework, each individual chases the prey from its own position (see also Moll and Tomasello 2007a). This event clearly is a group activity or group action, because the chimpanzees are "mutually responsive" as they coordinate their behaviors with that of the others in space and time (see also Melis et al. 2006). But what seems to be missing is the "togetherness" or "jointness" that distinguishes shared cooperative activities from other sorts of group actions.

This interpretation is strongly supported by studies that have investigated chimpanzees' abilities to cooperate in experimental settings. In one study, Warneken et al. (2006) tested three juvenile human-raised chimpanzees with a set of four different cooperation tasks. In two of these tasks, a human tried to engage the chimpanzee to cooperate in order to solve a problem (e.g., extracting a piece of food from an apparatus). In the other two tasks, the human tried to engage the ape to play a social game. The authors looked at two things: the chimpanzees' level of behavioral coordination and the chimpanzees' behaviors in the so-called interruption periods, in which the human suddenly stopped participating in the activity. The results were very consistent: in the problem-solving tasks, chimpanzees coordinated their behaviors quite well with that of the human, as shown by the fact that they were mostly successful in bringing about the desired result, as, for instance, extracting the piece of food from the apparatus. However, they showed no interest in the social games, and so the level of coordination in these tasks was low or absent. Most important was what happened when the human suddenly interrupted the activity. In none of the tasks did a chimpanzee ever make a communicative attempt to reengage the partner. Such attempts were missing even in cases in which they should have been highly motivated to obtain the desired result, as in the problem-solving task involving food. The absence of any efforts by the chimpanzees to reengage their human partner is crucial: it shows that the chimpanzees did not cooperate in the true sense, since they had not formed a joint goal with the human. If they had been committed to a joint goal, then we would expect them, at least in some instances, to persist in trying to bring it about and in trying to keep the collaboration going.

For humans, the situation is different from very early on in ontogeny. Warneken et al. (2006) conducted an analogous study with 18- and 24-month-old human children. Unlike the chimpanzees, children cooperated quite successfully and enthusiastically not only in the problem-solving tasks, but also in the social games. For example, these infants enjoyed playing a "trampoline" game together, in which both partners had to simultaneously lift up their sides of a small trampoline with their hands, such that a ball could bounce on it without falling off. Most importantly, when the adult stopped participating at a certain point during the activity, every child at least once produced a communicative attempt in order to reengage him. In some cases, the children grabbed the adult by his arm and drew him to the apparatus. The older children of 24 months of age also often made linguistic attempts to tell the recalcitrant partner to continue. Unlike the chimpanzees, we thus find in human infants the ability to cooperate with joint commitment to a shared goal: the children "reminded" the recalcitrant partner of their shared goal and expected him to continue in order to achieve it. There was even some evidence that the children already understood the normativity behind the social games and the way they "ought to be played." For example, in one of the games, they always used a can in order to catch a toy when it came falling out of one end of a tube after their partner had thrown it in from the other end. They could have also caught it with their hands, but they preferred to do it the way it had been demonstrated to them. This implies that they perceived the can as a constitutive element of the game, and they wanted to play the game the way it "ought" to be played. The chimpanzees, on the other hand, never used the can in order to catch the toy – if they engaged in the game at all, they simply used their hands. It, thus, seems that human infants by the age of 18 months, in contrast to apes, are able to jointly commit to a shared goal.

Role reversal. The second criterion for cooperation, as we define it, is role-taking. True cooperation should involve that the partners perform reciprocal roles and also understand them, in the sense that they coordinate their actions and intentions with the possibility of reversing roles. This form of role-taking would suggest that each partner represents the entire collaboration, its shared goal and reciprocal roles, holistically from a "bird's eye view" instead of just from within whatever role they happen to be taking at the moment. One study purporting to show role reversal in chimpanzees is that of Povinelli et al. (1992). In that study, chimpanzees were trained in one of two roles of a cooperative hiding game with a human. Some chimpanzees were trained in the role of a communicator, who indicated to the human where a piece of food was located. The other chimpanzees were trained in the complementary role of the "operator," who extracted the food from the location indicated by the human. When the chimpanzees had learned their initial role to criterion, a role switch was initiated and the question was whether the chimpanzees would spontaneously reverse the roles. One of the chimpanzees, whose initial role was that of the communicator, was immediately successful as operator after the switch. But the problem is that this individual most likely comprehended human indicating gestures before the study – as this animal had extensive interactions with humans. The two individuals that switched to be a communicator also seemed to reverse the roles effectively, as they were reported to provide the human with cues about the location of the food fairly quickly. However, the problem in this case is that it is not clear that the chimpanzees actually produced any communicative signals at all, but instead the humans simply interpreted their natural bodily orientation to the food.

A better controlled investigation of role-reversal skills in chimpanzees was done by Tomasello and Carpenter (2005) with the same three young human-raised chimpanzees which participated in Warneken et al.'s (2006) study. In this study, a human demonstrated to the chimpanzee various actions with each of four pairs of objects. For each pair of objects, one functioned as a "base" and the other as an "actor." The human then demonstrated to the chimpanzee how the two, the actor and the base, are put together. For instance, she put a "Tigger" figure on a plate and "Winnie the Pooh" figure in a little toy car. Then E gave the actor (e.g., Tigger) to the chimpanzee and held out the base (the plate) towards the chimpanzee, thus offering that the chimpanzee put the actor on the base to complete the act. If chimpanzees did not perform the role of putting the actor on the base spontaneously, E encouraged them to do so by vocalizing and, and if they still did not respond, by helping them put the actor on the base. To test for role reversal, E then handed the chimpanzee the base and held out the actor to see whether she would spontaneously offer the base. Two of the three chimpanzees held out the base object at some point. But, crucially, none of these responses occurred spontaneously, and more importantly, in none of these responses was the holding out of the base accompanied by a look to E's face. A look to the partner's face while holding out the object is a key criterion of "offering" used in all studies with human infants (Bates 1979; Camaioni 1993). Thus, in Tomasello and Carpenter's (2005) study, there was no indication that the chimpanzees offered the base to the human, and so there were no acts of role reversal.

An analogous study with human infants of 12 and 18 months of age was conducted by Carpenter et al. (2005). As in the study with the chimpanzees, situations were set up in which an adult did things like hold out a basket in which the infant was asked to place a toy. After the infant complied, in the test for role reversal, the adult placed the basket within the infant's reach and held up the toy herself. Impressively, even some of the 12-month olds spontaneously held out the basket for the adult while at the same time looking to her face, presumably in an anticipation of her placing the toy inside. Thus, the infant's handing behaviors, in contrast to those of the chimpanzees, were clearly the acts of offering learned through role reversal. It, thus looks as though chimpanzees, in contrast to young human children, do not fulfill either of the first two criteria of cooperation: sharing a joint goal and understanding the roles of a joint activity in some general way.

Mutual support. The third criterion is that, if needed, the partners of a joint cooperative activity help one another do their part successfully. This criterion is inherently linked to the other two: the commitment to the joint (and not just an individual) goal implies a responsibility not just for the successful completion of one's own role, but also to some degree for that of the other participants, and so helping them fulfill their goal (or, in some instances even replacing them) is an integral part of true collaboration. In two recent studies, chimpanzees did not take an opportunity to "help" another individual to obtain food (Silk et al. 2005; Jensen et al. 2006). But food is a resource over which apes are used to compete, and so helping might be better investigated in situations that do not revolve around food. Given our interest in helping as a constituent of collaboration, the most important form of helping is "instrumental helping," in which one individual helps another instrumentally to achieve a behavioral goal. We know of only one study investigating instrumental helping in nonhuman primates. Warneken and Tomasello (2006) had three human-raised juvenile chimpanzees watch a human attempt, but fail to achieve, different kinds of individual goals. Reasons for her failure were that her desired objects were out of reach, that she ran into physical obstacles, clumsily produced wrong results, or used ineffective means. The chimpanzees helped the human in some cases. However, the range of situations in which they helped was very limited: only when the adult effortfully reached but failed to grasp objects did the chimpanzees help by fetching them for her.

An analogous study was conducted with 18-month-old human infants, who also saw an adult fail to reach her goals for the same reasons (Warneken and Tomasello 2006). In this study, infants as young as 18 months of age helped the adult in various scenarios: for instance, they spontaneously removed physical obstacles that hindered the adult (e.g., they opened a cabinet so that the adult could place books inside) and showed him means that they knew were effective to bring about the intended result. It thus seems that, even though some helping behavior can be found in nonhuman primates, only human infants display helping actions in a variety of situations, providing whatever help is needed in the given situation.

What we conclude from these experimental studies is that, despite their group hunting in the wild, chimpanzees do not have "we-intentionality" (see Bratman 1992; Searle 1995; Tuomela 2002). They do not form a joint commitment to a shared goal and they do not perform reciprocal roles in the true sense as they do not generally understand roles from a bird's eye view, in the same representational format. Finally, they seem to be limited in their abilities to help another individual – which is a necessary prerequisite to engage in cooperative activities narrowly defined. Human infants and young children, in contrast, have this we-intentionality and act cooperatively from at least 14 to 18 months of age. They "remind" their partner of the joint commitment to a shared goal, as they reengage her when she suddenly interrupts the activity (Warneken et al. 2006; Warneken and Tomasello 2007); they begin to reverse and understand their roles as early as 12 months of age (Carpenter et al. 2005); and they help others in the fulfillment of their individual roles in various ways by at least 14–18 months (Warneken and Tomasello 2006, 2007).

16.3.2 Cooperative Communication

A related domain, which also requires some form of cooperation is communication. As noted above, chimpanzees usually perform poorly in experiments that require some understanding of cooperative communication. Here, we address this issue in more detail by first looking at nonhuman primates' own production of communicative gestures, and then at their comprehension of such gestures produced by others.

Chimpanzees gesture to one another in different contexts. Some of these gestures are clearly intentional, in the sense that they are not just triggered by certain environmental conditions, but used flexibly to do such things as elicit play in the other (by an "arm-raise") or to request nursing (by a "touch-side"). That these gestures are indeed used flexibly is illustrated by a number of phenomena, for instance, the fact that visual gestures are only used in instances in which the recipient is visually oriented towards the sender (e.g., Tomasello et al. 1997a; Kaminski et al. 2004). One might think that if chimpanzees can gesture flexibly and understand some things about visual perception (see Call and Tomasello 2008), they should also use gestures to direct another chimpanzee's attention to a certain event or object by pointing. There are certainly occasions in which it would be very helpful if one ape pointed for another ape to indicate the locus of some relevant event. It must, therefore, seem somewhat surprising that, in fact, there has not been a single reliable documentation of any scientist in any part of the world of one ape pointing for another. But captive apes which have had regular interactions with humans point for their human caretakers in some situations. Leavens and Hopkins (1998, 2005) conducted a study with chimpanzees in which a human experimenter placed a piece of food outside of the ape's reach and then left. When another human came in, the chimpanzees pointed to the food so that the human would get it for him (pointing was usually done with the whole hand, but some points were produced with just the index finger; see also Leavens et al. 2004). Human-raised chimpanzees have also been found to point to humans in order to obtain access to locations where there is food (Savage-Rumbaugh 1990), and some orangutans point for humans to the location where they can find a hidden tool, which they will then hopefully use to obtain food for the orangutans (Call and Tomasello 1994).

We thus find that apes do sometimes point for humans – given that they have had some contact with humans in the past. Importantly though, they use this manual gesture imperatively only. That is, they point for humans either in order to obtain a desirable object from them directly, as in the studies by Leavens and Hopkins (1998, 2005), or indirectly by requesting from the human to provide the necessary conditions for them to get the object themselves, as in Savage-Rumbaugh's (1990) study. It, thus seems that what the apes have learned from their experience with humans is that the human will help them, and that they can use the pointing gesture instrumentally in order to make him help them. They, thus, "use" the human as a "social tool" in order to get things they otherwise could not get, and they have learned that pointing gets this tool to work (the term "social tool" was first used by Bates et al. 1975). However, no ape has ever been observed to point for another ape or for a human declaratively – that is, just for the sake of sharing attention to some entity or event, or to inform others cooperatively, as humans often do.

Liszkowski et al. (2004, 2006) have shown in a series of experiments that even when they first begin to point at around 1 year of age, human infants do this with a full range of different motives – including the motive to share attention and interest. In one study (Liszkowski et al. 2004), an adult reacted differently towards infants' points, and the infants' responses to the adult reaction were investigated. The main finding was that if the adult did not jointly attend to the event with the infant (by alternating gaze between infant and event and commenting on it) – but instead either (1) just "registered" the event without sharing it with the infant or (2) only looked and emoted positively to the infant while ignoring the event – the infants were dissatisfied and tried to correct the situation. In contrast, in the joint attention condition, infants appeared satisfied with the response. Using the same basic methodology, Liszkowski et al. (2006) found that beyond the classic distinction of imperative and declarative pointing, 12-month-olds point for others also to inform them about things that are relevant for them. In that study, they directed an adult's attention to the location of an object for which that person was searching. What this suggests is that in human ontogeny, pointing is used from the very beginning not just in order to obtain certain objects via helpful adults as social tools, but with the motivation to help/inform others or to just jointly attend to things in the world with them.

The question is thus why apes do not point to share interest and inform others as human infants do from very early in development (see also Tomasello 2006). They clearly have the necessary motor abilities to do so. And again, it would surely be useful if they spatially indicated important events for one another. So why do they not do it? To answer this question, one needs to look at apes' understanding of pointing. One of the main paradigms that has been used to assess chimpanzees' comprehension of pointing is the Object Choice task. In the task designed by Tomasello et al. (1997b), one human, the hider, hides a piece of food for the ape in one of several containers. Then another human, the helper, shows the ape where it is by tilting the container so that she can look inside and see the food. After this "warm-up," the hider again places a piece of food in one of the containers, but now the helper indicates the location of the food for the ape by pointing at the baited container with his index finger (or by gazing at it). Variations of this method involve other kinds of communicative cues (Call and Tomasello 2005) and a trained chimpanzee instead of a human as the provider of the cue (Itakura et al. 1999). The results were the same in all these studies: the apes performed poorly, that is, they chose the correct container at chance level. They often followed the human's point (or gaze cue) to the container with their eyes, but they did not make any inferences from there about the location of food. That is, they cannot use or exploit the information that is conveyed to them via the pointing gesture – they do not know what it means. When following the human's point with their gaze, all they perceive is a useless bucket. To understand that the point is not directed at the bucket as such, but at the bucket qua location or qua container of a desired object, the apes would need to understand something about cooperation or communication. They would need to understand that the other is trying to communicate to them something that might be relevant for the achievement of their goal. In other words, an understanding of the meaning of the pointing gesture presupposes a more general understanding that others might want to help or inform us about the things which they assume are relevant for our purposes. And this understanding obviously goes beyond the apes' social-cognitive skills.

The view that the challenge of the Object Choice task does, indeed, lie in its cooperative structure is supported by recent studies using a competitive version of the task. In one version, Hare and Tomasello (2004), instead of pointing to the baited container, reached unsuccessfully for it. Superficially, this reaching behavior is very similar to the pointing gesture: the human's hand is oriented towards the container in which the food is hidden (the difference being that when pointing, only the index finger is stretched out, whereas in the case of reaching, all fingers point at the container). However, the chimpanzees' response in the reaching version was very different, as they successfully retrieved the food from the correct container. The reason for this must be that, even though the two tasks are superficially highly similar, their underlying structure is very different. Our interpretation is that in the case of reaching, the chimpanzees just need to perceive the goal-directedness of the human's reaching action and 'see' that there must be something desirable in the container. This task can, thus be solved with some understanding of the individual intentionality of the reaching action. In contrast, to understand pointing, the subject needs to understand more than the individual goal-directed behavior. She needs to understand that by pointing towards a location, the other individual attempts to communicate to her where a desired object is located; that the other tries to inform her about something that is relevant for her. So the ape would need to understand something about this directedness towards itself 'this is for me!' and about the communicative intention behind the gesture in order to profit from it. Apparently, apes do not understand that the cue is "for them" – used by the other in a helpful, informative and communicative way. Even though they are quite skillful in understanding intentional behavior that is directed at objects in the world (see Tomasello et al. 2005, for a review), they do not understand communicative intentions, which are intentions that are not directed at things or behaviors but at another individual's intentional states (with the embedded structure: "I intend for you to know that I intend for you x").

In order to explain why the apes fail to understand communicative intentions, one needs to broaden the perspective and focus on what we call the "joint attentional frame." The joint attentional frame or common ground (Clark and Brennan 1991) is what gives a pointing gesture its meaning – it is what "grounds" the communication in the shared space of meaning. To illustrate the point, imagine you are walking down the aisle of a hardware store and all of a sudden a stranger looks at you and points to a bucket standing in one of the shelves. You see the bucket, but, with a quizzical look on your face, look back at the stranger, because you do not know what is going on. The reason why you do not know what is going on is that you lack a joint attentional frame with the stranger, which would give the point its meaning. The pointing as such, in this frameless scenario, does not mean anything. But if, instead, you are walking down the same aisle with a friend because you are looking for a bucket to use for cleaning purposes, and your friend points out the bucket to you, you would know immediately what he means: "Here is one!" The presence of the joint attentional frame, which could be described by something like "we are searching for a bucket," grounds the point in the ongoing activity and gives it its meaning. Another possible scenario could be that you and your friend are looking for anything that is made of a certain kind of plastic because you like it so much. In this case, your friend's point would have a different meaning, namely something like: "Here is an item which is made of that plastic that you like so much!" The referent of the pointing gesture thus varies as a function of the joint attentional frame in which the pointing is anchored. One can imagine an endless number of joint attentional frames for the same basic scenario, with the referents of the pointing gesture being, for instance, "item with texture of kind x," "item which is similar to that other item we just saw," and so forth. The pointing gesture does not just indicate some spatial location, but instead it already contains a certain perspective from which the indicated object or location is to be viewed. And the perspective is carried by the joint attentional frame.

Humans can read pointing gestures based on joint attentional frames from as early as 14 months of age. Behne et al. (2005) found that 14-month-olds choose the correct container in the Object Choice task significantly above chance, thus demonstrating that they understand the pointing gesture cooperatively. Infants also know that the validity of a joint attentional frame is limited to those people who share it. Liebal et al. (2009) had 18-month-old infants clean up with an adult by picking up toys and putting them in a basket. At one point, the adult stopped and pointed to a ring toy, which infants then picked up and placed in the basket, presumably to help clean up. However, when the adult pointed to this same toy in this same way but in a different context, infants did not pick up the ring toy and put it in the basket; specifically, when the infant and adult were engaged in stacking ring toys on a post, children ignored the basket and brought the ring toy back to stack it on the post. The crucial point is that in both conditions the adult pointed to the same toy in the same way (and everything else in the room was the same), but the infant extracted a different meaning in the two cases – based on the two different joint attentional frames involved, and the jointness is, indeed, crucial here. Thus, in a control condition, the infant and adult cleaned up exactly as in the shared clean-up condition, but then a second adult who had not shared this context entered the room and pointed towards the ring toy in exactly the same way as the first adult in the other two conditions. In this case, infants did not put the toy away into the basket, presumably because the second adult had not shared the cleaning context with them. Rather, because they had no shared frame with this adult, they seemed most often to interpret the new adult's point as a simple invitation to note and share attention to the toy.

We, thus find that apes "communicate" individualistically, to get others to do things, and without joint attentional frames to ground the communicative intentions in a preexisting space of shared meaning. Human infants from as early as 14 months of age, on the other hand, communicate cooperatively – to share interest in things and inform others of things – and they construct and participate in joint attentional frames, which give cooperative gestures their meaning. Without a foundation in cooperative communication of this type, human language is not even thinkable (Tomasello 2008).

16.3.3 Cultural Learning

Human behavioral traditions have a cumulative history, with some of them showing a kind of "ratchet effect" of accumulating complexity over time (Tomasello et al. 1993). There is no convincing demonstration of the ratchet effect or any other form of cumulative cultural evolution for chimpanzees or any other nonhuman animals.

The explanation for this difference involved four components. First, although chimpanzees learn much socially (see for example the recent work of Whiten and colleagues as summarized in Whiten in press, this volume), humans seem to be more focused on actions than are chimpanzees, who are mainly focused on outcomes and goals. Humans are better and more accurate social learners: they are cultural learners. This special focus on actions enables them to socially learn activities from others in a much more accurate fashion, which not only contributes to the ratchet effect over time, but also enables the acquisition of cultural conventions that are only arbitrarily related to any causal relations in the world (such as linguistic symbols), since in this case faithful copying of actions is required.

Second, humans rely on teaching as a complement to their natural skills of social and cultural learning. Gergely and Csibra (2006) have recently elaborated an account explaining why the existence of relatively "opaque" cultural conventions (there is no causal structure or else it is difficult to see this structure) requires that human adults be specifically adapted for pedagogy toward children and human children be specifically adapted for recognizing when adults are being pedagogical (what Tomasello et al. 1993, called "instructed learning"). Engaging with others in this way is a kind of shared intentionality relying on cooperative and communication, in which the learner trusts the information given by the teacher. There has been no systematic study of chimpanzees engaged in anything resembling teaching since the observations of Boesch (1991), which have multiple interpretations.

Third, humans imitate one another not simply when they are aimed at acquiring more effective behavioral strategies in instrumental situations, but they also imitate for purely social reasons – to be like others. The tendency of human beings to follow fads and fashions and to conform are well known and well documented, and the proposal here, following Carpenter (2006), is that this represents a different and important motivation for social learning that may produce qualitatively different behaviors. For example, human infants have a greater tendency than do chimpanzees for copying the unnecessary "style" of an instrumental action (Carpenter and Tomasello, unpubl. data), and in acquiring linguistic conventions, children are not just driven by communicative efficacy but also by the desire to do it the way the others do it (Tomasello 2003). This analysis would also explain why children in the studies cited above sometimes imitated poor demonstrators when it would have been to their advantage to ignore them, and, in general, why children copy the actual actions of others more readily than do other apes. This so-called "social function" of imitation (Uzgiris 1981), the urge to be like others, is clearly an important part of human culture and cultural transmission.

Finally, human culture persists and has the character it has, not just because human children do what others do, but also because adults expect and even demand that they behave in certain ways: children understand that this is not just the way that something *is* done, but rather the way it *should* be done. This normative judgment is another aspect of shared intentionality, as it is essentially a judgment based on the perspective of the group – how "we" do things. In a recent study, Rakoczy et al. (2008) found that 3-year-old human children not only copied the way that others did things, but when they observed a third party doing them in some other way they objected and told them they were doing it "wrong" – that is not how "one" does it. Kelemen (1999) has also shown that young children learn very quickly that a particular artifact is for a particular function, and other uses of it may be considered "wrong" – this is not how "we" use this artifact. This normative dimension to human cultural traditions serves to guarantee their faithful transmission across generations in a way that supports further ratcheting up in complexity across historical time.

It may very well be, then, that it is these processes and aspects (cultural learning, teaching, normativity) that give human cultural traditions their extraordinary stability and cumulativity over time. Integral in all of these is a kind of social engagement depending on skills and motivations for shared intentionality.

16.4 Joint Attention and Perspective

It is, thus clear that human infants, before they are fully participating members of a culture, already have a special motivation for sharing experiences with other persons, and they possess special skills for creating with others joint goals, joint intentions, and joint attention. They learn from others in unique ways as well.

However, our claim goes further. Our claim is that participation in interactions involving shared intentionality transforms human cognition in fundamental ways. Most importantly, it actually creates new forms of cognitive representation, specifically, perspectival or dialogic cognitive representations (see also Tomasello 1999; Tomasello et al. 2005). In understanding and internalizing an adult's intentional states, including those directed towards her, at the same time she experiences her own intentional states towards the other, the child comes to conceptualize the interaction simultaneously from both the first and third person's perspective (see Barresi and Moore 1996) – forming a bird's eye view of the collaboration in which both commonalities and differences are all comprehended within a single representational format. Such perspectival representations are necessary not only for supporting cooperative interactions online, but also for the creation and use of certain kinds of cultural artifacts, most importantly linguistic and other kinds of symbols, which are socially constituted and bi-directional in the sense of containing simultaneously the perspective_s_of_both_speaker and listener (see Mead 1934).

These perspectival cognitive representations pave the way for later uniquely human cognitive achievements. Importantly, following Harris (1996), Tomasello and Rakoczy (2003) argued and presented evidence that coming to understand false beliefs – the fact that someone else's cognitive perspective about a state of affair is different from what I know to be true – depends on children's participation over a several year period in perspective-shifting discourse. In such linguistic discourse, including such things as misunderstandings and requests for clarification, children experience regularly that what another person thinks is often different from what they think, and the understanding of false beliefs – which, in almost everyone's account, is fundamental to mature human social cognition – is apparently unique to humans (Call and Tomasello 2008). And at age 4, children not only come to understand that others might hold false beliefs, they develop a sophisticated understanding of perspectives more generally. That is, they appreciate that different people might see or conceptualize a given event or object in different ways and also, that one and the same person can view or construe an object differently at different times. This ability to simultaneously "confront" perspectives becomes manifest in a variety of tasks besides the standard false belief task. For example, children now understand that one and the same object might (1) look like a rock but really be a sponge (the so-called appearance-reality distinction, Flavell et al. 1986), (2) be both an animal and a rabbit (as shown by their acceptance of alternative labels for a given object, Doherty and Perner 1998), and (3) be seen right-side-up from one perspective but upside-down from another (the so-called level 2 perspective-taking, Masangkay et al. 1974).

We would argue that young children come to understand and operate with the concept of perspective only after first experiencing the sharedness of attention on one and the same thing (see also Barresi and Moore 1996). From thereon, they can later begin to understand the differences in perspectives which converge on the shared target or object of interest. The foundations again can be found in infancy.

Evidence for this comes from a series of studies in which infants must determine what an adult has experienced and has not experienced. Tomasello and Haberl (2003) had 12- and 18-month-old infants play with an adult with two toys in turn. Before a third toy was brought out by an assistant, the adult left the room. During her absence, the infant played with the third toy together with the assistant. Finally, all three toys were held in front of the infant, at which point the adult returned into the room and exclaimed excitement followed by an unspecified request for the infant to give her *that* toy (without indicating by gazing or pointing which specific toy she was attending to). Surprisingly, infants of both ages selected the toy the adult had not experienced (the one which was new for her). In order to solve this task, infants had to understand (1) that people get excited about new, not familiar things and (2) which of the toys was new for the adult and which she was already familiar with from previous experience.

In this study, infants knew what was familiar for the adult after they had participated with her in joint attention around two of the objects (but not the third). This suggests the possibility that infants need to attend to another person's experience in joint attention with her in order to register the other as knowing the object in question. And this is what was basically found in the two studies by Moll et al. (2007), Moll and Tomasello (2007b). Following the basic procedure of Tomasello and Haberl (2003), 14- and 18-month-old infants either (1) became familiar with the first two objects in a joint attentional frame together with the adult or (2) simply witnessed the adult become familiar with the known objects individually. In each case, infants themselves became equally familiar with all three objects, as in the original study. The result was that infants knew which of the three objects was new for the adult, and thus captured her attention only when they had explored the known objects in a joint attentional format with her. They could not make this distinction when they had just witnessed her exploring them on her own, outside of a joint attentional frame. The shared attention to the known objects thus highlighted the fact that the third object was not jointly experienced. It is, thus inside of joint attentional frames, that infants first begin to realize differences in people's experiences and perceptions. This early understanding of other's experiences is the foundation for the later developing understanding of divergent perspectives on one and the same thing – an understanding which is, just like the ability to jointly engage with others, uniquely human.

Our argument is thus that basically all species-unique aspects of human cognition reflect their cooperative roots in fundamental ways. The ability to take the perspective of others – which spawns the understanding of false beliefs, perspectival cognitive representations, and collective/institutional reality – is only possible for organisms that can participate in social interactions involving shared intentionality, especially involving joint attention. Let us be very clear on this point. Participation in these interactions is critical. A child raised on a desert island would have all of the biological preparations for participation in interactions involving shared intentionality, but because the child did not actually participate in such interactions, she would have nothing to internalize into perspectival cognitive representations. Ontogeny in this case is critical.

16.5 From Collaboration to Culture

We, thus find that human infants in their second year of life are much more skilled, and much more motivated, than are great apes at participating in collaborative problem solving and cooperative communication, and their skills of social/cultural learning have unique qualities as well. Following Tomasello et al. (2005), our claim is that the reason for this difference is that human infants are biologically adapted for social/cultural interactions involving shared intentionality. Even at this tender age, human infants already have special skills for creating with other persons joint goals, joint intentions, and joint attention, and special motivations for helping and sharing with others – and for communicating with and learning from others within these special interactions as well.

It would seem that we are still a long way from such things as governments and religions and sciences and other large-scale cultural institutions. But actually we are not. These kind of cultural institutions arose only very recently in human evolution, after, and partially as a result of, the agricultural revolution when people began living in large cities. There, many new demands arose for collaborating with strangers and in much larger groups and across longer spans of time than previously. Our simple proposal is that the collaborative and communicative skills we see in young children form the necessary foundation for beginning to participate in such large-scale collaboration. Virtually no one believes that there were any genetic events at the agricultural revolution that led to the unprecedented population explosion and flowering of cultural institutions associated with that great event. So what is required for scaling up small group collaboration into large group collaboration is mainly certain sociological conditions, as a first step, and then cultural-historical processes over time and generations in which such institutions could be formed.

What we have called the cultural intelligence hypothesis, or the Vygotskian intelligence hypothesis, reflects the idea that the incredible complexities of modern human culture and its institutions are the result of a qualitatively new process that arose in human evolution. Although there may be some cultural transmission in some animal species, nothing like the human creation of cultural artifacts and institutions – which ratchet up in complexity over time and within which children's ontogeny proceeds and on which it depends – takes place in other species. This is because human beings collaborate with, communicate with, and learn from their groupmates based on unique skills and motivations for shared intentionality and cultural learning. We do not believe that the human gap can be explained by any appeal to individual cognitive skills, but rather it can only be explained by the social–cultural processes for which they are specially adapted and within which each generation of modern humans has evolved.

References

Barresi J, Moore C (1996) Intentional relations and social understanding. Behav Brain Sci 19:107–154

Bates E (1979) The emergence of symbols: cognition and communication in infancy. Academic, New York

Bates E, Camaioni L, Volterra V (1975) The acquisition of performatives prior to speech. Merrill Palmer Q 21:205–226

Behne T, Carpenter M, Tomasello M (2005) One-year-olds comprehend the communicative intentions behind gestures in a hiding game. Dev Sci 8:492–499

Boesch C (1991) Teaching among wild chimpanzees. Anim Behav 41:530–532

Boesch C (2005) Joint cooperative hunting among wild chimpanzees: taking natural observations seriously. Behav Brain Sci 28:692

Boesch C, Boesch H (1989) Hunting behavior of wild chimpanzees in the Taï-National-Park. Am J Phys Anthropol 78:547–573

Boesch C, Boesch-Achermann H (2000) The chimpanzees of the Taï forest: behavioural ecology and evolution. Oxford University Press, Oxford

Bratman ME (1992) Shared cooperative activity. Phil Rev 101:327–341

Call J, Tomasello M (1994) The production and comprehension of referential pointing by orangutans. J Comp Psychol 108:307–317

Call J, Tomasello M (2005) What chimpanzees know about seeing revisited: an explanation of the third kind. In: Eilan N, Hoerl C, McCormack T, Roessler R (eds) Joint attention. Oxford University Press, Oxford, pp 45–64

Call J, Tomasello M (2008) Does the chimpanzee have a theory of mind? 30 years later. Trends Cogn Sci 12:187–192

Camaioni L (1993) The development of intentional communication: a re-analysis. In: Nadel J, Camaioni L (eds) New perspectives in early communicative development. Routledge, New York, pp 82–96

Carpenter M (2006) Instrumental, social, and shared goals and intentions in imitation. In: Rogers SJ, Williams J (eds) Imitation and the development of the social mind: lessons from typical development and autism. Guilford, New York, pp 48–70

Carpenter M, Tomasello M, Striano T (2005) Role reversal imitation and language in typically developing infants and children with autism. Infancy 8:253–278

Clark HH, Brennan SE (1991) Grounding in communication. In: Resnick LB, Levine JM, Teasley SD (eds) Perspectives on socially shared cognition. APA Books, Washington, DC, pp 127–149

Doherty M, Perner J (1998) Metalinguistic awareness and theory of mind: just two words for the same thing? Cogn Dev 13:279–305

Flavell JH, Green FL, Flavell ER (1986) Development of knowledge about the appearance-reality distinction. Monogr Soc Res Child Dev 51(Serial No 212):i–v 1–87

Gergely G, Csibra G (2006) Sylvia's recipe: the role of imitation and pedagogy in the transmission of cultural knowledge. In: Enfield NJ, Levinson SC (eds) Roots of human sociality: culture, cognition, and interaction. Berg, Oxford, pp 229–255

Hare B, Tomasello M (2004) Chimpanzees are more skillful in competitive than in cooperative cognitive tasks. Anim Behav 68:571–581

Harris P (1996) Desires, beliefs, and language. In: Carruthers P, Smith PK (eds) Theories of theories of mind. Cambridge University Press, Cambridge, pp 200–222

Herrmann E, Call J, Hernández-Lloreda MV, Hare B, Tomasello M (2007) Humans have evolved specialized skills of social cognition: the cultural intelligence hypothesis. Science 317:1360–1366

Herrmann E, Call J, Hernández-Lloreda M, Hare B, Tomasello M (in press) The structure of individual differences in the cognitive abilities of children and chimpanzees. Psychological Science

Itakura S, Agnetta B, Hare B, Tomasello M (1999) Chimpanzees use human and conspecific social cues to locate hidden food. Dev Sci 2:448–456

Jensen K, Hare B, Call J, Tomasello M (2006) What's in it for me? Self-regard precludes altruism and spite in chimpanzees. Proc Roy Soc Lond B 273:1013–1021

Kaminski J, Call J, Tomasello M (2004) Body orientation and face orientation: two factors controlling apes' begging behavior from humans. Anim Cogn 7:216–223

Kelemen D (1999) Functions, goals and intentions: children's teleological reasoning about objects. Trends Cogn Sci 12:461–468

Leavens DA, Hopkins WD (1998) Intentional communication by chimpanzees: a cross-sectional study of the use of referential gestures. Dev Psychol 34:813–822

Leavens DA, Hopkins WD (2005) Multimodal concomitants of manual gesture by chimpanzees (*Pan troglodytes*): influence of food size and distance. Gesture 5:73–88

Leavens DA, Hostetter AB, Wesley MJ, Hopkins WD (2004) Tactical use of unimodal and bimodal communication by chimpanzees, *Pan troglodytes*. Anim Behav 67:467–476

Liebal K, Behne T, Carpenter M, Tomasello M (2009) Infants use shared experience to interpret pointing gestures. Dev Sci 12:264–271

Liszkowski U, Carpenter M, Henning A, Striano T, Tomasello M (2004) Twelve-month-olds point to share attention and interest. Dev Sci 7:297–307

Liszkowski U, Carpenter M, Striano T, Tomasello M (2006) 12- and 18-month-olds point to provide information for others. J Cogn Dev 7:173–187

Masangkay ZS, McCluskey KA, McIntyre CW, Sims-Knight J, Vaughn BE, Flavell JH (1974) The early development of inferences about the visual percepts of others. Child Dev 45:357–366

Mead GH (1934) Mind, self, and society. University of Chicago Press, Chicago, IL

Melis AP, Hare B, Tomasello M (2006) Chimpanzees recruit the best collaborators. Science 311:1297–1300

Moll H, Tomasello M (2007a) Cooperation and human cognition: the Vygotskian intelligence hypothesis. Phil Trans Roy Soc B 362:639–648

Moll H, Tomasello M (2007b) How 14- and 18-month-olds know what others have experienced. Dev Psychol 43:309–317

Moll H, Carpenter M, Tomasello M (2007) Fourteen-month-olds know what others experience only in joint engagement. Dev Sci 10:826–835

Povinelli DJ, Nelson KE, Boysen ST (1992) Comprehension of role reversal in chimpanzees: evidence of empathy. Anim Behav 43:633–640

Rakoczy H, Warneken F, Tomasello M (2008) The sources of normativity: young children's awareness of the normative structure of games. Dev Psychol 44:875–881

Savage-Rumbaugh ES (1990) Language as a cause-effect communication system. Philos Psychol 3:55–76

Searle JR (1995) The construction of social reality. Free Press, New York

Silk JB, Brosnan SF, Vonk J, Henrich J, Povinelli DJ, Richardson AS, Lambeth SP, Mascaro J, Schapiro SJ (2005) Chimpanzees are indifferent to the welfare of unrelated group members. Nature 437:1357–1359

Tomasello M (1999) The cultural origins of human cognition. Harvard University Press, Cambridge, MA

Tomasello M (2003) Constructing a language: a usage-based theory of language acquisition. Harvard University Press, Cambridge, MA

Tomasello M (2006) Why don't apes point? In: Enfield NJ, Levinson SC (eds) Roots of human sociality: culture, cognition and interaction. Berg, New York, pp 506–524

Tomasello M (2008) Origins of human communication. MIT, Cambridge, MA

Tomasello M, Carpenter M (2005) The emergence of social cognition in three young chimpanzees. Monogr Soc Res Child Dev 70:1–132

Tomasello M, Haberl K (2003) Understanding attention: 12- and 18-month-olds know what is new for other persons. Dev Psychol 39:906–912

Tomasello M, Rakoczy H (2003) What makes human cognition unique? From individual to shared to collective intentionality. Mind Lang 18:121–147

Tomasello M, Kruger AC, Ratner HH (1993) Cultural learning. Behav Brain Sci 16:495–552

Tomasello M, Call J, Gluckman A (1997a) Comprehension of novel communicative signs by apes and human children. Child Dev 68:1067–1080

Tomasello M, Call J, Warren J, Frost GT, Carpenter M, Nagell K (1997b) The ontogeny of chimpanzee gestural signals: a comparison across groups and generations. Evol Comm 1:223–259

Tomasello M, Carpenter M, Call J, Behne T, Moll H (2005) Understanding and sharing intentions: the ontogeny and phylogeny of cultural cognition. Behav Brain Sci 28:675–735

Tuomela R (2002) Collective goals and communicative action. J Phil Res 27:29–64

Uzgiris IC (1981) Two functions of initation during infancy. Int J Behav Dev 4:1–12

Vygotsky LS (1978) Mind in society: the development of higher psychological processes. Harvad University Press, Cambridge, MA

Warneken F, Tomasello M (2006) Altruistic helping in human infants and young chimpanzees. Science 311:1301–1303

Warneken F, Tomasello M (2007) Helping and cooperation at 14 months of age. Infancy 11: 271–294

Warneken F, Chen F, Tomasello M (2006) Cooperative activities in young children and chimpanzees. Child Dev 77:640–663

Whiten A. Coming of age for cultural Panthropology. In: Lonsdorf E, Ross S, Matsuzawa T (eds) The mind of the chimpanzee. University of Chicago Press, Chicago, IL (in press)

Chapter 17
The Evolution and Development of Human Social Cognition

David F. Bjorklund, Kayla Causey, and Virginia Periss

Abstract Humans' ancestors experienced intense selection pressures to develop enhanced social-cognitive abilities, facilitating the coevolution of an extended childhood, larger brain, and increased social complexity. This chapter describes the emergence of human social-cognitive abilities from an evolutionary developmental perspective, focusing on the importance of social interaction and epigenetic inheritance. The development of shared attention and referential communication, empathy, social learning, and theory of mind is discussed as it occurs in human children, and research demonstrating the importance of parent–child interactions and individual differences in maternal behavior during the development of these abilities is highlighted. A discussion of how these abilities are expressed in mother-reared and enculturated chimpanzees is also included and indicates that these animals possess substantial social-cognitive competencies, which, under certain rearing conditions, can be modified to resemble a more *Homo sapiens* way of thinking. This is strongly suggestive that our common ancestor with chimpanzees also possessed the neurological plasticity to adapt its behavior and cognition in response to changes in environmental conditions.

17.1 Introduction

Biologists since Darwin have believed that there is a continuity of mental functioning in evolution, so that the cognition of a species such as *Homo sapiens* should share many features with species with which they recently shared a common ancestor. In a quest both to understand human evolution and the cognition of extant animals, scientists have assessed the cognitive gap between humans and other

D.F. Bjorklund (✉), K. Causey, and V. Periss
Department of Psychology, Florida Atlantic University, Boca Raton, FL, USA
e-mail: dbjorklu@fau.edu, klabeth@mac.com, vperiss@gmail.com

P.M. Kappeler and J.B. Silk (eds.), *Mind the Gap*,
DOI 10.1007/978-3-642-02725-3_17, © Springer-Verlag Berlin Heidelberg 2010

primates. In fact, the size of the gap between humans and our great-ape relatives has been a point of contention among scientists and laypeople for decades, perhaps centuries. Some view chimpanzees, for instance, as "almost human," whereas others see them as just "clever animals." How people view the great apes is also related to how they view humans, as "next to angels" (literally or figuratively) or simply as "the thinking animal."

The urge to either increase or decrease the cognitive and evolutionary distance between apes and humans varies both with one's scientific and philosophical (and, perhaps, political and theological) viewpoints. The scientific debates hinge on what unique attributes evolved, over the past 5–8 million years when modern humans last shared a common ancestor with *Pan troglodytes* and *Pan paniscus*, in the line that led to *H. sapiens*. Identifying the set of selective pressures that is the best candidate for the driving force in human cognitive evolution has also provoked much discussion. Many have been nominated, and we believe there is no single domain that can be pointed to as "the" cause of human evolution. However, we, as many others (e.g., Humphrey 1976; Alexander 1989; Dunbar 1995) have concluded that human intellectual evolution was driven by the need to deal with conspecifics. As such, human social cognition can be viewed as the more primary form of thought, with *H. sapiens*' impressive suite of intellectual abilities being essentially the derivative of cognition initially evolved to cooperate and compete with fellow hominids. Humans obviously share many social-cognitive abilities with other social primates, and so, even for those that are unique to humans, we are able to see their roots in the behaviors of our close genetic relatives.

There are many different aspects of human social cognition and its development, but at their core is the understanding that other people are *intentional agents* – beings whose behavior is based on knowledge and desires, and whose actions are deliberate in achieving a goal (i.e., they do things "on purpose"; see Bandura 2006; Tomasello and Carpenter 2005). Although signs of intentional representation appear late in the first year, children's understanding of others as intentional agents develops over childhood, culminating in the ability to pass false-belief tasks around 4 years of age, the benchmark for attaining theory of mind.

Although most aspects of social cognition follow a regular developmental pathway, there are substantial individual differences in the rate at which some contents are acquired or the level they are expressed. Given the wide range of social environments in which humans live, it is necessary that children's social cognition be flexible, matched to the demands of their particular culture. Although there are a host of potential factors that contribute to individual differences in social cognition, parents likely play a particularly important role. Mothers (and, sometimes, fathers) are children's first social partners, and humans' extended period of dependency requires prolonged parental care. Although peers and other adults in a community certainly influence the children's development (e.g., Harris 1995; Bronfenbrenner and Morris 2006) – increasingly so as they become more independent from their parents, beginning around 6 or 7 years of age – the foundation of social cognition is laid down during the preschool years, and it is this time when parents' influences tend to be greatest (e.g., Scarr 1993).

The focus of this chapter is on the development of social cognition in human children. In addition to describing the typical course of development, we also examine the influence that parental behaviors have on the development of social-cognitive abilities of humans as well as on the social-cognitive abilities of chimpanzees (*P. troglodytes*). Chimpanzees provide the best model for what the social cognition of our ancestors may have been like. Like humans, chimpanzees and bonobos (*P. paniscus*) have big brains, an extended juvenile period, and live in socially complex communities, conditions that some have proposed were necessary ingredients for the evolution of the human mind (e.g., Bjorklund and Rosenberg 2005; Bjorklund 2006). We should, thus, see hints of some of the social-cognitive abilities that we see in young humans in our close genetic relatives. Moreover, just as individual differences in rearing environment affect the course of social cognition in children, they may also influence the development of social cognition in young apes. We, therefore, include in our review evidence that human rearing (enculturation) of chimpanzees can alter ape cognition in a more human-like fashion. *Enculturation* in this context refers to apes being reared by humans much as human children are reared. This typically includes language, encouraging shared attention, and deliberate teaching (Call and Tomasello 1996).

We start our review with the early-developing foundational social-cognitive abilities of shared attention, referential communication, and empathy. We then examine more advanced forms of social cognition, specifically, social learning and theory of mind.

17.2 Shared Attention and Referential Communication

Shared (or *joint*) *attention* involves a triadic interaction between the child, another person, and an object. For example, parents often draw their child's attention to an object by pointing or gazing at the object, a form of *referential communication*, which indicates that the "pointer" understands that he or she sees something that the observer does not.

17.2.1 Development

Although parents may engage in this type of behavior from the earliest days of a baby's life, it takes infants a while to fully catch on. However, infants are oriented to social interactions from birth and "hit the ground running," so to speak, regarding the acquisition of information through social interaction. For instance, neonates orient to the human face and quickly learn to seek their mothers' faces (Feldman and Eidelman 2004). By 2 months, infants begin to engage in dyadic exchanges, reciprocating the social patterns of a partner in an intimate one-on-one context (e.g., Lavelli and Fogel 2005), and by 2 or 3 months, they can recognize self-produced, biological motion (Bertenthal et al. 1987) and soon turn to look in the same direction of another (Tomasello et al. 2003). These behaviors are necessary for shared attention, which is typically first seen in infants around 9-months of age and increases over the next year. For example, 12-month olds will point to inform others about unknown events (Liszkowski et al. 2007), and between 12 and 18 months, infants acquire the ability to reference others' eye gaze as a cue to direct their own attention (Brooks and Meltzoff 2002) and to point to objects to direct an adult's attention to something he or she is searching for (Liszkowski et al. 2006). Between 18 and 24 months of age, toddlers use eye gaze along with other directional cues, such as pointing and head orientation, for word learning and social referencing (Poulin-Dubois and Forbes 2002). These findings indicate that beginning about 9 months of age, infants view other people as *intentional agents*, an understanding that other peoples' behavior is based on their goals and intentions (see Tomasello 1999; Tomasello et al. 2007).

By the end of the first year of life, infants extend shared attention to social referencing to guide their behavior when encountering ambiguous events (e.g., Feinman 1991; Vaish and Striano 2004). Infants can make use of a parent's facial expression, tone of voice, gestures, or combinations of these sources to determine their actions in an uncertain situation (Hornik et al. 1987; Vaish and Striano 2004). In these contexts, infants exploit the triadic interaction between themselves, an object, and an adult to derive information from the adult's responses to the object or situation. Differences in the amount and quality of the information that is transmitted by the mother may have significant, and sometimes dire, consequences for her child. Moreover, as a situation in which infants must infer the emotional response of an individual, social referencing may provide the stepping-stone necessary to develop empathy.

By being drawn into the social relationship and taking the perspective of the parent, children are able to extend their attention beyond the dyad to the object being referenced. This ability to take the perspective of others is at the core of social cognition, and, as such, deficits in shared attention are taken as early indicators of autism (e.g., Adamson et al. 2001).

17.2.2 Parental Effects

While the development of shared attention seems highly canalized, a supportive social partner is required and this is typically a child's parent, frequently his or her mother. As such, characteristics of the parent may lead to individual differences in parent–infant interactions.

For example, dysphoric (depressed mood) mothers spend less time engaging in shared-attention activities with their infants than nondysphoric mothers (Goldsmith and Rogoff 1997). But it is not only the quantity of shared attention that can be affected, but also the quality. For example, Deák et al. (2008) reported that 15- and 21-month-old infants were more likely to look at a parent (mostly mothers) who elicited their attention verbally and to follow the shifting gaze of a parent who also pointed or used directing verbalizations, indicating that differences in maternal behaviors can influence how effective parents are in attracting and redirecting their child's attention.

Individual differences in parental behavior that lead to variations in dyadic interactions may also have effects on infants' later social-cognitive development. For instance, Carpenter et al. (1998b) found that the amount of time 9-month-old infants spent in shared attention with their mothers and a mother's use of language containing some reference to an object predicted infants' communication skills at 16 months. Other research found that 12-month olds' initiations of, and responses to, shared attention predicted their social competence at 30 months, even after accounting for infant temperament, cognition, language, and demographic variables (Van Hecke et al. 2007). Taken together, these findings underscore the important role of early social interaction in the development of later social-cognitive abilities.

17.2.3 Abilities of chimpanzees

Although chimpanzees, and even monkeys, will follow the gaze of another individual in some contexts (Bering and Povinelli 2003; Bräuer et al. 2005), most researchers argue that there is no evidence that either mother-reared or enculturated chimpanzees engage in shared attention (Tomasello and Carpenter 2005; Herrmann et al. 2007). Others disagree, noting, for example, that captive chimpanzees point to food only in the presence of a caretaker (Leavens et al. 2005). Moreover, evidence that wild chimpanzees will scratch themselves in an exaggerated way to direct the action of a grooming partner (Pika and Mitani 2006) suggests that even apes in the wild may engage in some form of referential communication, albeit a type that is centered on a chimpanzee's body and not on distant objects.

In contrast to mother or nursery-reared chimpanzees, clear evidence of referential pointing is observed in each of the great ape species for enculturated individuals: orangutans (*Pongo pygmaeus*, e.g., Miles 1990), gorillas (*Gorilla gorilla*, e.g., Patterson 1978), chimpanzees (*P. troglodytes*, e.g., Povinelli et al. 1992), and bonobos (*P. paniscus*, e.g., Savage-Rumbaugh et al. 1986).

17.3 Empathy

As evolutionary primatologist Sarah Hrdy (1999, p. 392) wrote: "*What makes us humans rather that just apes is the capacity to combine intelligence with articulate empathy*." Generally, empathy refers to the ability to recognize, perceive, and feel the emotion of another and requires the ability to take the perspective of another. Here, we distinguish between empathy that is insightful (cognitive empathy), sympathy (emotional recognition), and emotional contagion (emotional empathy). When considering its development, it is important to consider different types of empathy as reflecting different underlying mechanisms. For example, recognizing that Mom is angry based on her facial, vocal, or postural expression can be characterized as *sympathy*. The sympathizer can recognize the emotion and may even be moved to change it, but lacks insight into the nature or cause of the emotion. Sympathy (emotion recognition) is less sophisticated than the ability to recognize that Mom is angry because she got cut off in traffic (understanding the object of emotion) and is running late (nonpsychological context) and is quick to anger, in general (character traits). These insights are considered truly *empathic* because the observer gets an immediate sense of what it would be like "in Mom's shoes" and understands the perspective of Mom (cognitive empathy). It is also important to distinguish empathy from *emotional contagion,* an automatic or unconscious process by which the observer shares the emotions of another, but that does not require insight on the part of the observer (Preston and de Waal 2002).

17.3.1 Development

There is evidence that empathy is rooted in an early ability to associate the behaviors, vocal cues, and facial expressions of another with the proprioceptive feelings that occur when producing the corresponding movements or expressions oneself (e.g., Singer et al. 2004). For example, even newborn infants display emotional contagion, engaging in contagious crying in response to the cries of other newborns, although not to the cries of older infants or to their own cries when played back to them (Martin and Clark 1982; Dondi et al. 1999).

Cognitive empathy, in contrast, is described as a secondary, or self-conscious, emotion and emerges in the second half of the second year of life (Lewis 1993). For instance, in one study, researchers examined 12- to 24-month-old children's responses to the distress of other people (Zahn-Waxler et al. 1992). They reported that empathic responses increased with age, such that 2-year-old children often made facial expressions or other gestures indicative of sadness, tried to comfort distressed people, and sometimes sought information about the person's distress (What's wrong?). Bischof-Köhler (1988, 1991) proposed that cognitive empathy, insofar as it involves emotional insight, is a mechanism that aids understanding of another's belief or desire state by allowing the observer to vicariously share in this state. In other words, fully developed empathy requires theory of mind, while emotional contagion does not. As such, true empathic insight relies on self-recognition, the ability to make a clearcut distinction between the emotional domains of self and other, which develops typically in the latter half of the second year. Thus, much of the "empathic" behaviors observed in infants can be explained as resulting from emotional contagion.

In a series of experiments, Bischof-Köhler (1988, 1991, 1994) demonstrated that self-objectification (measured by mirror self-recognition) and empathy (objectified as prosocial behavior) emerge almost simultaneously in development: only those children capable of passing the "rouge test" for self-recognition tried to help a person in need who demonstrated grief, whereas nonrecognizers stayed indifferent.

Bischoff-Köhler's depiction of empathy development is stage-like: initially, infants are victims of emotional contagion, but emergence of self-objectification serves as a switch that "turns on," or allows, empathy. The process requires that infants orient to emotional cues early on and their ability to do so is partly due to the manner in which their caregivers direct them to relevant information.

17.3.2 Parental Effects

Kuchuk et al. (1986) showed 3-month-old infants static pictures of faces that varied on the intensity of their smiling expression. Infants were more sensitive to the subtle variations of smiles in the pictures if their mothers more often directed attention to their own faces when smiling. Importantly, this relationship was strongest for mothers who displayed relatively low levels of smiling, suggesting that it is not the number of smiles that a mother shows her infant, but the *quality* of the dyadic interaction during these experiences that affects infants' ability to discriminate between subtly different expressions. Kuchuk et al. suggested that events that are made salient during atypical experience might be more distinctive to infants, increasing their sensitivity to corresponding subtle differences as a result.

In support of this, although most infants look longer at sad faces relative to happy faces (Field et al. 1998), infants of depressed mothers tend to experience disproportionately high levels of exposure to sad, angry, or neutral faces, and as a result seem to be desensitized to sad expressions and do not display the typical looking-time preference for sad faces but rather prefer smiling faces (Striano et al. 2002). This and related work with maltreated children (e.g., Pollak and Kistler 2002; Pollak and Sinha 2002) suggest that young children possess mechanisms that allow them to become sensitized to particular types of emotional expressions and heighten their sensitivity to other types. For example, it would be advantageous for maltreated children to quickly recognize anger in order to avoid negative consequences, but quickly detecting sad faces would not provide any additional advantage to children of depressed mothers (de Haan et al. 2004).

While the existing data suggest that the quality of early mother–child interactions affects children's differential attention to emotional expressions, we view this ability as a stepping stone from emotional contagion toward fully developed empathy that requires the ability to take the perspective of another, and presumably to treat the other individual as an intentional agent, whose emotions affect his or her goals and behaviors. This is often referred to as possessing a "theory of mind," and we discuss maternal effects on these abilities later in this chapter.

17.3.3 Abilities in Chimpanzees

As we noted, cognitive empathy seems to require the ability of the observer (or empathizer) to take the perspective of another in order to identify with or to understand another's feelings. However, Preston and de Waal (2002) proposed that simpler forms, such as emotional contagion and sympathy, are observed in a wide range of social animals.

At least one study provides evidence that enculturated apes will spontaneously provide help to a familiar human. In a study by Warneken and Tomasello (2006), 18- and 24-month-old human children helped an adult when he or she, for example, accidentally dropped a marker on the floor or when misplacing a book on a stack. They failed to help (e.g., retrieve the fallen marker or book) when the adult intentionally threw the marker on the floor or placed the book beside the stack. Warneken and Tomasello reported similar results for three human-reared chimpanzees, although only when the human was reaching unsuccessfully for an object, not for other types of tasks. These young, enculturated apes displayed evidence of "helping" behavior, something not previously observed in unrelated chimpanzees. A more recent controlled study reported that semi-free ranging and nonenculturated chimpanzees engaged in some helping behaviors to both humans and conspecifics, suggesting that the roots of altruism may be part of chimpanzee, as well as human, nature (Warneken et al. 2007).

De Waal (1997, 2005) has argued that great apes display more advanced forms of empathy under certain circumstances, including cognitive empathy, and he provides substantial anecdotal evidence in support of this. For example, de Waal (2005) relates the episode in which Binti Jua, an 8-year-old female gorilla, helped a 3-year-old boy who fell into the primates' cage at the Chicago Brookfield Zoo. In an equally provocative example, a long-time researcher at the Stuttgart Zoo introduced her newborn baby to the bonobos, upon which the alpha female disappeared for a short time and returned with her own newborn (de Waal 1997).

Yet, in laboratory settings, chimpanzees seem to be indifferent to the welfare of others. For example, chimpanzees will not provide food to other familiar but unrelated chimpanzees, even though there is no material cost to themselves (Silk et al. 2005, see also Jensen et al. 2006).

17.4 Social Learning

Social learning is broadly defined as occurring in a situation "*in which one individual comes to behave similarly to another*" (Boesch and Tomasello 1998), and is further differentiated based on presumed underlying mechanisms (see Tomasello and Call 1997; Tomasello 1999).

Consider the various ways in which a young child might go about learning as a consequence of observing his of her sister push a chair over to the refrigerator, climb on top, and retrieve the cookies hidden above. Perhaps the simplest form of social learning is local enhancement, in which an individual demonstrates an increased interest in the particular location at which he or she observed another individual performing an action (in this case, the refrigerator). As a result, the child moves to that location and independently discovers a useful outcome (cookies).

In a similar process, stimulus enhancement involves an observer's increased interest in manipulating a particular object (a chair) as opposed to moving to a particular location. Mimicry involves learning about the causal actions of an individual, as opposed to the object or goal, and replicating those actions without any insight as to why those actions are being performed or that a goal is even present (pushing the chair to the refrigerator and climbing on it just because your sister did it – without knowing ahead of time that it helps you reach the cookies on top). Goal emulation involves recognizing that a particular goal exists in the environment (there are desirable cookies on top of the refrigerator) and setting to reach that goal by one's own means (climbing on the counter to reach the top of the refrigerator rather than climbing on a chair). Imitation involves understanding and reproducing the goal (get cookies) and the means by which it was achieved (push the chair over to the fridge and climb up on it).

While imitation might be the most sophisticated form of social learning someone can demonstrate, perhaps the most sophisticated component of the learning process is teaching, which involves the teacher modifying his or her behavior in order that the "student" acquires new knowledge. We define teaching following Tomasello et al. (1993a), as requiring that the learner appreciate the perspective of the teacher and that the teacher be sensitive to the knowledge, motivations, and emotions of the learner.

17.4.1 Development

Although neonates will match the facial expression of a model (Meltzoff and Moore 1977), infants are learning nothing new via such matching behavior, and thus, this likely does not reflect social learning (Jacobson 1979; Bjorklund 1987). Rather, infants and caregivers are engaging in social mirroring that helps foster social interaction between the infant and caregiver (Byrne 2005). Such early facilitation of social interaction is important for establishing patterns of later social interaction and, as a result, social learning. In fact, Heimann (1989) found that neonatal "imitation" of tongue protrusions positively predicted the levels of mother–infant social interaction 3 months later.

Infants' social learning clearly increases over the first year of life, although it is difficult to ascertain with certainty if infants are merely mimicking an adult model or engaging in other more sophisticated forms of social learning. Early in the second year of life, however, infants are sensitive to the model's *intentions*, as reflected by their selective imitation of a model's intended behavior rather than his or her actual behavior (Meltzoff 1995; Carpenter et al. 1998a). For example, in one study 14- to 18-month-old infants watched adults perform complex behavior sequences, some of which appeared intentional as reflected by the model's vocal behavior and others that, based on what the model said, were accidental. When they were later given the chance to imitate the model, the infants reproduced twice as many "intentional" as "accidental" behaviors (Carpenter et al. 1998b).

Although young children's social learning is flexible, most research indicates that they are more apt to engage in mimicry or imitation, copying relatively precisely the actions of a model, rather than goal emulation. For example, Horner and Whiten (2005) presented 3- and 4-year-old children with an opaque puzzle box and demonstrated a series of three actions, two of which were necessary and one of which was not, to retrieve a gummy bear from inside the box. The children copied all three of a model's actions to retrieve a gummy bear. However, when presented with an identical *transparent* puzzle box, children still copied all three actions, even when the irrelevancy of some actions was readily observable. These findings suggest that children blindly imitate the behavior of a model (see also Nagell et al. 1993). Young children may be motivated to mimic the actions of others in order to maintain a social interaction, and such social motivation may be as important, or more so, as an understanding of intentions in some contexts of social learning. Relatedly, young children's "overimitation" may be due to the belief that all of an adult's actions are goal-directed, making imitation of those actions a reasonable course to take (Lyons et al. 2007).

Despite a tendency toward "overimitation," even 14-month olds will engage in goal emulation rather than mimicry or imitation in some situations. For example, extending an earlier study by Meltzoff (1988), Gergely et al. (2002) had 14-month-old infants watch as a model used her head to press a button to turn on a light in one of two conditions: (a) her hands were free, as in the earlier Meltzoff study, or (b) her hands were wrapped in a blanket, that is, her hands were occupied and thus not available to turn on the light. Most of the infants (69%) in the "hands-free" condition used their heads to turn on the light, just as Meltzoff had found. But the pattern was reversed in the "hands-occupied" condition. Now, most of the infants (79%) used their hands to turn on the light. That is, when there was a reason why the model did not use her hands (they were wrapped in a blanket), the babies focused on the goal (turn on the light), not the means (use your head), and used their hands as the most efficient way to turn on the light. When no such reason was available, they copied the model's behavior exactly, reflective of imitative learning. Gergely and colleagues referred to this as rational imitation. Other researchers have similarly shown that young children's social learning is flexible, with children between 12- and 26-months of age displaying both imitation (focusing on the means of a model, as well as the ends) and goal emulation (focusing only on the ends), depending on the context (e.g., Carpenter et al. 2005; Nielsen 2006). This is important because it indicates that children possess the plasticity to modify their learning based on the type of information they receive.

17.4.2 Parental Effects

Parents' role in social learning begins early. For example, newborns not only copy the facial gestures of an adult but also provoke imitation from them and demonstrate different patterns of heart-rate change that correspond to these different behaviors (increasing heart rate when imitating and decreasing heart rate when provoking, Nagy 2006). Thus, although infants are not engaging in any form of social learning *yet*, these behaviors serve a unique function during the early months of life. When members of a dyad match each other's behavior, they are engaging in a nonverbal dialog, expressing mutual identification and maintaining mutual attention between the two interactants.

Mothers and infants continue this reciprocal imitative relationship over the course of infancy. For instance, Masur and Rodemaker (1999) observed mothers and their infants during free play or bath time when the babies were 10-, 13-, 17-, and 21-months old. Rates of imitation were about one episode per minute, with mothers imitating infants more than infants imitating mothers. And unlike neonatal imitation or mutual imitation described by Piaget (1962), these older infants are learning many new things through social learning. For example, in a study in which parents were asked to keep diaries of their children's imitative behavior, 12-, 15-, and 18-month olds learned, on average, one or two new behaviors a day simply by watching (Barr and Hayne 2003).

Being engaged in imitation and social learning may facilitate infants' and young children's abilities to learn that others are "like me" and aid children in the acquisition of perspective-taking abilities (Meltzoff 2005). As children learn from and imitate a model's behavior, they become self-aware and better able to reason about the relationship between their own behaviors, knowledge, and desires and the behaviors, knowledge, and desires of others.

17.4.3 Abilities in Chimpanzees

Chimpanzees are impressive social learners. They have been found to transmit nongenetic information across generations, including forms of greeting, grooming, and foraging, a characteristic of culture (e.g., Whiten et al. 1999; Whiten 2007). Orangutans (van Schaik et al. 2003) and cetaceans (Rendell and Whitehead 2001; Bender et al. 2009) have also been observed to transmit information across generations, although no species does so with the ease, fidelity, and to the extent that humans do.

It is unequivocal that chimpanzees engage in various forms of social learning, including goal emulation, in which new behaviors are acquired (e.g., Call et al. 2005; Horner and Whiten 2005). For example, we described earlier a study in which 3- and 4-year-old children imitated irrelevant actions to open a puzzle box to obtain a treat (Horner and Whiten 2005). In contrast, chimpanzees given the same task used goal emulation, imitating only relevant actions (see also Nagell et al. 1993). There is less unequivocal evidence that chimpanzees engage in imitation, in which the observer must understand the goal of the model and reproduce important aspects of the model's behavior to achieve that goal (see Tomasello and Call 1997; Whiten et al. 2004).

An exception to this seems to be for enculturated apes that have demonstrated imitation both immediately after viewing a model's behavior (e.g., Tomasello et al. 1993a; Buttelmann et al. 2007) and following a significant delay (e.g., Tomasello et al. 1993b; Bering et al. 2000; Bjorklund et al. 2002). For example, Buttelmann et al. (2007) used a procedure similar to that described earlier by Gergely et al. (2002), in which enculturated chimpanzees observed a human model using an unusual body part to operate an apparatus (such as using his forehead to turn on a light), both with his hands free and occupied. Similar to Gergely et al.'s findings with 14-month-old human infants, the enculturated chimpanzees imitated rationally, using the unusual body part when the model's hands were free and using the more convenient hands when the model's hands had been occupied.

With respect to teaching, there have been a handful of observations of chimpanzees in the wild, all involving foraging behaviors of mothers with their infants (e.g., Boesch 1991; Greenfield et al. 2000). For example, Boesch observed mother chimpanzees making exaggerated movements while cracking nuts when in the presence of their infants, which he interpreted as teaching. More recently, Lonsdorf (2006) observed mother and infant chimpanzees with respect to the activity of termite fishing and reported that the amount of time mothers spent alone or with maternal family members (which is highly correlated with time spent on termite fishing) was related to their offspring's skill at specific components of termite fishing, suggesting that the young animals were, indeed, learning these skills from their mothers. However, she reported no evidence of active facilitation of the offsprings' actions by the mothers, suggesting that teaching, though possibly practiced by some mothers, is not the primary means by which young chimpanzees learn foraging skills (Bering and Povinelli 2003).

To our knowledge, there is no evidence from controlled studies that either laboratory-reared or enculturated chimpanzees engage in teaching one another. There is anecdotal evidence, however, that some enculturated apes may engage in teaching. For example, Fouts (1997) reported that the language-trained and enculturated chimpanzee, Washoe, taught her adopted son, Loulis, about 50 signs, which the younger animal used to communicate both with Washoe and with his human caretakers.

17.5 Theory of Mind

Advanced forms of social cognition in modern humans, such as cheater detection and the negotiation of social contracts, are argued to have their basis in belief-desire reasoning (Cosmides 1989). This reasoning involves understanding that one's actions and the actions of others are based on what one knows and what one wants, and can differ across time, and that one's own knowledge and desires can differ from others. Possessing the ability to reason in this way is often referred to as having a *theory of mind* under the assumption that people develop a theory of how their own and others' minds work.

17.5.1 Development

While the foundations of theory of mind are observed earlier, belief-desire reasoning is not reliably observed in children younger than 4 years old, as measured by passing false-belief tasks (see Wellman et al. 2001). The false-belief task is used to measure children's awareness that an individual can believe something that is not actually true. This is assessed in the *unexpected-transfer* task (Wimmer and Perner 1983), sometimes referred to as the *Sally-Anne task* (Baron-Cohen et al. 1985). Children watch as an item (a cookie, for instance) is hidden in one location. Sally and Anne also watch, and then Sally leaves the room. Anne then moves the cookie from one location to another. Children are then asked where Sally will look for the cookie. Most children under 4 years of age say that Sally will look in the new location, where it is actually hidden, whereas most children 4 years of age and older understand that Sally has a false belief and that she will look for it at its original hiding location.

A second frequently used task to assess false belief is the *Smarties* task, in which a researcher presents children with a box of Smarties (candy familiar to British children), for example, and then reveals that it really contains something else, pennies for instance. Children are considered to pass this task if they understand that they falsely thought there were Smarties in the box before or that someone else who has not seen what is in the box will believe there are Smarties in the box, not pennies (Hogrefe et al. 1986).

Although most 3-year-old children cannot solve false-belief tasks, they have some understanding that other people have desires and knowledge different from their own (see Wellman and Liu 2004). For example, in one study, Repacholi and Gopnik (1997) tested 14- and 18-month-old infants for their preference for two types of food (Pepperidge Farm Gold-fish and raw vegetables). The infants then watched as a woman tasted the food, expressing happiness for one type and disgust for the other. When the woman then requested food from the infants, the 18-month olds gave the woman the type of food she preferred, regardless of their own preferences, presumably recognizing that their likes were different from those of another person. In contrast, the 14-month olds gave the woman the type of food *they* liked, regardless of what food the woman liked.

The developmental patterns observed in variants of false-belief tasks have been found universally (e.g., Avis and Harris 1991; Tardif and Wellman 2000; Sabbagh 2006), although the timetable for passing false-belief tasks may be slower for children in some cultures than others (see Lillard 1998; Wellman et al. 2006; Liu et al. 2008 for a discussion of cultural variations in theories of mind).

17.5.2 Parental Effects

While findings are fairly robust regarding the universal developmental trajectory of theory of mind in humans (e.g., Wellman et al. 2001), individual differences related to factors in the child's environment significantly impact the rate of acquisition and sophistication of these abilities (Carpendale and Lewis 2004). Among them are quality of attachment, parenting styles, parent–child communication (e.g., Carpendale and Lewis 2004), language skills (e.g., Milligan et al. 2007), maternal warmth, the extent to which mothers use mental state talk (i.e., talking about what they and their children are thinking) (e.g., Ruffman et al. 2006), and family size and structure (e.g., Perner et al. 1994; Jenkins and Astington 1996).

For example, Ruffman and colleagues showed that mothers' frequent use of mental-state utterances increases children's likelihood of passing a subsequent false-belief task (Ruffman et al. 2002; Ruffman et al. 2006). In related work (Taumoepeau and Ruffman 2006, 2008), mothers' use of desire versus belief terms was measured when their children were 15, 24, and 33 months old. The researchers reported that mothers increased their talk about belief/knowledge during this time, while talk about desire/emotional states remained the same. Mothers' talk about beliefs and knowledge preceded their children's use of such terms and was a strong predictor of children's social understanding at 33 months. Together, these findings suggest that mothers who, at some level, monitor their children's theory-of-mind development and fine-tune their own linguistic behaviors in response, facilitate the acquisition and sophistication of belief-desire reasoning.

Other research has shown a significant relation between individual differences in preschool children's performance on a theory-of-mind scale and performance of their parents on an adult theory-of-mind task (Sabbagh and Seamans 2008). Although it is impossible at this point to determine the extent to which this transgenerational effect is attributed to genetics or to ways in which parents interact with their children, given the impact of parental behavior on theory-of-mind development (e.g., Carpendale and Lewis 2004; Ruffman et al. 2006), it would seem that at least some of this relationship can be attributed to individual differences in parental behavior.

In addition to parental effects, siblings also seem to influence children's theory-of-mind development (e.g., Jenkins and Astington 1996; Ruffman et al. 1998). Specifically, having an older, not a younger, sibling is associated with better theory of mind (Ruffman et al. 1998). One interpretation of this finding is that competition with an older sibling hastens the development of theory-of-mind skills in the younger child (Cummins 1998).

17.5.3 Abilities of Chimpanzees

Chimpanzees and other great apes have been shown to deal in deception on occasion, for example, leading a more dominant group member to a location where food is not hidden and then running to the actual source of the food, or inhibiting a distinctive cry during orgasm while copulating with a favorite female so as to not have to share the estrous partner with others (see Whiten and Byrne 1988), actions suggestive of theory of mind. In a recent experiment involving competition for food between mother-reared chimpanzees and humans, chimpanzees were able to deceive the human by selectively reaching through tunnels to retrieve food so that the human competitor could either not see the chimpanzee's hand or could not hear anything when the tunnel was opened, behaviors indicative of deception, as well as inferring that eyes (and ears) possess "knowledge" (Melis et al. 2006). These are sophisticated social behaviors that require substantial control of one's actions and the inhibition of prepotent responses, but they do not, in and of themselves, require an understanding of what another being "knows."

Other research supporting the idea that chimpanzees understand that the "eyes have knowledge" has been reported in a food-competition task (Hare et al. 2000, 2001). In these studies, lower-ranking chimpanzees would retrieve food placed in a room only if it were placed out of sight of a higher-ranking animal. This implies that chimpanzees understand that a conspecific that is looking at something "sees" that thing (i.e., has knowledge of it), and they can use this information adaptively, in this case to retrieve the food (or not). This ability is more limited than it is in humans, however. For example, in other research, chimpanzees had to make a "begging response" to one of two human caretakers who stood on either side of a treat placed on an out-of-reach table in front of the chimpanzee. One caretaker's eyes were occluded (e.g., due to wearing a blindfold, having a bucket over her head), whereas the other's were not (Povinelli and Eddy 1996; Reaux et al. 1999). The chimpanzees chose randomly, suggesting an ignorance of the role of seeing (and eyes) in "knowing."

These various research projects, plus anecdotal reports of impressive social-cognitive performance of chimpanzees (e.g., de Waal 1982; Fouts 1997), suggest that these animals possess at least the rudiments of theory of mind. However, there is no evidence that chimpanzees, either mother-reared or enculturated, pass false-belief tasks, at least under controlled conditions (Call and Tomasello 1999; Herrmann et al. 2007). Nonetheless, even if the social-cognitive abilities of chimpanzees are less than those of human preschool children (e.g., Herrmann et al. 2007), research cited in this and earlier sections (e.g., Horner and Whiten 2005; Buttelmann et al. 2007; Warneken et al. 2007) makes it clear that chimpanzees have some understanding of the psychological states of other individuals. The task of researchers is to determine which states these are and the extent of their understanding (Tomasello et al. 2003).

17.6 Conclusion

We have included only a partial list of the development of important human social-cognitive abilities. Omitted from the list are autobiographical memory, language, cognitions associated with attachment, as well as general intelligence, all of which are influenced by individual differences in parental, mainly maternal, behavior (see Bjorklund et al. 2009). We hope that our brief description of the development of several aspects of social cognition makes it clear that parents shape the course of social intelligence in their offspring. Parents, however, do not create these tendencies in their children out of whole cloth, but serve as a supportive environment for the emergence of species-typical and universal patterns of social cognition. Yet, these abilities are not observed in other species, at least to the extent they are in humans, and they must have evolved sometime in human phylogeny following the separation of the line that leads to modern-day humans from those of contemporary chimpanzees and bonobos. However, as our review of chimpanzee social cognition indicates, these animals possess substantial social-cognitive competencies, which, under certain rearing conditions, can be modified to resemble a more *H. sapiens* way of thinking. This is strongly suggestive that our common ancestor with chimpanzees also possessed the neurological plasticity to adapt its behavior and cognition in response to changes in environmental conditions.

References

Adamson LB, McArthur D, Markov Y, Dunbar B, Bakeman R (2001) Autism and joint attention: young children's responses to maternal bids. J Appl Dev Psychol 22:439–453

Alexander RD (1989) Evolution of the human psyche. In: Mellers P, Stringer C (eds) The human revolution: behavioural and biological perspectives on the origins of modern humans. Princeton University Press, Princeton, NJ, pp 455–513

Avis J, Harris PL (1991) Belief-desire reasoning among Baka children: evidence for a universal conception of mind. Child Dev 62:460–467

Bandura A (2006) Toward a psychology of human agency. Perspect Psychol Sci 1:164–180

Baron-Cohen S, Leslie AM, Frith U (1985) Does the autistic child have a 'theory of mind'? Cognition 21:37–46

Barr R, Hayne H (2003) It's not what you know, it's who you know: older siblings facilitate imitation during infancy. Int J Early Years Educ 11:7–21

Bender CE, Herzing DL, Bjorklund DF (2009) Evidence of teaching in Atlantic spotted dolphins (*Stenella frontalis*) by mother dolphins foraging in the presence of their calves. Anim Cogn 12:43–53

Bering JM, Povinelli DJ (2003) Comparing cognitive development. In: Maestripieri D (ed) Primate psychology: bridging the gap between the mind and behavior of human and nonhuman primates. Harvard University Press, Cambridge, MA, pp 205–233

Bering JM, Bjorklund DF, Ragan P (2000) Deferred imitation of object-related actions in human-reared juvenile chimpanzees and orangutans. Dev Psychobiol 36:218–232

Bertenthal BL, Proffitt DR, Kramer SJ (1987) Perception of biomechanical motions by infants: implementation of various processing constraints. J Exp Psychol Hum Percept Perform 13:577–585

Bischof-Köhler D (1988) Über den Zusammenhang von Empathie und der Fähigkeit, sich im Spiegel zu erkennen. Schweiz Z Psychol 47:147–159

Bischof-Köhler D (1991) The development of empathy in infants. In: Lamb ME, Keller H (eds) Infant development. Perspectives from German speaking countries. Lawrence Erlbaum Associates, Hillsdale, pp 245–273

Bischof-Köhler D (1994) Self object and interpersonal emotions. Identification of own mirror image, empathy and prosocial behavior in the 2nd year of life. Z Psychol Z Angew Psychol 202:349–377

Bjorklund DF (1987) A note on neonatal imitation. Dev Rev 7:86–92

Bjorklund DF (2006) Mother knows best: epigenetic inheritance, maternal effects, and the evolution of human intelligence. Dev Rev 26:213–242

Bjorklund DF, Rosenberg JS (2005) The role of developmental plasticity in the evolution of human cognition: evidence from enculturated, juvenile great apes. In: Ellis BJ, Bjorklund DF (eds) Origins of the social mind: evolutionary psychology and child development. Guilford Press, New York, pp 45–75

Bjorklund DF, Yunger JL, Bering JM, Ragan P (2002) The generalization of deferred imitation in enculturated chimpanzees (*Pan troglodytes*). Anim Cogn 5:49–58

Bjorklund DF, Grotuss J, Csinady A (2009) Maternal effects, social cognitive development, and the evolution of human intelligence. In: Maestripieri D, Mateo JM (eds) Maternal effects in mammals. Chicago University Press, Chicago pp. 292–321

Boesch C (1991) Teaching among wild chimpanzees. Anim Behav 41:530–532

Boesch C, Tomasello M (1998) Chimpanzee and human cultures. Curr Anthropol 39:591–614

Bräuer J, Call J, Tomasello M (2005) All great ape species follow gaze to distant locations and around barriers. J Comp Psychol 119:145–154

Bronfenbrenner U, Morris PA (2006) The bioecological model of human development. In: Lerner RM (ed) Handbook of child psychology, vol 1, Theoretical models of human development. Wiley, Hoboken, NJ, pp 793–828

Brooks R, Meltzoff AN (2002) The importance of eyes: how infants interpret adult looking behavior. Dev Psychol 38:958–966

Buttelmann D, Carpenter M, Call J, Tomasello M (2007) Enculturated chimpanzees imitate rationally. Dev Sci 10:F31–F38

Byrne RW (2005) Social cognition: imitation, imitation, imitation. Curr Biol 15:R498–R499

Call J, Tomasello M (1996) The effect of humans on the cognitive development of apes. In: Russon AE, Bard KA, Parker ST (eds) Reaching into thought: the minds of the great apes. Cambridge University Press, New York, pp 371–403

Call J, Tomasello M (1999) A nonverbal false belief task: the performance of children and great apes. Child Dev 70:381–395

Call J, Carpenter M, Tomasello M (2005) Copying results and copying actions in the process of social learning: chimpanzees (*Pan troglodytes*) and human children (*Homo sapiens*). Anim Cogn 8:151–163

Carpendale JIM, Lewis C (2004) Constructing an understanding of mind: the development of children's social understanding within social interaction. Behav Brain Sci 27:79–96

Carpenter M, Akhtar N, Tomasello M (1998a) Fourteen- through 18-month-old infants differentially imitate intentional and accidental actions. Infant Behav Dev 21:315–330

Carpenter M, Nagell K, Tomasello M (1998b) Social cognition, joint attention, and communicative competence from 9 to 15 months of age. Monogr Soc Res Child Dev 63:1–143

Carpenter M, Call J, Tomasello M (2005) Twelve- and 18-month-olds copy actions in terms of goals. Dev Sci 8:F13–F20

Cosmides L (1989) The logic of social exchange: has natural selection shaped how humans reason? Studies with the Wason selection task. Cognition 31:187–276

Cummins DD (1998) Social norms and other minds: the evolutionary roots of higher cognition. In: Cummins DD, Allen CA (eds) The evolution of mind. Oxford University Press, New York, pp 30–50

de Haan M, Belsky J, Reid V, Volein A, Johnson MH (2004) Maternal personality and infants' neural and visual responsivity to facial expressions of emotion. J Child Psychol Psychiatry 45:1209–1218

de Waal FBM (1982) Chimpanzee politics: power and sex among apes. Jonathan Cape, London

de Waal FBM (1997) Bonobo the forgotten ape. University of California Press, Los Angeles

de Waal FBM (2005) Our inner ape. Riverhead Books, New York

Deák GO, Walden TA, Kaiser MY, Lewis A (2008) Driven from distraction: how infants respond to parents' attempts to elicit and re-direct their attention. Infant Behav Dev 31:34–50

Dondi M, Simion F, Caltran G (1999) Can newborns discriminate between their own cry and the cry of another newborn infant? Dev Psychol 35:418–426

Dunbar RIM (1995) Neocortical size and language. Behav Brain Sci 18:388–389

Feinman S (1991) Bringing babies back into the social world. In: Lewis M, Feinman S (eds) Social influences and socialization in infancy. Plenum Press, New York, pp 281–325

Feldman R, Eidelman AI (2004) Parent-infant synchrony and the social-emotional development of triplets. Dev Psychol 40:1133–1147

Field T, Pickens J, Fox NA, Gonzalez J, Nawrocki T (1998) Facial expression and EEG responses to happy and sad faces/voices by 3-month-old infants of depressed mothers. Br J Dev Psychol 16:485–494

Fouts R (1997) Next of kin: my conversations with chimpanzees. William Morrow, New York

Gergely G, Bekkering H, Király I (2002) Rational imitation in preverbal infants. Nature 415:755

Goldsmith DF, Rogoff B (1997) Mothers' and toddlers' coordinated joint focus of attention: variations with maternal dysphoric symptoms. Dev Psychol 33:113–119

Greenfield PM, Maynard AE, Boehm C, Schmidtling EY (2000) Cultural apprenticeship and cultural change: tool learning and imitation in chimpanzees and humans. In: Parker ST, Langer J, McKinney ML (eds) Biology, brains, and behavior: the evolution of human development. School of American Research Press, Santa Fe NM, pp 237–277

Hare B, Call J, Agnetta B, Tomasello M (2000) Chimpanzees know what conspecifics do and do not see. Anim Behav 59:771–785

Hare B, Call J, Tomasello M (2001) Do chimpanzees know what conspecifics know? Anim Behav 61:139–151

Harris JR (1995) Where is the child's environment? A group socialization theory of development. Psychol Rev 102:458–489

Heimann M (1989) Neonatal imitation, gaze aversion, and mother-infant interaction. Infant Behav Dev 12:495–505

Herrmann E, Call J, Hernández-Lloreda MV, Hare B, Tomasello M (2007) Humans have evolved specialized skills of social cognition: the cultural intelligence hypothesis. Science 317: 1360–1366

Hogrefe G-J, Wimmer H, Perner J (1986) Ignorance versus false belief: a developmental lag in attribution of epistemic states. Child Dev 57:567–582

Horner V, Whiten A (2005) Causal knowledge and imitation/emulation switching in chimpanzees (*Pan troglodytes*) and children (*Homo sapiens*). Anim Cogn 8:164–181

Hornik R, Risenhoover N, Gunnar M (1987) The effects of maternal positive, neutral, and negative affective communication on infant responses to new toys. Child Dev 58:937–944

Hrdy SB (1999) Mother nature: maternal instincts and how they shape the human species. Ballantine Books, New York

Humphrey NK (1976) The social function of intellect. In: Bateson PPG, Hinde RA (eds) Growing points in ethology. Cambridge University Press, Cambridge, pp 303–317

Jacobson SW (1979) Matching behavior in the young infant. Child Dev 50:425–430

Jenkins JM, Astington JW (1996) Cognitive factors and family structure associated with theory of mind development in young children. Dev Psychol 32:70–78

Jensen K, Hare B, Call J, Tomasello M (2006) What's in it for me? Self-regard precludes altruism and spite in chimpanzees. Proc Roy Soc Lond B 273:1013–1021

Kuchuk A, Vibbert M, Bornstein MH (1986) The perception of smiling and its experiential correlates in three-month-old infants. Child Dev 57:1054–1061

Lavelli M, Fogel A (2005) Developmental changes in the relationship between the infant's attention and emotion during early face-to-face communication: the 2-month transition. Dev Psychol 41:265–280

Leavens DA, Hopkins WD, Bard KA (2005) Understanding the point of chimpanzee pointing. Epigenesis and ecological validity. Curr Dir Psychol Sci 14:185–189

Lewis M (1993) Self-conscious emotions: embarrassment, pride, shame, and guilt. In: Lewis M, Haviland JM (eds) Handbook of emotions, 2nd edn. Guilford, New York, pp 623–636

Lillard A (1998) Theories behind theories of mind. Hum Dev 41:40–46

Liszkowski U, Carpenter M, Striano T, Tomasello M (2006) 12- and 18-month-olds point to provide information for others. J Cogn Dev 7:173–187

Liszkowski U, Carpenter M, Tomasello M (2007) Pointing out new news, old news, and absent referents at 12 months of age. Dev Sci 10:F1–F7

Liu D, Wellman HM, Tardif T, Sabbagh MA (2008) Theory of mind development in Chinese children: a meta-analysis of false-belief understanding across cultures and languages. Dev Psychol 44:523–531

Lonsdorf EV (2006) The role of the mother in the acquisition of tool-use skills in wild chimpanzees. Anim Cogn 9:36–46

Lyons DE, Young AG, Keil FC (2007) The hidden structure of overimitation. Proc Natl Acad Sci USA 104:19751–19756

Martin GB, Clark RD III (1982) Distress crying in neonates: species and peer specificity. Dev Psychol 18:3–9

Masur EF, Rodemaker JE (1999) Mothers' and infants' spontaneous vocal, verbal, and action imitation during the second year. Merrill Palmer Q 45:392–412

Melis AP, Call J, Tomasello M (2006) Chimpanzees (*Pan troglodytes*) conceal visual and auditory information from others. J Comp Psychol 120:154–162

Meltzoff AN (1988) Infant imitation after a 1-week delay: long-term memory for novel acts and multiple stimuli. Dev Psychol 24:470–476

Meltzoff AN (1995) What infant memory tells us about infantile amnesia: long-term recall and deferred imitation. J Exp Child Psychol 59:497–515

Meltzoff AN (2005) Imitation and other minds: the 'like me' hypothesis. In: Hurley S, Chater N (eds) Perspectives on imitation: from neuroscience to social science, vol 2. MIT, Cambridge, MA, pp 55–77

Meltzoff AN, Moore MK (1977) Imitation of facial and manual gestures by human neonates. Science 198:75–78

Miles HL (1990) The cognitive foundations for reference in a signing orangutan. In: Parker ST, Gibson KR (eds) 'Language' and intelligence in monkeys and apes. Cambridge University Press, Cambridge, pp 511–539

Milligan K, Astington JW, Dack LA (2007) Language and theory of mind: meta-analysis of the relation between language ability and false-belief understanding. Child Dev 78:622–646

Nagell K, Olguin RS, Tomasello M (1993) Processes of social learning in the tool use of chimpanzees (*Pan troglodytes*) and human children (*Homo sapiens*). J Comp Psychol 107:174–186

Nagy E (2006) From imitation to conversation: the first dialogues with human neonates. Infant Child Dev 15:223–232

Nielsen M (2006) Copying actions and copying outcomes: social learning through the second year. Dev Psychol 42:555–565

Patterson FG (1978) Linguistic capabilities of a lowland gorilla. In: Peng FCC (ed) Sign language and language acquisition in man and ape: new dimensions in comparative pedolinguistics. Westview Press, Boulder, CO, pp 161–201

Perner J, Ruffman T, Leekam SR (1994) Theory of mind is contagious: you catch it from your sibs. Child Dev 65:1228–1238

Piaget J (1962) Play, dreams and imitation in childhood. Norton, New York

Pika S, Mitani JC (2006) Referential gestural communication in wild chimpanzees (*Pan troglodytes*). Curr Biol 16:R191–R192

Pollak SD, Kistler DJ (2002) Early experience is associated with the development of categorical representations for facial expressions of emotion. Proc Natl Acad Sci USA 99:9072–9076

Pollak SD, Sinha P (2002) Effects of early experience on children's recognition of facial displays of emotion. Dev Psychol 38:784–791

Poulin-Dubois D, Forbes JN (2002) Toddlers' attention to intentions-in-action in learning novel action words. Dev Psychol 38:104–114

Povinelli DJ, Eddy TJ (1996) What young chimpanzees know about seeing. Monogr Soc Res Child Dev 61:1–152

Povinelli DJ, Nelson KE, Boysen ST (1992) Comprehension of role reversal in chimpanzees: evidence of empathy? Anim Behav 43:633–640

Preston SD, de Waal FBM (2002) Empathy: its ultimate and proximate bases. Behav Brain Sci 25:1–72

Reaux JE, Theall LA, Povinelli DJ (1999) A longitudinal investigation of chimpanzees' understanding of visual perception. Child Dev 70:275–290

Rendell L, Whitehead H (2001) Culture in whales and dolphins. Behav Brain Sci 24:309–382

Repacholi B, Gopnik A (1997) Early understanding of desires: evidence from 14 and 18-month-olds. Dev Psychol 33:12–21

Ruffman T, Perner J, Naito M, Parkin L, Clements WA (1998) Older (but not younger) siblings facilitate false belief understanding. Dev Psychol 34:161–174

Ruffman T, Slade L, Crowe E (2002) The relation between children's and mothers' mental state language and theory-of-mind understanding. Child Dev 73:734–751

Ruffman T, Slade L, Devitt K, Crowe E (2006) What mothers say and what they do: the relation between parenting, theory of mind, language and conflict/cooperation. Br J Dev Psychol 24:105–124

Sabbagh MA (2006) Neurocognitive bases of preschoolers' theory-of-mind development: integrating cognitive neuroscience and cognitive development. In: Marshall P, Fox N (eds) The development of social engagement: neurobiological perspectives. Oxford University Press, New York, pp 153–170

Sabbagh MA, Seamans EL (2008) Intergenerational transmission of theory-of-mind. Dev Sci 11:354–360

Savage-Rumbaugh S, McDonald K, Sevcik RA, Hopkins WD, Rubert E (1986) Spontaneous symbol acquisition and communicative use by pygmy chimpanzees (*Pan paniscus*). J Exp Psychol Gen 115:211–235

Scarr S (1993) Biological and cultural diversity: the legacy of Darwin for development. Child Dev 64:1333–1353

Silk JB, Brosnan SF, Vonk J, Henrich J, Povinelli DJ, Richardson AS, Lambet SP, Mascaro J, Schapiro SJ (2005) Chimpanzees are indifferent to the welfare of unrelated group members. Nature 437:1357–1359

Singer T, Seymour B, O'Doherty J, Kaube H, Dolan RJ, Frith CD (2004) Empathy for pain involves the affective but not sensory components of pain. Science 303:1157–1162

Striano T, Brennan PA, Vanman EJ (2002) Maternal depressive symptoms and 6-month-old infants' sensitivity to facial expressions. Infancy 3:115–126

Tardif T, Wellman HM (2000) Acquisition of mental state language in Mandarin- and Cantonese-speaking children. Dev Psychol 36:25–43

Taumoepeau M, Ruffman T (2006) Mother and infant talk about mental states relate to desire language and emotion understanding. Child Dev 77:465–481

Taumoepeau M, Ruffman T (2008) Stepping stones to others' minds: maternal talk relates to child mental state language and emotion understanding at 15, 24, and 33 months. Child Dev 79:284–302

Tomasello M (1999) Having intentions, understanding intentions, and understanding communicative intentions. In: Zelazo RD, Astington JW, Olson DR (eds) Developing theories of intention. Lawrence Erlbaum Associates, Mahwah, NJ, pp 63–75

Tomasello M, Call J (1997) Primate cognition. Oxford University Press, New York

Tomasello M, Carpenter M (2005) The emergence of social cognition in three young chimpanzees. Monogr Soc Res Child Dev 70:1–132

Tomasello M, Kruger AC, Ratner HH (1993a) Cultural learning. Behav Brain Sci 16:495–552

Tomasello M, Savage-Rumbaugh S, Kruger AC (1993b) Imitative learning of actions on objects by children, chimpanzees, and enculturated chimpanzees. Child Dev 64:1688–1705

Tomasello M, Call J, Hare B (2003) Chimpanzees understand psychological states – the question is which ones and to what extent. Trends Cogn Sci 7:153–156

Tomasello M, Carpenter M, Liszkowski U (2007) A new look at infant pointing. Child Dev 78:705–722

Vaish A, Striano T (2004) Is visual reference necessary? Contributions of facial versus vocal cues in 12-month-olds' social referencing behavior. Dev Sci 7:261–269

Van Hecke AV, Mundy PC, Acra CF, Block JJ, Delgado CEF, Parlade MV, Meyer JA, Neal AR, Pomares YB (2007) Infant joint attention, temperament, and social competence in preschool children. Child Dev 78:53–69

van Schaik CP, Ancrenaz M, Borgen G, Galdikas B, Knott CD, Singleton I, Suzuki A, Utami SS, Merrill M (2003) Orangutan cultures and the evolution of material culture. Science 299: 102–105

Warneken F, Tomasello M (2006) Altruistic helping in human infants and young chimpanzees. Science 311:1301–1303

Warneken F, Hare D, Melis AP, Hanus D, Tomasello M (2007) Spontaneous altruism by chimpanzees and young children. PLoS Biol 5:e184. doi:10.1371/journal.pbio.0050184

Wellman HM, Liu D (2004) Scaling of theory-of-mind tasks. Child Dev 75:523–541

Wellman HM, Cross D, Watson J (2001) Meta-analysis of theory-of-mind development: the truth about false belief. Child Dev 72:655–684

Wellman HM, Fang F, Liu D, Zhu L, Liu G (2006) Scaling of theory-of-mind understandings in Chinese children. Psychol Sci 17:1075–1081

Whiten A (2007) Pan African culture: memes and genes in wild chimpanzees. Proc Natl Acad Sci USA 104:17559–17560

Whiten A, Byrne RW (1988) Tactical deception in primates. Behav Brain Sci 11:233–273

Whiten A, Goodall J, McGrew WC, Nishida T, Reynolds V, Sugiyama Y, Tutin CE, Wrangham RW, Boesch C (1999) Cultures in chimpanzees. Nature 399:682–685

Whiten A, Horner V, Litchfield CA, Marshall-Pescini S (2004) How do apes ape? Learn Behav 32:36–52

Wimmer H, Perner J (1983) Beliefs about beliefs: representation and constraining function of wrong beliefs in young children's understanding of deception. Cognition 13:103–128

Zahn-Waxler C, Radke-Yarrow M, Wagner E, Chapman M (1992) Development of concern for others. Dev Psychol 28:126–136

Chapter 18
Deceit and Self-Deception

Robert Trivers

Abstract Deception is a universal feature of life, at all levels and in all relationships both within species and between species, inside individuals and outside, with strong effects on both deceiver and deceived. Being detected often results in a sharp reversal of fortune for the deceiver thereby intensifying selection to deceive successfully. In encounters between human strangers, nervousness, signs of control and of cognitive load can all serve as cues of deception but cognitive load appears to be the most important. Self-deception is defined as hiding true information from the conscious mind in the unconscious, and is illustrated by classical experimental work. Selection to deceive can favor self-deception, the better to hide the deception and separately to reduce its cognitive costs. Four examples are described. There is a general tendency toward self-inflation in humans, the better to give off a positive image. Conscious thought suppression, studied via fMRI, shows that one area of the brain has been coopted to suppress memory formation elsewhere in the brain. When people reach age 60, they fail to attend to negative social reality, and this old-age positivity may give immune benefits. Across primates there is a strong positive association between relative size of the neocortex and frequency of deceptive acts in nature. If the relationship holds within species, we may expect relatively intelligent humans to be prone to self-deception. There is such a thing as imposed self-deception, in which we act out the system of self-deception of another. Likewise, there is parasitized self-deception in which our system of self-deception makes us more vulnerable to deception by others. Con artists are given as an example. One could model the evolution of deceit and self-deception as a multiplayer game, which can then be analyzed mathematically, modeled via computer simulations or tested experimentally. One promising possibility is a variant of the Ultimatum Game, in which deception and detection of deception are permitted and given quantitative values.

R. Trivers
Department of Anthropology, Rutgers University, Rutgers, 131 George St, NJ, USA
e-mail: trivers@rci.rutgers.edu

P.M. Kappeler and J.B. Silk (eds.), *Mind the Gap*,
DOI 10.1007/978-3-642-02725-3_18, © Springer-Verlag Berlin Heidelberg 2010

18.1 Introduction

In our own species, deceit and self-deception are often two sides of the same coin – best seen together. If by deception, we limit ourselves to consciously propagated deception – lies – we will miss the larger category of unconscious deception, including active self-deception. On the other hand, if we look at self-deception and fail to see its connection to deception, we will miss its major function. The central claim is that self-deception evolves in the service of deception (1) the better to avoid its detection and (2) to reduce the immediate cognitive costs. In the first case, the self-deceived person fails to give off the cues that go with consciously mediated deception, thus escaping detection. In the second, the actual process of deception is rendered less expensive by keeping part of the truth in the unconscious mind. But this turns out to be a very complicated matter. Suppressing the truth may have short-term cognitive benefits as well as almost immediate immunological costs, or for that matter, longer-term cognitive ones. In addition, deception and self-deception are also being directed at you by others, so that there is such a thing as imposed self-deception, in which you act out the system of self-deception of another person, presumably often to your own detriment. In this account, we will try to fuse the two topics of deceit and self-deception into a single coherent account, with special attention to the associated costs and benefits, that is, with their selective effects.

18.2 Deception is Everywhere, at All Levels of Life

Deception (misleading others) is a very deep feature of life. It occurs at all levels and, it would seem, by any means possible. It tends to hide from view and is, by its nature, difficult to see and to study. Self-deception is even worse, hiding itself more deeply in our own unconscious minds. It is, thus, important to build up our logic carefully, with a full view of the enormous variation that has evolved for various reasons (Trivers 2000). Unfortunately, there has been very little formal theory on the subject, so I will limit myself to making a series of general comments, with special focus on cognitive load and self-deception.

When I say that deception occurs at all levels of life, I mean that viruses practice it, bacteria do, so do plants, and so do the insects preying on plants, and a wide range of other animals. It is everywhere. Even *within* our genomes deception may flourish, as selfish genetic elements use deceptive molecular techniques to over-reproduce at the expense of the larger genome (Burt and Trivers 2006). Or, when a selfish paternal orientation collides with an oppositely oriented maternal one (Haig 2002). Deception infects all the fundamental relationships in life, parasite and host, predator and prey, plant and animal, male and female, neighbor and neighbor, parent and offspring (including mother and fetus), and even the relationship of an organism to itself.

Viruses and bacteria actively deceive to gain entry in to their hosts, by mimicking body parts for example. Or, as in HIV, the virus deceives by changing coat proteins so often as to make mounting an enduring defense almost impossible. Predators gain from being invisible to their prey or resembling items attractive to them while prey gain by being invisible to their predators or mimicking items noxious to them, e.g., poisonous species or their predator's predator.

Deception within species requires only imperfect (<1) degrees of relatedness between individuals (so that their self-interests are not identical) and imperfect information (so that the other party can be fooled). Both are easily satisfied. Clonal species are rare in nature and imperfect degrees of relatedness the rule (under outbreeding, 1/2, 1/4 down to near-zero). Perfect information is impossible. Deception can allow you to steal or induce the transfer of food and other resources, engage in extra-pair copulations undetected, manipulate your parents, your mate, your offspring, your neighbors – even the maternal (or paternal) half of yourself.

And deception always takes the lead, while detection of deception plays catch-up. As has been said regarding rumor, the lie is halfway around the world, before the truth goes to work. When a new deception shows up, it starts rare in a world that lacks a proper defense. As it increases in frequency, it selects for such defenses in the victim, so that eventually its spread will be halted by the appearance and spread of countermoves, but new defenses can always be bypassed and new tricks invented.

The adaptive potential of deception is chronically overlooked by those with an attachment to the truth. Bill Gates told the world confidently, in 2004, that the problem of spam "would be solved by 2006" (N.Y. Times 12/06/06). He saw that defenses could easily be erected against the set of spamming devices then in use but he could not imagine that these defenses could easily be bypassed while yet newer forms of spamming were continually being invented. Spam is now at least ten-fold as frequent as it was in 2004, and correspondingly costly. One inevitable cost is the destruction of true information by spam-detectors too stringently set, including by oneself. This is a universal problem in animal discrimination. Greater powers of discrimination will inevitably increase the so-called false negatives, rejecting something as false which is, in fact, true.

It always amuses me to hear economists saying that the costs of deceptive excesses in our economy (so-called "white" crime robbery) will be naturally checked by "market forces." Are these the same forces that force us to add one unit to every price we see in order to know the true price? Why should the human species be immune to the general rule that where selection is strong, deception can be generated that extracts a substantial net cost every generation. Consider the following.

Deception is such an important feature of life that it can entrain the evolution of entire groups of organisms, as well as the evolution of specialized deceitful morphs within species. For example, the Phasmatodea, or stick insects, is a group that has given itself over to the imitation of either sticks (~3,000 species) or leaves (~30 species) (Markle 2007). In the case of sticks, there is apparently a tremendous evolutionary pressure to produce a long, thin (stick-like) body, even if this forces the individual to forego the benefits of bilateral symmetry. Thus, to fit the internal organs into a diminishing space, one of two organs have often been sacrificed, only one kidney, one ovary, one testis, etc. This shows that selection for successful deception has been powerful enough not only to remold the external shape of the creature but also to remold its internal organs as well – even when this is otherwise disadvantageous to the larger organism, as loss of symmetry, in principle, must usually be.

Likewise, selection for deception has been strong enough to mold morphs that are obligately committed to deception; that is, morphological forms whose strategy depends entirely on deception of others. A classic example occurs in the blue-gill sunfish, where a specialized male form has evolved that mimics a female in appearance and behavior, being 1/6th the size of a territorial male and roughly the size of an actual female (for a recent reference see Stoltz and Neff 2006). This female-mimic seeks out a territorial male, permits himself to be courted, and responds enough to keep the other male interested, so that when a true female spawns the pseudo-female is ready nearby, along with the territorial male, to fertilize the eggs. It is as if the territorial male imagines he is in bed with two females when, in fact, he is in bed with one female and one male. The two kinds of males appear to be distinct forms that never turn into each other. To have persisted for so long their long-term reproductive success must be identical; that is, the deceiver is doing exactly as good as the deceived – and this equality must, in turn, have been produced by frequency-dependent selection. That is, when the female-mimic is relatively rare, he will do relatively well; when common, less so.

18.3 Detection of Deception Often Leads to Negative Consequences, Including Punishment

In our own species, we hardly need convincing that when our deception is detected, we may receive some harsh feedback, a beating, a public humiliation, a lost love, or at the very least, a withdrawal of some trust and affection. This appears to be true of many other animals. That is, in insects and birds, deceivers are often punished aggressively, and most often by those whose status is being threatened by the deceiver (for some examples see Trivers 1985; Møller 1987; Hauser 1992; Tibbetts and Dale 2004). If we take it as generally true that detection of deception often leads to an unfortunate reversal in fortune, then a deceiver will be under pressure when deceiving and these signs of pressure are available for detection by others. Indeed, the lie detector test is based on this fact.

But here a caution is worth mentioning. The lie detector test continuously measures a series of physiological parameters in response to a series of posed questions designed to detect unusual signs of stress in particular places. For example, the most useful device is the "guilty knowledge" test, which suddenly introduces some pertinent fact known only to the guilty (was the gun red?). Here, *any* deviation from the norm – high reactivity or suppressed reactivity – shows guilty knowledge.

Daily life is very different from this. At one extreme, such as intimate relations between husband and wife, or partner and partner, each may indeed have a detailed behavioral template of the other, honest and dishonest, against which to evaluate the ongoing behavior, while at the other extreme, perfect strangers are interacting with no prior knowledge of each other.

18.4 Cognitive Load is a Key Factor in the Detection of Deception in Humans

In anonymous or infrequent interactions, behavioral cues can not be read against a background of known behavior, so more general attributes of lying must be used.

Three such attributes have been emphasized (Vrij 2004):

1. *Nervousness*: because of the negative consequences of being detected, including being aggressed against, nervousness can reveal deception (DePaulo et al. 2003)
2. *Control*: in response to concern over appearing nervous (or concentrating too hard), people exert control, trying to suppress both, but this requires additional effort, with possible side effects, and there is the danger of over-acting, over-control, a planned or rehearsed impression
3. *Cognitive load*: lying is cognitively demanding: you must suppress the truth and construct a falsehood, one that will not contradict anything known by the listener, or likely to be known, you must tell it in a convincing manner and in such a way that you can remember the story. This takes time and concentration, both of which may give off secondary cues.

The most recent work suggests that cognitive load is the critical variable among the three, with a minor role for control and very little for nervousness. At least this seems to be true in real criminal investigations as well as experimental situations designed to mimic them (Mann and Vrij 2006). People who are lying have to think too hard and this causes several effects, some of which are opposite to those of nervousness.

Consider, for example, blinking. We blink our eyes more often when nervous, but we blink them less under increasing cognitive load (e.g., while solving arithmetic problems). Recent studies of deception suggest that we blink less when deceiving, that is, cognitive load rules (Vrij 2004; Vrij et al. 2006). Nervousness makes us fidget more but cognitive load has the opposite effect. Again, contra usual expectation, people fidget less in certain deceptive situations. Again consistent with cognitive load effects, men use less hand gestures while deceiving and both sexes show longer pauses while they speak.

The effects of control are illustrated by pitch of voice and (separately) displacement activities. Deceivers tend to have higher-pitched voices (DePaulo et al. 2003).

This is a natural consequence of any effort to suppress behavior by becoming more rigid. Tensing up the body inevitably raises the pitch of voice.

Another effect of suppression is the production of displacement activities. As classically described in other animals, these are irrelevant activities often seen when two opposing motivations are simultaneously aroused. Since neither impulse can express itself, the blocked energy easily activates irrelevant behavior, such as a twitch. For this reason, displacement activities in primates are a reliable indicator of stress (Troisi 2002).

Nervousness is almost universally cited as a factor associated with deception both by those trying to detect deception as well as by those trying to avoid detection, yet surprisingly enough, it is one of the weaker factors predicting deception. This is, perhaps, precisely because we are conscious of our nervousness so that mechanisms of suppression may be almost as well developed as the nervousness itself. The point about cognitive load is that there is no escape. If it is cognitively expensive to lie, there is no obvious way to reduce the expense. Mechanisms of denial and repression may serve to reduce immediate cost, but alas, as we shall see, with ramifying costs later on.

18.5 What is Self-Deception?

Before going any further it would be useful to define self-deception, and give an example. Some people, especially philosophers, imagine that self-deception is a contradiction in terms: how can the self deceive itself, does that not require that it knows what it does not know? But the contradiction is easily resolved by defining the self as the conscious mind, so that self-deception occurs when the self is being kept in the dark, when the larger organism preferentially keeps true information out of consciousness, and when it misleads the conscious mind. Sometimes, this may involve activities of the conscious mind itself, which arrange – e.g., via active memory suppression – for later mental states to be biased in a particular way but usually the processes are unconscious and bias conscious mentation in a great variety of ways (see Gilbert 2006). So the key to defining self-deception is that true information is preferentially excluded from consciousness and, if held at all, is held in (varying degrees of) unconsciousness.

A very dramatic example of self-deception was demonstrated experimentally almost 30 years ago, with true and false information simultaneously stored within individuals but showing a strong bias toward the true information hidden in the unconscious mind. Gur and Sackeim (1979) also showed that they could manipulate an individual's self-deception – so defined – by the simple device of increasing or decreasing the individuals' opinion of themselves.

The experiment was based on a simple fact of human biology: we are physiologically aroused by the sound of the human voice but more so to the sound of our own voice (as reproduced, for example, on a tape recorder). We are unconscious of these effects. Thus, you can play a game of self-recognition, in which people are asked whether a voice is their own or not (conscious self-recognition) at the same time recording whether unconscious self-recognition was achieved, via higher arousal.

Here is how it worked. People were matched for age and sex and asked to read the same paragraph from Thomas Kuhn's "Structure of Scientific Revolutions," then these recordings were chopped into 2, 4, 6, 12, and 24 s segments, and a master tape was created consisting of a mixture of own and other voices. Meantime, the individual is hooked up to a machine measuring his or her galvanic skin response (GSR), which is normally twice as great for hearing one's own voice as hearing someone else's. People are asked to press a button to indicate if they think the recording is of self and another button to indicate how sure they are.

Several interesting things were discovered. First, some people denied their own voices some of the time, this was the only kind of mistake they made, and they seemed to be unconscious of making it (when interviewed later, only one was aware of having made this kind of mistake), yet the skin had it correct, that is, showed the large increase in GSR expected upon hearing one's own voice. By contrast, another set of people heard themselves talking when they were not – they projected their voice – and half were aware later that they had sometimes made this kind of mistake, but the skin once again had it correct. There were two other categories, those who never made mistakes and those who made both kinds, sometimes even fooling their skin; but in what follows, we neglected these two categories.

This is unconscious self-recognition shown to be superior to conscious recognition, but Gur and Sackheim also showed that one could affect the tendency toward self-denial or projection by manipulating the person's opinion of himself/herself. Made to feel bad by a poor score (in fact, randomly administered) on a pseudo-exam, individuals started to deny their voices more often. Made to feel good by a good score, individuals started to project their voices. It was as if self-presentation was expanding under success and contracting in response to failure.

Another interesting feature – never analyzed statistically – was that deniers also showed the highest GSRs to all stimuli. It was as if they were primed to respond quickly, to deny the reality and get it out of sight, while inventing the reality (projecting) seemed a more relaxed enterprise. Perhaps, reality that needs to be denied is more threatening than is the absence of reality one wishes to construct – or, in any case, denial can be dealt with more quickly.

Although this example is nice in showing clear, strong effects, it is not the only paradigm for self-deception. It is not necessary, for example, for the true information to be simultaneously stored. We act very early in information-generating processes to produce biased results. For example, we will choose not to read factual articles on a position we oppose (e.g., that marijuana consumption has negative personal effects) but to read those consistent with our views (e.g., marijuana consumption is beneficial). Somewhere in our body may be stored facts regarding preferential attention but not the content of the missed articles.

Alas! except for a couple of trivial articles quibbling with aspects of Gur and Sackeim's work (e.g., Douglas and Gibbons 1983), no follow-up work has appeared. How do those who make no mistakes differ from those who deny or project? And why do some individuals make both kinds of errors? Are deniers really more aroused, in general, than the other categories of people? And which voices of one's own do we deny and which expropriate from others, and why? Unfortunately, we have no answers to these or a host of related questions. In part, this reflects the difficulty of the work itself, very painstaking and demanding, especially, in the pre-computer era, but mostly it reflects the degree to which self-deception has not been seen as a major subject in psychology, for which this was a significant methodological breakthrough. In any case, it is a shame: Gur and Sackeim created a whole new line of work which then failed to develop. The fact that is has failed to develop is cited as evidence against the original work. One prominent social psychologist told me that the results were so uncertain in his field that unless there was at least one replication, no one paid attention to solitary results.

18.6 One Needs a Separate Theory of Reality

Since deceit and self-deception must always be judged against the truth, one needs a separate theory of reality that is reliable against which to test self-deception. Evolutionary theory, of course, pretends to provide exactly that. The dangers of a false theory of reality are illustrated by the failure of both Freud and Marx's theories of self-deception. Although Freud was able to describe such phenomena as denial, repression, and projection, his own unfounded theory of human development led him to deny a thing and project the other. For example, he denied that sexual advances from male relatives or step-relatives were a common problem for females, and projected on these women, the desire for exactly such encounters. An evolutionary approach is not congenial to the notion that women should have an inborn desire for sexual congress with close male relatives. Quite the contrary, but with a critical asymmetry, males are more likely to benefit genetically from such inbreeding than are the investing females. Here Freud's absence of any plausible view of human development allowed him to twist his self-deception argument any which way. Marx, in turn, provided an analysis of bourgeois deceit and self-deception but his naïve theory of inevitable economic evolution only encouraged socialist self-deception.

18.7 Is Self-Deception the Psyche's Immune System?

The immune analogy is very popular within psychology. The argument goes as follows. Just as our body is under constant threat from parasites, so is our psyche under threat from factors that reduce happiness. Hence, we have psychological defense mechanisms, just as we have immunological ones, the one to keep us healthy and disease-free, the other to keep us happy. In one formulation, people are seen as having a "*psychological immune system that defends the mind against unhappiness in much the same way that the physical immune system defends the body against illness*" (Gilbert 2006).

This is said to be an "unusually appropriate" analogy because immune response and the degree of psychological defensiveness both share the trait that too little is bad but so is too much. Alas, this is true of all biological systems. Otherwise, we would not have stable phenotypes. Too little oxygen is bad, and so is too much. Too little food is bad – as is too much. And right down the line, height, weight, salt, water, curvature of your left thumb, tendency to visit dentists, everything is bad in its extremes.

What then is the image of how this immune-like defensive system works? "We need to be defended – not defenseless or defensive – and thus our minds naturally look for the best view of things while simultaneously insisting that those views stick reasonably closely to the facts" (Gilbert 2006).

"Reasonability" is the operative word here, undefined and as elastic as you could want. We are seen as keeping ourselves happy in good part via self-deception: denial, projection, disassociation, and so on. We cook the facts, we bias the logic, we overlook the alternatives – in short, we lie to ourselves. Meanwhile, we apparently have a "reasonability center" that, by unknown criteria, determines just how far we will be permitted to protect our happiness via self-deception (without, for example, looking ridiculous to others). Why was evolution unable to produce a more sensible way of regulating such an important emotion as happiness? Something more reasonable?

Contrast the real immune system. It deals with a major problem common to all of life, that of parasites, organisms that eat us out from the inside. Parasites are often the major selection pressure every generation on their hosts, much stronger than that of predators, for example. The immune system uses a variety of direct reality-based molecular mechanisms to attack, disable, engulf, and kill a veritable zoo of invading organisms – thousands of species of viruses, bacteria, fungi, protozoa, and worms – themselves using techniques honed over hundreds of millions years of intense natural selection. The immune system also stores away an accurate and large library of previous attacks, with the appropriate counter-response now pre-programmed in advance.

The vertebrate immune system is at least 300 million years old and is extremely sophisticated (similar evidence is now emerging for the insect immune system). Several dozen cell types are produced and marshaled in a bewildering array of patterns by a wide range of neurochemicals. The system is also very costly, perhaps on the order of the brain itself. It generates several grapefruit volumes of new tissue every 2 weeks. When unnecessarily aroused through false antigens (e.g., sheep red blood cells), immune arousal compromises survival in nature, sometimes strongly. Because it is so costly, it also acts as a great reservoir of energy against which the larger system can borrow in times of need; for example, effects of stress and sex steroids act to depress immune function, presumably the energy is spent on more immediately pressing problems (whatever is causing the stress or sexual and aggressive opportunities).

What on earth does all of this have to do with the ego's need to defend itself against various "threats"? And what exactly are these a threat to? One's self-opinion? Why is that a matter of pressing survival value? Parasites threaten your life. And why adopt something as dubious as self-deception to solve this problem? Put another way, where does our immune system improve its function by lying to itself – why is there no analogy on this key point? Granted the selection pressures associated with social interactions have steadily increased in our lineage – perhaps, exploding with language – but self-image, self-esteem, and ego strength are surely a small part of all of this.

Is it possible that psychology has gotten all of this terribly wrong for so long by simply taking seriously an inside–outside approach to life – introspection will show us the way – in which we, in effect, choose our self-deceptions as building blocks in our theory? Social psychology has wedded itself to a thoroughly defensive view of self-deception, one that is itself congenial to an inflated self-perception: I am not lying to myself the better to deceive you but rather I lie to myself to defend myself from attacks on my personal integrity, my very happiness.

18.8 Self-Deception Helps Fool Others, While Reducing the Cognitive Cost of Doing So

Imagine two animals squaring off in a physical conflict. Each is assessing its opponent's self-confidence along with its own – variables expected to predict the outcome some of the time. It is easy to imagine that biased information flow within the individual can facilitate false self-confidence, which some of the time will pay for itself by fooling the opponent. Nonverbal self-deception can be selected in aggressive and competitive situations, the better to fool antagonists. Much the same could be said for male/female courtship relations. A male's false self-confidence may give him a boost some of the time. A biased mental representation can be produced, by assumption, without language.

The above is meant to demonstrate that in at least two contexts – aggressive conflict and courtship – selection for deception may easily have favored self-deception; that is, biased information flow within an organism to its consciousness, even when no language is involved. There are undoubtedly many other such contexts, for example, parent/offspring. But the simple fact is that language must have greatly expanded the opportunities for deceit and self-deception. If one great virtue of language is its ability to make true statements about events distant in space and time, then surely one of its social drawbacks is its ability to make similarly distant false statements, so less easily contradicted than statements about the immediate world. Once you have language, you have an explicit theory of self, social relationships and the world, ready to communicate to others. Numbers of new true assertions possible are matched by an infinitely greater number of false assertions, and so on.

A variety of ways in which self-deception helps deceive others has been described and many more await their day (reviewed briefly in Trivers 2000). What is also probably true is that self-deception reduces the immediate cognitive costs of deception. This alone should help hide the deception but it should also serve to permit the ongoing cognitive activities that would otherwise be blocked by a fully conscious effort to lie. The literature on cognitive dissonance suggests that conscious contradictions are costly and the mind acts to reduce dissonance in a variety of ways (Tavris and Aronson 2007). Yet exploring this point further is very difficult. We know that the brain consumes 20% of resting metabolic rate no matter what (Clark and Sokoloff 1999). With the exception of a slight increase during REM sleep, it does not increase under any known conditions, including hard mental work, schizophrenia, or an LSD trip. It does not decrease under depression or under any other state, except for a slight diminution in non-REM sleep. But the brain and body can borrow from the immune system and do so under a variety of conditions (e.g., stress, sex hormone increases, and thought suppression).

No one has been able to compare the cognitive costs of conscious deception with the same act performed unconsciously. We know that when white Americans with a strong implicit bias against black people (mostly unconscious, see below) are forced to interact briefly and pleasantly with a black experimenter, their performance on a subsequent Stroop test is compromised apparently because of increased cognitive load (Richeson and Shelton 2003). In the Stroop test, the subject must give the color in which a word denoting a color is written. Thus, the word "red" may be written in green ink, and the subject must say "green." Since one's first impulse is to read the word, this must be suppressed to give the correct answer. While viewing black faces, they also show activation of brain regions associated with cognitive control as if already preparing a public face (Gehring et al. 2003; Richeson et al. 2003). What we do not know, of course, is how conscious the various people are of their own negative feelings and how differences in degree of consciousness may be associated with variation in either cognitive control or cognitive load.

18.9 Four Examples of Self-Deception

Although we have few examples of self-deception with the simple clarity and power of Gur and Sackeim (1979), there is, in fact, an enormous literature on the subject, crossing several disciplines including social psychology, behavioral economics, animal behavior, neurophysiology, immunology, and the study of everyday life. I have chosen four examples. First is self-inflation, often measured verbally but now with new techniques at a deeper level. It appears to be very general in human life. Second is the neurophysiology of thought suppression, which combines spatial brain coordinates of ongoing mental activity (fMRI) with measures of success at suppressing material from consciousness, an activity known also to have immediate immune costs (Pennebaker 1997). Third is old-age positivity, a series of cognitive biases that creep in by age 60 so as to give a rosier view of the social world than is warranted. Does this produce immune benefits, especially useful in later life? Finally, the study of monkey and ape behavior in the wild suggests that deceptive behavior is more common per unit time in species with relatively large neocortexes (social brains). If this is true within species, we might expect brighter people to be, on average, more self-deceived, a possibility with some serious social implications.

18.10 Self-Inflation is the Rule in Life

Animal self-inflation routinely occurs in aggressive situations (size, confidence, color) as well as during courtship (same variables). Self-inflation is the dominant style in human psychological life (Greenwald 1980; Gilbert 2006), adaptive self-diminution appearing as an occasional variant (Hartung 1988). People routinely put themselves in the top half of positive distributions and the lower half of negative ones. Eighty percent of U.S. high school students place themselves in the top half of students in leadership ability, but, of course, for extreme examples of self-deception you can hardly beat academics: 94% of them in one survey place themselves in the top half of their profession. I plead guilty. Even when tied to a bed in some back mental ward of a hospital, I still believe I am performing better than half of my colleagues – and this is not only a comment on Rutgers University.

Subtler linguistic features of self-deception have been described. When describing a positive group effect, we adopt an active voice but when the effect is negative, we unconsciously shift to a passive voice: this happened and then that happened and then costs rained down on all of us (reviewed in Greenwald 1980). When in-group members do something positive, we tend to make a general statement, "she is a good person," just as we do when out-group members do something negative "he is wicked" but when an in-group member does something bad, we tend to describe it precisely, "she stepped on my toes," just as we do when an out-group member does something good, "she gave me directions to the Bahnhof" (Maass 1999).

A recent methodology permits a very striking result (Epley and Whitchurch 2008). With the help of a computer, individual photos were morphed either 20% toward attractive faces or 20% toward unattractive. Among other striking effects, when the individual tries quickly to locate his or her real face, its 20% positive or its 20% negative in a background sea of 11 faces of other people, people were quickest to spot the positive face (1.86 s), less so for the real face (2.08 s), and slowest for the ugly one (2.16 s). On average, increasing degree of attractiveness improved the speed of perception by 1/10th of a second out of 2 seconds. The beauty of this is that there has not been the usual verbal filter – what do you think of yourself? – but only a measure of speed of perception.

18.11 The Neurophysiology of Thought Suppression

One particular kind of self-deception – consciously mediated efforts at suppressing true information from consciousness – has been studied neurophysiologically in a most revealing way (Anderson et al. 2004). Different sections of the brain may have been coopted in evolution to suppress the activity of other sections in order to create self-deceptive thinking.

Consider, for example, the active conscious suppression of memory. At times, we actively attempt to suppress thoughts. In the laboratory, individuals are instructed to forget an arbitrary set of symbols that they have just learned. The effect of such efforts is highly variable, measured as the degree of forgetting achieved a month later and this variation is associated with variation in the underlying neurophysiology. The more highly the dorsolateral prefrontal area is activated during directed forgetting, the more it suppresses the ongoing activity in the hippocampus (where memories are typically stored) and the less is remembered a month later. The dorsolateral prefrontal area is otherwise often involved in overcoming cognitive obstacles and in prepotent motor activity, that is, preparing for physical activity (muscle movement). It is tempting to speculate that this area of the brain was coopted for the new function of suppressing memories because it was often involved in affecting other brain areas, i.e., activating behavior. It may be unrelated but whenever I experience an unwanted thought and act at once to suppress it, I often experience an involuntary twitch in my arms, as if trying to push something down (and out of sight).

18.12 Old-Age Positivity and Immune Function

There is a striking bias toward positive social memories and perceptions that sets in by age 60 (and perhaps somewhat earlier). At ages 20–30, the human shows no tendency to remember faces with positive expressions more often than those with negative ones, or to spend more time examining such pictures. But by age 60, a bias is apparent: positive faces are remembered more readily and they are attended to more carefully (Mather and Carstensen 2005). When a dot is presented on the side of a screen at which a positive face is presented, the dot is perceived more quickly if it succeeds the positive face (and less slowly if it succeeds a negative one, compared with neutral Mather and Carstensen 2003). This involves a measurable effect in the amygdala, where positive faces evoke a stronger response than negative in older people but not in younger people (Mather et al. 2004). Older people, compared to younger, are more likely to respond to a musically induced negative mood by preferentially looking at positive faces, as if actively inducing a positive mood (Isaakowitz et al. 2008).

Why show such a positivity bias? Half the problem is trivial. Young people will be wise to pay attention to reality, both positive and negative, the better to make the appropriate responses later; by old age, it hardly matters what you learn and since greater positive affect is associated with stronger immune response (Rosenkranz et al. 2003), you may trade grasp of reality for a boost in dealing with your main problem, that of parasites and cancer. A positivity bias sacrifices learning in the future concerning negative outcomes the better to enjoy strong immune function now. Grandchildren may admire gramps and grandma because nothing seems to faze them, but gramps and grandma might be living in a positivity-enveloped world, the better to deal with their internal enemies.

It is an interesting coincidence that although our implicit bias in favor of youth over old age hardly changes with age – from 20 to 70, we favor young over old – by our 40s, our explicit bias in favor of youth declines until at exactly 60 we start to prefer older to younger – like everyone else we implicitly associate youth with positive features, but we start preaching the opposite at the same time at which we ourselves display the old-age positivity bias (Nosek et al. 2002).

18.13 Are Intelligence and Self-Deception Correlated?

It is easy to imagine that intelligence and consciousness are two independent axes of human behavior, perhaps equally important, but uncorrelated. Thus, one can be bright and deluded or slow and honest with all combinations equally likely. Likewise, it is easy to imagine that the two axes are positively correlated. The smarter you are the less self-deception (greater consciousness). Your innate superiority in intellectual power can easily be turned back on your deceptive tendencies, so that you see through your lies and adjust appropriately. But what does that mean – you lie less or lie more?

I do not believe that degree of self-deception and intelligence are uncorrelated or that they are negatively correlated. I believe quite the opposite. Degree of consciousness and intelligence are positively correlated: brighter people are *more* likely to act deceptively and to practice self-deception. This increases the chance that the net effect of their actions will be negative instead of positive. This is, to put it mildly, an underemphasized underbelly of high intelligence. Of course, there are exceptions. It is not surprising that the academically less gifted are more likely to cheat (and thus act deceitfully) as indeed they are.

One line of evidence comes from monkeys and apes. The size of the neocortex – or better still, its relative portion of total brain size – is positively associated with use of tactical deception in nature, where tactical deception includes any kind of deception that can be seen to give an advantage. A large list of appropriate acts from nature was assembled from the primary scientific literature and used to solicit a still larger sample from active scientists. Study effort and group size were controlled as were taxonomic effects. Conclusion: since neocortex size is correlated with intelligence – including social intelligence – across a broad range of monkeys and apes, we know that deception occurs more often the smarter the species is. So, perhaps, does self-deception.

Another line of evidence comes from children. As children mature, they become increasingly intelligent and increasingly deceptive. This is not an accident. The very maturing capacity that gives them greater general intelligence also gives them greater ability to suppress their behavior and create a novel behavior. There is also a clear evidence that natural variation in intelligence, corrected for age, is positively correlated with deception in children (Lewis unpubl. data) using the peek/no peek–lie/not lie paradigm that has been used to such good effect (e.g., Crossman and Lewis 2006). Even health of the child at birth (as measured by a weighted sum of 32 perinatal factors) is positively correlated with lying.

Thus, if you wish to cherish a self-image that you are smarter than average or even that your group is, you may also need to imagine that you (and the group) are more prone to deceit and self-deception, with net effect on others uncertain.

18.14 Imposed Self-Deception

So far we have spoken of self-deception evolving in the service of the actor. This is the natural first step, but we are also highly sensitive to others, to their opinions, desires, actions, and so on. More to the point, we can be manipulated and dominated by them. This can result in self-deception being *imposed* on us by others (with varying degrees of force). Extreme examples are instructive. A captive may come to identify with his captor, an abused wife may take on the world-view of her abuser, molested children may blame themselves for the transgressions against them and the resulting misery. These are cases of imposed self-deception and if they are acting functionally from the standpoint of the victimized (by no means certain), then they probably do so by reducing conflict with the dominant individual. At least this is often the theory of the participants themselves: an abused wife may be deeply frightened and rationalizes acquiescence as the path least likely to provoke additional severe assaults, this is soon most effective if actually believed.

Let us consider another example of imposed self-deception, one with deeper social implications. It is possible to measure something called a person's "explicit" self-preference as well as an "implicit" one. The explicit simply asks people to state their preferences directly, e.g., for the so-called "black" people over "white" (to use the degraded language of the United States) where the actor is one or the other. The implicit measure is more subtle. It asks people to push a right-hand button for black *or* "good" stimuli (e.g., positive words) and left for white *or* bad ones – and then reverses everything, black or bad, white or good. We now look at latencies – how long does it take an individual to respond when they must punch white or bad versus white or good – and assume that shorter latencies (quicker responses) means the terms are, by implication, more strongly associated in the brain. Hence, the term "implicit association test" (IAT), invented only in 1998 (Greenwald et al.), has now generated an enormous literature, including (unusual for the social sciences) actual improvements in methodology (Greenwald et al. 2003). There are several websites that harvest enormous volumes of data over the internet (e.g., at Harvard, Yale, and the University of Washington), and these studies have produced some striking findings (Nosek et al. 2002).

For example, black and white people were similar in their explicit tendency to value self over other, blacks if anything more strongly so, but when it came to the implicit measures, whites were even more strongly in their own favor than they were explicitly, while blacks – on average – preferred white over black, not by a huge margin but, nevertheless, they preferred other to self (Nosek et al. 2002). This is most unexpected according to evolutionary theory, where self is the beginning (if not end) of self-interest.

This has the earmarks of an imposed self-deception – valuing yourself less than you do others – and it may come with some negative consequences. For example, priming black students for their ethnicity strongly impairs performance. This was indeed one of the first of what are now hundreds of demonstrations of this "priming" effect. Black and white undergraduates at Stanford arrived in a lab to take relatively difficult aptitude tests. In one situation, the students were simply given the exams; in the other, each was asked to give a few personal facts, one of which was their own ethnicity. Black and white scored equally well with no prime. With a prime, whites did slightly (but not significantly) better while blacks' scores plummeted by nearly 1/2. You can even manipulate one person's performance in opposite directions by giving opposing primes: Asian women perform better on math tests when primed with "Asian" and worse when primed with "woman." No one knows how long the effect of such primes endures but nor does anyone know how often a prime appears: how often is an African-American reminded that he or she is such? Once a day? Every half hour? Once a month? I think the number is somewhere between the second and the third.

The strong suggestions then is that it is possible for a historically degraded and/or despised minority group, now subordinate, to have an implicit self-image that is negative, to prefer other to self – indeed, oppressor to self – and to under-perform as soon as made conscious of the subordinate identity. This suggests the power of imposed or induced self-deception – some or, indeed, many subordinate individuals adopting the dominant stereotype regarding themselves. Not all of course, and the latter presumably more likely to oppose their subjugation since they are conscious of it. In any case, revolutionary moments often seem to occur in history when large numbers of individuals have a change in consciousness – regarding themselves and their status. Whether there is an accompanying change in IAT is another matter.

One more form of induced self-deception is worth mentioning. It is surprisingly easy to convince people to make false confessions to major crimes even though this may – and surprisingly often does – result in incarceration for long periods of time. All that is required is a susceptible victim and good old-fashioned police work applied 24/7: isolation of the victim from others, sleep deprivation, coercive interrogation in which denial and refutation are not permitted, false facts provided and hypothetical stories told (we have your blood on the murder weapon, perhaps you woke in a state of semi-consciousness and killed your parents without intending or being aware of it etc., with the implication that a confession will end their interrogation). People vary in in the susceptibility range they are to these pressures and in how much self-deception is eventually induced. Some go on to create false memories to back up their false confessions (Kassin 2008).

There is also a kind of self-deception that could be called imposed self-deception, but which could also be considered *defensive* self-deception. Consider an individual being tortured. The pain can be so great that something called "disassociation" may occur: the pain is separated off from other mental systems, presumably so as to reduce its intensity. As if the psyche or nervous system protects itself from severe pain by objectifying it, distancing it, and splitting it off from the rest of the system. One can think of this as being imposed by the torturer but also as a defensive reaction permitting immediate survival under most unfavorable circumstances. We know from many personal accounts that this is but a temporary solution and that the torture itself and utter helplessness against it endure as psychological and biological costs. There are, of course, more modest forms of "disassociation" from pain than that of torture, such as a mother distracting her child by tickling it.

18.15 Deceit and Self-Deception Seen as an Evolutionary Game

To model the evolution of deceit and self-deception in humans more exactly, I recommend pursuing the strategy that has proven so successful with reciprocal altruism and cooperation. That is, model opposing strategies as simple rules with specified costs and benefits in interaction with each other. This was first applied successfully to the evolution of reciprocal altruism by Axelrod and Hamilton (1981) modeled as a successive series of simultaneous moves with only two options for each of the two players, cooperate or defect, each combination with specified pay-offs (the so-called iterated Prisoner's Dilemma).

It was shown that a very simple strategy – tit-for-tat – beat out all others in computer tournaments and is a fairly robust strategy in evolutionary games: be cooperative on the first move, and imitate your partner's previous move on your next one. In short, reward cooperation with cooperation and punish defection with defection. An advantage of casting the problem in terms of games is that these can be played both mathematically and for real. When their patterns are convergent as they are here, we can have even greater confidence in the underlying logic.

Once the simple tit-for-tat strategy was described, certain problems were discovered that required modified strategies (briefly reviewed in Trivers 2005). For example, occasional errors can lock two tit-for-tatters into a most unfortunate situation, endless "vendettas" in which each reverses its move exactly out of synchrony with the other, never achieving simultaneous cooperation, thereby greatly reducing the success of the tit-for-tat strategy. "Generous tit-for-tat" solved this problem by allowing a tit-for-tatter occasionally (say 1/3rd of the time) to cooperate after a defection by the partner – small cost, larger gain under a variety of realistic conditions (Nowak and Sigmund 2004). Later, "win-stay, lose-shift" was shown to be superior still against a background of tit-for-tatters, defectors, and generous tit-for-tatters (Nowak and Sigmund 1993). Observer effects can also be modeled (Nowak and Sigmund 1998a,b), leading to "indirect reciprocity." In this case, my strategy towards you depends not only on what you have done to me but also on what you have been observed doing to others.

The simplest application of the above to deceit would be to treat it as a Prisoner's Dilemma. Two individuals can tell each other the truth (both cooperate) or lie (both defect) or one of each. There are two problems with this. One is that a critical new variable becomes important: who believes whom? If I believe you and you are lying, I suffer. If you lie and I disbelieve you, you suffer. By contrast, in the Prisoner's Dilemma, each individual knows after each reciprocal play how the other played (cooperate or defect) and tit-for-tat and its elaborations provide a simple reciprocal mechanism that can operate under the humblest of conditions – as in bacteria. The second problem is that with deception, there is no obvious reciprocal logic. If you lie to me, this does not mean my best strategy is to lie back to you – it usually means that my best strategy is to distance myself from you or punish you.

Karl Sigmund (pers. comm.) has suggested that it might be useful to adapt the Ultimatum Game to this problem. In the UG, a proposer offers a given split of (say) \$100: e.g., \$80 to self, \$20 to the responder. The responder, in turn, can accept the split, in which case the money is split accordingly or the responder can reject the offer, in which case neither receives anything. Often the game is played as a one-shot anonymous encounter, i.e., individuals play only once with others whom they do not know and with whom they will not interact in the future.

Sigmund argues as follows. Imagine a modified UG in which there are two possible pots (say \$100 and \$400) and both players know this. One pot is then randomly assigned to the proposer. Now let us say the proposer offers you \$40; this could represent 40% of the pot (in which case you should accept) or 10% (most people would reject). The proposer is permitted to lie and tell you that the pot is the smaller of the two when, in fact, it is the larger. You can trust the proposer or not but the key is that you are permitted to pay to find out the truth from a (disinterested) third party. (This measures the value you place in reducing your uncertainty regarding the proposer's honesty). If you then discover that the proposer lied, you should have a moral (or, at least, moralistic) motive to reject the offer, and the other way around, for the truth (all compared to uncertainty, i.e., not paying to find out). Note that from a purely economic point of view, there is no benefit in finding out the truth, since it costs money after which it may lead to an (otherwise) unnecessary loss of whatever is offered. In Sigmund's words: "how much would a responder be prepared to pay for reducing the uncertainty and go for a possibly inconvenient truth?" Note that the game can be played in real life with varying degrees of anonymity and also multiple times, as in the iterated Prisoner's Dilemma. As ability to discriminate develops, the other person will benefit more from your honesty (quickly seen as such) and suffer less from deception (spotted and discarded). When people are in greater need, they may be expected not to pay to find out the truth but rather to accept the offer whatever its relative size is.

When we add self-deception, a possible game quickly becomes very complicated. One can imagine actors who are stone-cold honest (cost: information given away, naive regarding deception by others), consciously dishonest to a high degree but with low self-deception (cost: higher cognitive cost and when detected), dishonest with high self-deception (more superficially convincing at lower immediate cognitive cost but suffering later defects and acting more often in the service of others, and so on). Without anything else to offer along these lines, I suggest that those talented at the mathematics of simple games or studying them via computer simulations might find it rewarding to define a set of characters along the lines just mentioned, and then assign variable quantitative effects so as to explore their combined evolutionary trajectory. Perhaps, results will be trivial and trajectories will depend completely on the relative quantitative effects assigned but more likely deeper connections will emerge, seen only when the coevolutionary struggle is formulated explicitly.

18.16 The Cost of Deconstructing Lies

The cost of seeing through deception is not trivial, and in some cases, it is substantial. Think of the daily drumbeat of propaganda emerging from the government in times of war or class warfare (2001–2006 in the U.S., for example). On a personal level, I was first exposed to this when I was about 6 years old. Having saved 6 dollars over the space of 2 months to buy a knife displayed in a window, I showed up, only to be told that I was 1 dollar short. Nonsense, I said, the sign outside says "$6." The shopkeeper took me to the sign and showed me that written after the 6 in very small letters was 98, i.e., almost 1 dollar. I was incredulous – and very angry – how did it make sense, I wanted to know, for him to misrepresent the true value of an item by subtracting two pennies so as to generate just the kind of mistake I had, in fact, made? He assured me that the practice was widespread. I soon confirmed that this was true. Almost all prices – gas, food, furniture – when spotted at a distance, appeared to be one full unit below what they actually were. For a couple of weeks I walked around in a daze, benumbed at the amount of unnecessary arithmetical calculation this system required: always adding a unit or two to the total in order to calculate the real value. How was it possible, I kept asking myself, that this was the system of posting prices that had developed?

And, I think, there was my mistake. This was not a rational system agreed upon by all actors – or what Jesus might have told us to do – this was the system that had actually developed over time. In a nutshell, honesty is not evolutionarily stable. It is easily displaced by deception which, in turn, forms a new equilibrium. Further, deception may be counter-selected but so may be a return to honesty, since honest valuations will often be devalued along with those that are hyped. That is, we will frequently add a unit to honest prices, decline to buy them, and the honest shopkeeper suffers.

So also with self-deception. Over evolutionary time, we have been driven downhill by selection such that a degree of self-deception is common to most or all of us. It has formed a new equilibrium such that honest people may come off as lacking, in part because others unconsciously compensate in their estimates for the expected degree of inflation. No inflation so we make them smaller than they really are. And, perhaps, the happiness thermostat in our body has been reset so as to assume a certain degree of self-deception.

References

Anderson MC, Ochsner KN, Kuhl B, Cooper J, Robertson E, Gabrieli SW, Glover GH, Gabrieli JDE (2004) Neural systems underlying the suppression of unwanted memories. Science 303:232–235

Axelrod R, Hamilton WD (1981) The evolution of cooperation. Science 211:1390–1396

Burt A, Trivers R (2006) Genes in conflict: the biology of selfish genetic elements. Belknap Press, Cambridge, MA

Clark DD, Sokoloff L (1999) Circulation and energy metabolism of the brain. In: Siegel GJ, Agranoff BW, Albers RW, Fisher SK, Uhler MD (eds) Basic neurochemistry: molecular, cellular and medical aspects. Lippincott-Raven, Philadelphia, pp 637–670

Crossman AM, Lewis M (2006) Adults' ability to detect children's lying. Behav Sci Law 24:703–715

DePaulo BM, Lindsay JJ, Malone BE, Muhlenbruck L, Charlton K, Cooper H (2003) Cues to deception. Psychol Bull 129:74–118

Douglas W, Gibbons K (1983) Inadequacy of voice recognition as a demonstration of self-deception. J Pers Soc Psychol 44:589–592

Epley N, Whitchurch E (2008) Mirror, mirror on the wall: enhancement in self-recognition. Pers Soc Psychol Bull 34:1159–1170

Gehring WJ, Karpinski A, Hilton JI (2003) Thinking about interracial interactions. Nat Neurosci 6:1241–1243

Gilbert D (2006) Stumbling toward happiness. Alfred A Knopf, New York

Greenwald AG (1980) The totalitarian ego: fabrication and revision of personal histories. Am Psychol 35:603–618

Greenwald AG, McGhee DE, Schwartz JLK (1998) Measuring individual differences in implicit cognition: the implicit association test. J Pers Soc Psychol 74:1464–1480

Greenwald AG, Nosek BA, Banaji MR (2003) Understanding and using the implicit association test: I. An improved scoring algorithm. J Pers Soc Psychol 85:197–216

Gur RC, Sackeim HA (1979) Self-deception: a concept in search of a phenomenon. J Pers Soc Psychol 37:147–169

Haig D (2002) Genomic imprinting and kinship. Rutgers University Press, New Brunswick, NJ

Hartung J (1988) Deceiving down: conjectures on the management of subordinate status. In: Lockard JS, Paulhus DL (eds) Self-deception: an adaptive mechanism?. Prentice Hall, Englewood Cliffs, NJ, pp 170–185

Hauser MD (1992) Costs of deception: cheaters are punished in rhesus monkeys (*Macaca mulatta*). Proc Natl Acad Sci USA 89:12137–12139

Isaakowitz DM, Toner K, Goren D, Wilson HR (2008) Looking while unhappy: mood-congruent gaze in young adults, positive gaze in older adults. Psychol Sci 19:848–853

Kassin SM (2008) False confessions: causes, consequences and implications for reform. Curr Dir Psychol Sci 17:249–253

Maass A (1999) Linguistic intergroup bias: stereotype perpetuation through language. Adv Exp Soc Psychol 31:79–121

Mann S, Vrij A (2006) Police officers judgements of veracity, tenseness, cognitive load and attempted behavioral control in real life police interviews. Psychol Crime Law 12:307–319

Markle S (2007) Stick insects: masters of defense. Lerner Publishing Group, Minneapolis

Mather M, Carstensen LL (2003) Aging and attentional biases for emotional faces. Psychol Sci 14:409–415

Mather M, Carstensen LL (2005) Aging and motivated cognition: the positivity effect in attention and memory. Trends Cogn Sci 9:496–502

Mather M, Canli T, English T, Whitfield S, Wais P, Ochsner K, Gabrieli JDE, Carstensen LL (2004) Amygdala responses to emotionally valenced stimuli in older and younger adults. Psychol Sci 15:259–263

Møller AP (1987) Social control of deception among status signaling house sparrows *Passer domesticus*. Behav Ecol Sociobiol 20:307–311

Nosek BA, Banaji MR, Greenwald AG (2002) Harvesting implicit group attitudes and beliefs from a demonstration website. Group Dyn 6:101–115

Nowak MA, Sigmund K (1993) A strategy of win-stay, lose-shift that outperforms tit-for-tat in the prisoner's dilemma game. Nature 364:56–58

Nowak MA, Sigmund K (1998a) Evolution of indirect reciprocity by image scoring. Nature 393:573–577

Nowak MA, Sigmund K (1998b) The dynamics of indirect reciprocity. J Theor Biol 194:561–574

Nowak MA, Sigmund K (2004) Evolutionary dynamics of biological games. Science 303:793–799

Pennebaker JW (1997) Opening up: the healing power of expressing emotions. Guilford Press, New York

Richeson JA, Shelton JN (2003) When prejudice does not pay: effects of interracial contact on executive function. Psychol Sci 14:287–290

Richeson JA, Baird AA, Gordon HL, Heatherton TF, Wyland CL, Trawalter S, Shelton JN (2003) An fMRI investigation of the impact of interracial contact on executive function. Nat Neurosci 6:1323–1328

Rosenkranz MA, Jackson DC, Dalton KM, Dolski I, Ryff CD, Singer BH, Muller D, Kalin NH, Davidson RJ (2003) Affective style and *in vivo* immune response: neurobehavioral mechanisms. Proc Natl Acad Sci USA 100:11148–11152

Stoltz JA, Neff BD (2006) Male size and mating tactic influence proximity to females during sperm competition in bluegill sunfish. Behav Ecol Sociobiol 59:811–818

Tavris C, Aronson E (2007) Mistakes were made (but not by *me*): why we justify foolish beliefs, bad decisions, and hurtful acts. Harcourt, New York

Tibbetts EA, Dale J (2004) A socially enforced signal of quality in a paper wasp. Nature 432:218–222

Trivers R (1985) Social evolution. Benjamin/Cummings, Menlo Park, CA

Trivers R (2000) The elements of a scientific theory of self-deception. Ann NY Acad Sci 907:114–131

Trivers R (2005) Reciprocal altruism: 30 years later. In: van Schaik CP, Kappeler PM (eds) Cooperation in primates and humans: mechanisms and evolution. Springer, Berlin, pp 67–83

Troisi A (2002) Displacement activities as a behavioral measure of stress in nonhuman primates and human subjects. Stress 5:47–54

Vrij A (2004) Why professionals fail to catch liars and how they can improve. Leg Criminol Psychol 9:159–181

Vrij A, Fisher R, Mann S, Leal S (2006) Detecting deception by manipulating cognitive load. Trends Cogn Sci 10:141–142

Chapter 19
Human Universals and Primate Symplesiomorphies: Establishing the Lemur Baseline

Claudia Fichtel and Peter M. Kappeler

> Lemur *and* Propithecus *are both socially intelligent and socially dependent. They are, however, hopelessly stupid towards unknown inanimate objects. In this branch of the primates, the basic qualities of primate society have evolved without the formal inventive intelligence of true monkeys.*
>
> Alison Jolly (1966a: 165–166)

Abstract Social behavior, culture, and cognition are domains where presumably most human universals exist. Identification of these derived human traits depends and relies on comparisons with other primates, notably chimpanzees. This approach can also be used to reconstruct primate and human behavioral evolution. Accordingly, traits found in both *Homo* and *Pan* can be inferred to have existed in their last common ancestor as well. By analogy, traits shared between humans and other primates can be traced back even further down on our family tree. Here, we look at the other side of human universals, i.e., behavioral and cognitive traits of the most basal living primates, which ought to represent the common primate legacy upon which later taxon-specific specializations were built. Specifically, we review studies investigating cognitive abilities and social behavior of the lemuriform primates of Madagascar. The Malagasy lemurs are particularly important for this purpose because they alone, among strepsirrhine primates, have evolved group-living, which characterizes most living haplorrhines. Even though lemurs have relatively smaller brains than New and Old World monkeys and great apes, their ability to solve problems that require technical intelligence is qualitatively on par with that of haplorrhines. In the domain of social intelligence, however, lemurs deviate from the

C. Fichtel (✉)
Department of Behavioral Ecology & Sociobiology, German Primate Center, Göttingen, Germany
e-mail: claudia.fichtel@gwdg.de

P.M. Kappeler
Department of Sociobiology/Anthropology, University of Göttingen, Göttingen, Germany
e-mail: pkappel@gwdg.de

P.M. Kappeler and J.B. Silk (eds.), *Mind the Gap*,
DOI 10.1007/978-3-642-02725-3_19, © Springer-Verlag Berlin Heidelberg 2010

better-known haplorrhine models (i.e., cercopithecines) in several respects. Most importantly, their behavioral strategies reflect an emphasis on within-group competition, rather than cooperation, which may represent lemur-specific adaptations to an ecologically unpredictable environment, rather than fundamental deficits in social intelligence. In any event, a broad comparative perspective including the best living models of the earliest gregarious primates can enrich reconstructions and other evolutionary analyzes of primate social behavior, including that of humans.

19.1 Introduction

Behavioral characteristics unique to *Homo sapiens* can only be identified as such by reference to a meaningful out-group. Because humans are members of the order Primates, this lineage provides the natural out-group for such comparisons. Primates, however, are a diverse group with hundreds of living species and 80 million years of evolutionary history, so that specific deviations from our basal evolutionary legacy may not be that evident. Specifically, our biological continuity with other animals is evident in those behavioral, morphological, and physiological traits that have a genetic basis and define our affiliation to vertebrates, mammals, primates, haplorrhines, catarrhines, and hominids. Our bodies and minds can therefore be seen as a complex puzzle made up of pieces we share with these various groups, interspersed with a few derived pieces. Traditionally, most comparative studies attempting to identify the derived human pieces of this puzzle have relied on the contrasts between humans and chimpanzees (*Pan troglodytes*). Sequencing of the chimpanzee genome revealed <1% difference with the human genome (Mikkelsen et al. 2005), and molecular clock studies pinpointed the last common ancestor of *Homo* and *Pan* at around 6 million years before the present (Bradley 2008). Chimpanzees (and bonobos) therefore provide the most immediate step back into our deep behavioral past and the most appropriate specific referents for comparative studies of human behavior (e.g., Boesch 2007; Whiten, this volume). This approach can also be used to reconstruct primate and human behavioral evolution. According to this logic, characteristics or traits found in both taxa today are assumed to have been already present in their most recent common ancestor, i.e., they represent plesiomorphies in the terminology of cladistics. In this terminology, human universals represent autapomorphies, i.e., derived traits that are unique to a terminal group.

Extending comparisons of behavioral traits beyond the obvious *Homo-Pan* contrasts has led to additional insights. For example, the demonstration of social and technical traditions in orangutans (*Pongo pygmaeus*) has led to the novel conclusion that such cultural abilities must have been shared by the last common ancestor of all great apes (van Schaik et al. 2003), which lived about 14 million years ago. Similarly, comparative studies of cognitive abilities of great apes and other haplorrhine primates revealed that New and Old World monkeys share many common features, and that only the level of performance varied among species (Amici et al. 2008). These examples highlight the importance of considering behavioral plesiomorphies in the analysis of potential human (or hominid) universals. In other words, characteristics attributed to the last common ancestor of chimpanzees and humans (or of the hominids) may, in fact, have an even longer evolutionary history. Thus, comparative analyses of human behavior limited to chimpanzees or other great apes risk drawing inaccurate conclusions by failing to explicitly recognize primate symplesiomorphies, i.e., character states that originated in an earlier common ancestor.

Evidence for the existence of precursors of cognitive and social traits in the behavioral domains, where human universals are most pronounced, has been produced for various New and Old World primates, including great apes (reviewed in Tomasello and Call 1997). Capuchin monkeys (*Cebus* spp.) exhibit remarkable technical intelligence (Visalberghi 1993), all too familiar socio–emotional responses (Brosnan and de Waal 2003), and local variation in cultural traditions among wild populations (Perry and Manson 2003). These observations question the existence of a possible deep behavioral and cognitive gap between hominins or hominids and all other primates. On the other hand, there are pronounced grade shifts in relative brain size among primate lineages (Martin and Harvey 1985; Dunbar and Shultz 2007), which may underlie qualitative differences in their social behavior and cognitive abilities (Deaner et al. 2007; Shultz and Dunbar 2007, Dunbar this volume). The cognitive and cultural capabilities of haplorrhine primates have been reviewed elsewhere (e.g., Tomasello and Call 1997; Whiten and van Schaik 2007). Here, we extend these comparisons to the most basal living primates: the strepsirrhine suborder. While recent strepsirrhines (lemurs and lorises) have their own distinctive evolutionary history and adaptations, they have retained a number of primitive features that almost certainly characterized the earliest primates (Yoder 2007). These living species, therefore, represent the legacy upon which all living primates have built their specific derived adaptations.

Malagasy lemurs (Lemuriformes) are particularly interesting in this context, because they are the only strepsirrhines that have evolved multi-male multi-female groups, like those that are characteristic of most haplorrhines. Group-living lemurs therefore represent the most appropriate models to establish the baseline for primate social intelligence and complexity, whereas cognitive abilities related to technical intelligence should be found independent of a particular social system. The specific goal of this chapter is to summarize and evaluate studies of social and technical cognitive abilities among lemurs. Even though Alison Jolly (1966b) established the importance of comparative studies of lemur social intelligence in the early days of primatology, subsequent research on lemur cognition and social communication has not been conducted with the enthusiasm and rigor that has characterized similar research on haplorrhines, and great apes in particular, in recent decades. Nevertheless, by bringing together some old, often overlooked studies and some more recent work, we can begin to sketch the outlines of lemur cognition and social behavior. This endeavor will help to put comparative work on human universals, and on hominid behavior more generally, into broader perspective, insofar as that a closer look at the basal living primates will provide baseline information about shared ancestral traits of all primates.

19.2 Lemurs

Based on genetic differences and several morphological features, primates can be divided into two suborders: strepsirrhines (lemurs and lorises) and haplorrhines (New and Old World monkeys including the great apes). The living lemurs represent the 100 or so endpoints of an adaptive radiation following a single successful colonization event of Madagascar during the Eocene (Karanth et al. 2005; Tattersall 2007). They can be grouped into five families and 15 genera, which, together with recently extinct taxa, exhibit almost the full range of diversity in social, ecological, and life-history adaptations found among all other primates (Richard and Dewar 1991). The majority of living lemurs are nocturnal and solitary or pair-living, but, according to recent genetic analyzes (Horvath et al. 2008), life in multi-male, multi-female groups has evolved independently in the Lemuridae (in the genera *Lemur*, *Eulemur*, *Hapalemur*, and *Varecia*) and Indridae (in the genus *Propithecus*). After controlling for body size and phylogenetic effects, lemur groups in both families are, on average, smaller than those of haplorrhines (Kappeler and Heymann 1996), and they are generally characterized by even adult sex ratios (Kappeler 2000). As in many haplorrhines, group-living lemurs are characterized by predominant female philopatry (Richard et al. 1993; Sussman 1992), diurnal activity (at least partially) (Kappeler and Erkert 2003; Erkert and Kappeler 2004), and their vocal repertoires sometimes include functionally referential calls (Fichtel and Kappeler 2002). Relative brain size of lemurs tends to be smaller than that of haplorrhines (Armstrong 1985; Dunbar 1998) and olfactory communication is used in a variety of behavioral contexts (Kappeler 1998; Pochron et al. 2005). Thus, the social systems (sensu Kappeler and van Schaik 2002) of lemurs exhibit a mixture of idiosyncrasies as well as convergences with those of other primates, but the basic pillars of sociality appear to be comparable.

Below, we summarize the results of a literature review of studies of lemur cognition and social behavior that bear relevance to the study of human behavioral universals. We divide this review into two sections that deal with technical and social intelligence, respectively. We do not attempt explicit and detailed comparisons with great apes or all other haplorrhines, and we refrain from extending comparisons to other mammalian orders. Instead, we aim to provide a concise summary of the socio-cognitive abilities of lemurs and other strepsirrhines that may contribute interesting baseline information about primate sociality for comparisons among other primates and mammals, including humans.

19.3 Technical Intelligence

19.3.1 Space and Objects

The most critical challenge to survival is the ability to deal effectively with (three-dimensional) space and objects, such as food, shelter, and predators. Physical cognition, i.e., the understanding of object features and their various spatial and causal interrelations, is presumably most adaptive in the contexts of foraging and locomotion. Independent of their social organization, most primates tend to remain within a particular home range. Cognitive abilities that enable animals to identify their position, to remember what is located where, and to travel efficiently between these sites, represent selective advantages (Anderson 1983; Gallistel 1989). Because successful foraging and efficient locomotion are general ecological problems, the relevant cognitive skills of lemurs are expected to be similar to those of other primates.

The cognitive abilities of lemurs in the context of spatial mapping and spatial memory have been the focus of experimental studies in captivity and, more recently, in observational and experimental studies in the wild. In Madagascar, gray mouse lemurs (*Microcebus murinus*) inhabit dry deciduous forests with pronounced seasonal fluctuations in food availability. During a long dry season, when food availability is low, mouse lemurs mainly rely on resources that are sparsely distributed but predictable in space, such as gum, secretions from colonial insects, and nectar (Dammhahn and Kappeler 2008). Field observations revealed that solitary mouse lemurs revisited stationary feeding sites more often than nonstationary feeding sites (Joly and Zimmermann 2007). Using an experimental approach, Lührs et al. (2009) set mouse lemurs a spatial memory task by confronting them with two different patterns of baited and non-baited artificial feeding stations. Mouse lemurs used spatial cues to relocate baited feeding stations and they were able to rapidly learn a new spatial arrangement. In a release experiment, they also exhibited high travel efficiency in directed movements, suggesting that their spatial memory is based on some kind of mental representation that is more detailed than a route-based network map (Lührs et al. 2009). The existence of a topological or route-based map has also been proposed for group movements of two group-living lemurs, Milne Edwards' sifakas (*Propithecus edwardsi*) and redfronted lemurs (*Eulemur fulvus rufus*, Erhart and Overdorff 2008). Because route-based mental representation of spatial relationships, straight-line travel, and efficient goal-directed movements between distant sites have been suggested for several haplorrhines (Boesch and Boesch 1984; Gallistel 1989; Garber 1988; Menzel 1991; Janson 1998; Noser and Byrne 2007), cognitive abilities in the context of spatial orientation do not appear to differ fundamentally between lemurs and other primates.

Another set of cognitive spatial skills is required to search for hidden food, as in object permanence experiments, or to trace invisible displacements of food hidden by an experimenter. When various lorisids, ring-tailed (*Lemur catta*), and brown lemurs (*E. fulvus*) were tested for their ability to find hidden food, only one loris failed to master the task (Jolly 1964a). Object permanence was studied in redfronted lemurs, mongoose lemurs (*E. mongoz)*, ring-tailed lemurs, and bamboo lemurs (*Hapalemur griseus*) (Deppe et al. 2009). Lemurs performed well above chance levels in tracking food that had been in clear view before being hidden (visible displacements). However, when lemurs were not allowed to search for up to 25 s, performance declined with increasing time-delay. They did not outperform chance levels in tracking food in invisible displacement tasks. Many haplorrhine primates solve visible displacement tasks, whereas the ability to perform invisible displacement has been demonstrated in apes, rhesus macaques (*Macaca mulatta*), and cotton top tamarins (*Saguinus oedipus*) (de Blois et al. 1998; Call 2001; Hauser 2001; Neiworth et al. 2003; Mendes and Huber 2004).

Primates often face the problem of getting food that is out of reach, such as fruits at the periphery of a branch. Thus, spatial understanding of objects that allows them to determine to which branch a fruit is connected to pull it closer or to use another branch as a tool to bring fruits into reach should be advantageous. Detour problems test this kind of understanding. Lemurs quite successfully mastered detour problems, in which food was impaled on a bent wire and the subject had to move it to the left or right and had to push or pull the food (Davis and Leary 1968). Although Old World monkeys were best at performing these tasks, lemurs did not differ from some New World monkeys, and squirrel monkeys failed the task entirely.

Maze experiments represent another type of detour problems, in which the spatial memory of subjects is investigated. Picq (1993, 2007) conducted radial maze experiments with captive mouse lemurs. In this experimental setup, subjects learned to choose one out of eight possible arms to get access to a reward; in this case, a nest-box. Mouse lemurs mastered this task quickly, and their learning curves matched those of New and Old world primates, including chimpanzees.

Because of their fast life histories, mouse lemurs are also well suited to address questions of aging in memory. Picq (2007) applied different visual and spatial discrimination as well as generalization tasks in an eight-armed radial maze. Young mouse lemurs were able to learn all tasks quickly; older mouse lemurs performed as well as the young ones in some tasks, but showed impairment in several other tasks, indicating that the acquisition of skills is not affected, but the shifts in attention from visual to spatial cues and, thus, the flexible use of acquired memories in novel situations as well as the generation of novel solution strategies were impaired. Similar maze experiments with haplorrhines revealed that basic spatial memory skills are comparable across taxa (Tomasello and Call 1997). Other tasks, in which spatial understanding of objects is investigated, such as patterned-string problems, in which subjects have to disentangle strings differently to get a reward, or mental rotations skills, have not been conducted with lemurs so far.

19.3.2 Tools and Causality

Primate foods are either immediately ready to eat or require manipulation before ingestion. Manipulation tasks vary from simple to complex: from just picking a fruit or leaf off the branch, through digging for roots, uncovering food items under leaves or tree barks, capturing mobile insects, using tools to open nuts, and modifying probes to fish for ants (van Schaik et al. 1999). Observations of wild lemurs revealed little evidence for object manipulation in a foraging context, except for the aye-aye (*Daubentonia madagascariensis*), which uses a thin, long, tap-scanning and probing middle finger to locate and extract insects embedded in trees or branches (Erickson 1995). Black (*E. macaco*) and brown lemurs sometimes manipulate millipedes vigorously, presumably in an effort to annoint their bodies with an insect repellent (Birkinshaw 1999; pers. observ.). No other manipulative interactions with food items by lemurs have been reported so far.

Studies of manipulative interactions with novel objects in captivity are more abundant. When various lorisids were confronted with different complex novel objects, they either stared at them and struck them with their hand (Jolly 1964a, b) or pushed, pulled, or even grasped them (Ehrlich 1970; Renner et al. 1992). Parker (1973, 1974) compared manipulation behavior with hands and mouth towards novel objects in ring-tailed and black lemurs with that of several haplorrhines, including great apes. He found that manipulative behavior was most variable in great apes, slightly less variable in macaques, and least variable in langurs, spider monkeys, and gibbons. Both lemur species were intermediate between the great apes/macaques and the other cluster of species. Because differences between groups could not be explained by hand anatomy, but by habitat use, i.e., a distinction between feeding specialists and generalists, Parker (1973) suggested that broad-niched opportunists need to develop more explorative behavior than specialists to adapt to the wide variety of circumstances in their habitat. A similar pattern of object manipulation variability was found in another study comparing 74 species of primates. Lemurs, marmosets, and leaf-eating monkeys showed less variable behavior than frugivorous and insectivorous Old World monkeys, but not folivores. Capuchin monkeys, as well as great apes, showed the most variable behaviors (Torigoe 1985). Thus, lemur manipulatory skills are roughly comparable with those of at least some New and Old World monkeys.

In addition, more complex object manipulation skills appear to exist in gray mouse lemurs. In a series of experiments, individuals first had to open a plastic box in three different ways to get access to a reward (Schilling 2007). In the second task, the reward was hidden in a cylindrical box sliding inside an opaque second box. Subjects were required to manipulate a string to move the inner box closer in order to reach and pull out the reward. All mouse lemurs learned the tasks rapidly and improved over time. In the third task, mouse lemurs were tested with a vertical mirror box presenting a mealworm hanging behind an opaque wall in such a way that the reward could only be obtained by learning to use its reversed image. All but one individual mastered this rather complex task, which may require some form of mental rotation.

The aye-aye, which has the largest relative brain size among strepsirrhines (Stephan et al. 1988), uses a unique form of percussive tap-foraging, during which insect larvae are extracted from wood by a series of coordinated actions with the elongated third digit (Miliken et al. 1991; Lhota et al. 2008). They are also able to open complex puzzle boxes and performed better than other lemurs on such a test (Digby et al. 2008). Opening of simple boxes has also been demonstrated in brown, black, and ring-tailed lemurs (Jolly 1964a, b; Kappeler 1987; Fornasieri et al. 1990; Anderson et al. 1992), though Fornasieri et al. (1990) stated that they showed "little comprehension" of the task.

Understanding of physical and causally relevant aspects of objects are prerequisites for using tools. Tool-use has been reported in several haplorrhine primates, with chimpanzees and orangutans exhibiting the most complex skills (Whiten et al. 1999; van Schaik et al. 2003; Moura and Lee 2004). In contrast, tool-use has not been reported from the wild for any strepsirrhine. However, there is one observation of aye-ayes manipulating an object that required some sort of sensorimotor intelligence that is also required for tool-use: they grasped and moved a liana over a branch under which they had been feeding to gain better access to a feeding site (Sterling 1994). These sensorimotor skills were examined in more detail in an experiment in which aye-ayes could use a simulated liana (rope) to get access to feeding cups fixed to the wall (Sterling and Povinelli 1999). However, they failed to move the rope horizontally close to the feeding cups, though they readily climbed up and down the simulated liana. The authors concluded that aye-ayes do not achieve comprehension to use tools, but rather may use trial-and-error learning to develop tool-use behavior.

Hence, the question arises whether strepsirrhines simply do not posses the underlying cognitive abilities to understand the functionality of objects for potential tool-use. In a recent study, Santos et al. (2005a) set up a series of experiments with brown and ring-tailed lemurs to address this question. They used and extended a design originally used with other haplorrhines (Hauser et al. 1999; Povinelli 2000; Fujita et al. 2003; Santos et al. 2006). In these experiments, lemurs were offered two cane-shaped tools to pull out-of-reach food items. In the first series of experiments, tools were identical and differed only in the orientation relative to the food reward, with one tool being more effective to reach the food. Lemurs had to choose the more effective tool and did so just as successfully as capuchins (Cummins-Sebree and Fragaszy 2001; Fujita et al. 2003). In the second experiment, lemurs were tested with novel tools differing from the originals in one dimension, to test whether they spontaneously attend to some of the features that are causally relevant for a successful pulling tool. Lemurs attended more to the sizes than to the colors of tools, but made no distinction between tools' shapes and textures. The next two experimental designs presented problems in which one of the tools had to be modified to access the food. In these tests, the authors used familiar, already successfully used tools, and unfamiliar tools. Lemurs did not prefer familiar over unfamiliar tools, indicating that they chose tools on the basis of features that were functionally relevant for the task. Thus, lemurs solved the can-pulling tasks like other tool-using haplorrhines, indicating that many primates share an ability to reason about basic functional properties of different objects, even if they do not use tools normally (see also Hauser et al. 2002; Spaulding and Hauser 2005). In contrast to this basic understanding of features of tools, regular tool users have a more sophisticated understanding of causal relationships between features of the tool and the problems they can solve with it (Martin-Ordas et al. 2008; Seed et al. 2009). Orangutans, for example, even spat water spontaneously into a transparent tube to get access to an out-of-reach peanut floating inside the tube (Mendes et al. 2007).

19.3.3 Features and Categories

All primates locate and manipulate objects, which they identify on the basis of certain observable features. However, in some cases, primates also identify objects on the basis of conceptual categories that go beyond direct perception (Tomasello and Call 1997). Basic discrimination learning of objects that vary either in shape, color, pattern, brightness, location, or sound has been demonstrated in a variety of lemurs and other strepsirrhines (reviewed in Ehrlich et al. 1976; Meador et al. 1987; Tomasello and Call 1997).

Learning sets are tests in which individuals may become better at discrimination when they solve different sets of similar problems over time and have learned to deal with a particular type of problem in general. Numerous studies on learning set phenomena have led to the consensus that successful problem-solving indicates the use of some type of abstract rule (Harlow 1949, reviewed in Fobes and King 1982). Tests of object discrimination learning sets have been conducted with bushbabies, lorises, black and ring-tailed lemurs (Stevens 1965; Cooper 1974; Ohta 1983; Ohta et al. 1984, 1987). A comparison across the primate order revealed that there are no taxonomic differences with respect to success in object discrimination tasks: "*after 200 problems, approximately 80% correct performance is achieved by species as different from another as black lemurs, chimpanzees, rhesus macaques, and gorillas*" (Tomasello and Call 1997).

Reversal learning paradigms investigate the ability to reverse a previously learned discrimination. Subjects first learn an object discrimination to get a reward before a previously nonrewarded object becomes rewarded. This paradigm is thought to reflect a subject's ability to form and use abstract rules or hypotheses (Rumbough 1970). When brown and ring-tailed, fork-marked (*Phaner* spp.), and mouse lemurs were subjected to this test paradigm, their performance was inferior to that of haplorrhines (Stevens 1965, reviewed in Rumbough 1997). However, more recent studies of mouse lemurs revealed that their reversal skills are comparable to those of haplorrhines (Picq 1993, 2007). Cross-modal transfer of objects from one perceptual domain to another also belongs to the kinds of tasks that go beyond stimulus-response associations. Only bushbabies (*Galago senegalensis*) among strepsirrhines were presented with this task, in which they were able to transfer learned responses from vision to audition (Ward et al. 1976).

The delayed response paradigm investigates a subject's memory or ability to maintain a perception of an item when it is no longer available. Typically, the subject sees a reward hidden in one of two locations, and after a certain delay, it may search for it. The few studies that applied this test paradigm to redfronted, mongoose, bamboo, ring-tailed, and ruffed lemurs (*Varecia variegata*), as well as to bushbabies, found that they were all inferior to haplorrhines in their performance (Harlow et al. 1932; Maslow and Harlow 1932; Jolly 1964a; Deppe et al. 2009).

Discrimination learning of relational categories involves a concept that can be learned only by comparing objects to one another and by inducing some relational difference (e.g., "larger than"). In oddity concept studies, subjects are presented three stimuli, two of which are the same. Subjects are rewarded for responding to the odd one; for example, in a "square–square–triangle" constellation. After that, some of the training subjects are confronted with three completely new objects (line–circle–line). In this paradigm, the inference is that subjects understand the concept of "odd." This concept allows animals, for example, to categorize environmental features such as different food items. Davis et al. (1967) confronted several New World and Old World monkeys as well as ring-tailed lemurs with oddity problems, and found that the performance of ring-tailed lemurs was inferior to most haplorrhines, but better than guenons. In summary, there are no qualitative differences in performance across major primate radiations in object discrimination learning set formation tasks. However, a few studies on strepsirrhines suggest that they do not seem to be very skilled in reversal learning and delayed response, but they do seem to have an understanding of oddity problems.

The ability to form categories of objects belonging to same or different classes is another task that provides insights into cognitive abilities of animals. In such tests, subjects are presented with many objects simultaneously and asked to sort them into groups on the basis of their similarities and differences. This is a demanding task because subjects are required to coordinate both the similarities and differences of multiple objects simultaneously and then to manipulate the objects in line with that understanding. The only study of such capacities in strepsirrhine primates reported remarkable skills in serial ordering of objects in ring-tailed lemurs (Merritt et al. 2007).

Many animals are also able to organize sequences in memory and retrieve ordered sequences without language (Sands and Wright 1980; Straub and Terrace 1981). For example, capuchin monkeys and rhesus macaques were able to select a series of photographs according to a consistent arbitrary order (D'Amato and Colombo 1989; Terrace et al. 2003). In this simultaneous chaining paradigm, a series of arbitrary stimuli (such as photographs) are presented simultaneously in random spatial position on a touch-sensitive monitor. Subjects are rewarded when they respond in a prespecified arbitrary order without error. This paradigm is particularly useful for cognitive studies because it investigates the internal representation of the sequence. Merritt et al. (2007) tested ring-tailed lemurs with such a paradigm. Ring-tailed lemurs were capable of learning three-, four-, and five-items lists. Moreover, these lemurs showed a remarkable similarity in accuracy and reaction time with that of capuchin and rhesus monkeys (D'Amato and Colombo 1989; Terrace et al. 2003).

19.3.4 Quantities

Primates also need to have an understanding of quantities to estimate food availability at different feeding patches or the number of opponents in a potential fight. There have been many studies demonstrating that monkeys and apes are able to judge the absolute and relative numerousness of objects (Tomasello and Call 1997; Beran and Beran 2004; Hanus and Call 2007; Evans et al. 2009). For example, it has been shown that anthropoids possess numerical representation that is modulated by Weber's law, such that as the numerical magnitude increases, a larger disparity is needed to obtain the same level of discrimination. By applying a search task in which grapes were placed into a bucket, Lewis et al. (2005) studied mongoose lemurs' numerical ability. They were able to differentiate numerosities that differed by 1:2, but not those that differed by 2:3 or 3:4. Thus, lemurs' understanding of numerosity also seems to be modulated by Weber's law. Nevertheless, lemurs' numerical discrimination seems to be inferior to that of New World and Old World monkeys; tamarin monkeys (*Saguinus* spp.) were able to differentiate sequences of syllables that differed by 1:2 and 2:3 but not the 3:4 ratios (Hauser et al. 2003), and rhesus macaques even discriminated numerosities that differed by a 4:5 ratio (Brannon and Terrace 2000).

Expectation about numerical events has been studied in ring-tailed, brown, mongoose, and ruffed lemurs (Santos et al. 2005b). By using looking techniques, they explored how lemurs represent small numbers of objects spontaneously in the absence of explicit training (see Hauser 2000 for review). Santos and her colleagues conducted experiments that were modeled after Wynn's violation of expectancy paradigm for human infants (Wynn 1992), and tested whether lemurs look longer when the number of objects revealed behind a screen differs from the number that should be there. They presented lemurs with two lemons that disappeared sequentially behind an occluder; lemurs looked longer at an unexpected outcome of only one lemon than at an expected outcome of two lemons. Similarly, lemurs looked longer at an unexpected outcome of three lemons than towards an expected outcome of two lemons. In addition, lemurs attended to the size of objects; they looked longer at an object twice the size of the original object than at an expected outcome of two objects of the original size. Thus, these lemurs understand the outcome of simple arithmetic operations of 1 + 1 events. These findings are in line with those in human infants (Wynn 1992; Feigenson et al. 2002), rhesus macaques (Hauser et al. 1996), and cotton-top tamarins (Uller et al. 2001). However, capuchin monkeys (*Cebus apella*) have been shown to be able to judge the quantity of 1–5 items in a sequentially presented food choice experiment (Evans et al. 2009). Moreover, great apes were able to differentiate quantities of up to ten items when items were presented simultaneously. However, sequential presentation of food items resulted in a correct judgment of only up to six items (Hanus and Call 2007). Furthermore, some chimpanzees, which were trained in lexical language skills, could judge of up to ten sequentially presented items correctly (Beran and Beran 2004). Thus, without training, haplorrhine primates are able to perform arithmetic operation of up to 6. Because newly hatched domestic chicks (*Gallus gallus*) are able to add and subtract up to five sequentially presented items, mental number representation might be present among many more vertebrates, however (Rugani et al. 2009).

19.4 Social Intelligence

There is a wealth of studies of social cognition in haplorrhine primates (Tomasello and Moll this volume, Cheney and Seyfarth this volume) from which the social brain hypothesis has been developed (Dunbar this volume). Strepsirrhines have not been well represented in this field of research – either because they are not interesting in this context due to their relatively small brain size (cf. Deaner et al. 2006) or because Jolly's (1966b) first impression of lemur intelligence has impeded subsequent research endeavors. However, there are some lemur studies that are relevant to assumptions and predictions of the social brain hypothesis, and show that group-living lemurs exhibit some interesting differences in their social lives from their haplorrhine cousins.

According to the social intelligence hypothesis, the challenges of living in social groups have favored the expansion and reorganization of the primate brain (Whiten and Byrne 1997; Dunbar and Shultz 2007; Silk 2007; Dunbar, this volume). Comparative studies of brain size among primates revealed that relative brain size correlates with several indices of social complexity, including group size (Dunbar 1995), number of females in the group (Lindenfors 2005), the frequency of coalitions (Dunbar and Shultz 2007), grooming clique size (Kudo and Dunbar 2001), the prevalence of social play (Lewis 2000), the frequency of tactical deception (Byrne and Corp 2004), and the frequency of social learning (Reader and Laland 2002). Below, we will summarize our current knowledge of lemur social complexity, focusing on group size and composition, the structure of social relationships (coalitions, cooperation, postconflict behavior, grooming networks), deception, social learning, and innovations, as well as communication. The aim of this review is not to be exhaustive, but rather to highlight the key differences and similarities.

19.4.1 Social Complexity and the Structure of Social Relationships

One way to test the social intelligence hypothesis experimentally is to examine whether species with complex social environments show unusual intelligence in nonsocial domains compared with closely related, less social species (Bond et al. 2003). Transitive inference (if $A > B$ and $B < C$, then $A > C$) is a form of deductive reasoning that has been suggested as one cognitive mechanism with which animals could learn the many relationships within their group's dominance hierarchy. This process, thus, bears relevance to the social intelligence hypothesis, which posits evolutionary links between various forms of social and nonsocial cognition. The relationship between social complexity and transitive reasoning has been studied in ring-tailed and mongoose lemurs (MacLean et al. 2008). The group-living ring-tailed lemurs and the pair-living mongoose lemurs showed similar transitive inference, indicating that both species possess similar fundamental cognitive abilities in this respect, obscuring potential effects of group size and complexity.

Because females are philopatric in many haplorrhine primates and form long-term social networks, the average number of females per group has an evolutionary impact on the development of large brains (Lindenfors 2005), and, hence, social intelligence. Lemur groups usually contain only one to five reproductive females, which are also philopatric (Kappeler 2000). Social networks, such as matrilineal dominance hierarchies, in which maternal kin occupy adjacent ranks and females form close and stable relationships, have been described for many Old World monkeys (Silk 2007). Similar bonds have not been observed among lemurid females (reviewed in Kappeler 1999), even though transitive dominance relationships are established in most, but not all lemur species (Kappeler 1993b; Pereira et al. 1990). In species with dominance relationships, all females dominate all males (Jolly 1966a; Richard 1987; Pochron et al. 2003). Reproductive opportunities seem to be more limited for lemurid females than for cercopithecine females, because, on average, only one or two females give birth per year in groups of most lemurid species (Overdorff et al. 1999; Kappeler 2000; Pochron et al. 2004). Targeted aggression by female group members towards close relatives, often adolescent females, resulting in severe injury or eviction, has been observed in captive and field settings in representatives of both Lemuridae and Indriidae (Vick and Pereira 1989; Pereira 1993; Barthold et al. 2009; Kappeler unpubl. data). Furthermore, infanticide by females has been observed in several lemur species (Andrews 1998; Jolly et al. 2000). Coalitionary defense of home ranges against neighboring groups indicates that competition between groups is also pronounced (Nunn and Deaner 2004; Benadi et al. 2008). Because lemurs live in a relatively harsh and unpredictable environment with pronounced seasonality (Wright 1999; Dewar and Richard 2007), ecological factors may have favored competitive, rather than cooperative tendencies in group-living lemurid females.

Overt cooperative behavior, another hallmark of social complexity (Silk and Boyd this volume), has only rarely been observed in lemurs. Coalitions of related redfronted lemur males have been observed to take over other groups (Ostner and Kappeler 2004), and ring-tailed lemur males sometimes migrate in pairs or trios (Jones 1983; Sussman 1992). Only a tiny fraction of agonistic interactions among females involve coalitionary support (Pereira and Kappeler 1997) even though joint territorial defense is common (see above). Solitary species exhibit a spatio-genetic structure characterized by spatial clustering of related females (Kappeler et al. 2002; Wimmer et al. 2002), which may facilitate cooperative behavior among relatives. For example, several gray mouse lemurs mobbed a snake that held a conspecific until it could escape (Eberle and Kappeler 2008), and communal breeding among closely related females with a high mortality risk may provide each of them with a form of family insurance (Eberle and Kappeler 2006).

Postconflict reconciliation is another important mechanism with which many haplorrhines deal with the disruptive social consequences of intragroup conflict on group cohesion (Aureli and de Waal 2000). Although reconciliation has also been described for other mammals (e.g., Cools et al. 2008), there is mixed evidence for lemurs. Reconciliation has been demonstrated in redfronted lemurs, albeit at relatively low levels, but it could not be demonstrated in ring-tailed lemurs, despite a clear dominance hierarchy and within-group kin structure (Kappeler 1993a). However, studies of other captive populations of ring-tailed lemurs found low levels or seasonal occurrence of reconciliatory behavior (Rolland and Roeder 2000; Palagi et al. 2005). Absence of reconciliatory behavior was reported for black lemurs (Roeder et al. 2002), whereas sifakas (*Propithecus verreauxi*) reconciled during the mating season (Palagi et al. 2008). Thus, in contrast to many haplorrhines, strategic use of affiliative interactions to foster social relationships is not pronounced among group-living lemurs.

Given their small group size, it is not surprising that lemur grooming networks are relatively small (Kudo and Dunbar 2001). Grooming cliques have been considered to be synonymous with coalition size, on the grounds that primates use grooming to reinforce the bonds on which coalitionary support is based (Seyfarth and Cheney 1984). This potential function of grooming has been studied in redfronted lemurs (Port et al. 2009). Here, the exchange of grooming bouts is highly reciprocal, but grooming is biased in favor of higher-ranking partners. In addition, aggression occurred at higher frequencies between classes of individuals that were characterized by nonreciprocal grooming, suggesting that grooming may serve as a means to reduce aggression in dyads with a high potential for conflicts. Thus, grooming might be exchanged for tolerance, suggesting that lemur grooming networks might form part of a biological market of the kind described for various Old World monkeys (Barrett et al. 1999; Henzi and Barrett 1999).

19.4.2 Tactical Deception and Related Skills

Neocortex size also predicts deception rate in primates (Byrne and Corp 2004). Deception of conspecifics is often thought to be evidence of considerable cognitive sophistication (Mitchell and Thompson 1986), and reflects very efficient learning ability and sensitivity to a wide range of social discriminations (Cheney and Seyfarth 1990; Byrne and Corp 2004; see also Trivers, this volume). Deception in lemurs seems to be rare. Deaner et al. (2006) tested ring-tailed lemurs with the classical deception paradigm of Menzel (1973), in which a subordinate was informed of the location of a hidden food item and was subsequently released into an enclosure simultaneously with an uniformed dominant female. Male ring-tailed lemurs did not reliably deceive the dominant female, which is not too surprising given the natural response of a male towards a female in a feeding context in this species. However, recent field observations at artificial feeding platforms within the home ranges of wild red-fronted lemurs suggested that some males and females of two different groups behaved as if they deceived other group members. Several times when a group passed the platforms at distances of about 150 m, some individuals sneaked away silently (redfronted lemurs usually produce grunts while locomoting!), ran quickly towards the platforms, and depleted them before uttering their long distance contact calls to reestablish contact with their group (Lennart Pyritz pers. comm.). Experiments in captive settings also indicated that brown and black lemurs seem to learn to deceive a human competitor (Genty and Roeder 2006; Genty et al. 2008).

The same authors also report on self-control behavior in brown and black lemurs (Genty et al. 2004). Self-control has been operationally defined as the ability to inhibit a natural tendency to reach for the greater of two amounts of foods (Anderson 2001). Self-control behavior in preschool children, i.e., the ability to delay gratification, has been related to later cognitive competencies (Mischel et al. 1989). Brown and black lemurs initially chose the larger array of food, but learned after a correction procedure to choose the smaller array of food, indicating that they show some form of self-control (Genty et al. 2004). Similarly, several New World and Old World monkeys, chimpanzees, and even children over 4 years old initially showed the tendency to select the larger array of a reward (Boysen and Berntson 1995; Silberberg and Fujita 1996; Anderson et al. 2000; Kralik et al. 2002). Only orangutans showed the spontaneous ability to understand the task (Schumaker et al. 2001).

Studies in a wide range of species, including apes, dogs, and goats (Tomasello et al. 1998; Call et al. 2003; Kaminski et al. 2005) showed that individuals follow the gaze of others. Such shared attention is thought to underlie a theory of mind and language acquisition (Tomasello and Moll this volume). Earlier studies reported that ring-tailed lemurs do not follow human gaze (Itakura 1996; Anderson and Mitchell 1999). However, a recent study, in which ring-tailed lemurs were equipped with a novel telemetric gaze-tracking system, showed that they preferentially gaze towards others, and follow other lemurs' gaze while freely moving and interacting in naturalistic social and ecological environments (Shepherd and Platt 2008). Moreover, Ruiz et al. (2009) demonstrated that brown and black lemurs use coorientation to find hidden food in an object-choice experiment. Lemurs were more likely to choose correctly after having looked in the same direction as the model, in this case a photograph of a conspecific, indicating that the adaptive value of gaze following might be a way of reading the attentional focus of others. Interestingly, other primates have been shown to coorient with humans (Tomasello et al. 1998; Bräuer et al. 2005), but failed to reliably select the correct location of the hidden food by using human cues (Call et al. 2000; Hare and Tomasello 2004), which might be due to the fact that object-choice tasks and coorientation have been tested separately. The integration of both tasks revealed that there is a connection between visual coorientation and foraging choice. These results, however, do not indicate that lemurs understand gaze as mental perspective taking of others. Objects or locations may simply become more salient for an observer, as a result of following another individual's attention to that object or location – a process that has been defined as "gaze-priming" (Ruiz et al. 2009). Thus, gaze-following or better gaze-priming ability is also present in strepsirrhines.

19.4.3 Social Learning and Innovations

The social intelligence hypothesis also invokes behavioral flexibility as a key advantage of enhanced brain size. Innovation and social learning allow animals to exploit the environment in new ways, and brain size seems to correlate with frequencies of innovation and social learning (Reader and Laland 2002). Feldman and Klopfer (1972) suggested that social learning, i.e., stimulus enhancement, may also play a role in object-choice performance in brown lemurs. Observations of predatory behavior on insects, small mammals, and birds in captive brown, black and ring-tailed lemurs also raised the question of whether such behavior may lead to the development of local traditions (Glander et al. 1985; Jolly and Oliver 1985). The first experimental study of the acquisition process of a novel behavior was conducted with ring-tailed lemurs (Kappeler 1987). Adult females, but not males, and juveniles acquired the novel behavior, and remembered it over several months. Similar experiments with brown, black, and ring-tailed lemurs also showed that novel behaviors, in this case opening a baited food-box, are learned socially (Fornasieri et al. 1990; Anderson et al. 1992). Social influences on feeding decisions involving familiar and novel food have been shown in black lemurs. In these experiments, the consumption of high-quality novel food was acquired individually, but the dominant female influenced the consumption rate of low-quality novel food (Gosset and Roeder 2001). Social influences have also been shown in the complex foraging behavior of aye-ayes. A comparison with ruffed lemurs revealed that aye-aye mothers co-fed and shared food with their infants and engaged in socially mediated learning more often than ruffed lemurs. As a consequence, ruffed lemurs showed less neophobia towards novel food and relied earlier on their own foraging decisions (Krakauer 2005).

Spontaneous innovation of a novel behavior has been reported for strepsirrhines only once so far. Semi-free-ranging ring-tailed lemurs developed a new behavior, that is, immersing the tail in water and then sucking on the wet tail. Almost all group members acquired "drinking-from-tail" behavior, and individuals who did not acquire it were allowed to drink from the wet tail of animals which did (Hosey et al. 1997).

In a similar vein, variation in antipredator behavior between populations has been documented in sifakas (Fichtel and van Schaik 2006, Fichtel and Kappeler unpubl. data). In primates, the usage and comprehension of alarm calls, i.e., their association with predator-specific escape strategies, appear to be socially learned (Seyfarth and Cheney 1980; Fichtel 2008). Thus, alarm calls provide flexible behavioral mechanisms that allow animals to develop appropriate responses to local predators (Curio et al. 1978; Cook and Mineka 1989; Laland 2004). A comparison of three types of alarm calls and antipredator strategies in a semi-free-ranging and a wild population of Coquerels' sifakas (*Propithecus coquereli*) revealed that the captive and wild sifakas used their alarm calls in the same contexts, but exhibited similar behavioral responses in response to only two of the three calls. All members of the captive population, including a wild-caught individual, apparently associated the third alarm call with the presence of a raptor, whereas individuals of the wild population associated no specific threat with this particular call.

Similarly, a comparison of two wild populations of Verreaux's sifakas in habitats with a low and a high density of carnivores also revealed a different comprehension of the alarm calls given to these predators. Sifakas in the habitat with a high density of carnivores associated a predator-specific escape response with these alarm calls, whereas sifakas in the other habitat did not. This differential comprehension of alarm calls is likely to reflect the operation of social learning processes that caused changes in signal content due to changes in the set of predators to which these two populations have been exposed (Fichtel and van Schaik 2006). Thus, social learning appears to be present in lemurs, whereas innovations and tool-use seem to be extremely rare, indicating that the innovative and tool-using anthropoids show greater flexibility in developing new behavior to exploit the environment.

19.4.4 Communication

The evolution of language is clearly one hallmark of humans. Vocal communication of nonhuman primates is very different from human language (Cheney and Seyfarth, this volume). Nonhuman primates have a relatively small repertoire of vocalizations, whose production is predominantly innate (Winter et al. 1973; Hammerschmidt et al. 2001). Although their vocal repertoire is limited, it can provide listeners with an open-ended, highly modifiable, and cognitively rich set of meanings (Cheney and Seyfarth, this volume). In some cases, such as alarm calls, the context eliciting a vocalization is narrowed down to the eliciting stimulus, in this case the type of predator or danger. However, vocalizations given in other contexts, such as during social interactions, depend on both the immediate social context and the history of interactions between particular individuals.

Several haplorrhine primates have been reported to produce acoustically distinct alarm calls for different types of predators, the so-called functionally referential alarm calls (Seyfarth and Cheney 2003). Group-living lemurs have developed different kinds of alarm call systems, from functionally referential alarm calls in ring-tailed lemurs (Macedonia 1990; Pereira and Macedonia 1991), to arousal-based alarm calls in ruffed lemurs (Macedonia 1990), and a mixed alarm call system in sifakas and redfronted lemurs (Fichtel and Kappeler 2002). The mixed alarm call system consists of functionally referential alarm calls for raptors and general alarm calls that are given in response to predators and other threats. Interestingly, the same sort of alarm call system has also been suggested to exist in some New World monkeys, i.e., saddleback tamarins (*Saguinus fuscicollis*), as well as white-faced and tufted capuchin monkeys (Fichtel et al. 2005; Kirchhof and Hammerschmidt 2006; Wheeler 2008). Nocturnal strepsirrhines do not seem to rely on early warning of predators, but produce general alarm calls that are primarily directed to predators or conspecifics (reviewed in Fichtel 2007). These calls may be the ancestral form of primate alarm calling.

As suggested by Cheney and Seyfarth (this volume), the vocalizations that baboons give during social interactions depend on both the immediate social context and the history of interactions between particular individuals. Although several lemurs are group-living, the usage and potential function of vocalizations during social interactions have not been studied yet. In the context of group coordination, some haplorrhine species produce a particular travel call to initiate group movements (reviewed in Boinski and Garber 2000). Sifakas converge in several fundamental proximate aspects of group coordination, but they do not use a particular call or other signals to initiate group movements (Trillmich et al. 2004).

Finally, several similarities seem to exist across primates, including humans, in acoustic features of the expression of the caller's arousal or emotional state. Specifically, primates use common principles, such as an energy shift towards higher frequencies, to encode basic emotions in vocalizations (Fichtel et al. 2001; Fichtel and Hammerschmidt 2002, 2003; Scheiner et al. 2002, 2006; Hammerschmidt and Jürgens 2007). Most of the basic emotions of humans appear to have deep phylogenetic roots, which extend back to the common ancestors of haplorrhine and steprshirrine primates (Fessler and Gervais, this volume).

However, in the domain of visual communication, i.e., facial expressions and gestures, strepsirrhines clearly differ from haplorrhine primates. Only a few facial expressions have been reported in aggressive contexts in ring-tailed lemurs (Pereira and Kappeler 1997) and during play (play-face) in sifakas (pers. observ.). The use of manual gestures seems to be almost absent, though several nocturnal lemurs (red-tailed sportive lemurs (*Lepilemus ruficaudatus*), mouse lemurs) use a shaking fist to threaten conspecifics or predators (pers. observ.). Strepsirrhine primates also exhibit some body gestures constituting conspicuous displays. Just to list a few of them, lorises, for example, raise their arms around their head while moving their body in cobra-like fashion when threatened (Charles-Dominique 1977), male ring-tailed lemurs exhibit conspicuous displays during their famous "stink-fights": while standing bipedally, they move their tail through the legs, "parfume" it with their antebrachial glands, and wave it in a stereotyped fashion in front of opponents (Jolly 1966a). Redfronted lemurs exhibit a friendly reciprocal arm-over display in which they put the proximal arm over the partner's back (Pereira and Kappeler 1997), and sifakas move their heads abruptly into the neck and back when aroused (Jolly 1966a). Nevertheless, in haplorrhine primates, gestural signals seem to be more flexible and subject to cultural transmission (Pollick and de Waal 2007; Fogassi and Ferrari 2007). For example, gestures in apes vary not only between social groups but also culturally between populations (Whiten et al. 1999; van Schaik et al. 2003; Pollick and de Waal 2007), leading to the hypothesis that the flexible use of gestures in combination with enhanced cognitive capacities played a crucial role in the evolution of human language (Arbib et al. 2008, Cheney and Seyfarth, this volume, Tomasello and Moll, this volume).

19.5 Discussion and Conclusions

This review indicates that primates are more heterogeneous with respect to aspects of social complexity and social intelligence than in the realm of physical intelligence. Even though only a few lemur species have been tested on various tasks, and the social structure of only a few species has been studied in detail in the wild, preliminary conclusions about the presence or absence of certain abilities and traits are beginning to emerge. More detailed comparisons of the level of cognitive performance have to await more tests with the same experimental paradigms in a larger number of lemur species, and more detailed comparisons of aspects of social structure related to social intelligence require additional studies of lemurs, New World monkeys, and colobines. On the basis of the available information, however, it is possible to begin characterizing the cognitive abilities of lemurs and to outline the cornerstones of their social complexity, but it remains difficult to separate traits that may have been present in the earliest primates and or haplorrhines from specific adaptations of lemurs that have evolved over the past 50 million years.

In most domains of technical cognition, in which experimental tests have been conducted, strepsirrhines seem to have the same sort of basic cognitive abilities as other primates, and the performance of lemurs was, in most cases, quantitatively not different from that of other primates (summarized in Table 19.1). In the domain of "space and objects," lemurs have a route-based mental representation of spatial relationships, show straight-line traveling, and efficient goal-directed movements between distant sites. They also search for hidden food and are able to solve detour problems. Lemurs manipulate objects less than most haplorrhines, but that might be due to the dominance of their olfactory sense and the less dexterous use of their hands. In discrimination learning tasks, they appear to be a bit slower and more error-prone, but in learning-set tasks, they are as skilled as other primates. They also seem to have some cross-modal skills, and sorting tasks are mastered just as well by lemurs as by New World and Old World monkeys. Though their numerical discrimination skills seem to be inferior, they understand the outcome of simple arithmetic operations. Tool-use and the associated abilities are a striking exception from the lack of fundamental differences from haplorrhines in this cognitive domain. Thus, if we simply consider whether strepsirrhines are able to perform a certain task, their cognitive abilities in physical domains are, by and large, comparable to those of New and Old World monkeys.

Table 19.1 Summary of studies of technical and social intelligence of lemurs and other strepsirrhines. The main categories in the left column are described in the text. The central columns summarize the names of species, in which the corresponding abilities or traits have been demonstrated

Categories	Examples	References
1. Technical intelligence		
1.1 Space and objects		
• Spatial memory	Mouse lemurs, redfronted lemurs, Milne Edwards Sifakas	Joly and Zimmermann (2007), Lührs et al. (2009), Erhart and Overdorff (2008)
• Hidden objects	Bushbabies, Bamboo lemurs, Brown lemurs, Mongoose lemurs, Redfronted lemurs, Ringtailed lemurs, Pottos	Jolly (1964a,b), Deppe et al. (2009)
• Invisible displacement	Bamboo lemurs, Mongoose lemurs, Redfronted lemurs, Ringtailed lemurs	Deppe et al. (2009)
• Detour problems, Bent wire, Maze experiments	Ringtailed lemurs, Mouse lemurs	Davis and Leary (1968), Picq (1993, 2007)
1.2 Tools and causality		
• Object manipulation	Bushbabies, Pottos, Mouse lemurs, Brown lemurs	Jolly (1964a,b), Parker (1973, 1974), Ehrlich et al. (1976), Torigoe (1985), Renner et al. (1992)
• Simple box	Ringtailed lemurs, Mouse lemurs, Black lemurs, Brown lemurs, Ringtailed lemurs	Kappeler (1987), Fornasieri et al. (1990), Anderson et al. (1992), Schilling (2007)
• Complex box	Mouse lemurs, Aye-Ayes	Schilling (2007), Digby et al. (2008)
Tool use		
• Wild, captivity	–	–
• Understanding of tools	Ringtailed lemurs	Santos et al. (2005a)
1.3 Features and categories		
• Learning sets	Bushbabies, Lorises, Black lemurs, Ringtailed lemurs	Stevens (1965), Cooper (1974), Ohta (1983), Ohta et al. (1984, 1987)
• Reversal learning	Brown lemurs, Ringtailed lemurs, Fork-marked lemurs, Mouse lemurs	Stevens (1965), Rumbough (1997), Picq (1993, 2007)
• Cross-modal transfer	Bushabies	Ward et al. (1976)
• Delayed response	Bushabies, Ringtailed lemurs, Ruffed lemurs	Harlow et al. (1932), Maslow and Harlow (1932), Jolly (1964a)
• Oddity	Ringtailed lemurs	Davis et al. (1967)
• Serial ordering	Ringtailed lemurs	Merritt et al. (2007)
• Quantities		
• Estimating numerosity	Mongoose lemurs	Lewis et al. (2005)
• Simple arithmetic operations of 1 + 1	Ringtailed lemurs, Brown lemurs, Mongoose lemurs, Ruffed lemurs	Santos et al. (2005b)

(*continued*)

Table 19.1 (continued)

Categories	Examples	References
2. Social intelligence		
2.1 Social complexity and structure of social relationships		
• Coalitions	Redfronted lemurs, Ringtailed lemurs	Ostner and Kappeler (2004), Sussman (1992), Jones (1983)
• Cooperation	Mouse lemurs	Eberle and Kappeler (2006, 2008)
• Post-conflict behavior	Redfronted lemurs, Ringtailed lemurs, Black lemurs, Sifakas	Kappeler (1993b), Rolland and Roeder (2000), Palagi et al. (2005), Roeder et al. (2002), Palagi et al. (2008)
• Grooming networks	Redfronted lemurs	Port et al. (2009)
• Dominance relationships	Ringtailed lemurs, Sifakas	Kappeler (1993b), Pochron et al. (2003)
2.2 Tactical deception and other related skills		
• Tactical deception	Redfronted lemurs, Ringtailed lemurs, Brown lemurs	L. Pyrritz pers. com., Deaner et al. (2006), Genty and Roeder (2006)
• Learning to deceive	Black lemurs	Genty et al. (2008)
• Self-control	Brown lemurs, Black lemurs	Genty et al. (2004)
• Gaze following	Ringtailed lemurs	Anderson and Mitchell (1999), Shepherd and Platt (2008), Ruiz et al. (2009)
2.3 Social learning and innovations		
• Social learning	Ringtailed lemurs, Brown lemurs, Black lemurs, Ruffed lemurs, Aye-aye	Feldman and Klopfer (1972), Glander et al. (1985), Jolly and Oliver (1985), Kappeler (1987), Fornasieri et al. (1990), Anderson et al. (1992), Gosset and Roeder (2001), Krakauer (2005)
• Innovations	Ringtailed lemurs	Hosey et al. (1997)
• Behavioral variation	Sifakas	Fichtel and van Schaik (2006)
2.4 Vocal communication		
• Functionally referential alarm calls	Ringtailed lemurs, Redfronted lemurs, Sifakas	Macedonia (1990), Pereira and Macedonia (1991), Fichtel and Kappeler (2002)
• Group coordination	Sifakas	Trillmich et al. (2004)
• Expressions of emotions	Redfronted lemurs	Fichtel and Hammerschmidt (2002)

On the basis of a meta-analysis of global cognition variables (detour, patterned string, invisible displacement, tool-use, reversal learning, oddity sorting, and delayed response), in which they ranked the performance of species, Deaner et al. (2006) concluded that strepsirrhines were inferior to most haplorrhines, but better than marmosets and talapoin monkeys (*Miopithecus talapopin*). This data set, however, included the performance of strepsirrhines in only four of the nine tasks that were compared across species. Tomasello and Call (1997) identified 15 paradigms in the domain of physical cognition to which various primates were subjected. Our review revealed that strepsirrhines were able to perform successfully in 12 of these paradigms, but the level of performance in some tasks was not up to par with haplorrhines. Nothing is yet known about strepsirrhines' ability to understand natural categories and the conservation of quantities. Thus, the grade shifts in brain size are not reflected by fundamental gaps in performance in these spatial and physical abilities among primates; but great apes, in particular, are superior on several tasks.

In the realm of social intelligence, lemurs exhibit a number of traits that differ from those described for the better-known haplorrhines, despite basic similarities in several aspects of social organization, such as the multi-male, multi-female composition of groups, the existence of dominance relations, and female philopatry. However, within-group coalitions, even between mothers and daughters, are extremely rare or absent altogether, postconflict reconciliation is also rare, but some basic exchange between grooming and other social commodities may exist. Very limited preliminary evidence suggests that some basal aspects of tactical deception exist and that lemurs can follow the gaze of conspecifics. Social learning abilities are more widespread among lemurs, but true innovations of novel behaviors are apparently rare. As demonstrated by the study of behavioral variation in the meaning of alarm calls among sifaka populations, however, more discoveries of innovations and variation among populations are likely, once more than one population is considered as the representative of its species. Finally, lemurs also vocalize with functionally referential vocalizations, exhibit coordinated group movements, and express their emotional status in structural features of their vocalizations. However, in the domain of visual communication, i.e., the use of gestures and facial expressions, strepsirrhine primates clearly differ from haplorrhines and use less variable signals.

With this information, the outlines of a proto-typical primate social structure and social cognition begin to emerge. Many basic features of social complexity exist, albeit often in rudimentary form, in lemurs, so that the observed variation among major primate radiations is primarily one in quantity, rather than in quality. It is striking that lemur social relationships differ most from the better-known haplorrhine models. Lemurs exhibit more similarities in this respect with New World primates (e.g., small group size, female competition) (Wright 1997). More detailed studies of additional New World monkeys, but also colobines, are required to establish cercopithecine monkeys with maternal rank inheritance as the typical haplorrhine reference for comparison with other primates (see also Strier 1994). It is, therefore, difficult to evaluate the observed differences in social structure between lemurs and haplorrhines. Because the traits where lemurs deviate most obviously are functionally related to intense within-group competition, they may represent lemur autapomorphies, rather than primate symplesiomorphies. These lemur idiosyncrasies are thought to reflect either adaptations to unusually harsh ecological conditions (Wright 1999) or an intermediate stage in a transition from pair-living to group-living (van Schaik and Kappeler 1996).

In the quest to identify human behavioral universals, a broader comparative perspective is useful. By acknowledging the biological continuity of some traits and abilities across the primate lineage, more focused comparisons and reconstructions among the various species of *Homo*, *Pan*, and their common ancestors are possible (Chapais, this volume). Moreover, by mapping social and cognitive variation on the full range of primate brain sizes, major grade shifts during primate evolution will be easier to recognize; for this, a more fine-grained data set that includes more stresirhhine species will be needed. Finally, lemurs should no longer be regarded as our embarrassing relatives, because their cognitive abilities and social complexity are not as utterly primitive as previously thought by some.

Acknowledgments We thank Joan Silk and an anonymous referee for very helpful comments on an earlier version of this manuscript.

References

Amici F, Aureli F, Call J (2008) Fission-fusion dynamics, behavioral flexibility, and inhibitory control in primates. Curr Biol 18:1415–1419

Anderson JR (1983) Optimal foraging and the traveling salesman. Theor Popul Biol 24:145–159

Anderson JR (2001) Self- and other-control in squirrel monkeys. In: Matsuzawa T (ed) Primate origins of human cognition and behavior. Springer, Tokyo, pp 330–347

Anderson JR, Mitchell RW (1999) Macaques but not lemurs co-orient visually with humans. Folia Primatol 70:17–22

Anderson JR, Fornasieri I, Ludes E, Roeder J-J (1992) Social processes and innovative behaviour in changing groups of *Lemur fulvus*. Behav Process 27:101–112

Anderson JR, Awazu S, Fujita K (2000) Can squirrel monkeys (*Saimiri sciureus*) learn self-control? A study using food array selection tests and reverse-reward contingency. J Exp Psychol Anim Behav Process 26:87–97

Andrews J (1998) Infanticide by a female black lemur, *Eulemur macaco*, in disturbed habitat on Nosy Be, North-Western Madagascar. Folia Primatol 69(Suppl 1):14–17

Arbib MA, Liebal K, Pika S (2008) Primate vocalization, gesture, and the evolution of human language. Curr Anthropol 49:1053–1076

Armstrong E (1985) Relative brain size in monkeys and prosimians. Am J Phys Anthropol 66:263–273

Aureli F, de Waal FBM (2000) Natural conflict resolution. University of California Press, Berkely

Barrett L, Henzi SP, Weingrill T, Lycett JE, Hill RA (1999) Market forces predict grooming reciprocity in female baboons. Proc Roy Soc Lond B 266:665–670

Barthold J, Fichtel C, Kappeler PM (2009) What is it going to be? Pattern and potential function of natal coat change in sexually dichromatic redfronted lemurs (*Eulemur fulvus rufus*). Am J Phys Anthropol 138:1–10

Benadi G, Fichtel C, Kappeler PM (2008) Intergroup relations and home range use in Verreaux's sifaka (*Propithecus verreauxi*). Am J Primatol 70:956–965

Beran MJ, Beran MM (2004) Chimpanzees remember the results of one-by-one addition of food items to sets over extended time periods. Psychol Sci 15:94–99

Birkinshaw CR (1999) Use of millipedes by black lemurs to anoint their bodies. Folia Primatol 70:170–171

Boesch C (2007) What makes us human (*Homo sapiens*)? The challenge of cognitive cross-species comparison. J Comp Psychol 121:227–240

Boesch C, Boesch H (1984) Mental map in wild chimpanzees: an analysis of hammer transports for nut cracking. Primates 25:160–170

Boinski S, Garber PA (2000) On the move: how and why animals travel in groups. University of Chicago Press, Chicago

Bond AB, Kamil AC, Balda RP (2003) Social complexity and transitive inference in corvids. Anim Behav 65:479–487

Boysen ST, Berntson GG (1995) Responses to quantity: perceptual versus cognitive mechanisms in chimpanzees (*Pan troglodytes*). J Exp Psychol Anim Behav Process 21:82–86

Bradley BJ (2008) Reconstructing phylogenies and phenotypes: a molecular view of human evolution. J Anat 212:337–353

Brannon E, Terrace HS (2000) Representation of the numerosities 1–9 by rhesus macaques (*Macaca mulatta*). J Exp Psychol Anim Behav Process 26:31–49

Bräuer J, Call J, Tomasello M (2005) All great ape species follow gaze to distant locations and around barriers. J Comp Psychol 119:145–154

Brosnan SF, de Waal FBM (2003) Monkeys reject unequal pay. Nature 425:297–299

Byrne RW, Corp N (2004) Neocortex size predicts deception rate in primates. Proc Roy Soc Lond B 271:1693–1699

Call J (2001) Object permanence in orangutans (*Pongo pygmaeus*), chimpanzees (*Pan troglodytes*), and children (*Homo sapiens*). J Comp Psychol 115:159–171

Call J, Agnetta B, Tomasello M (2000) Cues that chimpanzees do and do not use to find hidden objects. Anim Cogn 3:23–34

Call J, Bräuer J, Kaminski J, Tomasello M (2003) Domestic dogs (*Canis familiaris*) are sensitive to the attentional state of humans. J Comp Psychol 117:257–263

Charles-Dominique P (1977) Ecology and behaviour of nocturnal primates. Columbia University Press, New York

Cheney DL, Seyfarth RM (1990) How monkeys see the world. University of Chicago Press, Chicago

Cook M, Mineka S (1989) Observational conditioning of fear to fear-relevant versus fear-irrelevant stimuli in rhesus monkeys. J Abnorm Psychol 98:448–459

Cools AKA, van Hout AJM, Nelissen MHJ (2008) Canine reconciliation and third-party-initiated postconflict affiliation: do peacemaking social mechanisms in dogs rival those of higher primates? Ethology 114:53–63

Cooper H (1974) Learning set in *Lemur macaco*. In: Martin RD, Doyle GA, Walker AC (eds) Prosimian biology. University of Pittsburgh Press, Pittsburgh, PA, pp 293–300

Cummins-Sebree SE, Fragaszy D (2001) The right stuff: capuchin monkeys perceive affordances of tools. In: Burton G, Schmidt R (eds) Studies in perception and action, vol VI. Lawrence Erlbaum Associates, Hillsdale, NY, pp 89–92

Curio E, Ernst U, Vieth W (1978) Cultural transmission of enemy recognition: one function of mobbing. Science 202:899–901

D'Amato MR, Colombo M (1989) Serial learning with wild card items by monkeys (*Cebus apella*): implications for knowledge of ordinal position. J Comp Psychol 103:252–261

Dammhahn M, Kappeler PM (2008) Small-scale coexistence of two mouse lemur species (*Microcebus berthae* and *M. murinus*) within a homogeneous competitive environment. Oecologia 157:473–483

Davis RT, Leary RW (1968) Learning of detour problems by lemurs and seven species of monkeys. Percept Mot Skills 27:1031–1034

Davis RT, Leary RW, Stevens DA, Thompson RF (1967) Learning and perception of oddity problems by lemurs and seven species of monkeys. Primates 8:311–322

de Blois ST, Novak MA, Bond M (1998) Object permanence in orangutans (*Pongo pygmaeus*) and squirrel monkeys (*Saimiri sciureus*). J Comp Psychol 112:137–152

Deaner RO, van Schaik CP, Johnson V (2006) Do some taxa have better domain-general cognition than others? A meta-analysis of nonhuman primate studies. Evol Psychol 4:149–196

Deaner RO, Isler K, Burkart J, van Schaik CP (2007) Overall brain size, and not encephalization quotient, best predicts cognitive ability across non-human primates. Brain Behav Evol 70:115–124

Deppe AM, Wright PC, Szelistowski WA (2009) Object permanence in lemurs. Anim Cogn 12:381–388

Dewar RE, Richard AF (2007) Evolution in the hypervariable environment of Madagascar. Proc Natl Acad Sci USA 104:13723–13727

Digby LJ, Haley M, Schneider AC, Del Valle I (2008) Sensorimotor intelligence in aye-ayes and other lemurs: a puzzle box approach. Am J Primatol 70(Suppl 1):54

Dunbar RIM (1995) Neocortex size and group size in primates: a test of the hypothesis. J Hum Evol 28:287–296

Dunbar RIM (1998) The social brain hypothesis. Evol Anthropol 6:178–190

Dunbar RIM, Shultz S (2007) Understanding primate brain evolution. Phil Trans Roy Soc B 362:649–658

Eberle M, Kappeler PM (2006) Family insurance: kin selection and cooperative breeding in a solitary primate (*Microcebus murinus*). Behav Ecol Sociobiol 60:582–588

Eberle M, Kappeler PM (2008) Mutualism, reciprocity, or kin selection? Cooperative rescue of a conspecific from a boa in a nocturnal solitary forager the gray mouse lemur. Am J Primatol 70:410–414

Ehrlich A (1970) Response to novel objects in three lower primates: greater galago, slow loris, and owl monkey. Behaviour 37:55–63

Ehrlich A, Fobes JL, King JE (1976) Prosimian learning capacities. J Hum Evol 5:599–617

Erhart EM, Overdorff DJ (2008) Spatial memory during foraging in prosimian primates: *Propithecus edwardsi* and *Eulemur fulvus rufus*. Folia Primatol 79:185–196

Erickson CJ (1995) Feeding sites for extractive foraging by the aye-aye, *Daubentonia madagascariensis*. Am J Primatol 35:235–240

Erkert HG, Kappeler PM (2004) Arrived in the light: diel and seasonal activity patterns in wild Verreaux's sifakas (*Propithecus v. verreauxi*; Primates: Indriidae). Behav Ecol Sociobiol 57:174–186

Evans TA, Beran MJ, Harris EH, Rice DF (2009) Quantity judgments of sequentially presented food items by capuchin monkeys (*Cebus apella*). Anim Cogn 12:97–105

Feigenson L, Carey S, Spelke E (2002) Infants' discrimination of number vs. continuous extent. Cogn Psychol 44:33–66

Feldman DW, Klopfer PH (1972) A study of observational learning in lemurs. Z Tierpsychol 30:297–304

Fichtel C (2007) Avoiding predators at night: antipredator strategies in red-tailed sportive lemurs (*Lepilemur ruficaudatus*). Am J Primatol 69:611–624

Fichtel C (2008) Ontogeny of conspecific and heterospecific alarm call recognition in wild Verreaux's sifakas (*Propithecus v. verreauxi*). Am J Primatol 70:127–135

Fichtel C, Hammerschmidt K (2002) Responses of redfronted lemurs (*Eulemur fulvus rufus*) to experimentally modified alarm calls: evidence for urgency-based changes in call structure. Ethology 108:763–777

Fichtel C, Hammerschmidt K (2003) Responses of squirrel monkeys to their experimentally modified mobbing calls. J Acoust Soc Am 113:2927–2932

Fichtel C, Kappeler PM (2002) Anti-predator behavior of group-living Malagasy primates: mixed evidence for a referential alarm call system. Behav Ecol Sociobiol 51:262–275

Fichtel C, van Schaik CP (2006) Semantic differences in sifaka (*Propithecus verreauxi*) alarm calls: a reflection of genetic or cultural variants? Ethology 112:839–849

Fichtel C, Hammerschmidt K, Jürgens J (2001) On the vocal expression of emotion. A multi-parametric analysis of different states of aversion in the squirrel monkey. Behaviour 138:97–116

Fichtel C, Perry S, Gros-Louis J (2005) Alarm calls of white-faced capuchin monkeys: an acoustic analysis. Anim Behav 70:165–176

Fobes JL, King JE (1982) Measuring primate learning abilities. In: Fobes JL, King JE (eds) Primate behavior. Academic, New York, pp 289–326
Fogassi L, Ferrari PF (2007) Mirror neurons and the evolution of embodied language. Curr Dir Psychol Sci 16:136–141
Fornasieri I, Anderson JR, Roeder J-J (1990) Responses to a novel food acquisition task in three species of lemurs. Behav Process 21:143–156
Fujita K, Kuroshima H, Asai S (2003) How do tufted capuchin monkeys (*Cebus apella*) understand causality involved in tool use? J Exp Psychol Anim Behav Process 29:233–242
Gallistel CR (1989) Animal cognition: the representation of space, time and number. Annu Rev Psychol 40:155–189
Garber PA (1988) Foraging decisions during nectar feeding by tamarin monkeys (*Saguinus mystax* and *Saguinus fuscicollis,* Callitrichidae, Primates) in Amazonian Peru. Biotropica 20:100–106
Genty E, Roeder J-J (2006) Can lemurs learn to deceive? A study in the black lemurs (*Eulemur macaco*). J Exp Psychol Anim Behav Process 32:196–200
Genty E, Palmier C, Roeder J-J (2004) Learning to suppress responses to the larger of two rewards in two species of lemurs, *Eulemur fulvus* and *E. macaco*. Anim Behav 67:925–932
Genty E, Foltz J, Roeder J-J (2008) Can brown lemurs (*Eulemur fulvus*) learn to deceive a human competitor? Anim Cogn 11:255–266
Glander KE, Freed BZ, Ganzhorn JU (1985) Meat eating and predation in captive-born semi-free-ranging *Lemur fulvus* and caged *Lemur macaco*. Zoo Biol 4:361–365
Gosset D, Roeder J-J (2001) Factors affecting feeding decisions in a group of black lemurs confronted with novel food. Primates 42:175–182
Hammerschmidt K, Jürgens U (2007) Acoustical correlates of affective prosody. J Voice 21:531–540
Hammerschmidt K, Freudenstein T, Jürgens U (2001) Vocal development in squirrel monkeys. Behaviour 138:1179–1204
Hanus D, Call J (2007) Discrete quantity judgments in the great apes (*Pan paniscus, Pan troglodytes, Gorilla gorilla, Pongo pygmaeus*): the effect of presenting whole sets versus item-by-item. J Comp Psychol 121:241–249
Hare B, Tomasello M (2004) Chimpanzees are more skilful in competitive than in cooperative cognitive tasks. Anim Behav 68:571–581
Harlow HF (1949) The formation of learning sets. Psychol Rev 56:51–65
Harlow HF, Uehling H, Maslow AH (1932) Comparative behavior of primates. I. Delayed reaction tests on primates from the lemur to the orangutan. J Comp Psychol 13:313–343
Hauser MD (2000) Wild minds: what animals really think. Holt, New York
Hauser MD (2001) Searching for food in the wild: a nonhuman primate's expectations about invisible displacement. Dev Sci 4:84–93
Hauser MD, MacNeilage P, Ware M (1996) Numerical representations in primates. Proc Natl Acad Sci USA 93:1514–1517
Hauser MD, Kralik J, Botto-Mahan C (1999) Problem solving and functional design features: experiments on cotton-top tamarins, *Saguinus oedipus*. Anim Behav 57:565–582
Hauser MD, Santos LR, Spaepen GM, Pearson HE (2002) Problem solving, inhibition and domain-specific experience: experiments on cottontop tamarins (*Saguinus oedipus*). Anim Behav 64:387–396
Hauser MD, Tsao F, Garcia P, Spelke ES (2003) Evolutionary foundations of number: spontaneous representation of numerical magnitudes by cotton-top tamarins. Proc Roy Soc Lond B 270:1441–1446
Henzi SP, Barrett L (1999) The value of grooming to female primates. Primates 40:47–59
Horvath JE, Weisrock DW, Embry SL, Fiorentino I, Balhoff JP, Kappeler PM, Wray GA, Willard HF, Yoder AD (2008) Development and application of a phylogenomic toolkit: resolving the evolutionary history of Madagascar's lemurs. Genome Res 18:489–499
Hosey GR, Jacques M, Pitts A (1997) Drinking from tails: social learning of a novel behaviour in a group of ring-tailed lemurs (*Lemur catta*). Primates 38:415–422

Itakura S (1996) An exploratory study of gaze monitoring in non-human primates. Jpn Psychol Res 38:174–180

Janson CH (1998) Experimental evidence for spatial memory in foraging wild capuchin monkeys, *Cebus apella*. Anim Behav 55:1229–1243

Jolly A (1964a) Choice of cue in prosimian learning. Anim Behav 12:571–577

Jolly A (1964b) Prosimians' manipulation of simple object problems. Anim Behav 12:560–570

Jolly A (1966a) Lemur behavior. University of Chicago Press, Chicago

Jolly A (1966b) Lemur social behavior and primate intelligence. Science 153:501–506

Jolly A, Oliver WLR (1985) Predatory behavior in captive *Lemur* spp. Zoo Biol 4:139–145

Jolly A, Caless S, Cavigelli S, Gould L, Pereira ME, Pitts A, Pride RE, Rabenandrasana HD, Walker JD, Zafison T (2000) Infant killing, wounding and predation in *Eulemur* and *Lemur*. Int J Primatol 21:21–40

Joly M, Zimmermann E (2007) First evidence for relocation of stationary food resources during foraging in a strepsirrhine primate (*Microcebus murinus*). Am J Primatol 69:1045–1052

Jones KC (1983) Inter-troop transfer of *Lemur catta* males at Berenty, Madagascar. Folia Primatol 40:145–160

Kaminski J, Riedel J, Call J, Tomasello M (2005) Domestic goats, *Capra hircus*, follow gaze direction and use social cues in an object choice task. Anim Behav 69:11–18

Kappeler PM (1987) The acquisition process of a novel behavior pattern in a group of ring-tailed lemurs (*Lemur catta*). Primates 28:225–228

Kappeler PM (1993a) Reconciliation and post-conflict behaviour in ringtailed lemurs, *Lemur catta* and redfronted lemurs, *Eulemur fulvus rufus*. Anim Behav 45:901–915

Kappeler PM (1993b) Variation in social structure: the effects of sex and kinship on social interactions in three lemur species. Ethology 93:125–145

Kappeler PM (1998) To whom it may concern: the transmission and function of chemical signals in *Lemur catta*. Behav Ecol Sociobiol 42:411–4421

Kappeler PM (1999) Convergence and divergence in primate social systems. In: Fleagle JG, Janson CH, Reed K (eds) Primate communities. Cambridge University Press, Cambridge, pp 158–170

Kappeler PM (2000) Causes and consequences of unusual sex ratios among lemurs. In: Kappeler PM (ed) Primate males: causes and consequences of variation in group composition. Cambridge University Press, Cambridge, pp 55–63

Kappeler PM, Erkert HG (2003) On the move around the clock: correlates and determinants of cathemeral activity in wild redfronted lemurs (*Eulemur fulvus rufus*). Behav Ecol Sociobiol 54:359–369

Kappeler PM, Heymann EW (1996) Nonconvergence in the evolution of primate life history and socio-ecology. Biol J Linn Soc 59:297–326

Kappeler PM, van Schaik CP (2002) Evolution of primate social systems. Int J Primatol 23:707–740

Kappeler PM, Wimmer B, Zinner DP, Tautz D (2002) The hidden matrilineal structure of a solitary lemur: implications for primate social evolution. Proc Roy Soc Lond B 269:1755–1763

Karanth KP, Delefosse T, Rakotosamimanana B, Parsons TJ, Yoder AD (2005) Ancient DNA from giant extinct lemurs confirms single origin of Malagasy primates. Proc Natl Acad Sci USA 102:5090–5095

Kirchhof J, Hammerschmidt K (2006) Functionally referential alarm calls in tamarins (*Saguinus fuscicollis* and *Saguinus mystax*) – evidence from playback experiments. Ethology 112:346–354

Krakauer EB (2005) Development of aye-aye (*Daubentonia madagascariensis*) foraging skills: independent exploration and social learning. PhD-thesis, Duke University, Durham, NC

Kralik JD, Hauser MD, Zimlicki R (2002) The relationship between problem solving and inhibitory control: cotton-top tamarin (*Saguinus oedipus*) performance on a reversed contingency task. J Comp Psychol 116:39–50

Kudo H, Dunbar RIM (2001) Neocortex size and social network size in primates. Anim Behav 62:711–722

Laland KN (2004) Social learning strategies. Learn Behav 32:4–14

Lewis KP (2000) A comparative study of primate play behaviour: implications for the study of cognition. Folia Primatol 71:417–421

Lewis KP, Jaffe S, Brannon EM (2005) Analog number representations in mongoose lemurs (*Eulemur mongoz*): evidence from a search task. Anim Cogn 8:247–252

Lhota S, Jůnek T, Bartoš L, Kuběna AA (2008) Specialized use of two fingers in free-ranging aye-ayes (*Daubentonia madagascariensis*). Am J Primatol 70:786–795

Lindenfors P (2005) Neocortex evolution in primates: the 'social brain' is for females. Biol Lett 1:407–410

Lührs M-L, Dammhahn M, Kappeler PM, Fichtel C (2009) Spatial memory in the grey mouse lemur (*Microcebus murinus*). Anim Cogn . doi:10.1007/s10071-009-0219-y

Macedonia JM (1990) What is communicated in the antipredator calls of lemurs: evidence from playback experiments with ringtailed and ruffed lemurs. Ethology 86:177–190

MacLean EL, Merritt DJ, Brannon EM (2008) Social complexity predicts transitive reasoning in prosimian primates. Anim Behav 76:479–486

Martin RD, Harvey PH (1985) Brain size allometry: ontogeny and phylogeny. In: Jungers WL (ed) Size and scaling in primate biology. Plenum Press, New York, pp 147–173

Martin-Ordas G, Call J, Colmenares F (2008) Tubes, tables and traps: great apes solve two functionally equivalent trap tasks but show no evidence of transfer across tasks. Anim Cogn 11:423–430

Maslow AH, Harlow HF (1932) Comparative behavior of primates. II. Delayed reaction tests on primates at Bronx Park Zoo. J Comp Psychol 14:97–107

Meador DM, Rumbaugh DM, Pate JL, Bard KA (1987) Learning, problem solving, cognition, and intelligence. In: Mitchell G, Erwin J (eds) Comparative primate biology: behavior, cognition, and motivation. Alan R Liss, New York, pp 17–83

Mendes N, Huber L (2004) Object permanence in common marmosets (*Callithrix jacchus*). J Comp Psychol 118:103–112

Mendes N, Hanus D, Call J (2007) Raising the level: orangutans use water as a tool. Biol Lett 3:453–455

Menzel EW Jr (1973) Leadership and communication in young chimpanzees. In: Menzel EW Jr (ed) Precultural primate behavior. Karger, Basel, pp 192–225

Menzel CR (1991) Cognitive aspects of foraging in Japanese monkeys. Anim Behav 41:397–402

Merritt D, Maclean EL, Jaffe S, Brannon EM (2007) A comparative analysis of serial ordering in ring-tailed lemurs (*Lemur catta*). J Comp Psychol 121:363–371

Mikkelsen TS, Hillier LW, Eichler EE, Zody MC, Jaffe DB, Yang S-P, Enard W, Hellmann I, Lindblad-Toh K, Altheide TK, Archidiacono N, Bork P, Butler J, Chang JL, Cheng Z, Chinwalla AT, de Jong P, Delehaunty KD, Fronick CC, Fulton LL, Gilad Y, Glusman G, Gnerre S, Graves TA, Hayakawa T, Hayden KE, Huang X, Ji H, Kent WJ, King M-C, Kulbokas EJ, Lee MK, Liu G, Lopez-Otin C, Makova KD, Man O, Mardis ER, Mauceli E, Miner TL, Nash WE, Nelson JO, Pääbo S, Patterson NJ, Pohl CS, Pollard KS, Prüfer K, Puente XS, Reich D, Rocchi M, Rosenbloom K, Ruvolo M, Richter DJ, Schaffner SF, Smit AFA, Smith SM, Suyama M, Taylor J, Torrents D, Tuzun E, Varki A, Velasco G, Ventura M, Wallis JW, Wendl MC, Wilson RK, Lander ES, Waterston RH (2005) Initial sequence of the chimpanzee genome and comparison with the human genome. Nature 437:69–87

Miliken GW, Ward JP, Erickson CJ (1991) Independent digit control in foraging by the aye-aye (*Daubentonia madagascariensis*). Folia Primatol 56:219–224

Mischel W, Shoda Y, Rodriguez ML (1989) Delay of gratification in children. Science 244:933–938

Mitchell RW, Thompson NS (1986) Deception: perspectives on human and nonhuman deceit. SUNY Press, Albany

Moura AC, Lee PC (2004) Capuchin stone tool use in Caatinga dry forest. Science 306:1909

Neiworth JJ, Steinmark E, Basile BM, Wonders R, Steely F, DeHart C (2003) A test of object permanence in a new-world monkey species, cotton top tamarins (*Saguinus oedipus*). Anim Cogn 6:27–37

Noser R, Byrne RW (2007) Travel routes and planning of visits to out-of-sight resources in wild chacma baboons, *Papio ursinus*. Anim Behav 73:257–266

Nunn CL, Deaner RO (2004) Patterns of participation and free riding in territorial conflicts among ringtailed lemurs (*Lemur catta*). Behav Ecol Sociobiol 57:50–61

Ohta H (1983) Learning set formation in slow lorises (*Nycticebus coucang*). Folia Primatol 40:256–267

Ohta H, Ishida H, Matano S (1984) Learning set formation in ring-tailed lemurs (*Lemur catta*). Folia Primatol 43:53–58

Ohta H, Ishida H, Matano S (1987) Learning set formation in thick-tailed bush babies (*Galago crassicaudatus*) and comparison of learning ability among four species. Folia Primatol 48:1–8

Ostner J, Kappeler PM (2004) Male life history and the unusal adult sex ratios of redfronted lemur, *Eulemur fulvus rufus,* groups. Anim Behav 67:249–259

Overdorff DJ, Merenlender AM, Talata P, Telo A, Forward ZA (1999) Life history of *Eulemur fulvus rufus* from 1988–1998 in Southeastern Madagascar. Am J Phys Anthropol 108:295–310

Palagi E, Paoli T, Borgognini Tarli SM (2005) Aggression and reconciliation in two captive groups of *Lemur catta*. Int J Primatol 26:279–294

Palagi E, Antonacci D, Norscia I (2008) Peacemaking on treetops: first evidence of reconciliation from a wild prosimian (*Propithecus verreauxi*). Anim Behav 76:737–747

Parker CE (1973) Manipulatory behavior and responsiveness. In: Rumbaugh DM (ed) Gibbon and Siamang, vol 2, Anatyomy, dentition, taxonomy, molecular evolution and behavior. Karger, Basel, pp 185–207

Parker CE (1974) Behavioural diversity in ten species of nonhuman primates. J Comp Physiol Psychol 87:930–937

Pereira ME (1993) Agonistic interaction, dominance relation, and ontogenetic trajectories in ringtailed lemurs. In: Pereira ME, Fairbanks LA (eds) Juvenile primates: life history, development, and behavior. Oxford University Press, New York, pp 285–305

Pereira ME, Kappeler PM (1997) Divergent systems of agonistic behaviour in lemurid primates. Behaviour 134:225–274

Pereira ME, Macedonia JM (1991) Ringtailed lemur anti-predator calls denote predator class, not response urgency. Anim Behav 41:543–544

Pereira ME, Kaufman R, Kappeler PM, Overdorff DJ (1990) Female dominance does not characterize all of the Lemuridae. Folia Primatol 55:96–103

Perry S, Manson JH (2003) Traditions in monkeys. Evol Anthropol 12:71–81

Picq JL (1993) Radial maze performance in young and aged grey mouse lemurs (*Microcebus murinus*). Primates 24:223–226

Picq JL (2007) Aging affects executive functions and memory in mouse lemur primates. Exp Gerontol 42:223–232

Pochron ST, Fitzgerald J, Gilbert CC, Lawrence D, Grgas M, Rakotonirina G, Ratsimbazafy R, Rakotosoa R, Wright PC (2003) Patterns of female dominance in *Propithecus diadema edwardsi* of Ranomafana National Park, Madagascar. Am J Primatol 61:173–185

Pochron ST, Tucker WT, Wright PC (2004) Demography, life history, and social structure in Propithecus diadema edwardsi from 1986–2000 in Ranomafana National Park, Madagascar. Am J Phys Anthropol 125:61–72

Pochron ST, Morelli TL, Scirbona J, Wright PC (2005) Sex differences in scent-marking in *Propithecus edwardsi* of Ranomafana National Park, Madagascar. Am J Primatol 66:97–110

Pollick AS, de Waal FBM (2007) Ape gestures and language evolution. Proc Natl Acad Sci USA 104:8184–8189

Port M, Clough D, Kappeler PM (2009) Market effects offset the reciprocation of grooming in free-ranging red-fronted lemurs, *Eulemur fulvus rufus*. Anim Behav 77:29–36

Povinelli DJ (2000) Folk physics for apes: the chimpanzee's theory of how the world works. Oxford University Press, New York

Reader SM, Laland KN (2002) Social intelligence, innovation, and enhanced brain size in primates. Proc Natl Acad Sci USA 99:4436–4441

Renner MJ, Bennett AJ, Ford ML, Pierre PJ (1992) Investigation of inanimate objects by the greater bushbaby (*Otolemur garnettii*). Primates 33:315–327

Richard AF (1987) Malagasy prosimians: female dominance. In: Smuts BB, Cheney DL, Seyfarth RM, Wrangham RW, Struhsaker TT (eds) Primate societies. University of Chicago Press, Chicago, pp 25–33

Richard AF, Dewar RE (1991) Lemur ecology. Annu Rev Ecol Syst 22:145–175

Richard AF, Rakotomanga P, Schwartz M (1993) Dispersal by Propithecus verreauxi at Beza Mahafaly, Madagascar: 1984–1991. Am J Primatol 30:1–20

Roeder J-J, Fornasieri I, Gosset D (2002) Conflict and postconflict behaviour in two lemur species with different social organizations (*Eulemur fulvus* and *Eulemur macaco*)*:* a study on captive groups. Aggress Behav 28:62–74

Rolland N, Roeder J-J (2000) Do ringtailed lemurs (*Lemur catta*) reconcile in the hour post-conflict? A pilot study. Primates 41:223–227

Rugani R, Fontanari L, Simoni E, Regolin L, Vallortigara G (2009) Arithmetic in newborn chicks. Proc Roy Soc Lond B . doi:10.1098/rspb.2009.0044

Ruiz A, Gómez JC, Roeder J-J, Byrne RW (2009) Gaze following and gaze priming in lemurs. Anim Cogn 12:427–434

Rumbough DM (1970) Learning skills of anthropoids. In: Rosenblum LA (ed) Primate behavior: developments in field and laboratory research, vol 1. Academic, New York, pp 1–70

Rumbough DM (1997) Competence, cortex, and primate models: a comparative primate perspective. In: Krasnegor NA, Lyon GR, Goldman-Rakic PS (eds) Development of the prefontal cortex: evolution, neurobiology and behavior. Paul H Brookes, Baltimore, MD, pp 117–139

Sands SF, Wright AA (1980) Primate memory: retention of serial list items by a rhesus monkey. Science 209:938–940

Santos LR, Mahajan N, Barnes JL (2005a) How prosimian primates represent tools: experiments with two lemur species (*Eulemur fulvus* and *Lemur catta*). J Comp Psychol 119:394–403

Santos LR, Barnes JL, Mahajan N (2005b) Expectations about numerical events in four lemur species (*Eulemur fulvus*, *Eulemur mongoz*, *Lemur catta* and *Varecia rubra*). Anim Cogn 8:253–262

Santos LR, Pearson HM, Spaepen GM, Tsao F, Hauser MD (2006) Probing the limits of tool competence: experiments with two non-tool-using species (*Cercopithecus aethiops* and *Saguinus oedipus*). Anim Cogn 9:94–109

Scheiner E, Hammerschmidt K, Jürgens U, Zwirner P (2002) Acoustic analyses of developmental changes and emotional expressions in the preverbal vocalizations of infants. J Voice 16:509–529

Scheiner E, Hammerschmidt K, Jürgens U, Zwirner P (2006) Vocal expression of emotions in normally hearing and hearing-impaired infants. J Voice 20:585–604

Schilling A (2007) Cognitive capacities studied through object manipulation in the prosimian primate *Microcebus murinus*. Proceedings of the international prosimian conference, Ithala, p 67

Schumaker RW, Palkovich AM, Beck BB, Guagnano GA, Morowitz H (2001) Spontaneous use of magnitude discrimination and ordination by the orangutan (*Pongo pygmaeus*). J Comp Psychol 115:385–391

Seed AM, Call J, Emery NJ, Clayton NS (2009) Chimpanzees solve the trap problem when the confound of tool-use is removed. J Exp Psychol Anim Behav Process 35:23–34

Seyfarth RM, Cheney DL (1980) The ontogeny of vervet monkey alarm calling behavior: a preliminary report. Z Tierpsychol 54:37–56

Seyfarth RM, Cheney DL (1984) Grooming, alliances and reciprocal altruism in vervet monkeys. Nature 308:541–543

Seyfarth RM, Cheney DL (2003) Signalers and receivers in animal communication. Annu Rev Psychol 54:145–173

Shepherd SV, Platt ML (2008) Spontaneous social orienting and gaze following in ringtailed lemurs (*Lemur catta*). Anim Cogn 11:13–20

Shultz S, Dunbar RIM (2007) The evolution of the social brain: anthropoid primates contrast with other vertebrates. Proc Roy Soc Lond B 274:2429–2436

Silberberg A, Fujita K (1996) Pointing at smaller food amounts in an analogue of Boysen and Berntson's (1995) procedure. J Exp Anal Behav 66:143–147

Silk JB (2007) Social components of fitness in primate groups. Science 317:1347–1351

Spaulding B, Hauser MD (2005) What experience is required for acquiring tool competence? Experiments with two callitrichids. Anim Behav 70:517–526

Stephan H, Baron G, Frahm HD (1988) Comparative size of brains and brain components. In: Steklis HD, Erwin J (eds) Comparative primate biology. Alan R Liss, New York, pp 1–38

Sterling EJ (1994) Aye-ayes: specialists on structurally defended resources. Folia Primatol 62:142–154

Sterling EJ, Povinelli DJ (1999) Tool use, aye-ayes, and sensorimotor intelligence. Folia Primatol 70:8–16

Stevens DA (1965) A comparison of learning in rhesus monkeys, cebus monkeys, lemurs, and Burmese cats. PhD-thesis, University of Oregon

Straub RO, Terrace HS (1981) Generalization of serial learning in the pigeon. Anim Learn Behav 9:454–468

Strier KB (1994) Myth of the typical primate. Yrbk Phys Anthropol 37:233–271

Sussman RW (1992) Male life history and intergroup mobility among ringtailed lemurs (*Lemur catta*). Int J Primatol 13:395–413

Tattersall I (2007) Madagascar's lemurs: cryptic diversity or taxonomic inflation? Evol Anthropol 16:12–23

Terrace HS, Son LK, Brannon EM (2003) Serial expertise of rhesus macaques. Psychol Sci 14:66–73

Tomasello M, Call J (1997) Primate cognition. Oxford University Press, New York

Tomasello M, Call J, Hare B (1998) Five primate species follow the visual gaze of conspecifics. Anim Behav 55:1063–1069

Torigoe T (1985) Comparison of object manipulation among 74 species of non-human primates. Primates 26:182–194

Trillmich J, Fichtel C, Kappeler PM (2004) Coordination of group movements in wild Verreaux's sifakas (*Propithecus verreauxi*). Behaviour 141:1103–1120

Uller C, Hauser MD, Carey S (2001) Spontaneous representation of number in cotton-top tamarins (*Saguinus oedipus*). J Comp Psychol 115:248–257

van Schaik CP, Kappeler PM (1996) The social systems of gregarious lemurs: lack of convergence with anthropoids due to evolutionary disequilibrium? Ethology 102:915–941

van Schaik CP, Deaner RO, Merrill MY (1999) The conditions for tool use in primates: implications for the evolution of material culture. J Hum Evol 36:719–741

van Schaik CP, Ancrenaz M, Borgen G, Galdikas B, Knott CD, Singleton I, Suzuki A, Utami SS, Merrill M (2003) Orangutan cultures and the evolution of material culture. Science 299:102–105

Vick LG, Pereira ME (1989) Episodic targeting aggression and the histories of *Lemur* social groups. Behav Ecol Sociobiol 25:3–12

Visalberghi E (1993) Capuchin monkeys: a window into tool use in apes and humans. In: Gibson KR, Ingold T (eds) Tools, language and cognition in human evolution. Cambridge University Press, Cambridge, pp 138–150

Ward JP, Silver BV, Frank J (1976) Preservation of cross-modal transfer of a rate discrimination in the bushbaby (*Galago senegalensis*) with lesions of posterior neocortex. J Comp Physiol Psychol 90:520–527

Wheeler BC (2008) Selfish or altruistic? An analysis of alarm call function in wild capuchin monkeys, *Cebus apella nigritus*. Anim Behav 76:1465–1475

Whiten A, Byrne RW (1997) Machiavellian intelligence. II. Extensions and evaluations. Cambridge University Press, Cambridge

Whiten A, van Schaik CP (2007) The evolution of animal 'cultures' and social intelligence. Phil Trans Roy Soc B 362:603–620

Whiten A, Goodall J, McGrew WC, Nishida T, Reynolds V, Sugiyama Y, Tutin CEG, Wrangham RW, Boesch C (1999) Cultures in chimpanzees. Nature 399:682–685

Wimmer B, Tautz D, Kappeler PM (2002) The genetic population structure of the gray mouse lemur (*Microcebus murinus*), a basal primate from Madagascar. Behav Ecol Sociobiol 52:166–175

Winter P, Handley P, Ploog D, Schott D (1973) Ontogeny of squirrel monkey calls under normal conditions and under acoustic isolation. Behaviour 47:230–239

Wright PC (1997) Behavioral and ecological comparisons of neotropical and Malagasy primates. In: Kinzey WG (ed) New world primates: ecology, evolution, and behavior. Aldine de Gruyter, New York, pp 127–141

Wright PC (1999) Lemur traits and Madagascar ecology: coping with an island environment. Ybk Phys Anthropol 42:31–72

Wynn K (1992) Addition and subtraction by human infants. Nature 358:749–750

Yoder AD (2007) Lemurs. Curr Biol 17:R866–R868

Part VII
Innovation & Culture

Chapter 20
Ape Behavior and the Origins of Human Culture

Andrew Whiten

Abstract Identification of behavioral characteristics shared by humans and our closest primate relatives allow us to reconstruct the nature of our shared behavioral ancestry. I use this approach to infer the core features of social learning, traditions, and culture that characterized our ancestry before the evolutionary split that created chimpanzees and hominins. An extensive corpus of field observations and behavioral experiments has, in recent years, provided a substantial empirical basis through which to realize this approach to our cultural past. Features of culture shared by ourselves and chimpanzees, and thus likely to have been shared also by our common ancestor around 6 million years ago, include (1) the capacity to sustain different local cultures composed of multiple and diverse traditions, both technical and social; (2) related contents of such traditions, such as tool use; and (3) a portfolio of different social learning mechanisms, extending to both emulation and imitation, that are flexibly applied to acquire behavioral routines, with net adaptive benefit. These would have constituted a crucial platform from which our own unique and complex cultural nature evolved.

20.1 Introduction: The Culture Gap

Culture is apt to be high on anybody's list of the key cognitive and behavioral universals that separate us as humans from the rest of the animal kingdom. Some might put culture at the top of the list, concurring with the adage that "culture maketh man." Only our species has left an archeological record of cultural progress that stretches back over 2.5 million years to the earliest stone tools, and in the meantime has generated such rich cultural practices, from our languages to our

---

A. Whiten
Centre for Social Learning & Cognitive Evolution and Scottish Primate Research Group, School of Psychology, University of St. Andrews, St. Andrews, Scotland
e-mail: aw2@st-andrews.ac.uk

P.M. Kappeler and J.B. Silk (eds.), *Mind the Gap*,
DOI 10.1007/978-3-642-02725-3_20, © Springer-Verlag Berlin Heidelberg 2010

technologies, that the culture into which a person is born will explain vast swathes of their behavioral repertoire. Accordingly, human beings may behave in very different ways according to the local cultural practices into which they are inducted. Through these diverse cultures, humans have come to dominate almost every part of the planet. The competitive edge this has given us is underlined by the fact that our 6-billion-plus population is all too successfully exterminating our closest sister species, the other apes, whose total world population has dwindled to the equivalent of just one modest human township.

All this means that at first sight, the gap between the riches of human culture and anything remotely resembling them in the rest of the animal kingdom may appear unbridgeable. To the contrary, in this chapter, I will argue that a considerable variety of aspects of the social learning and culture that characterize our own species are at work in other primates, and some are yet more widely distributed in other groups of animals; more particularly, in our closest relatives, the apes, we find some features distinctively closer to human ones. The result is a substantial literature of empirical findings that bear directly on questions about the evolutionary roots of culture. This presents a big contrast with that other distinctive human universal, language, where there appears much less one can identify in other species to directly bridge the evolutionary gap. In the case of culture, we are more lucky: in recent decades, we have discovered an empirical goldmine in the present state of the natural world that permits many inferences about the ancient origins of social learning, traditions, and culture.

20.2 Some Definitions

In the animal behavior literature, there is a long-standing tendency to treat the terms "culture" and "tradition" as equivalents (and of course, they are often used interchangeably in talking of humans too), but among authors who do distinguish the two terms, "tradition" is the less contentious and so is a good place to start. "A distinctive behavior pattern shared by two or more individuals in a social unit, which persists over time and that new practitioners acquire in part through socially aided learning," the definition of a tradition offered by Fragaszy and Perry (2003: xiii) would likely be acceptable to most researchers. "Social learning" is a crucial ingredient and can be defined simply as learning from or through others, as distinct from individual learning, in which the learner is dependent on only their own resources. Numerous kinds of social learning have in turn been distinguished and defined (Whiten et al. 2004) and several will be discussed further below.

Some writers, typically because they recognize that there is so much more to human culture than is covered by the above definition of a tradition, insist on additional criteria before they are willing to talk of "culture" in nonhuman animals. Such criteria vary between writers (nicely illustrated in the range of approaches offered in Laland and Galef 2009), but examples include (1) features taken to be distinctively human forms of social transmission, notably imitation and teaching

(Galef 1992); (2) signs that individuals perceive and react to transgressions of cultural norms (Perry 2009); and (3) the existence of multiple traditions (Whiten and van Schaik 2007). In this chapter, I continue with this last approach, because when we talk of different human cultures, we are typically picking up on the existence of differences across multiple traditions, including communication patterns (language/dialect, forms of greeting), artifacts, and cuisine.

20.3 How to Trace the Origins of Culture

We have no time-machine, so all approaches to tracing the origins of culture are indirect and rely upon various kinds of inference. Nevertheless, a range of different methods, used in conjunction, has allowed us to build up a compelling picture of the past.

One such approach is to recover concrete remains, such as stone tools (Whiten et al. 2009a). This has yielded artifacts that, due to their material form, remain very much as they were a million years ago or more. However, we cannot witness the *behavior* of those who made them. This, of course, is exactly what we can do in the case of living primates. The question is, how can such present-day observations be used to reconstruct the past? Various approaches have been used, but the most robust has been called the comparative method. If we study a related group of species such as the great apes and find they all share a behavioral and/or cognitive characteristic, then we infer that this is the result of descent from a common ancestor that exhibited this character. This is the basis of the first of three components of my own approach to the evolutionary origins of culture, which I outline next.

20.3.1 The Comparative Method

In the example above, a characteristic shared by the great apes is attributed to the common ape ancestor that lived approximately 14 million years ago. But by zooming in to more closely related subsets of species, or alternatively zooming out to bigger groups of more distantly related taxa and applying the same principle, one can identify the characteristics of more recent and more distant common ancestors respectively, and thence trace the evolutionary history of the characteristic of interest – in our case, some aspect of culture. So, for example, characteristics shared by the two living species of chimpanzee (common chimpanzee and bonobo) would be attributed to their recent common ancestor living just over 2 million years ago, whereas characteristics shared by all primates will be attributed to their more ancient common ancestor of around 65 million years ago.

It is also possible that any taxa will share a characteristic because it evolved more than once, independently, but such convergent evolution is less likely than shared ancestry when we are looking at taxa that are genetically most closely related to each other. This and other considerations about using the comparative method to reconstruct ancestral states are described in more depth by Harvey and Pagel (1991) and Byrne (1995).

20.3.2 Dissecting Culture

Recent years have seen an enormous flowering of research on the subject of animal culture (Whiten and van Schaik 2007; Laland and Galef 2009). Too often, however, the over-simplistic question put in both academic and popular debates is whether such-and-such a species "has culture." The all-or-none assumption embodied in this habit is not helpful to the goal of a serious scientific analysis of the subject.

Instead, my colleagues and I have advocated the dissection of cultural phenomena into a set of elements, the distribution of which cannot be assumed to be correlated across species, as it would be if some species had all the shared features and the rest had none (Whiten et al. 2003; Whiten 2005, 2009). It will likely be more productive to investigate the extent of such correlations between features as an empirical question, acknowledging that the distribution of the various elements may be somewhat mosaic-like across the animal kingdom.

There will be many ways to construct taxonomies of such elements but in the papers cited above, I have begun to focus on three major aspects of culture: (1) the population-level distribution of cultural variations; (2) the behavioral contents of these; and (3) the underlying social transmission mechanisms. These, in turn, can be dissected into subelements that may also vary independently among species. The bulk of this chapter is concerned with these three major elements of culture as well as significant subelements.

20.3.3 Eclectic Methodological Approaches

The best ethology often combines an eclectic mixture of methodological approaches. Pure observation of natural populations is a crucial first step to map out the natural forms of the phenomena of interest, and in the case of culture, particularly to identify local variations in behavior likely to constitute traditions. But it is difficult, through pure observation, to identify the crucial ingredient of social learning, and even more difficult to discriminate particular forms, such as imitation: experimental and control conditions (the first of these typically permitting social learning, the second preventing it) have much greater power to identify the underlying learning processes and so offer a vital complement to observational approaches. Conducting such experiments in the field represents another ethological ideal, but experiments performed under the more readily controlled conditions of captivity may also play their part. This has proved very important in the case of social learning in primates, given the logistic and, sometimes, ethical constraints on experimentation in the wild.

20.4 Dissecting Culture and Tracing its Origins

In the core empirical sections of this chapter, I apply the approaches described above, discriminating multiple different aspects of culture through the eclectic mix of methods advocated, and reconstructing ancestral evolutionary states via the comparative method. However, to do so, with the thoroughness that the empirical material now at hand would permit, is beyond the scope or scale of the current chapter. In particular, to deal with each of the 14 elements of culture I distinguish, for each of the multiple shared ancestral states one could trace in our past (ape ancestor, primate ancestor, and so on: see Dawkins 2004, who dubs these common ancestors "concestors") is not feasible here. Given the topic of human universals, I therefore choose to focus on the concestor of ourselves and our nearest living relatives, the chimpanzees (*Pan* spp.). This concestor lived approximately 6 million years ago. Reconstructing its cultural life makes very clear that our unique and overblown capacity for culture, although vast when compared with the chimpanzee's, did not spring out of the blue; instead it evolved from an ancestral state that provided many different foundations for the cultural phenomena that evolved in later hominins. Of course, we do not simply take the chimpanzee as a model of our concestor; instead we infer the cultural nature of this concestor from the features of culture that the descendant taxa, *Pan* and *Homo*, can be shown to share.

Recent decades of research have identified "culture" in the sense of socially transmitted traditions, in a variety of mammals, birds, and fish (Laland and Hoppitt 2003; Whiten and Mesoudi 2008). My approach here is to raise the threshold criteria for cultural phenomena beyond this: to adopt as minimum threshold criteria the existence of traditions and/or social learning, but then to additionally pick out several key respects in which human culture goes *beyond* these starting points and assess in what ways, if any, chimpanzees do so also, with the implications for the concestral state outlined above. In principle, this analysis could be applied in any comparative analysis of different taxa, and for reconstructing the cultural life of corresponding concestors. Table 20.1 offers a simplified chart summarizing the analyses set out below.

20.4.1 The Large-Scale, Population-Level Patterning of Traditions

Multiplicity of traditions. Human cultures differ over time, and space, and these differences can typically be mapped in terms of traditions of many different kinds.

Table 20.1 The scope of culture in chimpanzees, humans, and our "concestor." The table lists features of culture (1) shared by chimpanzees and humans; (2) distinctive in humans. Twelve aspects of culture are discriminated within three broad categories. Each aspect represents one way in which "culture" extends beyond the mere existence of a tradition: see text for a fuller description. Each aspect is suggested by the scope of human culture, but there is some evidence bearing on each in our closest relative, the chimpanzee; these features thus offer an outline of the cultural nature of our common ancestor of approximately 6 million years ago

Hypothesis	Shared features	Distinctive in humans
1. Population-level patterning		
1.1 Multiplicity of traditions	Traditions very numerous compared with other species: over 40 in both species	Traditions have become so numerous as to be countless
1.2 Communities differing culturally in multiple ways	Each local culture is defined by a unique suite of traditions, in chimpanzees currently documented as between 9 and 24	Local cultures are distinguished by vast numbers of different traditions
1.3 Clustering of traditions through core ideas	Remains to be determined: some clusterings may be consistent with this	Occurs (e.g., LeVine 1984) but status is debated (Boyd et al. 1997)
1.4 Cumulative cultural evolution	Minimal at best, and disputed	Extremely elaborate and progressive (Tomasello 1999)
2. Content of cultures		
2.1 Physical (non-social)	Includes non-tool foraging techniques as well as tools fashioned and used for foraging, comfort, hygiene	Includes tool construction methods, material culture for hunting, trapping, clothing, medicine, shelters, and more
2.2 Social behavior	Includes social use of tools (e.g., leaf-clip in courtship), grooming conventions, possibly dialects	Includes language and other symbolic conventions, moral norms, ceremonies, and institutions
3. Social learning processes		
3.1 Copying	A portfolio of social learning mechanisms available including imitation and emulation, capable of recognizably faithful copying of sequences of action across repeated transmission	Higher fidelity copying of complex actions routine
3.2 Selective acquisition	Selective copying modulated by sensitivity to causal and intentional structure of tasks	In some contexts, less selective than apes, generating blanket copying ("over-imitation")
3.3 Conformity	Limited evidence for tendency to copy majority, even when alternatives known of	Strong conformity common, extending to ready acquisition of arbitrary conventions, such as gestures
3.4 Ratcheting	Cumulative social learning constrained, in part by great conservatism	Able to upgrade sophistication of repertoire by observational learning

(*continued*)

Table 20.1 (continued)

Hypothesis	Shared features	Distinctive in humans
3.5 Recognition of the copying process	Chimpanzees and other apes have been able to learn rule "Do-as-I-do"	Extended to intentional education and propaganda
3.6 Teaching	Minimal "scaffolding" in limited contexts at best; disputed	Now common in some contexts, but apparently rare in hunter-gatherers so likely rare until very recently everywhere

The culture of ancient Egypt, for example, differed from that of Egypt in more recent times in a host of ways that include forms of communication, tools, weapons, food, and religion, and just the same can be said of the contrast between Egyptian culture and other, very different, but contemporaneous cultures, such as that of, say, Scotland.

As noted above, traditions have been identified in many different animals too, but typically, this is just one tradition per species or per study. Black rats in Israeli pine forests have been shown to transmit special techniques for stripping seeds from pine cones across generations, but that is all (Terkel 1995). Many birds have been shown to pass on local dialects, but that is all (Catchpole and Slater 1995). In these cases, we can talk of a tradition as defined above, but not cultures in the richer sense defined by multiple traditions in our own species.

However, there are intermediate cases where multiple traditions have been identified and these are the most diverse in the species with which we shared the most recent common ancestor, the chimpanzee. The pooling of decades of recordings from long-term field studies across Africa has revealed over 40 behavior patterns that vary locally in ways that appear inconsistent with either genetic or straightforward environmental explanations, and so have been classed as putative traditions (Whiten et al. 1999, 2001; Whiten 2005). In many cases, this inference is strengthened by data on the intense observation of the skills of their elders by the youngsters, sometimes extending to correlations between generations in the style of technique adopted (Lonsdorf 2005). A considerable variety of behavioral domains are implicated, including tool use, foraging techniques, social behavior, grooming methods, and courtship gambits. This suggests that the richness of human culture, while unmatched by any other species, did not spring from nowhere but instead evolved from a state that already represented unusual cultural complexity compared with other animals. The later finding that orangutans show very nearly as many different traditions, encompassing social and material domains (van Schaik et al. 2003), indicates that this state of affairs likely characterized the common ancestor of all the great apes.

Further studies in recent years have indicated multiple traditions in other taxa, although typically no more than a handful in each case. Killer whale communities, for example, can be differentiated not only by their vocal repertoires but also by their very different hunting targets (salmon versus seals) and social organization (Rendell and Whitehead 2001). Bower birds in different localities exhibit bower styles that differ in multiple respects, such as overall architecture and the nature and color of objects chosen for decoration (Madden 2008). Whether, as more such long-term studies accumulate, other taxa will be found to show the cultural complexity of apes remains to be seen.

In all these studies, however, it is difficult through pure observation in the field to establish that the variations are truly due to social transmission and not other alternatives, such as genetics or environmental constraints. In recent years, the chimpanzee field data have, therefore, been complemented by an extensive series of social learning experiments of a particular kind – "diffusion experiments" – which are explicitly designed to test for the spread of new behaviors through populations (Whiten and Mesoudi 2008). In all but one of six such studies with chimpanzees, all concerning different artificial foraging tasks, clear evidence of transmission occurred in the most ambitious cases, including the spread of different behavioral variants across a series of three groups with high fidelity, perhaps, in part, because this employed the most complex foraging routines employed in these studies to date (Whiten et al. 2007). However, the latter study relied on us allowing one group to watch another in an adjacent compound, a scenario not possible in the wild, because of intercommunity hostility (see Crofoot and Wrangham, this volume). Instead, transmission is thought to occur in the wild though intercommunity transfer of individuals, typically females. Simulating this experimentally would therefore be a valuable next step. Nevertheless, the studies to date demonstrate that chimpanzees have a capacity for cultural transmission consistent with the interpretation of behavioral variations in the wild as representing extensive cultural variation.

Communities differing culturally in multiple ways. A related but different discovery has been that each chimpanzee community expresses a unique profile of behavioral variants that define it culturally, implying a qualitative similarity to the human case in this respect (Fig. 20.1). Of course, in quantitative terms, the difference remains huge: on present evidence, the number of traditional behaviors

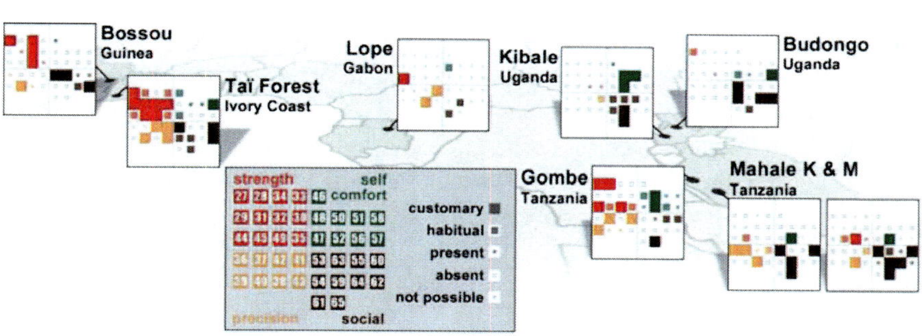

Fig. 20.1 The putative cultures of wild chimpanzees (after Whiten 2005). "Customary" acts are those typical in a community, "habitual" are less frequent yet consistent with social learning. Each community displays its own profile of such local behavioral variants, providing evidence of a unique culture for each locality. Numbers identify behavior patterns in the catalog attached to Whiten et al. (1999) and illustrated at http://culture.st-and.ac.uk/chimp

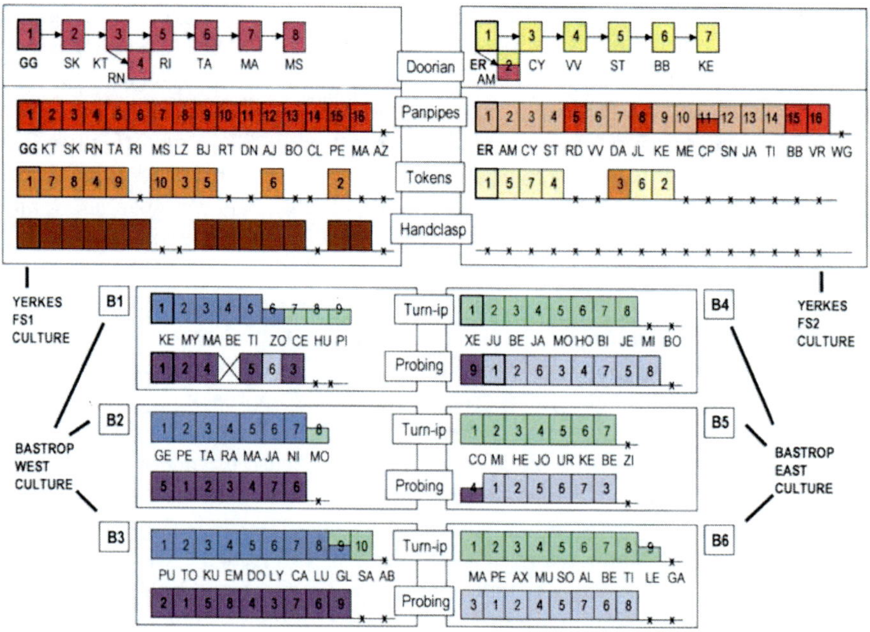

Fig. 20.2 The results of diffusion experiments in captive chimpanzees. Each *rectangle* represents a chimpanzee with two-character ID codes. Tasks, named in the center, were available in both local populations named on either side, but different techniques, color coded here, were seeded in one individual, marked here as No. 1, in each population. The Doorian experiment was run as a "transmission chain," as indicated by the *arrows*; all other experiments involved "open diffusion," with no predetermination of potential order of transmission. At Bastrop, transmission extended from groups B1 to B2 and B3, and from B4 to B5 and B6. Handclasp grooming spread spontaneously in the FS1 population. Numbers represent order of acquisition for each task. For further explanation, see text and Whiten et al. (2007). These studies demonstrate the capacity of chimpanzees to sustain multiple-traditions cultures, consistent with the interpretation of regional variations among wild chimpanzees summarized in Fig. 20.1

of any one chimpanzee community may typically lie in the teen numbers, whereas for humans, such variations can be numbered in the thousands – indeed, considering all the differences between any two different languages alone, it may be more apt to say the variations are "countless." The captive diffusion experiments with chimpanzees have essentially replicated in controlled conditions, the multiple-tradition cultures inferred to exist in the wild (Fig. 20.2), although on a smaller scale (Whiten et al. 2007).

Whether the great apes are truly as different from other species, as this picture currently suggests, or whether studies of other species will eventually identify similar complexity, remains to be seen. This is a young science. One recent study that underlines this caution concerns the peculiar behavior of stone-handling in Japanese macaques, where the case for cultural transmission is strong because the behavior appears so manifestly functionless (and thus unlikely to be shaped by environmental conditioning), and careful documentation has traced its emergence and steady spread from its earliest manifestations (Huffman 1996). The most recent studies have charted the variations between groups in numerous aspects of the way the stones are handled, generating a chart of behavioral variations approaching the complexity of that drawn for the great apes (Leca et al. 2007). One difference that remains, however, is that in the macaques these are all just minor variations on the same behavior, stone-handling, whereas those of chimpanzees span all major modes of behavior.

Core cultural clustering. The analyses of the two sections above view culture as made up of relatively particulate items – traditions that can be enumerated. This is consistent with the controversial idea that culture can be analyzed in terms of cultural inheritance units analogous to genes, dubbed "memes" by Dawkins (1976).

Early cultural anthropology largely embraced this approach, but later in the twentieth century anthropologists stressed the interconnectedness of elements of culture – "an organization of ideas rather than aggregate of independent traits" (LeVine 1984). Precursors of this in a nonverbal creature might be sought in the existence of correlated clusters of behavioral variants that would implicate core cognitive orientations influencing a whole suite of different behaviors. Possible examples of this in chimpanzees are discussed by Whiten et al. (2003) and include the tendency for some communities to tackle varied problems in their environment through the application of tools, whereas in other communities this orientation appears markedly lacking. Possibly the tool-rich cultures may be underlain by a global cognitive orientation to solve new problems technologically, a hypothesis that could perhaps be tested by field experiments presenting such problems to tool-rich and tool-poor communities. Compared to the data-rich comparisons that can be made in the two sections above, what can be said on this topic in nonhuman species remains presently only exploratory and will require more systematic investigations by field primatologists, in future. That said, just how interconnected are the components of humans culture remains a controversial issue by itself (Boyd et al. 1997); many elements do behave more like memes, spreading independently of others, a familiar contemporary example being the spread of mobile telephone use through very varied existing cultures worldwide.

Cumulative cultural evolution. By common consensus, the principal feature that distinguishes human culture from what we see in any nonhuman is cumulation. Generation by generation, cultures build on what went before so that in the later generations, achievements go massively beyond the earlier ones. We see this very clearly in the technological sphere, where we can trace the process from the earliest crude stone tools to the sophisticated technologies of today. For particular cultural items like the gun or the computer, we can trace cultural phylogenetic trees that chart diversification and progressive complexification over time and space, in many ways analogous to biological evolution (Mesoudi et al. 2006). Tomasello (1999) has described the progressive nature of this aspect of culture as a "ratchet effect."

At best, there are only very small signs of such ratcheting in apes, although we are often hampered in examining this because the earliest cultural phases are lost to us.

For example, chimpanzees in the Goualougo region in central Africa use a tool set to harvest subterranean termites, first using a stout stick to puncture a deep hole down into the ground, then withdrawing this and inserting a more delicate stem, the end of which they fray, such that it then collects a better harvest of termites (Sanz et al. 2004). This is unlikely to have emerged as a unitary brilliant invention, more likely developing step-wise from the use of a single probing tool, as with chimpanzees elsewhere, but we have no direct data on this.

However, most of the many traditions documented for chimpanzees show little evidence of cumulative culture, and cumulation is what has created the enormous gulf between present-day human and nonhuman cultures, from technology to language. We do not know why this difference arose. One explanation may simply be that chimpanzees could benefit little from more advanced cultures in the niche they have so successfully exploited for millions of years; by contrast, cumulative culture was a crucial ingredient in our ancestors' creation of the socio-cognitive niche of hunting and gathering, that allowed them to survive and thrive in the drastically changed African habitats that they engaged with (Whiten 1999a).

A different (although not necessarily mutually exclusive) possibility is suggested by the results of the first systematic experimental investigations of the question. Marshall-Pescini and Whiten (2008a) first exposed young, wild-born, sanctuary-living chimpanzees to an artificial foraging task that required sliding aside a small flap and dipping a stick-tool in the device to obtain small amounts of honey. In the first phase of the experiment, youngsters were shown to learn this socially, through watching a familiar human model the skill. In a second phase, the youngsters saw the same model progress to a more complex action, poking their stick in a hole to release a catch, so that the existing dipping action could now be used to lever the whole top of the device open and obtain a rich mixture of honey and nuts inside. If the chimpanzees would learn this, we would have a modest case of cumulative cultural evolution. However, they did not: they stuck to the existing, less productive skill they had mastered. This was not because the more complex routine was too difficult, for some control subjects who had not done the first simple dipping action did perform the more complex one. A similar conservatism was demonstrated in another recent study, in which chimpanzees who had learned to employ a tool to obtain out-of-reach food failed to copy a more productive technique some of their own group invented (Hrubesch et al. 2009). Together, these studies suggest that chimpanzees may exhibit a degree of conservatism that inhibits them from observationally learning to adopt approaches that are better than their existing habitual ones, an essential requirement for cumulative culture (Whiten et al. 2009b).

In sum, significant similarities and differences between chimpanzees and humans have been established dealing with the large-scale patterning of socially learned behavioral variations in time and space, described in the four sections above. These suggest that the cultural nature of our concestor would have allowed them to sustain cultures constituted by multiple behavioral variations, but with very limited capacity for cumulative change compared with the human case.

20.4.2 The Contents of Cultures

Whatever the similarities and differences in the population-level patterning described in the prior section, the actual behavioral content may vary, and this provides a second source of human-ape comparisons. For example, the content of human but not ape cultures includes language and religion, whereas chimpanzee courtship gambits that include various routines of noisy vegetation manipulation are not used by humans (Whiten et al. 2001). Such comparisons remain to be completed in any detail and so, by comparison with the foregoing (20.4.1) and following (20.4.3) sections, the treatment here remains relatively cursory. However, this is not to minimize its potential significance: as suggested by the examples above, major differences between human and nonhuman culture exist at the level of content and there are important shared features too.

Physical (nonsocial) content. This includes both material culture and other forms of nonsocial behavior. Under the heading of material culture, a major feature shared by humans and chimpanzees is the use and limited fashioning of tools, together with the functional contexts of their use, which range across the procurement of dietary items (including water and both vegetable and animal sources gained through dipping, probing, fishing, pounding, and stabbing techniques), "hygiene" such as dabbing and wiping feces, blood, and semen, and protection from elements, as in hat and seat-making (McGrew 1992; Whiten et al. 2001; Pruetz and Bertolani 2007). Not all of these should necessarily be attributed to our concestor – some specific instances will surely have evolved in the interim period in chimpanzees, as they so clearly have for humans – but the general capacity for learning a diversity of forms of tool use was likely there in our concestor, and indeed in the concestor of all the great apes, judging by the technological repertoire of orangutans (van Schaik et al. 2003).

Where did the differences begin? One important step taken by our hominin ancestors but not by chimpanzees is the construction of tools, by fitting together subcomponents. However, archeological evidence for this becomes available only much later than another that appears beyond chimpanzees but is still in the destructive mode: the fashioning of symmetrical stone tools (Acheulian blades: Mithen 1999; Whiten et al. 2003), discussed in detail later in this chapter.

Other shared aspects of content include complex, non-tool foraging techniques, and, perhaps, even medicinal plant use, which can require crucial discrimination between medicinal and poisonous elements, such as pith versus other parts of the plant (Whiten 2006).

Social behavioral content. On the basis of current published evidence, forms of social behavior that are socially learned are less common than the nonsocial ones outlined above. They include the categories of courtship gambits and physical social interaction, although in both of these the precise forms of behavior differ between chimpanzees and humans. Examples in chimpanzees include the A-frame "grooming hand-clasp" in which each of two grooming chimpanzees grasps one hand of the other above their heads (McGrew and Tutin 1978) and the more recently described "social scratch" that takes different forms in different communities (Nakamura et al. 2000). Of course, an important category of social behavior is communication, but here there is relatively little evidence for a role of social learning in chimpanzees, with early signs of dialect differences giving way to skepticism (Mitani et al. 1999). Indeed, apart from the courtship gestures noted above, the most obvious ways in which humans and chimpanzees differ in the social realm concern the socially learned vocal and gestural repertoire that became so enormously extensive in humans. The social content of our concestral culture may, thus, have involved only a relatively restricted set of physical routines and gestures, compared with the technical content described in the last section.

In humans, additionally, social institutions represent communally agreed forms of behavior, that community members are expected to abide by (Runciman 2001). They include such phenomena as marriage, moral norms, and ceremonies. Although such practices could be learnable to some extent just by observation, it is difficult to see how they can be established in the first place without linguistic negotiation, which would put them outside the range of nonhuman apes. The most likely candidates for shared features here, or at least potential precursors of them, may be "policing behavior," in which powerful individuals "keep the peace" in groups by, for example, breaking up fights by lower-ranking individuals as if they are contravening some unwritten, preferred, peaceable norms (de Waal 1989). However, we do not know if such policing is itself socially acquired; the closest evidence bearing on this is that the tendency towards peacemaking in the sense of reconciling after fights has been shown through cross-fostering experiments with monkeys to be influenced significantly by social learning (de Waal and Johanowicz 1993).

20.4.3 Social Learning Processes

The same category of behavioral content and the same spatio-temporal distribution of the behavior in the population could be generated by a variety of different social learning processes, so these provide the third principal basis on which grades of culture may be compared. In the human case, learning may be in a verbal mode, including listening to narratives, taking instruction, or engaging in dialog, and comparative primate research appears likely to cast relatively little light on the origins of these linguistic processes. Decades of research have focused instead on identifying and discriminating numerous forms of nonverbal, observational learning as well as different forms of teaching.

Copying. "To ape" means to copy or imitate. Several observational and experimental studies through the last century offered results apparently consistent with this image. However, in the 1980s and 1990s, a series of experimental results that did not support this picture combined with a series of powerful methodological critiques of the field to yield a much more skeptical view; that imitation in apes (as well as culture in the wild) remained to be convincingly demonstrated (Tomasello and Call 1997). Instead, it was suggested that what sometimes looked superficially like imitation was actually the ape learning about the affordances of some aspect of the physical world – such as, that sticks can help rake in food – a process labeled as "emulation."

In more recent studies, the pendulum has swung back somewhat, insofar as the fidelity of transmission of alternative behavior patterns has remained high in what might be thought of as the more demanding context of diffusion experiments that involve multiple transmission steps (Whiten et al. 2007). Numerous factors might potentially explain the differences in experimental results that litter the literature in this area, together with the different conclusions that have been drawn from them. For example, the relatively negative results have generally been obtained in the context of individual testing, in which isolated (yet highly socially oriented) chimpanzees may not demonstrate their true potential, which they do in the context of diffusion studies conducted in the group context. At present, we cannot be sure. But we are faced with the classic asymmetry of positive and negative results – if the results are positive in showing observers matching the actions of a model they watched, then the animals have demonstrated they have a capacity to copy; if the results are negative, with no copying, then this might be due to many possible factors, some of them merely artifactual and misleading. Of course, negative results are as important as positive ones – they are crucial to any comparative science – but they need to be replicated and tested repeatedly before we can acquiesce in concluding that the capacity at stake is lacking.

The roles of imitative and emulative processes have been further tackled recently through "ghost" experiments, in which the model is removed from the scene and the objects normally manipulated by the model are made to go through their motions by a "ghostly" subterfuge of the experimenter pulling them with fishing line. Applying this procedure to a tool-based foraging task that had earlier been shown to diffuse through a group with significant fidelity, Hopper et al. (2007) obtained the surprising result that chimpanzees could learn nothing of this by just watching the tool do its job as if moved by a ghostly hand – effectively just what an individual learning by emulation is supposed to be focusing on. In a follow-up experiment with a much simpler task that involved merely pushing a door to left or right to obtain the food hidden behind it, chimpanzees did show evidence of emulation on their first attempt, matching the direction they saw the door move in, but they quickly abandoned this in later trials (Hopper et al. 2008). By contrast, they were highly faithful to the direction in which they saw another chimpanzee push the door, persisting in this over repeated trials. Being able to actually copy another ape thus seems particularly important to them.

A very different source of evidence consistent with this came from a study that recently provided some of the first robust experimental evidence that chimpanzees will learn how to use a stone hammer to crack nuts (one of the putative local traditions of wild chimpanzees) through observing a proficient conspecific. In this study, the observer chimpanzee was several times seen to move its arm in sympathy

Fig. 20.3 Matching the actions of an expert. Two frames from a video record of an experiment assessing the role of social learning in the acquisition of nut-cracking using a stone hammer. Here, a naïve young chimpanzee (*top*) watches the performance of an already-proficient nut-cracker. The two frames illustrate the way in which the observer spontaneously mirrored the actions of the nutcracker, moving its arm up (*first frame*) and down (*second frame*) repeatedly in synchrony with tool-user. The video clip, in supplementary information to Marshall-Pescini and Whiten (2008b) can be viewed at http://dx.doi.org/10.1037/0735-7036.122.2.186.supp

with that of the model wielding the hammer (Fig. 20.3), going through the motions witnessed before ever having grasped the hammer stone (Marshall-Pescini and Whiten 2008b: video viewable in the electronic supplementary information to that paper). Similar sympathetic movements in humans (as in watching sports on the TV) have been interpreted as the effects of "mirror neurons" (Dijksterhuis 2005; Gallese 2007) that have the special property of firing both when performing and watching the same behavior, and on these occasions escape the inhibition to which they must normally be subject while watching others (otherwise, the observer would helplessly mimic all the other individual is doing). There is evidence that such neurons are involved in imitation in humans (Iacoboni et al. 2005), so the serendipitous finding in our nutcracking study may have profound implications for similarities in the processes underlying how chimpanzees and humans match conspecifics' actions.

Imitation has now been demonstrated experimentally in a wide range of species, including birds (e.g., pigeons and budgerigars) (Zentall 2004). Chimpanzees, like humans, have additionally been shown, experimentally, to copy such complexities as tool use and the sequencing of constituent subcomponents actions (Whiten et al. 2004). Something approximating these levels of imitative competence would accordingly be expected of our concestor.

Selective acquisition. Imitation by infants less than 2 years old has been shown to be already "rational" insofar as they will copy an adult performing a bizarre act like bending to switch on a light with their forehead, but only if the adult has their hands free: if the hands are enmeshed in clothing so that butting the light is simply the best thing to do, infants who then approach the task with their own hands free will eschew imitation and instead use their hands to press the switch (Gergely et al. 2002). A similar effect has recently been shown in chimpanzees (Buttelmann et al. 2007). In other experiments, chimpanzees who witnessed a component of a complex task that could be seen to be causally irrelevant eschewed copying that component, although they copied it in a version of the task where the materials were all opaque, so causally irrelevant events could not be directly perceived (Horner and Whiten 2005).

Together these studies indicate that chimpanzees do not blindly or mindlessly "ape" what they witness; their observational learning incorporates appraisal of both relevant and irrelevant components. Intriguingly, we found this was less true of young children tested in the experiment contrasting visible versus invisible causally irrelevant elements (Horner and Whiten 2005): children copied with high fidelity in both conditions, a tendency we interpreted as consistent with a more thoroughgoing assimilation of their cultural environments. Under just what conditions children are discriminating (as in the Gergely study cited above) or instead imitating in "mindless" blanket fashion is the subject of current studies. Developmental psychologists have begun to describe the latter tendency as "over-imitation" and have found that in certain contexts, children appear unable to inhibit it even when strongly encouraged to do so (Lyons et al. 2007). More generally, although our work has demonstrated a considerable capacity for moderately faithful copying by chimpanzees, it nevertheless remains the case that a substantial corpus of direct, comparative studies have shown children to copy with consistently greater fidelity (Whiten et al. 2004; Call et al. 2005).

Conformity. One mark of humans' cultural proclivity is conformity. Social psychologists have provided ample studies showing effects of the kind in which an experimental subject is exposed to a group who make judgments that fly in the face of what the subject can perceive with their own eyes (such as that a line A is longer than B, when it must be obvious to the subject that line B is significantly longer), yet, the subject goes along with the majority view: they conform (abjectly, we might say) (Asch 1956).

In a recent social diffusion experiment with chimpanzees, in which alternative foraging techniques were seeded in each of two separate groups, we found that some chimpanzees discovered the alternative technique for themselves; nevertheless, when tested again 2 months later, there was a significant tendency for these innovators to "return to the fold," conforming to what a majority in their group were doing (Whiten et al. 2005). We can describe this as conformity to a local "norm" in the statistical sense, without requiring that chimpanzees conceive of a norm, as such. Thus, the conformity we recorded may indicate the importance of cultural transmission to chimpanzees, as it does in the human case. However, Galef and Whiskin (2008) have followed up our study by showing a similar case of conformity in rats. Although this concerns only the rats' choice amongst two foods, rather than acquiring the complex, tool-based foraging technique that our study was concerned with, this suggests that the basic phenomenon of conformity may be more widespread in the animal kingdom. Conformity may take many forms and occur in different grades, so a systematic program of research on this topic is now warranted (Laland 2004).

In humans, conformity may be further shaped by active sanctions imposed by a group on any who do not conform (Hill 2009). Even without such sanctions, there may be a recognition that such-and-such is a proper or correct way to behave. In studies in which we have exposed young children to members of another group seeded with an alternative technique, we repeatedly hear the refrain "but that's not how you do it!." There is an early sense of how things "ought" to be done (see Tomasello et al. 2005 for an extended analysis of this and related aspects of social cognition). This sense of "ought" is difficult to test for in nonverbal apes, but in any case, clear signs of sanctions against nonconformers have not been documented. Indeed, it is not easy to say why chimpanzees should care if one of their group members does things a different way – which in turn raises the intriguing question of just why this has become so significant for humans, and when and why it may have done so.

Ratcheting. As noted in Sect. 20.4.1, cumulative cultural evolution is a distinctive characteristic of human culture, notably absent or minimal in nonhuman species. The recent experiments we described in that section, showing that chimpanzees offered an opportunity to ratchet up from their existing foraging technique to a superior one performed by a group mate, failed to do so (Hrubesch et al. 2009; Marshall-Pescini and Whiten 2008a), should be cross-referenced here because this really comes into the category of social learning mechanisms. This appears such a crux in what makes human culture different to that of other apes, that it is to be hoped that this will be a growth area in future research.

Recognition of the copying process. Preschool-age children become aware of when they are and are not copying others, such that the process of cultural acquisition may become to this extent, self-reflective. We see a sign of this "meta-cognition" in the ability of apes to learn the rule to attempt to copy another individual in the "Do-as-I-do" paradigm of imitation research and it contrasts with the failure of several attempts to train monkeys to do this (Whiten et al. 2004). In other words, apes appear to grasp the "concept" of imitation in a special way, as do children, which suggests that in some way, their cultural acquisition mechanisms are operating at this higher cognitive level.

Teaching. There are few signs of teaching in apes (Whiten 1999b). This may appear a striking contrast with the human case, but, in fact, anthropologists have remarked on the lack of teaching also in hunter-gatherer societies (Draper 1976; Hewlett and Cavalli-Sforza 1986): a serious question-mark thus remains about the significance of teaching for human culture in all but very recent historical times. This is discussed further by Whiten et al. (2003). It may be that among animals, behavior that can be called teaching, or perhaps more accurately "scaffolding" supporting the acquisition of difficult skills (Whiten 1999a), is to be seen in predatory species in which the young must make a big leap in skill to become independent foragers, such as in meerkats (Thornton and Raihani 2008).

20.5 Conclusions

What kind of being, then, was the human/chimp concestor? The comparative evidence reviewed above suggests that it was a significantly cultural creature. Although culture was to become inordinately more complex as our own line later evolved, this analysis, nevertheless, suggests that the repertoire of our concestor provided a significant cultural "platform," that makes the origins of our unique cultural nature much less mysterious than could any analysis blind to the evolutionary past.

Summarizing the key cultural features of our concestor based on the foregoing analyses, we would infer that while there would be little or no significant teaching occurring, observational learning would have been important in their lifestyle. This would have included a portfolio of social learning processes, including both emulative and imitative components permitting the acquisition from existing cultures of a whole repertoire of useful behaviors like foraging techniques. These would have extended to complex, sequentially structured actions of the kinds illustrated by the use of tool sets in wild chimpanzees, and incorporated into our diffusion experiments that have so robustly demonstrated the capacity for multiple cases of such behavior patterns to be spread by social learning. Social learning would have been sufficiently strong to create a degree of conformity to local norms of behavior, sophisticated enough for the mind to hold some concept of what it is to copy others, and capable of sustaining different local cultures defined by multiple and diverse traditions. Whiten and van Schaik (2007) have suggested that this adaptive complex may help explain the relatively large brain size of great apes, with culture requiring cognitive sophistication and culture, in turn, making tradition-acquirers smarter (the "cultural intelligence hypothesis": Whiten and van Schaik 2007).

Note that although the above analysis might superficially appear to amount to saying our ancestor was culturally like a chimpanzee, this is not the nature of our methodological enterprise. To take the chimpanzee as a straightforward representative of our ancestor – what has been called a "referential model" for our past – would be wrong. We did not evolve from a chimpanzee, any more than did chimpanzees evolve from humans. The crucial approach is instead to identify shared features of chimpanzees and humans and attribute them to the common ancestor. Derived features unique to either chimpanzee or humans do not figure in this, so the ancestor we are reconstructing was not simply like a chimp. Having said that, however, it was likely more like a chimpanzee than a human, one important reason for this being that its brain was still chimpanzee-sized, and thus only about a third of that of the cultural creature we have become.

When we turn to the other side of the coin and ask where the cultural differences between the concestor and our later ancestors began, the principal feature of interest must be the capacity for cumulative culture. This becomes evident in the archeological record as Oldwan tools gave way to the symmetry and sophistication of Acheulian blades and yet more sophisticated technology that was built step-wise on what had been achieved before (Mithen 1999). The common explanation for the emergence of cumulative cultural evolution, among both leading comparative psychologists (Tomasello 1999) and leading anthropologists (Boyd and Richerson 1996), has been the emergence of a capacity for imitation, which during the 1990s was argued to be lacking in nonhuman species such as the chimpanzee. As will have been evident from the foregoing review, more recent research with chimpanzees has demonstrated considerable powers for rather high-fidelity transmission of behavior, which suggests it was not a lack of imitative capacity that constrained cumulation. It is true that children are more imitative than chimpanzees, but the point is that chimpanzees, nevertheless, have ample copying ability to sustain cultures constituted of many complex behaviors. My colleagues and I have, therefore, suggested that the key to the rise of cumulation was instead the evolution of a greater degree of intelligence, associated with encephalization, that permitted more sophisticated innovations such as the Acheulian industry (Whiten et al. 2003). This is consistent with the analyses of archeologists such as Mithen (1999), who points out that Acheulian artifacts are sufficiently complex that they must have been acquired through considerable imitative capacity, yet there was minimal cumulation at this stage, with no change over millennia: ergo, imitation does not beget cumulation. The conservatism of social learning that we have identified in recent experiments with chimpanzees suggests an additional factor that may limit cumulation: chimpanzees and our concestor, we infer, could acquire much culture through observational learning, but the crystallization of well-worn habits soon constrained the ability to assimilate any new innovations that emerged. With our triple-sized ape brains, that is what we humans became supremely able to do.

Acknowledgments The author was supported by a Royal Society Leverhulme Trust Senior Research Fellowship during the preparation of this work.

References

Asch SE (1956) Studies of independence and conformity. I. A minority of one against a unanimous majority. Psychol Monogr 70:1–70

Boyd R, Richerson PJ (1996) Why culture is common, but cultural evolution is rare. Proc Br Acad 88:77–93

Boyd R, Borgerhoff-Mulder M, Durham WH, Richerson PJ (1997) Are cultural phylogenies possible? In: Weingart P, Mitchell SD, Richerson PJ, Maasen S (eds) Human by nature: between biology and the social sciences. Lawrence Erlbaum Associates, Mahwah, NJ, pp 355–386

Buttelmann D, Carpenter M, Call J, Tomasello M (2007) Enculturated chimpanzees imitate rationally. Dev Sci 10:F31–F38

Byrne RW (1995) The thinking ape: evolutionary origins of intelligence. Oxford University Press, Oxford

Call J, Carpenter M, Tomasello M (2005) Copying results and copying actions in the process of social learning: chimpanzees (*Pan troglodytes*) and human children (*Homo sapiens*). Anim Cogn 8:151–163

Catchpole CK, Slater PJB (1995) Bird song: biological themes and variations. Cambridge University Press, Cambridge

Dawkins R (1976) The selfish gene. Oxford University Press, Oxford, reprinted in 1989

Dawkins R (2004) The ancestor's tale: a pilgrimage to the dawn of life. Weidenfeld & Nicolson, London

de Waal FBM (1989) Peacemaking among primates. Harvard University Press, Cambridge, MA

de Waal FBM, Johanowicz DL (1993) Modification of reconciliation behaviour through social experience: an experiment with two macaque species. Child Dev 64:897–908

Dijksterhuis A (2005) Why we are social animals: the high road to imitation as social glue. In: Hurley S, Chater N (eds) Perspectives on imitation: from neuroscience to social science, vol 2. MIT, Cambridge, MA, pp 207–220

Draper P (1976) Social and economic constraints on child life among the !Kung. In: Lee RB, DeVore I (eds) Kalahari hunter-gatherers: studies of the !Kung San and their neighbours. Harvard University Press, Cambridge, MA, pp 199–217

Fragaszy DM, Perry S (2003) The biology of traditions: models and evidence. Cambridge University Press, Cambridge

Galef BG Jr (1992) The question of animal culture. Hum Nat 3:157–178

Galef BG Jr, Whiskin EE (2008) 'Conformity' in Norway rats? Anim Behav 75:2035–2039

Gallese V (2007) Before and below 'theory of mind': embodied simulation and the neural correlates of social cognition. Phil Trans Roy Soc B 362:659–669

Gergely G, Bekkering H, Király I (2002) Rational imitation in preverbal infants. Nature 415:755

Harvey PH, Pagel MD (1991) The comparative method in evolutionary biology. Oxford University Press, Oxford

Hewlett BS, Cavalli-Sforza LL (1986) Cultural transmission among Aka pygmies. Am Anthropol 88:922–934

Hill K (2009) Animal 'cultures'? In: Laland KN, Galef BG Jr (eds) The question of animal culture. Harvard University Press, Cambridge, MA, pp 269–287

Hopper LM, Spiteri A, Lambeth SP, Schapiro SJ, Horner V, Whiten A (2007) Experimental studies of traditions and underlying transmission processes in chimpanzees. Anim Behav 73:1021–1032

Hopper LM, Lambeth SP, Schapiro SJ, Whiten A (2008) Observational learning in chimpanzees and children studied through 'ghost' conditions. Proc Roy Soc Lond B 275:835–840

Horner V, Whiten A (2005) Causal knowledge and imitation/emulation switching in chimpanzees (*Pan troglodytes*) and children (*Homo sapiens*). Anim Cogn 8:164–181

Hrubesch C, Preuschoft S, van Schaik CP (2009) Skill mastery inhibits adoption of observed alternative solutions among chimpanzees. Anim Cogn 12:209–216

Huffman MA (1996) Acquisition of innovative cultural behaviors in non-human primates: a case study of stone handling, a socially transmitted behavior in Japanese macaques. In: Galef BG Jr, Heyes C (eds) Social learning in animals: the roots of culture. Academic, Orlando, pp 267–289

Iacoboni M, Molnar-Szakacs I, Gallese V, Buccino G, Mazziotta JC, Rizzolatti G (2005) Grasping the intentions of others with one's own mirror neuron system. PLoS Biol 3:e79. doi:doi/10.1371/journal.pbio.0030079

Laland KN (2004) Social learning strategies. Learn Behav 32:4–14

Laland KN, Galef BG Jr (2009) The question of animal culture. Harvard University Press, Cambridge, MA

Laland KN, Hoppitt W (2003) Do animals have culture? Evol Anthropol 12:150–159

Leca J-B, Gunst N, Huffman MA (2007) Japanese macaque cultures: inter- and intra-troop behavioural variability of stone-handling patterns across 10 groups. Behaviour 144:251–281

LeVine RA (1984) Properties of culture: an ethnographic view. In: Schweder RA, LeVine RA (eds) Culture theory: essays on mind, self, and emotion. Cambridge University Press, Cambridge, pp 67–87

Lonsdorf EV (2005) Sex differences in the development of termite-fishing skills in the wild chimpanzees, *Pan troglodytes schweinfurthii*, of Gombe National Park, Tanzania. Anim Behav 70:673–683

Lyons DE, Young AG, Keil FC (2007) The hidden structure of overimitation. Proc Natl Acad Sci USA 104:19751–19756

Madden JR (2008) Do bowerbirds exhibit cultures? Anim Cogn 11:1–12

Marshall-Pescini S, Whiten A (2008a) Chimpanzees (*Pan troglodytes*) and the question of cumulative culture: an experimental approach. Anim Cogn 11:449–456

Marshall-Pescini S, Whiten A (2008b) Social learning of nut-cracking behaviour in East African sanctuary-living chimpanzees (*Pan troglodytes schweinfurthii*). J Comp Psychol 122:186–194

McGrew WC (1992) Chimpanzee material culture: implications for human evolution. Cambridge University Press, Cambridge

McGrew WC, Tutin CEG (1978) Evidence for a social custom in wild chimpanzees. Man 13: 234–251

Mesoudi A, Whiten A, Laland KN (2006) Towards a unified science of cultural evolution. Behav Brain Sci 29:329–383

Mitani JC, Hunley KL, Murdoch ME (1999) Geographic variation in the calls of wild chimpanzees: a reassessment. Am J Primatol 47:133–151

Mithen S (1999) Imitation and cultural change: a view from the Stone Age, with specific reference to the manufacture of handaxes. In: Box HO, Gibson KR (eds) Mammalian social learning: comparative and ecological perspectives. Cambridge University Press, Cambridge, pp 389–399

Nakamura M, McGrew WC, Marchant LF, Nishida T (2000) Social scratch: another custom in wild chimpanzees? Primates 41:237–248

Perry S (2009) Are non-human primates likely to exhibit cultural capacities like those of humans, and if so, how can this be demonstrated? In: Laland KN, Galef BG Jr (eds) The question of animal culture. Harvard University Press, Cambridge, MA, pp 247–268

Pruetz JD, Bertolani P (2007) Savanna chimpanzees, *Pan troglodytes verus*, hunt with tools. Curr Biol 17:412–417

Rendell L, Whitehead H (2001) Culture in whales and dolphins. Behav Brain Sci 24:309–382

Runciman WG (2001) From nature to culture, from culture to society. In: Runciman WG (ed) The origin of human social institutions. Oxford University Press, Oxford, pp 235–254

Sanz C, Morgan D, Gulick S (2004) New insights into chimpanzees, tools and termites from the Congo Basin. Am Nat 164:567–581

Terkel J (1995) Cultural transmission in the black rat: pine-cone feeding. Adv Study Behav 24:119–154

Thornton A, Raihani NJ (2008) The evolution of teaching. Anim Behav 75:1823–1836

Tomasello M (1999) The cultural origins of human cognition. Harvard University Press, Cambridge, MA

Tomasello M, Call J (1997) Primate cognition. Oxford University Press, Oxford

Tomasello M, Carpenter M, Call J, Behne T, Moll H (2005) Understanding and sharing intentions: the origins of cultural cognition. Behav Brain Sci 28:675–735

Toth N, Schick K (2007) Overview of paleolithic archeology. In: Henke W, Tattersall I (eds) Handbook of paleoanthropology. Springer, Heidelberg, pp 1943–1963

van Schaik CP, Ancrenaz M, Borgen G, Galdikas BFM, Knott CD, Singleton I, Suzuki A, Utami SS, Merrill M (2003) Orangutan cultures and the evolution of material culture. Science 299:102–105

Whiten A (1999a) Parental encouragement in *Gorilla* in comparative perspective: implications for social cognition and the evolution of teaching. In: Parker ST, Miles HL, Mitchell RW (eds) The mentality of gorillas and orangutans. Cambridge University Press, Cambridge, pp 342–366

Whiten A (1999b) The evolution of deep social mind in humans. In: Corballis MC, Lea SEG (eds) The descent of mind: psychological perspectives on hominid evolution. Oxford University Press, Oxford, pp 173–193

Whiten A (2005) The second inheritance system of chimpanzees and humans. Nature 437:52–55

Whiten A (2006) The significance of socially transmitted information for nutrition and health in the great ape clade. In: Wells JCK, Laland KN, Strickland SS (eds) Social information transmission and human biology. CRC Press, London, pp 119–136

Whiten A (2009) The identification of culture in chimpanzees and other animals: from natural history to diffusion experiments. In: Laland KN, Galef BG Jr (eds) The question of animal culture. Harvard University Press, Cambridge, MA, pp 99–124

Whiten A, Mesoudi A (2008) Establishing an experimental science of culture: animal social diffusion experiments. Phil Trans Roy Soc B 363:3477–3488

Whiten A, van Schaik CP (2007) The evolution of animal 'cultures' and social intelligence. Phil Trans Roy Soc B 362:603–620

Whiten A, Schick K, Toth N (2009a) The evolution and cultural transmission of percussive technology: integrating evidence from paleoanthropology and primatology. J Hum Evol in press

Whiten A, McGuigan N, Marhsall-Pescini S, Hopper LM (2009b) Emulation, imitation, overimitation and the scope of culture for child and chimpanzee. Phil Trans R Soc 364:2417–2428

Whiten A, Goodall J, McGrew WC, Nishida T, Reynolds V, Sugiyama Y, Tutin CEG, Wrangham RW, Boesch C (1999) Cultures in chimpanzees. Nature 399:682–685

Whiten A, Goodall J, McGrew WC, Nishida T, Reynolds V, Sugiyama Y, Tutin CEG, Wrangham RW, Boesch C (2001) Charting cultural variation in chimpanzees. Behaviour 138:1481–1516

Whiten A, Horner V, Marshall-Pescini S (2003) Cultural Panthropology. Evol Anthropol 12:92–105

Whiten A, Horner V, Litchfield CA, Marshall-Pescini S (2004) How do apes ape? Learn Behav 32:36–52

Whiten A, Horner V, de Waal FBM (2005) Conformity to cultural norms of tool use in chimpanzees. Nature 437:737–740

Whiten A, Spiteri A, Horner V, Bonnie KE, Lambeth SP, Schapiro SJ, de Waal FBM (2007) Transmission of multiple traditions within and between chimpanzee groups. Curr Biol 17:1038–1043

Zentall TR (2004) Action imitation in birds. Learn Behav 32:15–23

Chapter 21
The Coevolution of Genes, Innovation, and Culture in Human Evolution

Richard McElreath

Abstract Much adaptive human behavior is much too complex to be invented by any single individual in his lifetime. Such complex behavior can be learned and maintained in human populations, however, because our species possesses evolved psychological abilities for acquiring and modifying complex behavior and knowledge. One puzzle surrounding the origin of human behavior, with its strong reliance on socially transmitted knowledge, is how natural selection can favor costly abilities for complex social learning before the existence of complex behavior to be learned. The finding of special-purpose social learning abilities in other apes has only sharpened this puzzle – if other apes are good at imitation, is the key difference between ourselves and chimpanzees instead rates of innovation? In this chapter, I explore this puzzle by considering the simultaneous coevolution of both social learning ability and individual innovation. When one allows both innovation and the accuracy of social learning to evolve independently of one another, natural selection can favor increased investment in social learning, but only if it first favors increased innovation. However, once social learning evolves to high accuracy, high innovation rates are no longer needed, and natural selection favors reduced investment in innovation. Thus, the debate about whether innovation or imitation defines the gap between humans and other apes may be misstated. Instead, the emergence of human culture may have required the coevolution of both kinds of learning.

R. McElreath
Department of Anthropology and Graduate Groups in Ecology, Animal Behavior and Population Biology, University of California, Davis, CA 95616
e-mail: mcelreath@ucdavis.edu

P.M. Kappeler and J.B. Silk (eds.), *Mind the Gap*,
DOI 10.1007/978-3-642-02725-3_21, © Springer-Verlag Berlin Heidelberg 2010

21.1 Introduction

Wheat is one of humanity's great inventions. Coming in great variety, locally adapted to microclimates, it converts energy into a form people can use to make more people. It is also nearly wholly dependent upon people for its survival – like other grains, wheat seed does not easily break from the grass, but instead stays firm, stuck to a "tough" rachis, waiting for the farmer or machine to remove them all at once. A sensible wild grain instead disperses seed to the wind. Other domestic species are similar; they have partly outsourced their reproduction to humans, in exchange for security. Domestic cattle – whatever you think of their intelligence – have done quite well, compared with their extinct wild ancestors.

All of the human species' domesticated inventions represent a transfer of information. Information about the environment – how to extract energy from it, how to resist drought, how to make harvest easier for a human farmer – makes its way into wheat's genes, during artificial selection. Generations of human farmers have shuttled information about the world, the world's pests, and their own preferences into the plant's genome, creating a technology that "knows" about the world we live in. Of course the farmer must have some extra information, in order to profit from wheat. But a tremendous amount of information must be contained in the plant's genome, and it accumulated there over many generations.

When I say "information" here, I mean that if we knew how to interpret it, the sequence of bases in wheat DNA would tell us new things about the environment and how to adapt to it. However, there are more formal definitions of information that suggest that natural selection accomplishes a similar feat, building information about the environment into our own genome (Frank 2009). The fit the Darwin observed between organisms and their environments reflects this flow of information. Each organism embodies a recent step in a long chain of information commerce, between the environment and the population of genes. This commerce is not without friction, of course. Sexual reproduction and recombination interfere directly with selections ability to describe the environment. But over time, natural selection manages to adapt organisms to the environment, by differentially favoring alleles.

Of course real environments fluctuate and vary. The planet we live in now is quite different than that of the Pleistocene, and spatial variation from pole to pole is at last as great as temporal change. As a result, some of the information that organisms accumulate about the environment is meta-information, information about information in the environment. Whenever a seed assays available moisture and postpones germination as a result, the plant is employing this kind of accumulated meta-information. In less philosophical language, it learns.

Learning is a kind of phenotypic plasticity, a condition in which information in the genome teaches the organism to respond to information in the environment. Instead of natural selection building in a direct description of the environment, there is instead meta-information about variation in the environment. This meta-information might be an array of phenotypes that are triggered as the organism receives information from the environment, during its development. This is what the seed does, when it "decides" whether or not to germinate. The meta-information can also be an exploration strategy, however, so that instead of the organism's genome containing information about, say, the location of a water source, the genome contains information about how to find a water source. This information is still relevant to only some environments, and is therefore information about these environments, because different exploration strategies – different ways of learning – are better or worse under different conditions.

These are two kinds of information about the environment: information built into the genome by generations of natural selection and information we acquire during our individual lifetimes, interpreted in light of information in the genome. But there is also an important third sort of information that some species make use of. Much of the knowledge that most farmers employ to manage wheat accumulated over many generations, but it is not contained in anyone's genome, at least not in any simple sense. Instead, farmers inherit each generation the accumulated culture of farming. This information is fit to the environment, just as other human traditions can exhibit amazing adaptation. But no individual in the course of his or her lifetime could accumulate it. Instead, it has taken many generations to develop, in a way similar to how information over many generations accumulated in wheat's genome. In the case of wheat, humans built the information directly into the plant. In the case of other elements of culture, humans built the information into human brains, and later books and other forms of storage that human brains can access. This information is also often meta-information, providing strategies for solving specific problems as well as strategies for learning in itself.

In this chapter, my aim is to provide an introduction to, and an example of, theory development within the evolution of this kind of accumulated cultural information and the genetic information that makes its accumulation possible. Unlike other apes, humans rely upon accumulated bodies of adaptive information – culture – that do not reside in the genome, but nevertheless do depend upon information in the genome for their continuity and pattern of evolution. In order to understand why humans are the only ape to have crossed this "gap" and become so committed to and dependent upon socially transmitted complex adaptations, we must understand both the genetical origins of the psychology that makes cultural evolution possible – the information-about-information that resides in our genomes – and the behavioral origins of accumulated socially transmitted information-about-information culture.

How did cultural evolution evolve? Our goal is to ultimately understand why the human genome, contrasted with those of other apes, has learned to learn about the environment in a way that generates complex behavioral adaptations that rival the complexity of those produced directly by natural selection. Addressing this question brings up a number of puzzles. I focus on two closely related concepts.

Accumulated culture is a poor guide to the origins of accumulation. First, complex cultural adaptations like boats and agriculture appear obviously worth learning. However, when cultural abilities were first evolving, these fancy, accumulated bodies of information must not have yet existed. Explaining the origins of domesticated wheat cannot reference contemporary highly domesticated varieties, but must instead reference wild varieties with their inconvenient wild characteristics. Similarly, we cannot explain the origin of human culture with reference to contemporary cultural adaptations. The information that has to be built into the genome, in order for a child or adult to acquire complex accumulated culture, would seem to pre-require the complex culture. How can culture get started, if there is little culture yet to acquire (Boyd and Richerson 1996)?

Culture makes humans evolvable, but that is not why culture evolved. Second, while culture makes human societies highly evolvable – they can quickly (in genetic time) generate complex local adaptations to a large number of novel circumstances – this evolvability is not easy to understand as the original reason for the evolution of cultural learning. Quite simply, the first cultural learners would not have benefitted from the evolvability. The long-term population-level benefits of complex accumulated culture seem obvious. Once fancy human culture existed, it allowed us to adapt to every terrestrial environment (except Antarctica) and accumulate powerful bodies of information like quantum mechanics. The fitness of any contemporary human has been buoyed by many past generations of accumulated extraction of information from the environment and through transmission and curation of this information. In this way, cumulative culture generates important and powerful group advantages. But these group benefits are not easily understood as the reasons for humans' evolving cultural abilities. Instead, evolutionary ecologists would rather attempt to explain the origins of cultural capacities with individual benefits accruing to individual learners. Thus, instead of marveling at the adaptedness of accumulated human culture, our task is to understand how psychological abilities driven by individual selective advantage can build group benefits as a by-product (Boyd and Richerson 1985).

The approach I take in this chapter is to first review the gene–culture or dual-inheritance approach to human evolution. I present relevant preexisting theory on the problem of getting cumulative culture started. I use this theory then to introduce a new model that considers the simultaneous coevolution of all three kinds of information transfer that I have outlined here: (1) the incorporation of meta-information – information about how to use information adaptively – into the genome through natural selection, (2) the adaptive use of direct environmental information through individual learning, and (3) the accumulation and transmission of environmental information across generations, outside of the genome. I will show that allowing for all three of these dynamics simultaneously illuminates one potential path across the cultural "gap," whereby cumulative culture can get started, despite our first puzzle above. In the process, however, the evolutionary dynamics seem to cover their own tracks, hiding the initial changes that make the crossing possible. Finally, by allowing social learning to enhance innovation in ways other than merely allowing one to start where others left off, the model can produce levels of behavioral adaptation much greater than are possible if innovation and cumulative social learning are considered orthogonal psychological abilities.

21.2 The Gene–Culture Framework

One way to view the development of the body of theory known as gene–culture or dual-inheritance theory (Cavalli-Sforza and Feldman 1981; Boyd and Richerson 1985; Durham 1991) is to think of it as a revolutionary idea that suggests that human evolution cannot be understood in the same purely genetical way that works for other organisms. One could view it this way, but I do not.

In evolutionary ecology, there is a long and successful tradition of modeling organic evolution as an interaction among information contained in genes, the structure of the population, and the state of the environment. In this framework, the only heritable variation is contained in genes, and so evolution is described as a change in this variation over time.

Or is it? In the simplest models, it is true that the only evolving aspects of the population are genes. But in slightly more complex models that consider population structure, gene frequencies cease to be the only evolving information. Now the distribution of genes, age and sex structure, and local population densities can all evolve and exert very strong influences on the future changes in gene frequencies. These systems cannot be reduced to gene-only descriptions – additional information about the population and environment is needed to understand and predict change. Applied mathematicians are keenly aware of this fact, because we must define "state variables" for each of these evolving bits of information.

Routinely, genes are not the only state variables, even in culture-free models. Gene–culture models, therefore, dwell well within this successful tradition in population biology. There is not necessarily anything unusual about theorizing and modeling the idea that extra-genetic information is required to adequately describe a population, even when our focus is on genetical evolution.

If gene–culture models are special in any way, it is that the extra-genetic information directly influences individual phenotype in a way very similar to the action of genes. The basic issue is to identify the minimal requirements for representing evolution of phenotype in a species. For example, we could construct a very simple genetic model in which the change (Δ) in the frequency of an allele, p, is a function of environmental state, E. This system would have a single recursion:

$$\Delta p = F(p, E),$$

where the function $F(p, E)$ is to be specified depending upon what model of adaptation to the environment we might choose. It might be that E has little effect on individuals with different alleles, or it might be that E favors one over the others. It might be that E is fluctuating, so that selection favors different alleles at different times. The change might depend upon p itself, as it does in the example of sickle-cell anemia and other cases of overdominance. But nowhere do we allow in such a system for E itself to evolve in response to p.

The scientific question is whether such models are sufficient to model the evolution of a given organism's phenotype. For example, some moths imprint on the plants they graze upon as larvae. When they are adults, they will seek out these same plants to lay their eggs. The information about which plant to seek is not coded in the genome, even though the strategy for imprinting is. In this case, if we only know genotypes and the state of the environment, we could not predict the behavior of organisms in the next time period. We also need to know the distribution of imprinted memories among moths. In such a case, we need at least one more equation:

$$\Delta p = F(p, q, E),$$
$$\Delta q = G(p, q, E),$$

where q is the frequency of some learned variant (an imprinted plant, say), and $G(p, q, E)$ a function telling us how learning responds to environment, E, and its own previous state, q, and the frequency of an allele, p.

This all sounds rather complex. And it can be. However, when important parts of a phenotype are acquired during development and depend upon previous phenotypes, some system like this is useful for understanding how the organism evolves. Unless we think the existing behavior could be predicted solely from knowing the environment and the distribution of genes, at some point evolutionary models may incorporate the dynamics of behavioral inheritance. No heroic assumptions are required for behavioral inheritance to exist: if portions of phenotype depend upon the phenotypes of other individuals, then weak or strong inheritance of behavior can exist. In the long run, in a given model, it might turn out that behavioral dynamics have little effect on the outcome. In others, it will make a huge difference.

Cultural evolutionary models (as well as niche construction models, see Odling-Smee et al. 2003) can model just the nongenetic behavioral dynamics, as if q above did not depend upon p, as well as joint dynamics of a coupled gene–culture system (Cavalli-Sforza and Feldman 1981; Boyd and Richerson 1985; Durham 1991). In each case, however, the structure of the model is decided by the question of interest. There are no global models that encompass all questions about evolution. This is why we call them "models." While a few axiomatic mathematical theories do exist in biology – the Price equation being the most famous – for the most part, formal evolutionary models are attempts to understand the consequences of assumptions and to explore the sufficient or necessary conditions for a given outcome. With such models, we can study which kinds of strategies and population structures can possibly produce a phenomenon – these are possible sufficient conditions. We can also study which assumptions can be omitted, still producing a phenomenon – these are then not necessary conditions.

The demonstrable success of the modeling strategy across the sciences recommends it well. While most of the work in evolutionary anthropology and evolutionary psychology is concerned with more proximate phenomena than the population dynamics of joint gene–culture systems, understanding both the details of the psychological differences between humans and other apes and the different population dynamics of human and ape societies will be necessary, before we have a satisfactory set of answers for how humans evolved, both in relation to other primates and broader trends in animal societies.

21.2.1 What Does It Mean to Say that Culture Is an Inheritance System?

While evolutionary principles are equally applicable to almost any dynamical system, many researchers approach models of cultural transmission and evolution via an analogy with genetical evolution. Analogies are often useful, but can disguise important differences. This has rightly led some to be concerned about the strength of the gene–culture analogy (Sperber 2000). If cultural variants are not discrete, are prone to "mutation," and are strongly affected by learning biases, then is it useful to speak of "transmission" of culture at all?

While I have no particular attachment to the term "transmission," the answer is definitively "yes." Even if all the above is true, culture can still be an evolving system that leads to cumulative adaptation. This does not mean that evolved psychology has no role to play in how culture evolves (on the contrary, psychology has a huge role to play in understanding culture), but it does mean that dismissing cultural evolution on the basis of imperfection of the genetic analogy is unwarranted.

Many people – enthusiasts of the "meme" approach and critics alike – seem to have been persuaded by Richard Dawkins' abstract statements on what is required for adaptive evolution to occur. In *The Extended Phenotype* (1982), he argued that any successfully replicating entity must exhibit (1) longevity, (2) fecundity, and (3) fidelity. The entity must last long enough (longevity) to make copies of itself (fecundity) that are reasonably similar to it (fidelity). Some have interpreted this to mean that anything with high mutation rates cannot be a successful replicator. Thus, if cultural ideas change in the process of social learning, the conclusion is that they do not constitute an evolving system at all (see citations in Henrich and Boyd 2002). Similarly, if cultural variants are continuous and blended entities, then they never exactly replicate, and again cannot produce adaptive evolution.

These conclusions are unfounded. Read very generally, Dawkins' conditions are necessary and sufficient – there must be some heritability for adaptive evolution to occur. However, there are many ways to produce heritable variation. So in the strict sense many people have read them, while Dawkins' conditions are sufficient, they are definitely not necessary. Reverse-engineering DNA may tell us how inheritance *can* work, but it does not tell us how it *must* work. Henrich and Boyd (2002) examine the problems with this reverse-engineering in greater depth (see also Henrich et al. 2008).

It is understandable that there is confusion about what is needed for adaptive evolution – even textbooks are confused. Before the union of genetics and Darwinism, most biologists, including Darwin, thought that inheritance was a blending process: offspring were a mix of parental phenotypes. Darwin was troubled by Jenkin's (1867) argument that natural selection could not produce adaptations, because inheritance would quickly deplete the variation that natural selection relies upon. Fisher's (1918) argument reconciling genetics with continuous phenotypic variation purportedly rescued Darwin. Many textbooks repeat this version of the history, reinforcing the notion that low-error discrete entities like genes is a necessary condition for adaptive evolution. But in reality both Jenkin's argument and those who think Fisher saved Darwin are simply wrong: blending inheritance can preserve variation, and particulate inheritance is neither necessary nor sufficient to preserve variation. Maynard Smith (1998) has a chapter that examines this problem, still not entirely resolved within modern population genetics (see also Barton and Keightley 2002).

21.2.2 Is Extra-Genetic "Inheritance" Common?

I think this history lesson teaches us that our verbal models of even genetical evolution are sorely lacking. If so, then, our informal appreciation of nongenetic influences on behavior is also a suspect. In many baboons, females inherit dominance rank from their mothers and sisters (Silk and Boyd 1983). In these species, fitness is strongly affected by this extra-genetic inheritance: any female adopted at birth into a high-ranking matriline would be better off than if she were adopted into a low-ranking matriline. And this female will have her dominance rank before she fights a single member of her social group. Dominance is heritable, has important effects on fitness, and yet the mechanism of inheritance is at least partly nongenetic. The rules of how this inheritance works are complicated and very unlike genes. It probably depends upon the composition of one's own matriline, the composition of the entire social group, and local resource density and feeding competition. And yet no primatologist could completely understand baboon biology without taking this complicated extra-genetic pedigree into account. Its existence may lead females to strive for rank because of its downstream consequences, in addition to its immediate resource access effects (Boyd 1982; Leimar 1996, Watts, this volume).

Extra- or "epigenetic" (Maynard Smith 1990) systems like this are increasingly recognized: everywhere biologists look, they find hints of inheritance systems either built on top of genes or built from entirely different mechanisms. If the key question is what mechanisms account for heritable phenotypic differences among organisms, then the answer appears to be "many." Jablonka and Lamb's *Evolution in Four Dimensions* (2005) mounts the empirically rich argument that heritable differences in many species are due to the action of several inheritance systems (genetic, epigenetic, behavioral, and symbolic), sometimes interacting, sometimes acting in parallel.

If one thinks about cell division for a moment, it is obvious that processes other than the replication of DNA are needed to explain how it works. Organelles need to be copied (Sheahan et al. 2004), and the genetic code itself needs to be copied (and this is not contained in the DNA, nor could it be). Beyond cell division, adult phenotypes depend upon imprinting and other forms of learning that may channel the environments offspring are exposed to (a kind of niche construction – Odling-Smee et al. 2003). And finally, most biologists believe that DNA was certainly not the first form of hereditary biological material (Szathmáry and Maynard Smith 1995). Thus, some inheritance systems must be able to sometimes create complementary and even usurping inheritance systems.

In light of these plausible "inheritance systems," it appears that human culture may not be so special or surprising at all, in the sense of being a nongenetic system of inheritance. Organisms as diverse as *Arabidopsis* (a small plant related to mustard that is a favorite of geneticists), common fruit flies, and single-celled microscopic animals, such as paramecia, exhibit heritable differences due at least in part to mechanisms other than the sequence of nucleotides in their DNA. The existence of social learning as a system of inheritance and adaptation that functions in complement to DNA may turn out to be unremarkable.

There will always be aspects of human behavior and evolution that can be usefully modeled as culture-free. These are, after all, just models: all of them are wrong, but some are more useful than others. To someone who makes formal models of evolutionary systems, the question that we must answer is what qualitatively different phenomena we miss when we represent human (or any other organism's) evolution with just state variables for its alleles. If we sometimes require state variables for early childhood experience, imprinting, or behaviors acquired via social learning, to make useful models of our own evolution, then attempts to construct culture-free models are simply scientifically inadequate. As with each of the possible systems above (e.g., Maynard Smith 1990; Jablonka and Lamb 1991; Pál and Miklós 1999), the specific dynamics and consequences of cultural learning may be rather unique and very important for understanding both micro- and macroevolution.

21.3 Simultaneous Evolution of Innovation and Social Learning

The literature on gene–culture evolution often presents individual and social learning as alternatives. At some point in its lifetime, an organism is forced to choose between relying upon individual experience or socially acquired information. This dichotomy obviously does not imply that social learning is free of inference or unguided by individually learned theory. But evolutionary models almost always engage the strategic and population levels, not the psychological. At the level of abstraction of our models, these psychological platitudes are granted, and different strategic uses of information are the focus.

So while evolutionary models do treat individual learning and social learning as strategic alternatives, I know of no theory that treats them as orthogonal influences on behavior. Even in the simplest sort of model, individual and social learning interact at the population level and across generations (Rogers 1988). These interactions are antagonistic, in some cases. Social learning parasitizes innovation, in the same way that students who cheat on tests parasitize (or attempt to parasitize) the students who study. To understand the pattern of behavior that emerges from these models, we have to appreciate that individuals are relying differentially upon both kinds of information.

Consider the genetic analogy again. Evolutionary theorists routinely speak of mutation and selection as separate "forces." Each is governed in part by unique parameters, within an evolutionary model. And yet no one charges theorists with assuming that mutation and selection do not interact. Indeed, there is still active theory development concerning the interaction of mutation and selection in determining patterns of genetic variation (Barton and Keightley 2002).

Similarly, gene–culture models have considered how individual and social learning interact, over an organism's lifetime, as well as over evolutionary time. Boyd and Richerson (1985) devoted considerable space to the adaptive interaction of individually and socially acquired information, *guided variation*. There are other, more subtle kinds of interactions. An adaptive synergy between individual innovation and social learning can arise from conditional effort in innovation (Boyd and Richerson 1996). Suppose that an organism can assess the efficacy of the information it acquires via social learning. The assessment can function through an initial trial or from analysis. If the organism judges that the behavior it acquired socially is below some threshold of efficacy, the organism has the option of then spending additional time innovating.

A second kind of adaptive synergy is when information gathered through individual learning can be build onto information gathered from social learning. When this is possible, the two kinds of learning interact to produce accumulations of information across generations (Boyd and Richerson 1996). There has been much less theory developed in the cumulative culture case. Previous models demonstrate that there may be a fitness valley between simple, inaccurate social learning that cannot accumulate complex behavior and more accurate social learning that can. Boyd and Richerson (1996) developed both a discrete behavior and truly continuous behavior versions of their model. In both cases, there were values of the parameters for which cumulative cultural learning could not invade when rare but was stable when common. However, neither of these models allowed individual learning to co-evolve with social learning. Innovation rate was a fixed parameter, and this lack of feedback hid some interesting dynamics.

The model in the remainder of this chapter treats the simultaneous dynamics of innovation – an organism's investment in acquiring new adaptive information directly from the relevant environment – and cumulative social learning – and organism's investment in acquiring complex information from other individuals.

21.3.1 What Is Learning For?

Levins (1968) produced what is probably the first broad formal analysis of the evolution of simple phenotypic plasticity. He asked us to imagine a large number of different environmental states which may vary through time and across space. Suppose there is a unique allele that is optimal in each state. Now imagine an alternative strategy that, instead of providing a fixed phenotype, assesses the environment and attempts to change phenotype to match it. Levins extracted several general conclusions from analyzing models of this type.

1. Both spatial and temporal variation reduce fitness, compared to a uniform, static environment. Call this fitness cost the "naive cost." It is what an unlearning, unresponsive organism pays, in the presence of environmental change.
2. An optimal strategy that either bet-hedges against change or learns can be found for any pattern of spatial/temporal change, but this optimal strategy cannot remove the entire naive cost.
3. A learning strategy itself imposes a cost on the organism, either because of sampling effort or processing costs. Call this cost the "learning penalty." Therefore in order for learning to evolve, it must reduce the naive cost by an amount greater than the learning penalty.

These generalizations have remaining quite robust to the specific forms of models.

Spatial and temporal variation do generate different results, however. Germane to the arguments in this chapter, when the environment varies through time, natural selection will favor a bet-hedging strategy that maximizes the geometric mean fitness of the organism, over the environmental states it experiences. Temporal variation does not maintain polymorphic fixed phenotypes, in the absence of learning. Spatial variation, on the other hand, may favor polymorphism in a species, depending upon the details.

The rest of this chapter builds a model that allows for both investments in individual exploration and, later, the ability to copy behavior and strategy from other individuals. In this section, I'll construct the individual learning and innovation core of the model. I'll analyze this core, before laying on social learning and analyzing the simultaneous dynamics of both cumulative social learning and innovation. By presenting the model in this way, I hope to lose fewer readers and better explain how cultural learning alters the outcomes.

Consider the evolution of simple learning, a form of phenotypic plasticity that uses information in the environment during an individual's development to alter phenotype, whether it is morphology or behavior. The evolution of the strategy of learning incorporates information into the genome that biases learning to be adaptive, rather than self-destructive.

Let's represent an individual's "genotype" with $d > 0$. The notion is that a number of regulatory and other genes combine to produce this continuous genotype that influences the amount of exploration and innovation. Individuals search and gather and process information about the environment so that each gains $q = d$ units of adaptive knowledge, in the current environment. These units translate into fitness, π (q), with diminishing returns. I have explored a number of specific diminishing returns functions, but found all of them to produce the same qualitatively behavior. The easiest to analyze is:

$$\pi(q) = \frac{bq}{\beta + q} = \frac{bd}{\beta + d},$$

where $b > 0$ determines the strength of selection, and $\beta > 0$ is a parameter that determines the rate at which marginal fitness benefits decline. When β is large, $\pi(q_i)$ is approximately linear, and there are no diminishing returns to knowledge. When instead β is small, the fitness benefits of increasing knowledge diminish rapidly. Because of the specific form of the function π, β turns out to be exactly the value of q_i that produces half of the maximum value of π, $b/2$. That is, $\pi(\beta) = b/2$.

Suppose there are a very large number of environmental states. In each state the environment could take, different phenotypes are favored. Each generation, there is a chance u that the environment changes to another random state. Since the variation here is stochastic, in the absence of phenotypic plasticity, a bet-hedging strategy will evolve that pays Levins' naive cost of variability. Let fitness after paying this naive cost be w_0.

Individuals can do better than this baseline, by attempting to learn the current state of the environment and use information from it to reduce the naive cost of variation. An individual who invests d in learning pays a cost cd, the learning penalty. Since investment in learning is continuous, this cost scales with it. As the fitness benefits of environmental knowledge, q, have diminishing returns, eventually the marginal benefits and costs of learning equal. At this point, selection will favor no further investments in plasticity.

These assumptions give us the following fitness, for an individual with genotype d:

$$w(d) = w_0 + \pi(d) - dc.$$

We find the evolutionarily stable investment in individual learning, $\hat{d}$, by solving $\partial w/\partial d = 0$ for d. This yields:

$$\hat{d} = \sqrt{\beta b/c} - \beta. \tag{21.1}$$

This is greater than zero, provided $b/c > \beta$, in which case selection favors learning. If this condition is not met, however, selection favors instead the bet-hedging fixed strategy that suffers the full naive cost of variation.

Are we ready yet to answer the question: what is learning for? According to this model, learning allows an organism to recoup fitness lost to temporal environmental variation. Note that I have assumed so far that this is an entirely asocial process. Fitness is not frequency dependent and there is no learning from conspecifics. In the next section, however, I add the possibility of social learning to the model. Then learning can be for building complex adaptations that fit the environment beyond the amount q.

21.3.2 Adding Cumulative Social Learning

Many organisms are capable of phenotype plasticity. All primates – and indeed all mammals – are capable of individual learning of the kind modeled above. In novel circumstances, animals employ search strategies that may allow them to adaptively exploit new environments. One of the best understood of these is foraging in rats – rats explore trash eagerly, but sample in small amounts and remember and avoid foods that make them ill (Galef 1996). This strategy allows rats to exploit varied urban environments, but it is possible because information about how to search for and use relevant information has been built into their genome by natural selection.

Animals also sometimes exhibit specialized adaptations for using conspecifics as cues of adaptive behavior. When Norway rats smell food on the muzzle of another rat, they are more likely to eat that same food (Galef 1996). Information in the rat's genome makes this possible, by directing attention to odors on conspecifics and enhancing memory of socially-encountered foods.

In humans, the motivations and psychological adaptations that we might call "social learning" involve symbolic communication, abstraction, and substantial individual practice. Speech is a good model – while substantial social input is necessary for any human to learn the speech patterns of his or her community, a lot of individual practice with sounds is needed, because the inputs (sounds) are quite different than the information that an individual eventually needs to encode in order to produce them (motor memory). Every individual has a differently shaped vocal tract, and so in order to "imitate" another speaker, all of us had to experiment with sound production. Likewise, acquiring a complex skill like hunting or agriculture may require years of instruction and practice. Readers who have learned to play a musical instrument may find it to be a rich source of intuitions about the assumptions of this model. Playing the cello takes many years of individual practice, but this practice is much more effective when guided by a master cellist. A lone cellist may eventually attain the skill of a master, after many years of individual effort, but it is much easier to match or surpass the master, if the master provides instruction or simply allows observation. The purely "social" component of social transmission may be quite small, in terms of the time it occupies. But very little transmission, if any, is possible without the social component.

Begin with the model of individual innovation presented above. Assume now that there is another set of loci that influence an individual's ability and motivation to learn socially. The "genotype" at those loci is represented by s, and an individual with $s > 0$ can successfully copy a fraction s of the adaptive behavior displayed by an adult from the previous generation. In order to separate innovation and social learning, I restrict $s < 1$, such that social learning will never accidently generate behavior that is more adaptive that what was observed. Investments that increase s may be attentional – improvements in studying and representing the behavior of other individuals – or motivational – increases in the extent to which goals and ways of achieving goals are open to social input. In both cases, greater investments in time or ability to acquire complex behavior from others results in the eventual acquisition of a larger portion of previously innovated behavior. If s is large enough, innovations generated over several generations may accumulate, generating behavior more complex than any individual innovation could in a single lifetime. If s remains low, however, then no amount of innovation will result in these complex behaviors, because each generation has to re-invent too much.

Increasing the accuracy of social learning is, however, costly. It costs the learner energy for upkeep and use of the psychology that makes social learning possible, and it costs the learner time in observation, practice, and missed opportunities to enhance fitness in other activities. I represent the total cost of social learning of this kind by ks. The more an individual invests in accurate copying of information, the more the individual pays.

With these assumptions, we can define a new fitness expression, now for a rare (mutant) individual investing d in innovation and s in social learning, in a population in which the common type invests d^* and s^* in each, respectively.

$$w(d, s, d^*, s^*) = w_0 + \pi(q_t) - dc - ks,$$

where q_t is the individual's behavioral phenotype, after both social transmission and innovation. Because learned information can be maintained across generations now, q will depend upon the amount of accumulated adaptive behavior in the population. This in turn depends upon the common phenotype, d^*, s^*, and the rate at which the environment changes and renders previously innovated behavior non-adaptive. The correct expression for q_t is:

$$q_t = (1 - u_t)sq' + d,$$

where u_t is a random variable taking the value 1 or 0, depending upon whether the environment changed last generation (with probability u) or not (probability $1 - u$), respectively. The symbol q' defines a recursion for the dynamics of behavior that is transmitted across generations. The behavior available to learn socially depends upon the common genotype, not that of the individual whose fitness we are modeling. The dynamics of behavior from one generation to the next are defined by:

$$q' = (1 - u_t)s^*q + d^*,$$

where q above is the average behavioral phenotype in the previous generation. Because u_t and u_{t-1} are random variables, there is no equilibrium amount of adaptive behavior in the population. Instead, q is reset to zero after each change in the environment and then begin climbing until the next change. One could assume instead that a proportion of adaptive information is retained across changes in the environment, but all this does is reset q to some minimum, rather than zero. There still will never be a stable value of q across generations. To cope with this kind of stochastic system, we solve for the mean of the stationary distribution of q. While there is no equilibrium, in a linear system like this one, the distribution of q across generations will eventually settle down. This is the system's stationary distribution. We can compute the mean of the stationary distribution, by taking expectations across generations and solving for $\hat{q}$, the mean of the stationary distribution. Doing this yields:

$$\hat{q} = \frac{d^*}{1 - s^*(1 - u)}.$$

This expression tells us that the mean level of adaptive behavior increases with increases in d^* and s^*, but decreases as u increases. It is helpful to consider some limiting cases. Suppose for example that the population has yet to evolve any effective social learning, $s^* = 0$. Then the average level of adaptive behavior will be $\hat{q} = d^*$. No adaptive behavior accumulates beyond what individuals can learn for themselves. Now suppose instead that $s^* = 1$. Now $\hat{q} = d/u$ – if u is small enough, substantial adaptive behavior will accumulate, because social learning is very (unrealistically) accurate.

Of course d^* and s^* are evolving genotypes. In order to analyze the simultaneous dynamics of innovation and social learning, we need to substitute $\hat{q}$ into the fitness expression:

$$\begin{aligned}\hat{w}(d, s, d^*, s^*) &= w_0 + \pi((1-u)s\hat{q} + d) - dc - ks,\\ &= w_0 + \pi\left(\frac{(1-u)sd^*}{1-s^*(1-u)} + d\right) - dc - ks.\end{aligned}$$

Note that the adaptive behavior available for the mutant individual to acquire depends upon the population genotypes d^* and s^*, while the accuracy of her own social learning and power of her own innovation depend upon the individual genotypes d and s. In this way, the invading genotype plays against the population in game theoretic fashion. Our goal is to find the values of d^* and s^* that cannot be invaded by any other values d and s, respectively.

21.3.3 Joint Dynamics of Innovation and Cumulative Social Learning

Before deriving the un-invadable values of innovation and social learning, it is useful to summarize the combined, two-dimensional, dynamics of this model. This system can evolve to two qualitatively different outcomes. First, social learning may increase when rare and evolve until its theoretical maximum. Second, social learning may be unable to invade when rare. Which of these two outcomes is realized depends upon the amount of innovation favored, when social learning is rare. If innovation is cheap, for example, then enough of it might be favored when social learning is absent. Social learning will then increase from $s^* = 0$, because there is complex information in the population worth copying. Once social learning begins to increase, however, selection favors less innovation, because of the diminishing fitness returns on knowledge. Eventually innovation may fall to the same level it was at, before social learning invaded. However, social learning remains high in the population. Once social learning can get a start from initially high innovation levels, it can invade.

Any potential evolutionarily stable values of d^* and s^* are found where $\partial w/\partial d|_{d,s = d^*,s^*} = 0$ and $\partial w/\partial s|_{d,s = d^*,s^*} = 0$. Call the evolutionarily stable values

$\hat{d}$ and $\hat{s}$. Setting $d^* = \hat{d}$, $s^* = \hat{s}$ and solving these equations for $\hat{d}$ and $\hat{s}$ yields one possible equilibrium for d^*:

$$\hat{d} = \left(\sqrt{\beta b/c} - \beta\right)(1 - \hat{s}(1 - u)). \tag{21.2}$$

Again note that innovation does not always evolve. If $\beta > b/c$, the expression above is negative and no innovation is favored by natural selection. The possibility of social learning, however, affects the stable amount of innovation. As $\hat{s}$ increases, $\hat{d}$ decreases. If the environmental rate of change u is large, however, then the effect of social learning on reducing $\hat{d}$ is reduced.

Instead of having an equilibrium value, s^* can either decrease or increase until it reaches zero or one (or another theoretical upper limit). That is, $\hat{s} = 1$ or $\hat{s} = 0$, depending upon the parameters. The condition for social learning to increase from zero, and so invade a population, is given by $\partial w/\partial s|_{d=\hat{d},s=s^*} 0$. This reduces to:

$$u < 1 - \frac{k}{\sqrt{b\beta c} - \beta c}.$$

If the environment changes too quickly, social learning is never favored. But if the marginal cost, k, of social learning is low enough and fitness benefits of behavior do not diminish too rapidly, then social learning will invade and increase until its theoretical limit.

These expressions do not immediately reveal what is happening, however. It is easier to understand the behavior of this model, by visualizing the joint evolution of innovation and cumulative social learning. Figure 21.1 shows the phase diagram of this model, for two different sets of parameter values. In each plot, position along the horizontal axis represents the value of d^*, from zero to one. Position along the vertical axis represents s^*, also from zero to one. Arrows represent the direction and magnitude of change for the system, at each point. The point in each plot is the eventually evolutionarily stable combination of innovation and social learning, in each case. On the left, the cost of innovation is set high, but not so high as to prevent individual learning from evolving at all. The high cost, however, does prevent d^* from ever evolving to high enough values to provide enough adaptive behavior to be worth investing in accurate social learning. Therefore, where ever the system begins, selection will eventually reduce social learning to its minimum. Cumulative culture does not evolve in this case, although rather fancy behavior is invented each generation, because of the non-zero equilibrium value of d^*.

On the right, the cost of innovation is slightly reduced. Suppose the system begins in the lower-left corner, at d^*, $s^* = 0$. Now innovation can increase to a higher level than on the left, before the arrows turn the other way and selection no longer favors any increases in innovation. Innovation can reach a high enough level, in fact, that the behavior that is invented each generation is now worth copying through investments in accurate social learning. Therefore the system evolves

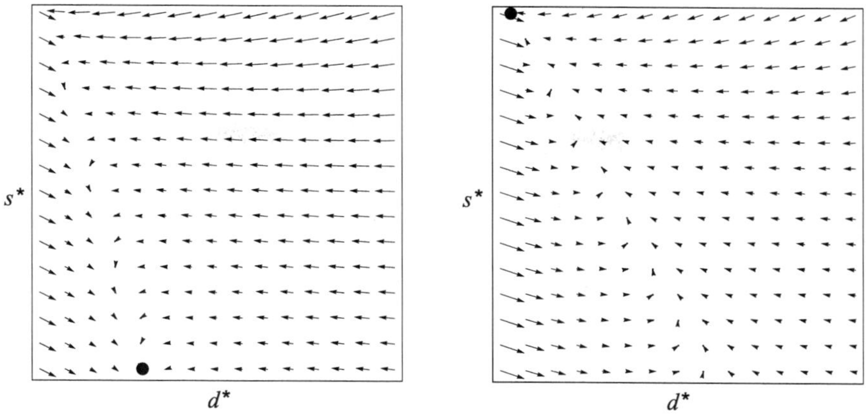

Fig. 21.1 Evolutionary dynamics of innovation and cumulative social learning. *Arrows* show the direction and magnitude of evolutionary change at each point in the possible state space of the population, defined by the average investment in innovation, d^*, and the average investment in social learning, s^*. *Black dots* in each panel show the only stable equilibrium in each case. In both panels, $b = 5$, $u = 0.05$, $\beta = 1$, $k = 1$. *Left panel*: $c = 3$. *Right panel*: $c = 2$. On the left, higher costs of innovation prevent individually learned information from reaching high enough levels for natural selection to favor cumulative social learning. Without much behavior worth copying, the system remains at a high level of innovation, but no imitation evolves. On the right, a slightly reduced cost of innovation leads initially to a higher investment in individual learning, a higher level of individually acquired behavior, and eventually to the invasion of social learning. As social learning increases, however, natural selection favors reduced investments in innovation, because of the diminishing fitness returns to knowledge. This system comes to rest where innovation is lower than the panel on the left, but social learning is highly accurate

towards the interior, favoring increasing amounts of social learning as it heads for the top of the figure. As selection favors social learning, however, it also favors less innovation (Expression 2). Thus the eventual equilibrium has highly accurate social learning ($\hat{s}$ is near one), but lower levels of innovation than the plot on the left. The behavior invented each generation is modest in comparison to the population in which social learning did not evolve. However, the mean level of adaptive behavior is twice as large. On the left, $\hat{q} = \hat{d} = 0.29$. On the right, $\hat{q} = 20$, $\hat{d} = 20$ $(0.029) = 0.58$.

21.3.4 How Much Cumulative Culture?

There is an irony lurking within the solution above, however. While the evolution of social learning appears to have resulted in higher levels of adaptive behavior, "culture," $\hat{q}$, in reality social learning has only provided a cheaper way to attain the same amount of adaptive behavior the population would have enjoyed, if it had relied entirely upon high levels of innovation. This is obvious, once we inspect the expression for stationary mean $\hat{q}$, at $\hat{d}$ $\hat{s}$. Let $\hat{\hat{q}}$ be the value of $\hat{q}$, evaluated at $d^* = \hat{d}$ $s^* = \hat{s}$. Then the expression for this expected average level of adaptive behavior is:

$$\hat{\hat{q}} = \frac{\sqrt{b\beta c} - \beta c}{c} = \sqrt{\beta b/c} - \beta.$$

Note that this expression does not contain $\hat{s}$. Therefore, it does not depend upon social learning at all. Furthermore, it is the same amount of adaptive behavior we would expect from the a-cultural model presented earlier (Expression 1)! Cumulative social learning has evolved, but it has failed in this model to produce long-term information gains beyond what would already have been possible using (highly advanced) innovation.

What is happening in the evolutionary economics is that lower costs of innovation allow behavior to reach a threshold that then allows social learning to invade. Once social learning invades, selection favors less innovation. Because behavior has diminishing returns, individuals do better by investing in an optimal mix of innovation and social learning. This optimal mix trades off the costs of innovation against the potentially unreliable benefits of socially learned behavior. Because individual benefit is driving the evolution of both innovation and cumulative culture in this model, selection does not necessarily maximize the group benefits of cumulative culture.

One way out of this unsatisfactory result is to note that we have only modeled a single domain of behavior. Social learning ability will be applied potentially to other domains with much lower relevant rates of environmental change. Consider that bows and arrows continue to function, even when climate changes substantially. Therefore different technologies and strategies experience different rates of change (u in the model). If fitness gains from more-slowly changing domains are important enough, then social learning will be pulled up to a higher level of accuracy, even in fast-changing domains, than would be optimal, if we consider those domains alone.

But this is a hand-waving argument. Are there other theoretical solutions that do not invoke large numbers of parameters that are poorly understood and potentially unmeasurable?

21.3.5 When Social Learning Enhances Innovation

How can we get selection to increase adaptive behavior beyond this selfish optimum? One way is by allowing social learning to improve the efficiency of innovation. This hypothesis is reasonable, if you believe that the psychological abilities that make cultural transmission possible also enhance an individual's ability to represent, remember, and explore new solutions. For example, language is a symbolic capacity that allows us to represent abstract systems, much like the model in this chapter. While language makes it possible to acquire complex behavior from other people, it also makes it possible to organize and transform information within one's own head. The dialog scientists carry on with themselves – sometimes out loud – suggests that at least some abilities that are possibly selected for enhancing social learning can simultaneously enhance imagination and innovation. If one of the things social learning does to human cognition is provide a quite open motivational and association system, so that we can remember arbitrary scripts and develop novel goals through communication, then any energies given to innovation may be able to tap these same abilities.

When we allow new synergy between social learning and innovation (so that social learning actually makes innovation cheaper), we have a new fitness expression:

$$w(d, s, d^*, s^*) = w_0 + \pi((1-u)s\hat{q} + d) - (1 - s/z)cd - ks.$$

The parameter $z > 1$ determines the amount of synergy. When $z = \infty$, the above reduces to the previous fitness expression and the result is unchanged. When z is small, however, there may be substantial cost reductions to innovation as social learning abilities increase. When $z = s$, innovation is effectively free (note that this is impossible, by the constraint that $z > 1$). The new steady state accumulated culture becomes:

$$\hat{q} = \frac{bz}{\sqrt{bcz(z-\hat{s})}} - \beta.$$

Figure 21.2 plots this expression over all possible values of $\hat{s}$, for two values of z. As $z \to \hat{s}$ from above, this quantity increases rapidly. In biological terms, as social

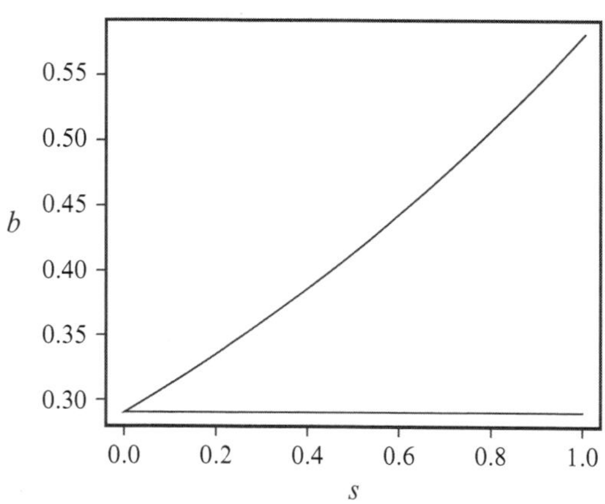

Fig. 21.2 The expected amount of adaptive behavior, Expression 3, as a function of the amount of cumulative social learning in the population, $\hat{s}$. *Horizontal line*: $b = 5$, $c = 3$, $\beta = 1$, $z = \infty$. *Sloped line*: $z = 3$. When social learning reduces the costs of innovation, the evolution of social learning leads to increases in adaptive behavior, beyond what innovation alone could provide. Otherwise, selection adjusts the amount of innovation so that the amount of adaptive behavior remains the same, whether social learning invades or not

learning increasingly makes innovation cheaper and more efficient, the steady state amount of adaptive behavior increases. In contrast, as $z \to \infty$, the above approaches the previous expression for $\hat{\hat{q}}$, $\sqrt{\beta b/c} - \beta$, resulting in no change in the amount of adaptive behavior as social learning increases.

21.4 Where Did Culture Come From?

Natural selection builds adaptive information into the genome. Learning gathers information about the environment, to be used by information in the gnome. Cumulative social learning takes information from behavior – whatever its source – and allows it to be stored and accumulated in human brains. The obvious adaptive utility of the products of this process – technologies and strategies too complex for any individual to invent in his or her own lifetime – make puzzling the gap between humans and other apes in this regard. If "culture" is such a great adaptive trick for genes to acquire, then why are not other apes similarly cultural? This question is additionally puzzling, given the evidence of at least proto-cultural social learning abilities in chimpanzees (see Whiten this volume).

The theory I have reviewed and developed in this chapter addresses the question of the origins of human cultural abilities. The first goal of the theory is to understand how natural selection on genes can fail to favor cumulative social learning and under what conditions it will lead to cultural evolution and accumulation. The second goal is to understand how the population-level adaptive benefits of this accumulation can appear, without these being the selective reasons for investments in learning.

21.4.1 Evolving Cultural Evolution

The first goal is addressed by the combined dynamics of innovation and social learning. When social learning allows accumulation and costs more and more as the complexity of what is copied or the accuracy with which it is copied increases, then a fitness value can appear between an a-cultural population and a cultural population (Boyd and Richerson 1996). If individual learning is effective enough, however, the model in this chapter suggests that it can provide a way around this valley. If selection favors improvements in innovation, independent of cumulative social learning, eventually there is complex behavior that – while not accumulated across generations – is nevertheless worth copying, because the costs of social learning are lower than those of innovation itself. Proximately, the lower costs of social learning may arise because innovation is an inherently harder activity. Many good ideas are hard to stumble upon, and much individually learned behavior takes a lifetime to assemble, despite not being a product of social learning. Once complex behavior is available, selection might favor acquiring it before any individual effort is made in innovation. Once this happens, selection trades off innovation against social learning, reducing the once-large innovation rate that was needed to cross the cultural valley. This may or may not be the right idea, to explain the gap between humans and other apes. But it points in the same direction as others have suggested (see for example Whiten this volume). Innovation and social learning are potentially co-adapted in humans, and explaining one may ultimately require an equally deep understanding of the evolution of the other.

In the end, the theory here required investments in social learning to make innovation more effective or cheaper. A number of other alternations could be made to the model, to modify the relationship between innovation and cumulative social learning. An obvious one is to allow the amount of adaptive behavior that is socially learned, rather than the efficiency of social learning itself, to enhance innovation. The idea is that previously evolved information may make future innovation easier, because it defines the relevant parts of the problem and provides tools to finding solutions. Much of how modern science works involves the development and dissemination of tools, not products. In this way, science is as much about building intellectual and technological solutions for discovery as it is about discovery itself.

Similarly, many of the social institutions and cooperative arrangements in modern societies enhance innovation. Governments actively structure patent law, so that more innovation is encouraged than would be individually optimal for firms. While patent law does not necessarily become more effective as knowledge accumulates, further enhancing innovation, there are other institutions which might. Division of labor and the exchange institutions that make it possible also enhance innovation, in two ways. First, division of labor carries with it the benefits of specialization. Economies of scale make innovation easier in each domain of behavior, and new information can be traded among specialists more easily than it can be independently discovered by all of them. Second, as culture accumulates, eventually the sum of what the population knows exceeds what any individual can learn, even with advanced social learning. The readers of this chapter are probably among the most educated people on the planet, and yet each is unlikely to be expert in more than one or two areas of science. Your author spent a decade learning to understand the intersection of anthropology and evolutionary ecology, and yet he still has little deep understanding of some branches of both anthropology and ecology. Like most scientists, he relies upon experts in other areas – combined with active skepticism and habits of thought – to keep track of relevant advances in neighboring fields. This division of labor allows knowledge in any particular domain – hunting large animals versus gathering palm fiber or processing medicinal plants versus childcare – to grow beyond the limits of individuals to learn and practice all domains.

21.4.2 Evolvability as a Side Effect

The second goal of this chapter has been to highlight the kind of theory that is required to understand the accumulation of socially-transmitted adaptive behavior, without the eventual highly-adaptive accumulations being the initial reason for the evolution of the psychology needed to make cumulative culture possible. Human societies are no doubt more adaptable than those of other apes – we have conquered (and according to many, ruined) nearly every environment on the planet, while other apes shrink in tropical refuges. Part of the explanation for our world dominance is the ability to generate complex, locally adapted behavior over generations (Richerson and Boyd 2005). Foragers in different parts of the world need quite different knowledge and strategy. The combination of innovation and social transmission makes local specialization and regional adaptability both possible.

And yet, selection does not favor costly social learning abilities, unless there is an immediate benefit to the organism. Our ancestors did not lug around brains capable of cumulative culture, because it would turn out to allow our species to dominate the planet. Instead, we have to seek short-term, individual fitness benefits in order to explain why an organism would cross the cultural gap. An acorn detects moisture when it decides whether or not to germinate, because acorns that were initially slightly sensitive to a moisture gradient produced more descendants. These descendants them had mutations that favored more sensitivity, until some rough optimum was reached. Selection favored every step, even though the eventual level of adaption was higher than the initial. Similarly, the theory in this chapter hypothesizes that cumulative social learning began as a way to avoid the costs of innovation. Especially as learned behavior becomes more complex, social learning allows an individual to rapidly acquire sensible locally-adapted behavior, saving time and energy for other activities. As each individual continues to add some continued improvement to what is learned socially, the average adaptiveness of behavior may increase over generations. However, selection favored each step along the way because of the benefits and costs at each step, not because of the population-level benefits that would eventually arise.

The specific model developed in this chapter suggests that one path to evolving cultural evolution lies in first getting selection to favor increases in innovation, as summarized just above. However, any successful theory of the evolution of evolvability must contend with this same challenge. Students of the evolution of development ("evo-devo") are fond of noting how animal body plans can make life very evolvable, over macro-evolutionary time. Developmental genes are organized in such a way as to make compartmentalized changes possible – the genome can make one set of limbs longer or even replace them with the genetic information for another specialized set (see Kirschner and Gerhart 1998). But while this source of evolutionary novelty may turn out to explain the very long term success of some groups of organisms (like bilaterally symmetric animals), it cannot be the reason the body plan arose in the first place. Higher-level selection, at the population or species level, can indeed explain the maintenance of such adaptations. A popular theory of the maintenance of sexual reproduction suggests that sex indeed makes populations more evolvable (Maynard Smith 1978).

References

Barton NH, Keightley PD (2002) Understanding quantitative genetic variation. Nat Rev Genet 3:11–21

Boyd R (1982) Density-dependent mortality and the evolution of social interactions. Anim Behav 30:972–982

Boyd R, Richerson PJ (1985) Culture and the evolutionary process. University of Chicago Press, Chicago, IL

Boyd R, Richerson PJ (1996) Why culture is common, but cultural evolution is rare. Proc Br Acad 88:77–93

Cavalli-Sforza LL, Feldman MW (1981) Cultural transmission and evolution: a quantitative approach. Princeton University Press, Princeton, NJ

Dawkins R (1982) The extended phenotype: the long reach of the gene. Oxford University Press, Oxford

Durham WH (1991) Coevolution: genes, culture, and human diversity. Stanford University Press, Stanford, CA

Fisher RA (1918) The correlation between relatives on the supposition of Mendelian inheritance. Trans Roy Soc Edinb 52:399–433

Frank SA (2009) Natural selection maximizes fisher information. J Evol Biol 22:231–244

Galef BG Jr (1996) Social enhancement of food preferences in Norway rats: a brief review. In: Heyes CM, Galef BG Jr (eds) Social learning and imitation: the roots of culture. Academic, San Diego, CA, pp 49–64

Henrich J, Boyd R (2002) On modeling cognition and culture: why cultural evolution does not require replication of representations. J Cogn Cult 2:87–112

Henrich J, Boyd R, Richerson PJ (2008) Five misunderstandings about cultural evolution. Human Nature 19(2):119–226

Jablonka E, Lamb MJ (1991) Sex chromosomes and speciation. Proc Roy Soc Lond B 243:203–208

Jablonka E, Lamb MJ (2005) Evolution in four dimensions: genetic, epigenetic, behavioral, and symbolic variation in the history of life. MIT, Cambridge, MA

Jenkin F (1867) The origin of species. North Brit Rev 46:277–318

Kirschner M, Gerhart J (1998) Evolvability. Proc Natl Acad Sci USA 95:8420–8427

Leimar O (1996) Life history analysis of the Trivers and Willard sex-ratio problem. Behav Ecol 7:316–325

Levins R (1968) Evolution in changing environments: some theoretical explorations. Princeton University Press, Princeton, NJ

Maynard Smith J (1978) The evolution of sex. Cambridge University Press, Cambridge

Maynard Smith J (1990) Models of a dual inheritance system. J Theor Biol 143:41–53

Maynard Smith J (1998) Evolutionary genetics, 2nd edn. Oxford University Press, Oxford

Odling-Smee FJ, Laland KN, Feldman MW (2003) Niche construction: the neglected process in evolution. Monographs in population biology 37. Princeton University Press, Princeton, NJ

Pál C, Miklós I (1999) Epigenetic inheritance, genetic assimilation and speciation. J Theor Biol 200:19–37

Richerson PJ, Boyd R (2005) Not by genes alone: how culture transformed human evolution. University of Chicago Press, Chicago, IL

Rogers AR (1988) Does biology constrain culture? Am Anthropol 90:819–831

Sheahan MB, Rose RJ, McCurdy DW (2004) Organelle inheritance in plant cell division: the actin cytoskeleton is required for unbiased inheritance of chloroplasts, mitochondria and endoplasmic reticulum in dividing protoplasts. Plant J 37:379–390

Silk JB, Boyd R (1983) Female cooperation, competition, and mate choice in matrilineal macaque groups. In: Wasser SK (ed) Social behavior of female vertebrates. Academic, New York, NY, pp 315–347

Sperber D (2000) An objection to the memetic approach to culture. In: Aunger R (ed) Darwinizing culture: the status of memetics as a science. Oxford University Press, Oxford, pp 163–173

Szathmáry E, Maynard Smith J (1995) The major evolutionary transitions. Nature 374:227–232

Part VIII
Conclusions

Chapter 22
Mind the Gap: Cooperative Breeding and the Evolution of Our Unique Features

Carel P. van Schaik and Judith M. Burkart

Abstract Humans are strikingly different from our close relatives, the great apes, in mind, behavior, and life history. We propose that the evolution of these derived features was a consequence of the adoption of cooperative breeding by early *Homo*. Among the species that adopted it, cooperative breeding generally produced changes in psychology toward greater prosociality and greater cognitive abilities. We propose that in our ancestors, the major energetic inputs to breeding females due to cooperative breeding explain the derived features of human life history and lifted energetic constraints on brain enlargement. Moreover, in combination with great-ape -level cognitive abilities, the cooperative-breeding psychology led to the evolution of many of the unusual socio-cognitive traits that we now celebrate as uniquely human: pedagogy, extensive cumulative culture, and cultural norms; intensive and nearly indiscriminate within-group cooperation and morality; a cooperative declarative communication system known as language; and full-blown theory of mind.

22.1 Introduction

Related species tend to share many features. Our species, *Homo sapiens*, is an African great ape. Our ancestors separated from the other apes a mere 6–8 million years ago (Glazko and Nei 2003). Hence, it would not be surprising if we shared many features with chimpanzees, bonobos and other great apes. Indeed, the similarities between humans and great apes generated by the research of primatologists are numerous, and their presence in humans does not require any other explanation

C.P. van Schaik (✉) and J.M. Burkart
Anthropological Institute and Museum, University of Zürich, Zürich, Switzerland
e-mail: vschaik@aim.uzh.ch, judith.burkart@access.uzh.ch

P.M. Kappeler and J.B. Silk (eds.), *Mind the Gap*,
DOI 10.1007/978-3-642-02725-3_22, © Springer-Verlag Berlin Heidelberg 2010

than that they have been present for a long time and apparently are not patently maladaptive in our species, allowing them to persist.

Every species also has derived features, not shared with its closest relatives, or else it would not be a separate species. Perhaps the most remarkable thing about humans from a comparative perspective is how many of those unique features arose. Humans are different enough from great apes that one could easily be misled into thinking that this biological relationship is irrelevant to understanding human nature – the path chosen by the human sciences for millennia. This dramatic departure is all the more striking since it is becoming ever more clear that it was not soon after the split with the other apes, around 6 million years ago, but only with the emergence of the genus *Homo*, roughly 2 million years ago, that many of these traits arose.

Here is a brief summary of the non-morphological features that are derived in humans relative to the great apes and that seem particularly relevant to us when it comes to understanding the processes that produced them (see also Flinn et al. 2005; Richerson and Boyd 2005; Burkart et al. in press). First, there are pronounced life-history differences with the other great apes. Humans have slower development (later age at sexual maturity) and a longer life span than our great ape relatives. At the same time, women also show higher birth rates, produce relatively larger neonates, which are nonetheless weaned much earlier and experience much earlier than expected cessation of reproduction, known as midlife menopause (Robson et al. 2006).

Second, our subsistence ecology became radically different from that of any other anthropoid. Hunting large game and gathering a limited set of plant resources requires learned, skill-intensive techniques and delayed processing (Kaplan et al. 2000), systematic sharing, especially of meat (Ridley 1996; Gurven 2004), and some degree of specialization, mainly by sex. This life style is based on extremely intense cooperation: high social tolerance and prosocial helping within social units, targeted largely toward kin, toward bonded non-kin, which we, nonetheless, surprisingly call relatives too, and affiliated non-relatives. This intense cooperation also finds expression in occasional, systematic violent between-group conflict (Gat, this volume). Cooperation extends to our social organization, which is based on long-term (monogamous or polygynous) pair bonds, which serve in part as economic units, and in which there is discreet sexual activity not limited to short periods of sexual attractivity, as in most other primate species.

Third, humans have far more elaborate and cumulative material culture than the great apes, involving complex artifacts and knowledge, but we also uniquely use symbols and build institutions based on them, and maintain cultural norms based on religiously informed normative values. Culture, therefore, plays a decisive role in both our ecological niche and our "groupishness" (group-serving behaviors).

Fourth, human cognition is distinguished by unusual physical and spatial intelligence, involving causal understanding, episodic memory, and long-term planning. Even more striking is our social understanding, involving mental perspective taking, and understanding and sharing of intentions. Humans uniquely use language to coordinate and plan activities, discuss reputations, and intentionally teach the young. Many of these cognitive abilities are ultimately based on the presence of another uniquely derived motivational feature, *shared intentionality*, i.e., the ability to participate with others in collaborative activities with shared goals and intentions (Tomasello and Rakoczy 2003; Tomasello et al. 2005). It is based on the desire to share emotional states and knowledge, which, in turn, is due to a prosocial (sharing and helping) motivation.

What could explain this drastic and rapid divergence? We believe these differences are linked, and are caused by the fact that our ancestors adopted cooperative breeding. Here, we will present the relevant comparative evidence to establish the basic credibility of this cooperative breeding hypothesis. Basically, the idea is that our ancestors were the first Old World primates to engage in extensive allomaternal care (cooperative breeding). Comparative data suggests that cooperative breeding installs a more prosocial psychology, which functions to support the more intensive cooperation in such species, and has immediate consequences for cognitive performance, and in some cases leads to larger brain size. In our ancestors, who had ape-like cognitive abilities, this fundamental change in attitudes led to a cascade of cognitive changes. This idea was first broached by Hrdy (1999), and then developed by Hrdy (2009) and independently by Burkart and van Schaik (in press; see also Burkart et al. 2007, in press; Burkart 2009).

22.2 Cooperative Breeding and Human Nature

22.2.1 Cooperative Breeding

Cooperative breeding is defined in rather different ways, depending on the taxonomic focus of the biologists using the term, and as a result there has been quite a lot of confusion in the literature (Hrdy 2009). For the present purpose, the presence of extensive allomaternal care, i.e., routine care by other individuals than the mother, suffices. Thus, fathers, grandmothers, older immature siblings, and aunts and uncles can all be allomothers. The energetic significance of cooperative breeding is that it directly or indirectly provides energy inputs to the mother, allowing her to reproduce more successfully than would otherwise be possible. Obviously, there is large variation in how many others contribute to rearing the young, how much they are involved, and what forms their caretaking takes.

There is no doubt that, in sharp contrast to any of the extant great apes, humans are cooperative breeders (Hrdy 2005, 2009; Mace and Sear 2005). The evidence is clear among foragers, but remains visible in most derived societies. First, among foragers, men bring in two-thirds of the calories on average (Marlowe 2003), even if provisioning is not always exclusive to the nuclear family, perhaps depending on their opportunities for costly signaling (Hawkes 1993). Second, grandmothers spend more time foraging, especially for difficult-to-process foods, such as tubers (Hawkes et al. 1989, 1998), and in many settled societies, grandmothers support

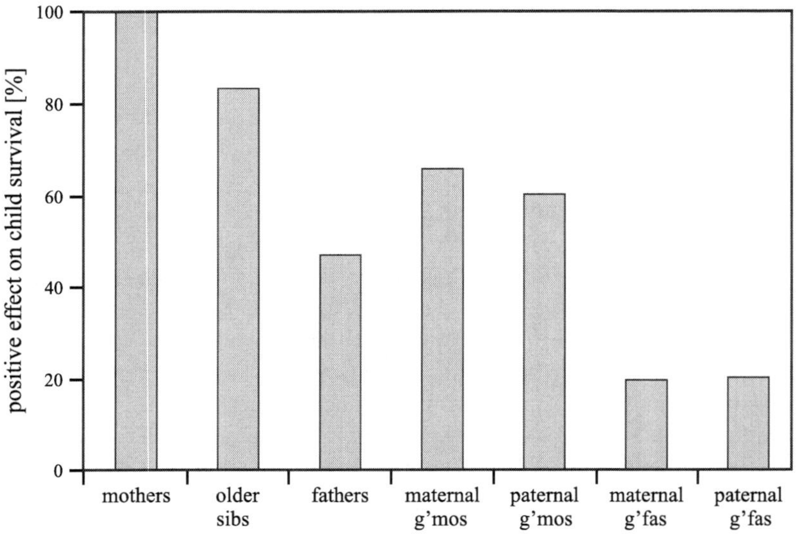

Fig. 22.1 Summary of positive effects of allomaternal care in humans (data from Sear and Mace 2008). Positive effects on child survival of different categories of care-givers in 45 studies of natural fertility populations. *g'mos* = grandmothers, *g'fas* = grandfathers

mothers in other ways. Third, among foragers and non-foragers, older siblings also play a major role in childcare and babysitting (Hawkes et al. 1995; Kramer 2005).

We now have quantitative estimates of the effect of this help on birth rates and infant performance in terms of growth and survival. Grandmothers have been shown to improve infant survival and maternal reproductive rates, not only among settled, natural fertility populations (Lahdenperä et al. 2004), but also among foragers (Blurton Jones et al. 2005). Other helpers also have a positive impact. Figure 22.1 summarizes the results of the review by Sear and Mace (2008).

Extensive allomaternal care can account for the derived features of human life history listed above, in particular the larger relative size of neonates, the much earlier weaning, and higher birth rates of humans relative to the other great apes (Robson et al. 2006), when the opposite would be expected given that brain size sets these developmental and reproductive parameters in primates (Barrickman et al. 2008) and human brains are about three times as large as those of the other great apes. Midlife menopause can also be regarded as an expression of cooperative breeding (Hawkes et al. 1998).

22.2.2 The Cooperative Breeding Hypothesis

Which features of a species are affected by the adoption of cooperative breeding? First and foremost, cooperative breeding creates a different social system, in which group members show extreme social tolerance and intensive cooperation, ranging from food sharing to collective action, in the form of communal predator mobbing or territory defense, and even to incipient forms of division of labor, such as sentinel duties (Burkart and van Schaik in press). This unusual social system requires an unusual psychology. Among cooperatively breeding primates and canids, we see a motivational predisposition (prosociality) that produces spontaneous assistance to others and empathic responses to signs of need, reflected in a concern for others and eagerness to share food and information with others and cooperation in a wide array of contexts, in some cases extending to active care for injured or ill group members (Burkart and van Schaik in press; Burkart et al. in press). Among callitrichid monkeys, cooperatively breeding primates, prosociality also extends to others in the group beyond infants, probably as a secondary development, or as a necessary corollary to cooperative breeding in *Callithrix jacchus* (Burkart et al. 2007), although the evidence is mixed for *Saguinus geoffroyi* (Cronin et al. 2005, in press; Cronin and Snowdon 2008). Capuchin monkeys (*Cebus* spp.) also show elements of cooperative breeding (Fragaszy et al. 2004); accordingly, they too show elements of spontaneous help toward others, although they were less indiscriminate (de Waal et al. 2008; Lakshminarayanan and Santos 2008). Other primates tested, including chimpanzees, do not show such prosociality (Silk 2007).[1]

The most relevant consequence of cooperative breeding for this chapter is that the cooperative-breeding psychology affects cognition, both directly and indirectly (Burkart 2009; Burkart et al. in press). Prosociality is expected to produce an immediate improvement in cognitive performance in two ways. First, it should improve the conditions for social learning, because it leads to higher social tolerance, increased attention to others, and active involvement of role models (teaching is above all a form of prosociality: Hoppitt et al. 2008). Second, prosociality should improve the coordination of activities, such as cooperation, in part through higher social tolerance, greater attentional biases toward others, and practice in the coordination of infant care. A review of the cognitive performance of cooperatively breeding callitrichids confirms this prediction: they outperform their independently breeding sister taxa, especially with respect to these aspects of social cognition, and less formal comparisons suggest a very similar pattern among carnivores (Burkart and van Schaik in press).

Indirectly, the great increase in opportunities for social learning improves the efficiency of use of brain tissue. The Cultural Intelligence hypothesis (Whiten and van Schaik 2007; van Schaik and Burkart in press) is based on the fact that the high energetic costs of growing and maintaining brain tissue imposes high obstacles to the evolution of larger brain size. Thus, any process that lowers these costs will lower the obstacles to selective benefits favoring increased brain size (Isler and

[1]There is large variation in reproductive skew in societies with extensive allomaternal care, from very high, where one breeding pair monopolizes mating (as in meerkats and callitrichid monkeys), to rather low, where all adult group members potentially breed (as in capuchin monkeys and humans). Likewise, species vary in which classes of helpers are the most important (siblings, males, grandmothers). How this variation affects prosociality remains to be examined.

van Schaik in press). Reliance on social learning rather than on individual exploration is perhaps the most obvious of such processes. Because the amount and quality of environmental inputs have been shown to affect the cognitive performance of adults (van Schaik and Burkart in press), most dramatically so in humans (Tomasello 1999), animals that can rely on social learning can achieve more with the same brain size. This leads to the prediction that the more species can rely on social learning to acquire their various skills, the better their cognitive performance (controlling for brain size) should be. As noted above, cooperative breeders can rely more on social learning than others. Over time, therefore, cooperative breeders may attain a larger equilibrium brain size than their independently breeding sister taxa, provided that improved cognitive abilities sufficiently enhance survival or reproduction. The latter prediction is, indeed, confirmed in preliminary analyzes of birds and mammals (K. Isler et al. unpubl. data).

22.2.3 Cooperative Breeding and Human Evolution

The cooperative breeding hypothesis argues that this breeding system installs a prosocial psychology in a species, which not only affects the nature of cooperation, but also cognition directly. Indirectly, it may improve cognitive abilities over evolutionary time. The influences of cooperative breeding on the derived aspects of human life history, behavior, and cognition can be arranged in three fundamental classes: (1) direct expressions or consequences of cooperative breeding, such as midlife menopause; (2) evolutionary consequences of cooperative breeding in the social or cognitive domain; and (3) side effects of the increase in brain size.

Some distinctive human features are merely an expression of cooperative breeding. Thus, facultative paternal care for infants and long lifespans following midlife menopause are simply a reflection of cooperative breeding (Hrdy 2009). It is also possible that pair-bonding, at least originally, was a direct expression of cooperative breeding. Grandmothering is most plausibly considered as an adaptation through which aging females achieve better fitness returns than when they were to continue to breed (Hawkes et al. 1998). This may have arisen in part because maternal mortality rises rapidly with age in humans (Temmerman et al. 2004; Pavard et al. 2008), probably more so than in other taxa, in part because human children rely for much longer after weaning on care (Williams 1957), making continued survival of the mother important. Nonetheless, the importance of grandmothers for the survival of grandchildren makes grandmothering a special case of allomaternal care. Cooperative breeding may also have affected our longevity because care for the sick and injured may reduce unavoidable mortality to lower levels than in independent breeders, such as great apes, thus enabling selection for the physiology underlying longer life (Kirkwood and Austad 2000).

The most obvious immediate evolutionary consequence of the adoption of cooperative breeding is the intense cooperation in humans, which is so different from that of other apes. Human cooperation must be underlain by a different psychological regulation system than found in great apes: prosociality makes active sharing and collective action possible, especially if aimed at all members of the core group (creating indirect reciprocity, if combined with concern for reputation). In humans, this prosocial attitude has been called strong reciprocity (Gintis 2000), and also encompasses Alexander's (1987) moral altruism. Although punishment of non-cooperators is rare in cooperative breeders (probably because non-cooperation is rare: Snowdon and Cronin 2007), strong reciprocity in humans also involves punishment of non-cooperators. However, the recent finding that altruistic (third-party) punishment is virtually absent among foragers (Marlowe et al. 2008; but see Wiessner 2009) suggests that it is a more recent development, necessitated by living in larger-scale societies. Thus, the psychology of altruism in humans living in small-scale foraging societies shows a strong similarity, perhaps reflecting convergence, to that found in cooperative breeders.

Cooperative breeding has also produced cognitive changes in humans, as it did in non-human cooperative breeders, by improving social learning and coordination of joint cooperative activities. Shared intentionality, i.e., the formation of shared goals and coordination of their actions in pursuit of these shared goals, forms the basis of the human ability for cooperative or joint problem solving, the origin of language, the presence of a full-fledged theory of mind (not just in competitive contexts), and our tendency to abide by social norms (Tomasello and Carpenter 2007). It also makes teaching intentional, and thus more effective, and thereby contributes to the presence of cumulative culture (Table 22.1). While its critical role in human cognition is now recognized, the evolution of shared intentionality

Table 22.1 The transitions from ape-like ancestral states to the human-derived states that can be explained by the adoption of cooperative breeding, and its psychological underpinnings (in particular, prosociality)

Ancestral state		Human state
Social learning and culture		
Observational social learning in apprenticeship	→	Joint attention and teaching (pedagogy)
Simple material culture	→	Cumulative material culture
Individual innovation	→	Cooperative problem-solving
Communication		
Vocal displays, suppressing information harmful to ego; imperatively used	→	Language (information donation), including much declarative use
Visual displays, suppressing information harmful to ego	→	Cooperative eyes (signaling of gaze direction), blushing, crying, and disgust face
Cooperation		
Direct, relationship-dependent reciprocity	→	Reputation-based indirect reciprocity (=indiscriminate within-group altruism, if all cooperate)
Dyadic obligations	→	Morality, religion
Unspecialized cooperation	→	Division of labor
Cognition		
Theory of Mind abilities, applied especially in competitive contexts	→	Shared intentionality and full-blown Theory of Mind, applied to coordinate cooperative activities

remained obscure. We suggest that prosociality gave rise to shared intentionality, because it allowed the nascent theory-of-mind abilities in our ancestors to be deployed to prosocial ends, and thus gradually enhanced.

Cooperative breeding can, therefore, be seen as underlying many of the cognitive differences with the great apes (see extensive discussion in Burkart et al. in press). Basically, while chimpanzees, and perhaps all great apes, meet many of the relevant cognitive preconditions for the evolution of human cognitive potential, they lack the motivational preconditions. In humans alone, these two components have come together, the cognitive component due to common descent, and the motivational component, resulting from the selection pressures associated with cooperative breeding (Table 22.1). Thus, the high social tolerance of mothers and eager apprentice attitude of the infants in great apes became the active teaching by parents and the full-blown system of pedagogy (Gergely et al. 2007). This change strongly facilitated cultural evolution. The prosocial attitude resulting from cooperative breeding also led to a fundamental change in communication, toward declarative use of communicative signals, which largely honestly reflected intentions and attitudes, thus enabling language evolution. By extending prosociality toward all in-group members, dyadic, relationship-dependent cooperation could become group-wide, indirect reciprocity and more intense collective action, later backed up my morality and religion.

Cooperative breeding can explain why these cognitive benefits could actually be expressed in increased brain size in the hominin lineage. Larger-brained organisms show strong reductions in maximum reproductive potential, r_{max} (Isler and van Schaik 2009), which is an estimate of a species' ability to recover from population crashes or to evolve new adaptations in the face of rapidly changing environmental conditions. This negative relationship indicates that each lineage has a maximum sustainable relative brain size, beyond which reproduction is so low that population extinction, and thus species extinction, becomes likely. Evolving organisms may therefore bump into a gray ceiling. In particular great apes are so large-brained that their demographic viability is severely compromised: they have the lowest r_{max} on record. Thus, further brain enlargement in great apes would be prohibited by its negative demographic consequences.

Given that this strong negative relationship is set by the energetic constraints of growth and reproduction imposed by larger brains, it is to be expected that it is relaxed among cooperative breeders (Hrdy 1999). Indeed, cooperative breeders have higher r_{max} than similar-sized independent breeders (Isler and van Schaik 2009), and tend to have larger brains than independent breeders, suggesting that the constraint on brain size has been lifted. Thus, if cooperative breeders experience clear fitness benefits from enhanced cognitive abilities, their brain size can increase beyond that possible for independent breeders. This process can account for the spectacular increase in brain size in the hominin lineage.

The third category of derived features in humans can be regarded as a by product of the large increase in brain size during hominin evolution. Brains are very costly tissues metabolically (Mink et al. 1981). According to the expensive brain hypothesis (Isler and van Schaik in press), brain size increases must be paid for either by increases in energy throughput (as indexed by basal metabolic rate) or by reduced energy allocation to production (i.e., growth and development as well as reproductive rates). There is good comparative evidence for these predictions among both mammals and birds (Isler et al. 2008; Isler and van Schaik in press). The reduced allocation to production, at least in precocial organisms, implies slower rate of development, controlling for body size, and reproduction once adult, unless the increased brain size can counteract these costs (more likely for reproduction than for development, when the brains are not yet mature). Again, these predictions hold up well in a large comparative sample of precocial mammals and birds. However, for selection to favor increases in brain size, the larger brains must compensate for the trend toward delayed and slower reproduction by making the organisms outlive (or, less plausibly, outreproduce) the smaller-brained variants. Increased longevity of larger-brained organisms is indeed observed (Ross and Jones 1999; Deaner et al. 2003; Isler and van Schaik in press).

The increased brain size can, therefore, account for the slower development of humans (Barrickman et al. 2008), and thus suggests that other, less parsimonious explanations are less likely to hold. Indeed, the developmental pace of humans corresponds to the value predicted for a non-human primate of our brain size (Isler and van Schaik in press). These effects can also explain the strong increase in human lifespan.

22.3 Cooperative Breeding in Hominins: When Did it Arise?

It may be next to impossible to develop a fully reliable estimate of the timing of the origin of cooperative breeding in the hominin lineage. It is, nonetheless, important to do this because we have argued that many other derived human features depend on the presence of prosociality, which therefore requires that cooperative breeding arose relatively early, i.e., before these other derived features arose. As a first step, we can bracket the timing of origin by examining the endpoint and starting point of the hominin lineage. Extant humans all descend from cooperative breeders, but it is also likely that Neanderthals, given their strong reliance on hunting and food sharing, were cooperative breeders, pushing the origins back at least as far as the ancestor of modern humans and Neanderthals. On the other hand, it is almost certain that the common ancestor of *Pan* and the hominins were not cooperative breeders, given the remarkable absence of any tendencies in that direction among all extant apes. Thus, we can surmise with some confidence that cooperative breeding arose somewhere between the earliest australopithecines and mid-Pleistocene *Homo*.

Several lines of argument suggest that the origin coincided with the emergence of *Homo erectus* (or *H. ergaster*) in East Africa around 1.8 Ma. First, there is evidence that these were the first hominins to acquire meat from large mammals (Foley 2001; Pobiner et al. 2008), which necessarily indicates the presence of cooperative hunting or at least cooperative defense of large carcasses. The large carcass size strongly suggests extensive and systematic food sharing, perhaps serving, as among extant hunter-gatherers, to even out the highly variable yields. Both these traits are essentially modern, suggesting the operation of similar psychological tendencies toward cooperation and food sharing, which, we suggest, are based on cooperative breeding.

Second, survival in the newly colonized savanna habitat presumably relied on skills acquired gradually during development, largely through social learning, involving some combination of tool-assisted hunting and processing of mammals and tool-based extraction of underground storage organs (O'Connell et al. 1999; Kaplan et al. 2000). However, resources harvested as efficiently by juveniles as adults, such as soft fruits, are much scarcer in savannas (Hawkes et al. 1995). As a result, immatures would have found it increasingly difficult to achieve a positive energy balance through their own foraging, unlike in all other primates, which would have produced increasingly long birth intervals, unless, of course, these offspring were provisioned by adults. This condition fits with those favoring cooperative breeding: helping is favored where successful dispersal is difficult, for whatever reason, and helping has a large positive impact on the immatures receiving it. The unpredictability of food supply in savanna habitats also may be a fairly common pacemaker for the evolution of cooperative breeding in birds (Rubenstein and Lovette 2007) and mammals (Clutton-Brock 2006) generally.

Third, *H. erectus* was the first hominid to colonize habitats outside Africa, which were different yet again from the habitats occupied earlier. Hrdy (2005, 2009) has argued convincingly that colonizing new habitats is facilitated by cooperative breeding, given the periods of scarcity encountered in a new habitat before novel solutions have been invented to deal with them.

Fourth, because female *H. erectus* were much larger in both body size and brain size than the females of the taxa that preceded them, an increased reproductive burden must have ensued. As a result, there must have been "*a revolution in the way in which females obtained and utilized energy to support their increased energetic requirements*" (Aiello and Key 2002). We suggest that this revolution included the emergence of shared care and provisioning, which therefore must have started around that time.

Finally, as we argued above, external energetic inputs (in particular through allomaternal care and provisioning, i.e., cooperative breeding) allow a taxon to break through its taxon-specific maximum viable brain size. This argument would put the beginning of cooperative breeding in the period directly preceding the rise of *H. erectus*, when hominin brain sizes clearly exceeded the great ape range for the first time (Schoenemann 2006). It is supported by reconstructions of the dental development of east African *H. erectus*, which suggest that this was the first taxon among the hominins to develop more slowly than the extant great apes (Dean et al. 2001). Among primates and other precocial mammals, longer maturation time is a direct consequence of increased brain size (Barrickman et al. 2008; Isler and van Schaik in press).

If this circumstantial evidence is accepted, it is consistent with the critical assumption that cooperative breeding preceded the gradual accumulation of the uniquely human features. Many of them would be expected to have arrived later, as brain size expanded further.

22.4 Discussion

Here, we address some obvious immediate possible objections. First, how does this idea relate to the other popular scenarios for the evolution of our unique features? Second, given that prosociality is now recognized as a major derived feature of humans, how does the cooperative breeding hypothesis relate to the various prominent hypotheses recently proposed to account for our unusual "groupishness?" Finally, what is known about the selective processes that favored the origin and guarantee the maintenance of cooperative breeding in general?

22.4.1 Relationship with Other Scenarios for Human Evolution

Producing reconstructions of human evolution, and the development of scenarios to explain the reconstructed course of events, has been a major pastime of paleoanthropologists from the very beginning. This is understandable, because otherwise there would be little point to attempting to find fossils of our ancestors. However, the reconstructions are often limited by the sparseness of the record and the limited reliability of paleontological "facts" (see the debates over age at death or aging and tooth wear: Hawkes and O'Connell 2005). Moreover, especially the earlier scenarios have been little more than fanciful teleological stories, explicitly or implicitly relying on notions such as progress and improvement (Cartmill 1993). Many were also simplistic silver bullet theories that attributed all of human evolution to one factor, such as bipedality or hunting. Subsequent ideas or reworkings of the old ones have been much more informed by evolutionary biology and comparative primatology, and generally focused on specific abilities or taxa, usually the origin of the genus *Homo* (e.g., Foley 2001).

Admittedly, the cooperative breeding hypothesis presented here (Hrdy 2009; Burkart et al. in press) is somewhat of a silver bullet theory for explaining *Homo*. What makes us so bold? The main reason is that this hypothesis does not postulate a single, exclusive force, but rather serves to provide the context that enables many other, previously identified mechanisms to operate, thus in turn raising the plausibility of these ideas, and sometimes even removing some of their weaknesses.

The predominant explanation for the evolution of human uniqueness has long been the hunting hypothesis (discussed in Cartmill 1993; Hawkes 2006). This hypothesis still has much to offer in a modified form, which puts the origin of hunting much later and no longer as a direct response to bipedality and the secondary altriciality of newborns as a result of the remodeling of the pelvis it induced. Life on the savanna opened up the hunting niche, which is skill-intensive, and thus requires an unusually large brain as well as long learning periods and hence delayed maturity. This system is only feasible if immatures are subsidized until reaching adulthood, and thus if there is parental provisioning ("embodied capital" hypothesis: Kaplan et al. 2000; Kaplan and Robson 2002). This proposal is therefore very similar to the cooperative breeding hypothesis, which explains why men, in particular, became hunters who brought back meat to share it with others. Men may engage in costly signaling to advertise their value as mates or allies (Hawkes 1993), but they generally also provision their families (Gurven 2004), and without prosociality and shared goals, the whole foraging ecology of gathering, hunting, processing (including cooking), and then systematic sharing with the family or the whole camp would quickly break down. Nonetheless, there are some differences. For instance, the embodied capital hypothesis does not consider the help provided by older siblings, other kin, or grandmothers. Most critically, it assumes that the need for skill learning is the limiting factor for the age at maturity, whereas the cooperative breeding hypothesis assumes that the slow development of humans can easily be accounted for by the large size of our brain, and not time needed to learn skills. This issue is subject to lively debate (but note that in species with altricial young, maturation time is not linked to brain size: Isler and van Schaik in press).

The grandmothering hypothesis aims to explain our derived life history features, in particular the presence of the long post-fertility lifespan shown by women (Hawkes et al. 1998). It argues that extended lifespan beyond the reproductive years were favored by selection because older women were more effective at helping rearing grandchildren than producing and rearing their own. It is, of course, fully compatible with the cooperative breeding hypothesis. However, neither the embodied capital nor the grandmother hypothesis haves systematically explored the consequences for the evolution of human psychology and cognition.

The controlled use of fire and the cooking of food it enabled have also been held responsible for the origin of many of our unique features (Wrangham et al. 1999). The hypothesis puts its origin at 1.9 Ma, around the origin of *H. erectus*, although many experts insist on far more recent dates. Regardless, cooking food makes the cooks extremely vulnerable to theft, and can only realistically have emerged in groups with social tolerance that went far beyond the level shown among great apes and which almost certainly systematically shared food. Moreover, managing fire is generally based on collective care and sharing. Thus, only hominins that bred cooperatively could have managed fire and cooked.

Unprecedented short-term variability in climate during the Plio-Pleistocene has been held responsible for the evolution of human cognitive and, especially, technological abilities (Potts 1998). It did so, not by selecting for particular genetically based adaptation but instead by selecting for enhanced phenotypic plasticity that generated adaptation to the new conditions through cognitive solutions. Although this idea can explain several observations (Potts 1998), one problem with it is why only hominins should have responded in this way. However, once cooperative breeding removed the obstacles to increased cognitive abilities and removed energetic constraints on brain size increases, the enhanced cognitive abilities, social learning capacity, and cooperative problem-solving skills envisaged by the cooperative breeding hypothesis allowed hominins to achieve adaptation through plasticity (cf. Hrdy 2005). Thus, the cooperative breeding hypothesis explains why variability-imposed selection could weigh in so heavily with the hominins without similarly affecting other synchronous and sympatric lineages.

On its own, human-like lethal between-group conflict (Alexander 1987; Flinn et al. 2005) has some difficulty accounting for the whole package of derived human traits. First, the kind of warfare seen among mobile foragers, in the form of raids, is rather similar to that seen among chimpanzees ("war below the military horizon": Crofoot and Wrangham this volume), which suggests that it cannot explain the radical reorganization of altruistic psychology that accompanied human evolution. Second, there is no good evidence for either saturated habitats or systemic human-like warfare until the last 15,000 years or so of human existence (Hrdy 2009), an observation consistent with estimates of Pleistocene population sizes (Pennington 2001). On the other hand, the cooperative breeding hypothesis can provide exactly the group-wide within-group prosociality required to sustain systematic warfare in humans, well beyond the level seen in chimpanzees, and difficult to achieve by independently breeding non-human primates (van Schaik et al. unpubl. data). Thus, cooperative breeding has ultimately made it possible for warfare to evolve to the level seen in humans, matching that seen among eusocial insects, which are, of course, obligate cooperative breeders.

The cooperative breeding hypothesis is less consistent with hypotheses that consider human behavior and cognition as driven by the need to deal with social complexity, including the various versions of the Machiavellian Intelligence hypothesis (Dunbar 2003). However, these hypotheses do not explain why different primate lineages differ so much in intelligence (van Schaik and Deaner 2003), or more specifically why humans became so different from the other apes, whose social complexity was comparable to that of early hominins (Rendall et al. 2007). Dunbar (1998) suggested that humans required larger brains and language, because their groups became too large and new mechanisms of group cohesion were needed. In the absence of data on hominin group sizes, this idea is nearly impossible to test (see Rendall et al. 2007).

22.4.2 Alternative Hypotheses for Human Prosociality

There has been much speculation to explain the "groupishness" of people, i.e., our tendency to be spontaneously altruistic toward all in-group members and to readily contribute to collective action. Traditional answers range from group selection and cultural group selection to reputation-based individual benefits. Yet, their relative merits remain subject of intense debate.

Group selection proposes that groups with altruists outcompete other groups and replace them with colonists from their own group. The first problem facing the group-selection explanation is that the frequent, lethal between-group competition it requires is probably far too recent to have overhauled our psychology, as argued above. Second, human groups are not usually as close to immigration as is required. Even if other groups are being eliminated, many of their members are absorbed into the dominant groups (Boyd and Richerson 2006), although perhaps not always (Bowles 2006).

Cultural group selection argues that groups that adopt prosocial social norms outcompete other groups, either in direct competition or simply by doing better in the struggle for life in hostile habitats or by surviving lean periods (Richerson and Boyd 2005). No genetic homogeneity is required and dispersal between groups is allowed, but what is needed is that individuals abide by the social (moral) norms of the group in which they find themselves (or select groups based on the reigning norms there). Cultural group selection is a very plausible model, supported by much evidence (Richerson and Boyd 2005), and it may explain much of the current variation observed among human societies, as well as the maintenance of social norms. However, Boyd and Richerson (2006) admit that it has difficulty explaining why hominins were amenable to adopt prosocial norms rather than other ones, or why they became so credulous or teachable. The cooperative breeding hypothesis posits that this attitude was there to begin with, removing this weakness to the cultural group selection hypothesis.

Much seemingly indiscriminate altruistic behavior in human groups may represent indirect reciprocity, in which actors of altruistic acts become recipients of altruistic acts by third parties because of the reputation gained from these altruistic acts (Milinski et al. 2002). However, while reputation may explain within-group altruism, it is disputed whether it can also account for group-level cooperation, i.e., collective action (Panchanathan and Boyd 2004). Moreover, since reputation effects can be far stronger in the presence of language (gossip), reputation may not yet have been a major factor in early *Homo*.

We have argued that cooperative breeding can account for the origin and maintenance of within-group prosociality and the tendency to engage in collective action in the small groups containing mostly relatives and bonded non-kin pairs that characterized humans for most of our evolutionary past. If that argument turns out to be valid, then one might reasonably insist that the explanation for this development in humans must be the same as for the other cooperative breeders, given that prosociality and generally also coordinated collective action (Burkart and van Schaik in press) are observed in many cooperative breeders (reviewed in Burkart and van Schaik in press). Cooperative breeders generally manage to prevent free-riding, and do so without obvious punishment (Snowdon and Cronin 2007), even when non-kin are involved, and remain prosocial even when groups also contain non-related helpers (Clutton-Brock 2002, 2006; Zahed et al. 2007; McDonald et al. 2009).

It is unlikely that any of the three standard explanations for the evolution of human prosociality discussed above can be applied directly to the cooperatively breeding callitrichids, carnivores, elephants, or birds. Explaining the evolution of cooperative breeding in general might, therefore, also solve the problem of the origin of prosociality in the small forager groups of our ancestors, which consisted of kin of variable relatedness and bonded partners. On the other hand, it is easy to see how cultural group selection could produce prosocial cultural norms in human groups, once cooperative breeding had established prosociality, which facilitated both cultural evolution and the establishment of social norms.

The maintenance of indiscriminate within-group prosociality in the larger groups that ensued when foragers began to settle after the adoption of agriculture, beginning at around 10,000 years ago, although perhaps earlier in some places (Johnson and Earle 2000), is a thorny problem, but one that is distinct from that of its origin and initial maintenance in mobile foragers. Accompanying these transitions, egalitarianism gave way to despotism, except within family units, and uniform within-group sharing disappeared, again except within families (Johnson and Earle 2000; Rowley-Conwy 2001), and specialized classes of norm enforcers (punishers) arose. Here, almost certainly uniquely human processes are needed to account for the maintenance of prosociality in these much larger units, which were not only more likely to contain distant kin, and later non-kin (after states were formed), but also routinely went to war with each other. This is probably when prosocial preaching in the form of organized non-animistic religions arose, and altruistic punishment by third parties became an important part of the package of strong reciprocity (Marlowe et al. 2008).

22.4.3 Why Did Cooperative Breeding Arise and Why was it Stable?

If cooperative breeding explains the origin of prosociality in the hominin lineage, the key question is how it arose. The details will probably always remain unknown, but it might be possible to develop a plausible scenario if we knew what conditions, in general, favor the evolution and maintenance of cooperative breeding. Unfortunately, there is no consensus in the behavioral ecology literature, but from it one can formulate a general condition. The origin is likely to lie in situations where helpers gain better fitness return from helping than from dispersing and attempting to join other units or found their own unit, or alternatively from trying to take over the natal unit, provided there are non-related adults available as mates (Russell 2004). This implies that cooperative breeding is likely where the following combination of factors applies: (1) effective dispersal is very difficult, for a variety of reasons, but at the same time (2) helping has a large impact on the fitness of the immatures receiving the help (which may happen for a variety of reasons: Clutton-Brock 2006). Subsequent changes in female reproductive capacity will increase the positive fitness impact of helping, making cooperative breeding increasingly more specialized and less opportunistic.

Kin selection can easily account for the origin, but once the system is established, we regularly see some non-kin in the groups, operating as helpers (Clutton-Brock 2002, 2006). The maintenance is, thus, difficult to explain against the risk of free-riding, be it by non-kin group members or by helpers that compete for a future breeding slot in the group. Theorists argue that free-riding is no threat in cooperative breeders whenever fitness is positively linked to group size and all group members share a stake in offspring survival, even the survival of offspring they are not related to (Kokko et al. 2001). Harming a group member would, thus, automatically diminish the group's prospect of persisting, and thus make free-riding not a viable option, even for non-kin. However, it is not clear how often these stringent conditions are met.

Neither is it clear why potential rivals remain largely prosocial, given that helping incurs some cost and by holding back in helping, pretenders might improve their competitive ability, and thus their chances of achieving the breeding position. The habitually tolerant and peaceful social relations among callitrichids are interrupted by aggression when disputes erupt over breeding status, which – as in most cooperative breeders – can be quite serious (Digby et al. 2007). The same thing seems to be true for other cooperative breeders, be they canids or small herpestids. For instance, meerkat helpers clearly pay a short-term cost for helping, which should affect their prospects of becoming a breeder, and thus creates some incentive for free-riding that would seem to exist. Yet, all individuals seem to help according to their ability (Clutton-Brock 2006). It is possible that there are constraints on the extent to which individuals are capable of free-riding just enough to gather its benefits without being attacked or evicted by others. Perhaps, the neuroendocrine state that produces prosociality cannot be combined simultaneously with some level of free-riding, unless the override is cognitively driven (cf. Wiltermuth and Heath 2009). Clearly, given its theoretical significance, it is an important issue for future research to examine how other cooperative breeders deal with the problem of free-riding and can maintain their prosociality. The solution to this problem may go a long way toward explaining the evolution of human prosociality.

22.5 Conclusion

In this chapter, we have examined the cooperative breeding hypothesis for the evolution of human cooperation and cognition, in order to assess whether it is viable enough to warrant more detailed evaluation. Having done this, we must now proceed to systematically test its many assumptions and predictions, in a range of cooperative breeders and in humans.

Acknowledgments We thank Charles Efferson, Kristen Hawkes, Sarah Hrdy, Karin Isler, and Maria van Noordwijk for valuable discussion or comments on the manuscript, and Peter Kappeler for the invitation to take part in the Freilandtage.

References

Aiello LC, Key C (2002) Energetic consequences of being a *Homo erectus* female. Am J Hum Biol 14:551–565

Alexander RD (1987) The biology of moral systems. Aldine de Gruyter, New York

Barrickman NL, Bastian ML, Isler K, van Schaik CP (2008) Life history costs and benefits of encephalization: a comparative test using data from long-term studies of primates in the wild. J Hum Evol 54:568–590

Blurton Jones NG, Hawkes K, O'Connell JF (2005) Older Hadza men and women as helpers: residence data. In: Hewlett BS, Lamb ME (eds) Hunter-gatherer childhoods. Aldine Transaction, New Brunswick, NJ, pp 214–236

Bowles S (2006) Group competition, reproductive leveling, and the evolution of human altruism. Science 314:1569–1572

Boyd R, Richerson PJ (2006) Culture and the evolution of the human social instincts. In: Enfield NJ, Levinson SC (eds) Roots of human sociality: culture, cognition and interaction. Berg, Oxford, pp 453–477

Burkart JM (2009) Socio-cognitive abilities and cooperative breeding. In: Röska-Hardy LS, Neumann-Held EM (eds) Learning from animals? Examining the nature of human uniqueness. Psychology Press, Hove, pp 123–140

Burkart JM, van Schaik CP (in press) Cognitive consequences of cooperative breeding in primates. Anim Cogn

Burkart JM, Fehr E, Efferson C, van Schaik CP (2007) Other-regarding preferences in a non-human primate: common marmosets provision food altruistically. Proc Natl Acad Sci USA 104:19762–19766

Burkart JM, Hrdy SB, van Schaik CP (in press) Cooperative breeding and human cognitive evolution. Evol Anthropol

Cartmill M (1993) A view to a death in the morning: hunting and nature through history. Harvard University Press, Cambridge, MA

Clutton-Brock TH (2002) Breeding together: kin selection and mutualism in cooperative vertebrates. Science 296:69–72

Clutton-Brock TH (2006) Cooperative breeding in mammals. In: Kappeler PM, van Schaik CP (eds) Cooperation in primates and humans: mechanisms and evolution. Springer, Berlin, pp 173–190

Cronin KA, Snowdon CT (2008) The effects of unequal reward distributions on cooperative problem solving by cottontop tamarins, *Saguinus oedipus*. Anim Behav 75:245–257

Cronin KA, Kurian AV, Snowdon CT (2005) Cooperative problem solving in a cooperatively breeding primate (*Saguinus oedipus*). Anim Behav 69:133–142

Cronin KA, Schroeder KKE, Rothewell ES, Silk JB, Snowdon CT (in press) Cooperatively breeding cottontop tamarins (*Saguinus oedipus*) do not donate rewards to their long-term mates. J Comp Psychol

de Waal FBM, Leimgruber K, Greenberg AR (2008) Giving is self-rewarding for monkeys. Proc Natl Acad Sci USA 105:13685–13689

Dean C, Leakey MG, Reid D, Schrenk F, Schwartz GT, Stringer C, Walker A (2001) Growth processes in teeth distinguish modern humans from *Homo erectus* and earlier hominins. Nature 414:628–631

Deaner RO, Barton RA, van Schaik CP (2003) Primate brains and life histories: renewing the connection. In: Kappeler PM, Pereira ME (eds) Life history and primate behavior. Chicago University Press, Chicago, pp 233–265

Digby LJ, Ferrari SF, Saltzman W (2007) Callitrichines: the role of competition in cooperatively breeding species. In: Campbell CJ, Fuentes A, MacKinnon KC, Panger M, Bearder SK (eds) Primates in perspective. Oxford University Press, New York, pp 85–106

Dunbar RIM (1998) The social brain hypothesis. Evol Anthropol 6:178–190

Dunbar RIM (2003) Evolution of the social brain. Science 302:1160–1161

Flinn MV, Geary DC, Ward CV (2005) Ecological dominance, social competition, and coalitionary arms races: why humans evolved extraordinary intelligence. Evol Hum Behav 26:10–46

Foley RA (2001) The evolutionary consequences of increased carnivory in hominids. In: Stanford CB, Bunn HT (eds) Meat-eating and human evolution. Oxford University Press, New York, pp 305–331

Fragaszy DM, Visalberghi E, Fedigan LM (2004) The complete capuchin: the biology of the genus *Cebus*. Cambridge University Press, Cambridge

Gergely G, Egyed K, Király I (2007) On pedagogy. Dev Sci 10:139–146

Gintis H (2000) Strong reciprocity and human sociality. J Theor Biol 206:169–179

Glazko GV, Nei M (2003) Estimation of divergence times for major lineages of primate species. Mol Biol Evol 20:424–434

Gurven M (2004) To give and to give not: the behavioral ecology of human food transfers. Behav Brain Sci 27:543–583

Hawkes K (1993) Why hunter-gatherers work: an ancient version of the problem of public goods. Curr Anthropol 34:341–361

Hawkes K (2006) Life history theory and human evolution: a chronicle of ideas and findings. In: Hawkes K, Paine RR (eds) The evolution of human life history. School of American Research Press, Santa Fe, pp 45–93

Hawkes K, O'Connell JF (2005) How old is human longevity. J Hum Evol 49:650–653

Hawkes K, O'Connell JF, Blurton Jones NG (1989) Hardworking Hadza grandmothers. In: Standen V, Foley RA (eds) Comparative socioecology: the behavioural ecology of humans and other mammals. Blackwell, Oxford, pp 341–366

Hawkes K, O'Connell JF, Blurton Jones NG (1995) Hadza children's foraging: juvenile dependency, social arrangements, and mobility among hunter-gatherers. Curr Anthropol 36:688–700

Hawkes K, O'Connell JF, Blurton Jones NG, Alvarez H, Charnov EL (1998) Grandmothering, menopause, and the evolution of human life histories. Proc Natl Acad Sci USA 95:1336–1339

Hoppitt WJE, Brown GR, Kendal R, Rendell L, Thornton A, Webster MM, Laland KN (2008) Lessons from animal teaching. Trends Ecol Evol 23:486–493

Hrdy SB (1999) Mother nature. A history of mothers, infants, and natural selection. Pantheon Books, New York

Hrdy SB (2005) Evolutionary context of human development: the cooperative breeding model. In: Carter CS, Ahnert L, Grossmann KE, Hrdy SB, Lamb ME, Porges SW, Sachser N (eds) Attachment and bonding: a new synthesis. 92nd Dahlem Workshop Report. MIT, Cambridge, MA, pp 9–32

Hrdy SB (2009) Mothers and others: the evolutionary origins of mutual understanding. Harvard University Press, Cambridge, MA

Isler K, van Schaik CP (2009) Why are there so few smart mammals (but so many smart birds)? Biol Lett 5:125–129

Isler K, van Schaik CP (in press) The expensive brain: a framework for explaining evolutionary changes in brain size. J Hum Evol

Isler K, Kirk EC, Miller JMA, Albrecht GA, Gelvin BR, Martin RD (2008) Endocranial volumes of primate species: scaling analyses using a comprehensive and reliable data set. J Hum Evol 55:967–978

Johnson AW, Earle T (2000) The evolution of human societies: from foraging group to agrarian state. Stanford University Press, Stanford

Kaplan H, Robson AJ (2002) The emergence of humans: the coevolution of intelligence and longevity with intergenerational transfers. Proc Natl Acad Sci USA 99:10221–10226

Kaplan H, Hill K, Lancaster JB, Hurtado AM (2000) A theory of human life history evolution: diet, intelligence and longevity. Evol Anthropol 9:156–185

Kirkwood TBL, Austad SN (2000) Why do we age? Nature 408:233–238

Kokko H, Johnstone RA, Clutton-Brock TH (2001) The evolution of cooperative breeding through group augmentation. Proc Roy Soc Lond B 268:187–196

Kramer KL (2005) Children's help and the pace of reproduction: cooperative breeding in humans. Evol Anthropol 14:224–237

Lahdenperä M, Lummaa V, Helle S, Tremblay M, Russell AF (2004) Fitness benefits of prolonged post-reproductive lifespan in women. Nature 428:178–181

Lakshminarayanan VR, Santos LR (2008) Capuchin monkeys are sensitive to others' welfare. Curr Biol 18:R999–R1000

Mace R, Sear R (2005) Are humans cooperative breeders? In: Voland E, Chasiotis A, Schiefenhövel W (eds) Grandmotherhood: the evolutionary significance of the second half of female life. Rutgers University Press, New Brunswick, NJ, pp 143–159

Marlowe FW (2003) A critical period for provisioning by Hadza men: implications for pair bonding. Evol Hum Behav 24:217–229

Marlowe FW, Berbesque JC, Barr A, Barrett C, Bolyanatz A, Cardenas JC, Ensminger J, Gurven M, Gwako E, Henrich J, Henrich N, Lesorogol C, McElreath R, Tracer D (2008) More 'altruistic' punishment in larger societies. Proc Roy Soc Lond B 275:587–590

McDonald PG, Kazem AJN, Wright J (2009) Cooperative provisioning dynamics: fathers and unrelated helpers show similar responses to manipulations of begging. Anim Behav 77:369–376

Milinski M, Semmann D, Krambeck H-J (2002) Reputation helps solve the 'tragedy of the commons'. Nature 415:424–426

Mink JW, Blumenschine RJ, Adams DB (1981) Ratio of central nervous system to body metabolism in vertebrates – its constancy and functional basis. Am J Physiol 241:R203–R212

O'Connell JF, Hawkes K, Blurton Jones NG (1999) Grandmothering and the evolution of *Homo erectus*. J Hum Evol 36:461–485

Panchanathan K, Boyd R (2004) Indirect reciprocity can stabilize cooperation without the second-order free rider problem. Nature 432:499–502

Pavard S, Metcalf CJE, Heyer E (2008) Senescence of reproduction may explain adaptive menopause in humans: a test of the 'mother' hypothesis. Am J Phys Anthropol 136:194–203

Pennington R (2001) Hunter-gatherer demography. In: Panther-Brick C, Layton RH, Rolwey-Conwy P (eds) Hunter-gatherers: an interdisciplinary perspective. Cambridge University Press, Cambridge, pp 170–204

Pobiner BL, Rogers MJ, Monahan CM, Harris JWK (2008) New evidence for hominin carcass processing strategies at 1.5 Ma, Koobi Fora, Kenya. J Hum Evol 55:103–130

Potts R (1998) Environmental hypotheses of hominin evolution. Yrbk Phys Anthropol 41:93–136

Rendall D, Notman H, Vokey JR (2007) Homologizing the mind. In: Dunbar RIM, Barrett L (eds) Oxford handbook of evolutionary psychology. Oxford University Press, Oxford, pp 59–70

Richerson PJ, Boyd R (2005) Not by genes alone: how culture transformed human evolution. University of Chicago Press, Chicago

Ridley M (1996) The origins of virtue: human instincts and the evolution of cooperation. Viking, New York

Robson SL, van Schaik CP, Hawkes K (2006) The derived features of human life history. In: Hawkes K, Paine RL (eds) The evolution of human life history. School of American Research Press, Santa Fe, pp 17–44

Ross C, Jones KE (1999) Socioecology and the evolution of primate reproductive rates. In: Lee PC (ed) Comparative primate socioecology. Cambridge University Press, Cambridge, pp 73–110

Rowley-Conwy P (2001) Time, change and the archaeology of hunter-gatherers: how original is the 'Original Affluent Society'? In: Panter-Brick C, Layton RH, Rowley-Conwy P (eds) Hunter-gatherers: an interdisciplinary perspective. Cambridge University Press, Cambridge, pp 39–72

Rubenstein DR, Lovette IJ (2007) Temporal environmental variability drives the evolution of cooperative breeding in birds. Curr Biol 17:1414–1419

Russell AF (2004) Mammals: comparisons and contrasts. In: Koenig WD, Dickinson JL (eds) Ecology and evolution of cooperative breeding in birds. Cambridge University Press, Cambridge, pp 210–227

Schoenemann PT (2006) Evolution of the size and functional areas of the human brain. Annu Rev Anthropol 35:379–406

Sear R, Mace R (2008) Who keeps children alive? A review of the effects of kin on child survival. Evol Hum Behav 29:1–18

Silk JB (2007) Empathy, sympathy, and prosocial preferences in primates. In: Dunbar RIM, Barrett L (eds) Oxford handbook of evolutionary psychology. Oxford University Press, New York, pp 115–126

Snowdon CT, Cronin KA (2007) Cooperative breeders do cooperate. Behav Process 76:138–141

Temmerman M, Verstraelen H, Martens G, Bekaert A (2004) Delayed childbearing and maternal mortality. Eur J Obstet Gynecol Reprod Biol 114:19–22

Tomasello M (1999) The cultural origins of human cognition. Harvard University Press, Cambridge, MA

Tomasello M, Carpenter M (2007) The emergence of social cognition in three young chimpanzees. Blackwell, Boston

Tomasello M, Rakoczy H (2003) What makes human cognition unique? From individual to shared to collective intentionality. Mind Lang 18:121–147

Tomasello M, Carpenter M, Call J, Behne T, Moll H (2005) Understanding and sharing intentions: the origins of cultural cognition. Behav Brain Sci 28:675–735

van Schaik CP, Burkart JM (in press) Why culture makes smart: the cultural intelligence hypothesis. Animal Cognition

van Schaik CP, Deaner RO (2003) Life history and cognitive evolution in primates. In: de Waal FBM, Tyack PL (eds) Animal social complexity. Harvard University Press, Cambridge, MA, pp 5–25

Whiten A, van Schaik CP (2007) The evolution of animal 'cultures' and social intelligence. Phil Trans Roy Soc B 362:603–620

Wiessner P (2009) Experimental games and games of life among the Ju/'hoan Bushmen. Curr Anthropol 50:133–138

Williams GC (1957) Pleiotropy, natural selection, and the evolution of senescence. Evolution 11:398–411

Wiltermuth SS, Heath C (2009) Synchrony and cooperation. Psychol Sci 20:1–5

Wrangham RW, Jones JH, Laden G, Pilbeam D, Conklin-Brittain NL (1999) The raw and the stolen: cooking and the ecology of human origins. Curr Anthropol 40:567–594

Zahed SR, Prudom SL, Snowdon CT, Ziegler TE (2007) Male parenting and response to infant stimuli in the common marmoset (*Callithrix jacchus*). Am J Primatol 70:84–92

Index